EcoProduction

Environmental Issues in Logistics and Manufacturing

Series editor

Paulina Golinska, Poznan, Poland

About the Series

The EcoProduction Series is a forum for presenting emerging environmental issues in Logistics and Manufacturing. Its main objective is a multidisciplinary approach to link the scientific activities in various manufacturing and logistics fields with the sustainability research. It encompasses topical monographs and selected conference proceedings, authored or edited by leading experts as well as by promising young scientists. The Series aims to provide the impulse for new ideas by reporting on the state-of-the-art and motivating for the future development of sustainable manufacturing systems, environmentally conscious operations management and reverse or closed loop logistics.

It aims to bring together academic, industry and government personnel from various countries to present and discuss the challenges for implementation of sustainable policy in the field of production and logistics.

More information about this series at http://www.springer.com/series/10152

Voratas Kachitvichyanukul
Kanchana Sethanan
Paulina Golinska-Dawson
Editors

Toward Sustainable Operations of Supply Chain and Logistics Systems

Springer

Editors
Voratas Kachitvichyanukul
Industrial and Manufacturing Engineering
Asian Institute of Technology
Khlong Luang, Pathumthani
Thailand

Paulina Golinska-Dawson
Engineering Management
Poznan University of Technology
Poznan
Poland

Kanchana Sethanan
Industrial Engineering
Khon Kaen University
Khon Kaen
Thailand

ISSN 2193-4614 ISSN 2193-4622 (electronic)
EcoProduction
ISBN 978-3-319-19005-1 ISBN 978-3-319-19006-8 (eBook)
DOI 10.1007/978-3-319-19006-8

Library of Congress Control Number: 2015941359

Springer Cham Heidelberg New York Dordrecht London

Printed on acid-free paper

Springer International Publishing AG Switzerland is part of Springer Science+Business Media (www.springer.com)

Preface

This book entitled "Toward Sustainable Operations of Supply Chain and Logistics Systems" aims to cover contemporary critical issues of logistics operations and supply chain management with special focus on sustainability. The supply chain operations have big impact on the natural environment. Nowadays, the procurement, production or distribution, and accompanying transportation operations or warehousing are significant on the global scale. Improvement of logistics operations by supply chain cooperation can results in reducing the burden to the natural environment. Reverse logistics has significantly developed in the last decade. Reverse network configuration, forward and reverse supply chain integration, and management of the recovery activities are examples of trends in logistics toward a more environmental friendly business practices.

The book covers both qualitative and qualitative aspects of sustainable supply chain and logistics operations management. The focus is placed on a multidisciplinary approach. This book presents viewpoints of the academic and the industry personnel on the challenges for implementation of sustainable police in logistics operations. Some best practices from different countries and industries are presented as reference for implementation of sustainable policies in the private enterprises and public sector's institutions. This book might be valuable to both academics and practitioners wishing to deepen their knowledge in the field of logistics operations and management with regard to sustainability issues.

The authors present in the individual chapters the results of the theoretical and empirical studies related to the following topics:

- Supply chain management,
- Sustainable logistics,
- Manufacturing and production logistics,
- Warehousing, distribution, and transportation problems,
- Metaheuristics and artificial technology methods,
- Mathematical Programming and Operations Research, and
- Statistical Techniques.

This book includes research contributions of geographically dispersed authors from Asia, Africa, Americas, and Europe. The high scientific quality of the chapters was assured by a rigorous double-blind review process implemented by the ICLS 2015 organizers.

This book provides an interesting composition of theoretical trends and practical applications. It presents selected papers of authors participating in the 10th International Congress on Logistics and SCM Systems (ICLS 2015) which is held in Chiang Mai, Thailand, during 1–4 July 2015. This Congress is organized by the Asian Institute of Technology and the International Federation of Logistics and SCM Systems (IFLS) to function as an international forum for leading academics, researchers, and practitioners to discuss and exchange ideas on the latest developments and to seek opportunities for collaboration among the participants as well as to promote excellence in the field. Although not all of the received chapters appear in this book, the efforts spent and the work done for this book are very much appreciated.

We would like to express our gratitude to the Board of the International Federation of Logistics and SCM Systems (IFLS) for the valuable contribution to the volume:

- Honorary Chairman—Prof. Karasawa, Yutaka, Kanagawa University, Japan
- Advisor—Prof. Katayama, Hiroshi, Waseda University, Japan
- Chairman—Prof. Lee, Young Hae, Hanyang University, Korea
- Senior Vice Chairman—Prof. Kachitvichyanukul, Voratas, Asian Institute of Technology, Thailand
- Vice Chairmen—Prof. Lai Kin Keung, City University of Hong Kong; Prof. Rim Suk-Chul, Ajou University, Korea; Prof. Wakabayashi Keizo, Nihon University, Japan and Wu Yen-Chen Jim, National Sun Yat-Sen University, Taiwan.

We would like to thank all reviewers whose names are not listed in the volume due to the confidentiality of the process. Their voluntary services and comments helped the authors to improve the quality of the manuscripts.

Voratas Kachitvichyanukul
Kanchana Sethanan
Paulina Golinska-Dawson

Contents

Part I
Supply Chain Management

A Supply Chain Network Design Considering Network Density

Kanokporn Rienkhemaniyom and Subramanian Pazhani

Abstract A geographical concentration of nodes within a supply chain or supply chain density is one of the supply chain network characteristics that may affect the resiliency of a supply chain to disruption. Some major cases include the disruption of the global PC supply chain due to the 1999 earthquake in Taiwan, the disruption of automotive supply chain due to the 2011 earthquake in Japan, and the disruption of hard disk supply chain due to the 2011 massive floods in Thailand. These disruptions were resulted from the high geographical concentration of suppliers and manufacturers. In this chapter, the optimal design of a supply chain is discussed with the objectives of maximizing the total profit and the density of the supply chain. A bi-criteria mixed-integer linear programming model (MILP) is formulated to determine the optimal locations of the facilities and the distribution of flows between facilities in the supply chain. A four-stage supply chain network model is developed, which consists of suppliers, manufacturing plants, warehouses, and retailers. The model is illustrated using a realistic example, and the results are discussed.

Keywords Supply chain network design · Bi-criteria MILP · Network density · Disruption · Resiliency

K. Rienkhemaniyom (✉)
Graduate School of Management and Innovation, King Mongkut's University of Technology Thonburi, Bangkok, Thailand
e-mail: kanokporn.rie@gmail.com; Kanokporn.rie@kmutt.ac.th

S. Pazhani
Harold and Inge Marcus Department of Industrial and Manufacturing Engineering, The Pennsylvania State University, University Park, PA 16802, USA
e-mail: sathish.subramanian84@gmail.com

V. Kachitvichyanukul et al. (eds.), *Toward Sustainable Operations of Supply Chain and Logistics Systems*, EcoProduction,
DOI 10.1007/978-3-319-19006-8_1

1 Introduction

Supply chain network design is an important strategic decision in supply chain management that enables firms to achieve their competitive strategies. Performances of any supply chain networks are usually to maximize customer expectations at minimum supply chain costs (Chopra 2003). Minimizing cost and maximizing customer responsiveness in the supply chain have resulted in globalization in search of cost minimization, access to new markets, and economies of scale. This is also further aggravated by customer's demand for variety of products with minimum delivery time. These have led to the increased complexity of supply chain functions and increased disruptions in the supply chains.

Network design in supply chains deals with the selection of the best location for the facilities in the supply chain network, numbers and capacities of facilities, and optimal product flows through the selected set of facilities (Melo et al. 2009; Subramanian and Ravindran 2014). Due to increased globalization of the firms' operations, supply chains have become more vulnerable to disruptions (Kungwalsong 2013; Subramanian and Ravindran 2014). Thus, supply chain disruption has become an important consideration while designing a supply chain network.

There are a number of empirical studies that have examined the relationship between supply chain characteristics and operational efficiency to the impact of disruptions. Wagner and Bode (2006) focused on the firms' dependence on customers and suppliers, the degree of single sourcing, and reliance on global supply sources. Craighead et al. (2007) revealed that density, complexity, node criticality, and capability of warning and recovery. Squire (2010) studied the impact of network characteristics to supply chains. Those characteristics include node criticality, node centrality, geographical distance, redundancy, number of nodes, and degree of distribution. Schmidt and Raman (2012) suggested that a higher rate of improvement in operating performance provokes the impact of internal disruptions but not external disruptions. Bode and Wagner (2015) studied the relationship between supply chain complexity and the frequency of supply chain disruptions. The authors consider three forms of supply chain complexity: horizontal complexity, which refers to the number of suppliers in each tier; vertical complexity, which refers to the number of tiers; and spatial complexity, which refers to the geographical dispersion among members within the network. The results indicated that all the three supply chain complexity types have a positive relationship with the frequency of supply chain disruptions. Kim et al. (2015) indicated that a network structure significantly relates to the probability of a network disruption. A network resiliency improves when the structural relationships in a network follow the power law.

In terms of quantitative approach, some of the supply chain design studies have included the topic of supply chain disruption into consideration in order to enhance both supply chain profit and supply chain resilience. Falasca et al. (2008) have extended the supply chain design characteristics' concept in Craighead et al. (2007) and proposed a quantitative measure for those characteristics. For instance, supply chain density can be measured as the number of nodes divided by the average

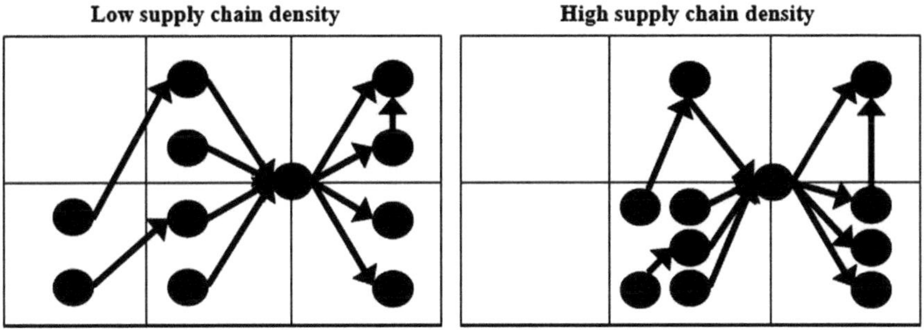

Fig. 1 Supply chain density (*Source* Falasca et al. 2008)

inter-node distance, as shown in Fig. 1. The supply chain is having a high-density level when a large number of nodes within a supply chain are clustered closely together. The authors indicated that the severity of a supply chain disruption is positively related to supply chain density.

To our knowledge, there is no supply chain network design model that explicitly integrates the supply chain density characteristics. In this paper, we propose a bi-criteria mixed-integer linear programming (MILP) for a supply chain network design problem with the objectives of maximizing the profit and density of the supply chain. We then use weighted fuzzy goal programming method to solve the bi-criteria problem.

The remainder of this article is organized as follows: Sect. 2 describes the problem definition. Section 3 presents a bi-criteria MILP model for designing a four-stage supply chain network and the solution technique. In Sect. 4, we present an illustrative example. We also present some managerial implications of the model. Section 5 presents conclusions and future research directions.

2 Problem Definition

The network design problem for a four-stage supply chain structure, considering maximizing profit density of the supply chain, is defined in this section. Let $S = \{1, 2 \ldots, n_S\}$ be the set of suppliers, $M = \{1, 2 \ldots, n_M\}$ be the set of manufacturing plants, $W = \{1, 2 \ldots, n_W\}$ be the set of warehouses, $C = \{1, 2, \ldots, n_C\}$ be the set of retailers, and $L = \{1, 2, \ldots, n_L\}$ be the set of capacity levels at the warehouses. The suppliers supply raw materials to manufacturing plants for producing products. The product flows from the manufacturing plant through a two distribution stages, managed by a single decision maker, i.e., centralized control system. The manufacturing plants produce products using the raw materials procured from the suppliers and distribute them to the retailers through a set of warehouses. Figure 2 shows the supply chain network considered in the study. Figure 3 shows the geographical dispersion of facilities in the supply chain.

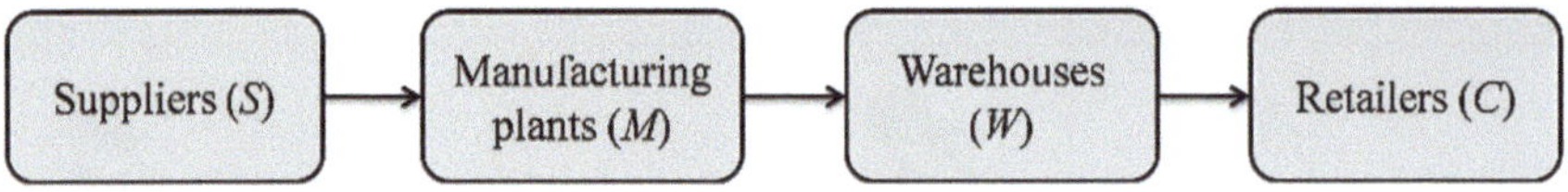

Fig. 2 Structure of the supply chain network

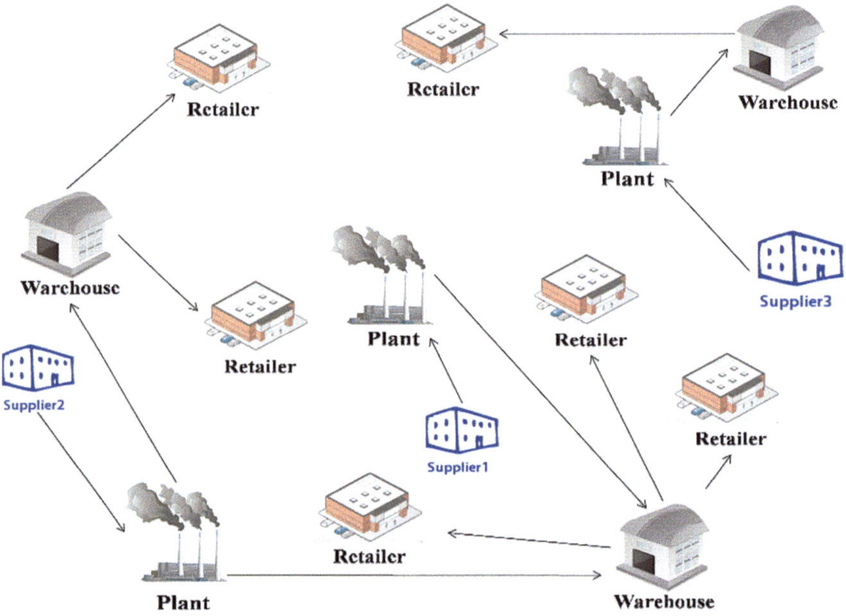

Fig. 3 Geographical dispersion of facilities in the supply chain

The main issues to be addressed in the bi-criteria model are to (i) determine the numbers and locations of suppliers, (ii) select the appropriate set of warehouse locations and capacity levels in the network, and (iii) determine the optimal manufacturing and distribution of products through the selected set of facilities. The conflicting objectives of the model are (1) maximizing supply chain profit and (2) maximizing supply density.

To quantify and model the supply density objective, we propose a function for calculating supply density based on the inter-stage distance and intra-stage distance. Note that a lower value of supply density implies that the nodes in the supply chain are clustered together. Thus, we maximize the supply density. This has implication of the resiliency of a supply chain. According to Squire (2010), accidental risk events are more likely to affect a node with few links compared to a hub node, therefore reducing the probability of disruption to a whole network. In other words, the impact of supply chain disruption should be reduced. Also, note that we consider the density function only for the supplier stage as this stage is the only external entity in the supply chain. Other stages are managed by a single decision maker (centralized control system).

3 Model Formulation

The proposed bi-criteria model for designing a four-stage supply chain network is presented in this section. The indices, input parameters, cost function components, and decision variables used to formulate the concerned supply chain network design problem are described below.

Indices

s	Index of candidate suppliers $s = 1, 2, \ldots, S$
m	Index of existing manufacturing plants $m = 1, 2, \ldots, M$
w	Index of candidate warehouse facilities $w = 1, 2, \ldots, W$
c	Index of existing retailers $c = 1, 2, \ldots, C$
l	Index of warehouse capacities $l = 1, 2, \ldots, L$
f	Index of facilities $f = 1, 2, \ldots, F$ $(F = S \cup M \cup W)$
i	Index of origin facilities $i \in S \cup M \cup W$
j	Index of destination facilities $j \in M \cup W \cup C$

Input parameters

cap_m	Production capacity at the manufacturing plant m, for $m \in M$
cap_w^l	Capacity of warehouse w of size l, for $w \in W$, $l \in L$
cap_s	Capacity of supplier s, for $s \in S$
d_c	Demand for products of retailer c, for $c \in C$
dis_{ij}	Distance between facilities i and j in the supply chain
msm	Minimum transportation quantity from suppliers to manufacturers

Cost function components

p_{sm}	Purchasing cost of raw material from supplier s by plant m, for $s \in S$, $m \in M$
tr_{mw}	Transportation cost per unit from plant m to warehouse w, for $m \in M$, $w \in W$
tr_{wc}	Transportation cost per unit from warehouse w to retailer c, for $w \in W$, $c \in C$
pc_m	Production cost of a product at plant m, for $m \in M$
np	Price of a product
f_w^l	Fixed cost of opening a warehouse w of capacity l, for $w \in W$, $l \in L$
ls_c	Lost sales cost of retailer c, for $c \in C$

Decision variables

QSM_{sm}	Quantity of raw materials purchased from supplier s by plant m, for $s \in S$, $m \in M$
QMW_{mw}	Quantity of products transported from plant m to warehouse w, for $m \in M$, $w \in W$
QWC_{wc}	Quantity of products transported from warehouse w to retailer c, for $w \in W$, $c \in C$
LD_c	Quantity of sales lost of retailer c, for $c \in C$

(continued)

(continued)

δ_w^l	$\begin{cases} 1, & \text{if warehouse } w \text{ is opened with size } l \\ 0, & \text{otherwise} \end{cases}$, for $w \in W$, $l \in L$
$S\alpha_{sm}$	$\begin{cases} 1, & \text{if supplier } s \text{ supplies raw materials to plant } m \\ 0, & \text{otherwise} \end{cases}$, for $s \in S$, $m \in M$
$S\beta_{ijm}, S\beta'_{ijm}$	$\begin{cases} 1, & \text{if supplier } i \text{ and supplier } j \text{ supplies to plant } m \\ 0, & \text{otherwise} \end{cases}$, for $(i,j) \in S$ and $(i \neq j)$, $m \in M$.

The following are the assumptions considered in this research:

- Demands of the retailers are deterministic. This assumption is reasonable, given that the proposed model is at the strategic level.
- The transportation cost of raw material from the supplier to the manufacturing plant is included in the raw material purchasing cost.
- The facilities in the entire supply chain (suppliers, manufacturing plants, warehouses) have capacity restrictions.

From the above notations, the bi-criteria network design problem can be formulated as follows.

3.1 Objective Functions

As mentioned in Sect. 1, two important and conflicting objective functions are considered in the formulation of the supply chain network design problem: (1) maximizing supply chain profit and (2) maximizing supply density.

3.1.1 First Objective: Maximizing Supply Chain Profit (Z_1)

- The supply chain profit comprises of the following cost components:

Revenue	$\text{np}\left(\sum_{w\in W}\sum_{c\in C} \text{QWC}_{wc}\right)$
Purchasing cost	$\sum_{s\in S}\sum_{m\in M} p_{sm}\text{QSM}_{sm}$
Production cost	$\sum_{m\in M} \text{pc}_m\left(\sum_{w\in W} \text{QMW}_{mw}\right)$
Transportation cost from plants to warehouses	$\sum_{m\in M}\sum_{w\in W} \text{tr}_{mw}\text{QMW}_{mw}$
Transportation cost from warehouses to retailers	$\sum_{w\in W}\sum_{c\in C} \text{tr}_{wc}\text{QWC}_{wc}$
Fixed cost of opening warehouses	$\sum_{l\in L}\sum_{w\in W} f_w^l \delta_w^l$
Lost sales cost	$\sum_{c\in C} \text{ls}_c\text{LD}_c$

- Therefore, the supply chain profit objective can be formulated as follows.

$$\begin{aligned}\text{Maximize } Z_1 = & -\left\{\text{np}\left(\sum_{w\in W}\sum_{c\in C}\text{QWC}_{wc}\right)\right\} \\ & -\left\{\sum_{s\in S}\sum_{m\in M}p_{sm}\text{QSM}_{sm} + \sum_{m\in M}\text{pc}_m\left(\sum_{w\in W}\text{QMW}_{mw}\right)\right. \\ & + \sum_{m\in M}\sum_{w\in W}\text{tr}_{mw}\text{QMW}_{mw} + \sum_{w\in W}\sum_{c\in C}\text{tr}_{wc}\text{QWC}_{wc} \\ & \left. + \sum_{l\in L}\sum_{w\in W}f_w^l\delta_w^l + \sum_{c\in C}\text{ls}_c\text{LD}_c\right\}\end{aligned}$$

3.1.2 Second Objective: Maximizing Supply Density (Z_2)

In this study, we propose a function for calculating based on the inter-stage distance and intra-stage distance. In this study, we propose a function for calculating supply density based on the inter-stage distance and intra-stage distance. Consider a supply chain with two stages, supplier and manufacturer. The inter-stage distance captures the distance between stages. For example, if the product is shipped from supplier 1 to manufacturer 3, then the inter-stage distance is the distance between supplier 1 and manufacturer 3. This measures the density of the suppliers in the supply chain. The intra-stage distance captures the distance within a stage. For example, if suppliers 1 and 3 supply products to manufacturer 1, then the intra-stage distance is the distance between suppliers 1 and 3. This measures the density of the suppliers within a stage in the supply chain. Therefore, the supply density objective, which is the density of the supply entity per unit of demand, can be formulated as follows.

$$\text{Maximize } Z_2 = \frac{1}{\sum_{c\in C} d_c}\left(\sum_{s\in S}\sum_{m\in M}\text{dis}_{sm}S\alpha_{sm} + \sum_{m\in M}\sum_{i\in S}\sum_{\substack{j\in S \\ i\neq j, i<j}}\text{dis}_{ij}S\beta_{ijm}\right)$$

The terms in the objective function Z_2 are described as follows. The two terms calculate the inter-stage and intra-stage distances for the supplier stage.

3.2 Constraints

3.2.1 Supplier Capacity

Each supplier s has a finite supply capacity, cap_s. Constraint set (1) ensures that the quantity of raw materials supplied by supplier s to all the manufacturing plants should be less than or equal to its capacity.

$$\sum_{m \in M} \text{QSM}_{sm} \leq \text{cap}_s \quad \forall\, s \in S \tag{1}$$

3.2.2 Inter-stage Flow

Constraint sets (2) and (3) determine the binary variable for the inter-stage flow between suppliers and manufacturers. Constraint set (2) ensures that if there is shipment between supplier s and manufacturer m, the binary variable is $S\alpha_{sm} = 1$. Constraint set (3) ensures that if there is no shipment between the stages, the binary variable is $S\alpha_{sm} = 0$. Constraint set (3) also ensures minimum shipment if there is shipment between supplier s and plant m.

$$\text{QSM}_{sm} \leq \text{cap}_s S\alpha_{sm} \quad \forall\, s \in S, m \in M \tag{2}$$

$$\text{QSM}_{sm} \geq \text{msm}\, S\alpha_{sm} \quad \forall\, s \in S, m \in M \tag{3}$$

3.2.3 Inter-stage Flow Binary Variable

Constraint sets (4) and (5) determine the binary variable for the intra-stage flow for the supplier stage. The right-hand-side term in constraint set (4) represents the product flow from suppliers i and j to manufacturer m. If there is product flow in both (supplier i − manufacturer m) and (supplier j − manufacturer m) links, the binary variable is $S\beta_{ijm} = 1$ and $S\beta'_{ijm} = 0$. If there is product flow in either one of the (supplier i − manufacturer m) and (supplier j − manufacturer m) links, then $S\beta'_{ijm} = 1$ and $S\beta_{ijm} = 0$. If there is no flow in both (supplier i − manufacturer m) and (supplier j − manufacturer m) links, then $S\beta_{ijm}$ and $S\beta'_{ijm} = 0$. Constraint set (5) ensures that either one of the following cases is true: $S\beta_{ijm} = 1$ and $S\beta'_{ijm} = 0$ or $S\beta_{ijm} = 0$ and $S\beta'_{ijm} = 1$ or $S\beta_{ijm} = 0$ and $S\beta'_{ijm} = 0$.

$$2S\beta_{ijm} + S\beta'_{ijm} = S\alpha_{im} + S\alpha_{jm} \quad \forall (i,j) \in S,\ i \neq j,\ i<j,\ m \in M \tag{4}$$

$$S\beta_{ijm} + S\beta'_{ijm} \leq 1 \quad \forall (i,j) \in S,\ i \neq j,\ i<j,\ m \in M \tag{5}$$

3.2.4 Production Capacity

Constraint set (6) is the production capacity constraints at the plants. The left-hand-side term of the constraint represents the sum of products transported to warehouses from plant m which should be less than or equal to its capacity.

$$\sum_{w \in W} \mathrm{QMW}_{mw} \leq \mathrm{cap}_m \quad \forall\, m \in M \tag{6}$$

3.2.5 Material Flow Between Suppliers and Plants

Constraint set (7) ensures that the quantity of raw material flowing into plant m is equal to the quantity of products flowing out of that plant to the warehouses.

$$\sum_{s \in S} \mathrm{QSM}_{sm} = \sum_{w \in W} \mathrm{QMW}_{mw} \quad \forall\, m \in M \tag{7}$$

3.2.6 Warehouse Capacity

Constraint set (8) ensures that the quantity of products flowing into warehouse w does not exceed its storage capacity, if the warehouse is selected. The left-hand-side term of the constraint is the sum of products flowing into warehouse w, and the right-hand-side term is the capacity of the selected warehouse. If a warehouse w is opened, constraint set (9) ensures that only one of the capacity levels is selected.

$$\sum_{c \in C} \mathrm{QWC}_{wc} \leq \sum_{l \in L} \mathrm{cap}_w^l \delta_w^l \quad \forall\, w \in W \tag{8}$$

$$\sum_{l \in L} \delta_w^l \leq 1 \quad \forall\, w \in W \tag{9}$$

3.2.7 Product Flows Between Warehouses and Retailers

Constraint set (10) ensures that the quantity of products flowing into warehouse w is equal to the amount of new products flowing out of that warehouse to the retailers.

$$\sum_{m \in M} \mathrm{QMW}_{mw} = \sum_{c \in C} \mathrm{QWC}_{wc} \quad \forall\, w \in W \tag{10}$$

3.2.8 Demand Requirement

Constraint set (11) represents the demand satisfaction constraints. The total quantity of products flowing into retailer c and the lost sales at the retailer c should be equal to the demand at that retailer.

$$\sum_{w \in W} \mathrm{QWC}_{wc} + \mathrm{LD}_c = d_c \quad \forall\, c \in C \tag{11}$$

3.2.9 Non-negativity and Binary Conditions

Constraints (12) and (13) describe non-negativity and binary conditions of the decision variables.

$$\mathrm{QSM}_{sm},\ \mathrm{QMW}_{mw},\ \mathrm{QWC}_{wc},\ \mathrm{LD}_c > 0 \tag{12}$$

$$\delta^{l}_{w}, S\alpha_{sm}, S\beta_{ijm}, S\beta'_{ijm} \in \{0, 1\} \tag{13}$$

3.3 Solution Method

In this section, we propose a weighted fuzzy goal programming (GP) model to solve the bi-criteria problem. In fuzzy GP, ideal values can be used as targets for the objectives (Masud and Ravindran 2008; Subramanian et al. 2013). In the proposed model, we minimize the sum of the weighted satisfaction degrees of the objectives. The proposed weighted fuzzy GP model is as follows:

$$\text{Minimize } w_1\mu_1 + w_2\mu_2$$

Subject to constraints (1)–(13)

$$\mu_1 = \left(\frac{Z_1^* - Z_1}{Z_1^* - Z_1^L}\right) \tag{14}$$

$$\mu_2 = \left(\frac{Z_2^* - Z_2}{Z_2^* - Z_2^L}\right) \tag{15}$$

where μ_1 and μ_2 are the satisfaction degrees of the profit and density objectives, Z_1^* is the ideal profit objective value, Z_2^* is the ideal density objective value, Z_1^L and Z_2^L are the lower bound values of profit and density objectives, w_1 is the weight for profit objective, and w_2 is the weight for supply density objective.

Note that $w_1 + w_2 = 1$. Also, note that Z_1 and Z_2 in the constraints (14) and (15) are the values of profit and density objectives and are calculated as shown in Sect. 3.1.

4 Illustrative Example

In this section, we present an illustrative example for designing a four-stage supply chain network that consists of suppliers, manufacturing plants, warehouses, and retailers. The supply chain network consists of the following entities:

- 20 candidate suppliers that can supply a raw material to manufacturing plants,
- 5 existing manufacturing plants that produce finished products,
- 25 candidate warehouse facilities that distribute finished products to the retailers. There are three possible capacity levels (sizes) of warehouses that could be built.
- 100 retailers, who face demand from the customers.

The supply chain entities are located in different regions, as summarized in Table 1. In this example, we consider six different regions according to the six continents of the world. Region 1 represents Africa; Region 2 represents Asia; Region 3 represents Europe; Region 4 represents North America; Region 5 represents Australia; and Region 6 represents South America. Figure 4 presents the geographical dispersion of those facilities in the six regions. We use a circle to represent a candidate supplier, a rectangle represents a manufacturing plant, a triangle represents a warehouse, and a diamond shape represents a retailer.

Table 1 Geographical locations of existing and candidate facilities

Region	Suppliers	Plants	Warehouses	Retailers
Region 1	S16	M3	W17	R65–R68
Region 2	S5, S8, S10, S11, S12, S15, S17, S18	M1, M2	W7, W9, W10, W11, W14, W15, W16, W19, W20, W21, W25	R25–R28, R33–R44, R53–R64, R73–R84, R97–R100
Region 3	S6, S7, S9, S14, S19	M4	W3, W4, W8, W13, W18, W22	R9–R16, R29–R32, R49–R52, R69–R72, R85–R88
Region 4	S4, S13, S14, S20	M5	W6, W12, W23	R21–R24, R45–R48, R89–R92
Region 5	S2	–	W2	R5–R8
Region 6	S1, S3	–	W1, W5, W24	R1–R4, R17–R20, R93–R96
Total facilities	20	5	25	100

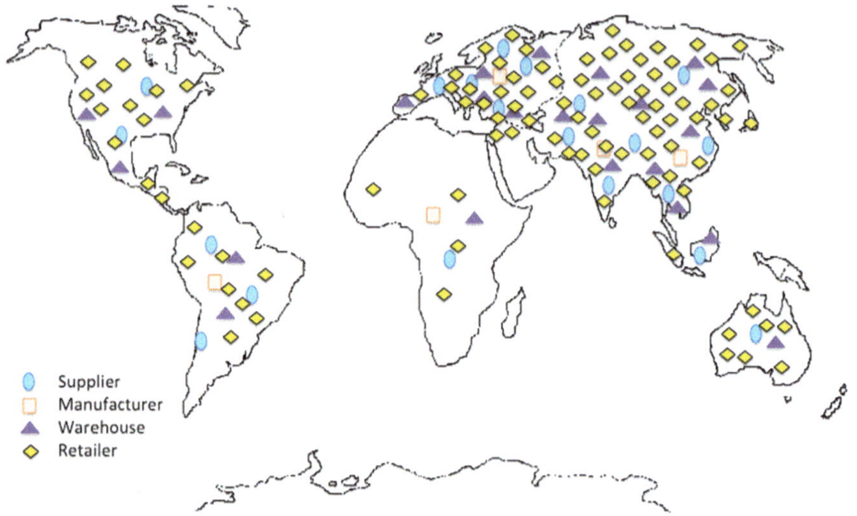

Fig. 4 Locations of candidate suppliers, plants, candidate warehouse facilities, and retailers

Table 2 Cost parameters (illustrative example)

Parameter	Variable	Setting
Total cost of a new product	pp	\$750
Profit margin		20 %
Price of new product	np	\$900
Purchasing cost of raw material	p_{sm}	~Unif (60, 65 %) * pp + (dis_{sm} /250)
Production cost of a new product	pc_m	~Unif (8, 12 %) * pp
Transportation cost per unit between plant and warehouse	tr_{mw}	~Unif (5.5, 6.5 %) * pp
Transportation cost per unit between retailer and warehouse	tr_{wc}	~Unif (8.5, 9.5 %) * pp

We model the cost parameters in the supply chain as a function of the product price. Table 2 shows the settings used to generate cost parameters.

The demands of the retailers are uniformly distributed between 500 and 700 units. Other parameters, including the purchasing cost, capacity of the supplier, capacity and production cost of the manufacturing plants, inter-stage distance between the supplier and the manufacturing plants, intra-stage distance between the suppliers, and capacity and fixed cost of warehouses, are available at https://www.dropbox.com/s/x8ehypv8o9zf660/Datafile_ICLS2015.docx?dl=0.

To solve the problem, the model is coded in Microsoft Visual C++ 6.0 and solved using ILOG Concert Technology with CPLEX 12.1 on a PC with Intel(R) Core (TM) 2 Duo processor with 2.8 GHz and 2.0 GB RAM.

4.1 Single-Objective Model Solution

In this section, we solve the problem considering the profit and density objectives separately as follows:

- Profit objective: Max Z_1, subject to constraints (1)–(13),
- Density objective: Max Z_2, subject to constraints (1)–(13).

The profit maximization objective has a tendency toward a centralized supply chain network by selecting 13 out of 20 suppliers in order to minimize the total cost. The supply chain profit is $13,248,680, representing the ideal profit value, and the supply density is 1.34. On the other hand, the density maximization objective has a tendency toward a decentralized supply chain network by selecting all 20 suppliers. The supply chain profit is $12,055,457, and the supply density value is 96.63, representing the ideal density value. More suppliers are selected when decision maker gives higher priority to the density objective, as shown in Table 3.

4.2 Bi-criteria Model Solution

In this section, we solve the bi-criteria model using fuzzy goal programming approach. The results confirm that the two objective functions, maximization of the supply chain profit and maximization of the supply chain density, are conflicting objectives. The increase in supply chain density leads to a decrease in supply chain profit and vice versa. The graphical representation of an efficient frontier is in Fig. 5.

Table 3 Single-objective model solutions

Region	Profit maximization		Supply density maximization	
	Selected suppliers	Selected warehouses	Selected suppliers	Selected warehouses
Region 1	S16	–	S16	–
Region 2	S8, S10, S11, S15, S17	–	S5, S8, S10, S11, S12, S15, S17, S18	W11
Region 3	S6, S7, S9, S14	W8	S6, S7, S9, S14, S19	–
Region 4	S4, S14	W12, W23	S4, S13, S14, S20	W12, W23
Region 5	–		S2	–
Region 6	S1		S1, S3	–
Profit value	$13,248,680 (ideal value)	$12,055,457		
Density value	1.34	96.63 (ideal value)		

Fig. 5 Efficient frontier between supply chain profit and supply density

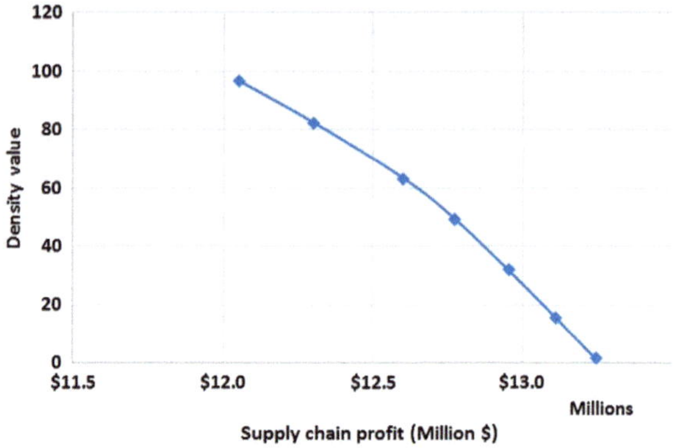

Table 4 Bi-criteria model solution

Region	Selected suppliers	Selected warehouses
Region 1	S16	–
Region 2	S5, S8, S10, S11, S12, S15, S17, S18	W19
Region 3	S6, S7, S9, S14, S19	W8
Region 4	S4, S13, S14, S20	W12
Region 5	S2	–
Region 6	S1, S3	–

For an illustrative purpose, we assume that the decision maker selects the solution, which has the profit of $12,954,399 and density of 31.85. The percentage deviation of the best compromise solution obtained corresponds to a 2.22 % decrease for the profit objective (from the ideal value of $13,248,690) and the increase for the density value from 1.34 to 31.85. Table 4 presents the selected suppliers and warehouses from the bi-criteria model. All 20 suppliers are selected to supply raw materials to the manufacturing plants for producing finished product. The material flows are presented in Table 5. Three warehouses (warehouses 8, 12, and 19) are selected to distribute the finished products to retailers. For the capacity level, warehouses 8 and 19 are selected in level size 3, while warehouse 12 is selected in level size 2. The product flows between manufacturing plants and the selected warehouses are presented in Table 6. All the retailers' demands are satisfied from the selected warehouses 8, 12, and 19.

4.3 Managerial Implications

This section discusses some managerial implications for the study. The model can be used by companies to design their supply chain network considering supplier density. From the numerical example, the model tends to select a fewer number of

Table 5 Product flows from the selected suppliers to manufacturing plants

Facilities	Plant 1	Plant 2	Plant 3	Plant 4	Plant 5	Total
Supplier 1	0	500	500	0	0	1000
Supplier 2	0	500	500	0	0	1000
Supplier 3	0	500	500	0	0	1000
Supplier 4	0	0	500	0	5795	6295
Supplier 5	0	500	500	0	0	1000
Supplier 6	0	500	0	5288	0	5788
Supplier 7	0	0	500	2116	0	2616
Supplier 8	0	6382	500	0	0	6882
Supplier 9	0	500	500	0	0	1000
Supplier 10	2767	500	500	0	0	3767
Supplier 11	2670	500	500	0	0	3670
Supplier 12	0	500	0	500	0	1000
Supplier 13	0	500	500	0	5188	6188
Supplier 14	0	500	2881	0	0	3381
Supplier 15	0	1427	500	500	0	2427
Supplier 16	0	500	6097	0	0	6597
Supplier 17	0	2953	500	0	0	3453
Supplier 18	0	500	0	0	0	500
Supplier 19	0	500	500	0	0	1000
Supplier 20	0	500	500	0	0	1000
Total	5437	18,262	16,478	8404	10,983	59,564

Table 6 Product flows from manufacturing plants to the selected warehouses

Facilities	Warehouse 8	Warehouse 12	Warehouse 19	Total
Plant 1	3143	0	2294	5437
Plant 2	0	0	18,262	18,262
Plant 3	16,478	0	0	16,478
Plant 4	0	8404	0	8404
Plant 5	0	8796	2187	10,983
Total	19,621	17,200	22743	59,564

suppliers when a decision maker gives higher priority to the profit objective (centralized supplier based). On the other hand, the model selects a larger number of suppliers when a decision maker gives higher priority to the density objective (decentralized supplier based). The centralized supplier-based profit objective provides lower cost but may result in high severity from a disruption. On the other hand, a decentralized supplier-based incurs extra cost of having additional suppliers, which offers a redundancy to the supply network. The model also gives the optimal product flows between stages. The efficient frontier shows the trade-off between supply chain profit and density.

5 Conclusion and Future Work

In current business scenario, companies strive to manufacture and distribute products in an efficient manner to gain competitive advantage. Network design is one of the important strategic decisions for a supply chain to operate efficiently and effectively. Due to increased globalization, supply chain network characteristics, such as density, which relates to the resiliency of a supply chain, should be considered while designing the supply chain network. In this paper, we proposed a bi-criteria MILP model that considers supply chain density as one of the decision criteria, along with maximizing the supply chain profit. The model determined the optimal location of suppliers, optimal location and capacities of warehouses, and the distribution of products in the supply chain network. We generate the efficient frontier representing the trade-off between supply chain profit and supply density objectives.

Scope for future work includes quantifying the resilience of the supply chain design solution from the proposed model and comparing it with the traditional supply chain design solution, which emphasizes on profit maximization objective (Rienkhemaniyom and Subramanian 2015). In this study, we considered the density function only for the supplier stage. The study can be extended by considering the total density across all the supply chain stages and evaluating the impact. Further, the mathematical model can also be extended to quantifying and incorporating other supply chain network characteristics, such as network complexity and node criticality, in order to have a better understanding of the supply chain resilience.

References

Bode C, Wagner SM (2015) Structural drivers of upstream supply chain complexity and the frequency of supply chain disruptions. J Oper Manag (accepted manuscript)

Chopra S (2003) Designing the distribution network in a supply chain. Transp Res Part E 39:123–140

Craighead CW, Blackhurst J, Rungtusanatham MJ, Handfield RB (2007) The severity of supply chain disruptions: design characteristics and mitigation capabilities. Decis Sci 38(1):131–156

Falasca M, Zobel CW, Cook D (2008) A decision support framework to assess supply chain resilience. In: Proceedings of the 5th international ISCRAM conference, Washington, DC, USA, pp 596–605

Kim Y, Chen Yi-Su, Linderman K (2015) Supply network disruption and resilience: a network structural perspective. J Oper Manag 33–34:43–59

Kungwalsong K (2013) Managing disruption risks in global supply chains. Unpublished PhD dissertation, Pennsylvania State University

Masud AM, Ravindran AR (2008) Multiple criteria decision making. In: Ravindran AR (ed) Operations research and management science handbook. CRC Press, Boca Raton

Melo MT, Nickel S, Saldanha-da-Gama F (2009) Facility location and supply chain management—a review. Eur J Oper Res 196:401–412

Rienkhemaniyom K, Subramanian P (2015) Resiliency analysis of supply chain network design. Working paper, King Mongkut's University of Technology Thonburi, Bangkok, Thailand

Schmidt W, Raman A (2012) When supply-chain disruptions matter. Harvard Business School (working paper)
Squire B (2010) Managing supply risks: understanding the impact of network characteristics. In: Ponis ST (ed) Managing risk in virtual enterprise networks. IGI Global
Subramanian P, Ravindran AR (2014) Design of closed loop supply chain networks. Int J Bus Anal 1(1):43–66
Subramanian P, Ramkumar N, Narendran TT, Ganesh K (2013) A bi-objective network design model for multi-period, multi-product closed-loop supply chain. J Ind Prod Eng 30(4):264–280
Wagner SM, Bode C (2006) An empirical investigation into supply chain vulnerability. J Purchas Supply Manag 12(6):301–312

Is Cash-to-Cash Cycle Appropriate to Measure Supply Chain Performance?

Seock-Jin Hong

Abstract In this chapter, the cash-to-cash cycle time (C2C) and financial performance metrics of seven selected industries (integrator, apparel, electronics, automobile, airline, logistics, and retail) are investigated with data taken from their annual reports from 2008 to 2012. The purpose is to study C2C whether it is appropriate to measure supply chain performance or not. The C2C characteristics are identified industry by industry with a focus on supply chain management (SCM) and logistics. The relationship between C2C and financial performances is evaluated using measurements of gross, net profit rate, revenue growth, current ratio, and return on asset (ROA). It is found that C2C is very powerful metric to understand supply chain strategies, but it is not a predictor of financial performance.

Keywords SCM performance · C2C · Financial performance · ROA · Profitability

1 Introduction

During the last few decades, there have been enormous changes in the global business environment. The world has been changing quickly due to the development of technologies spread of the internet, globalization and increasing mobility of freight and person. From a business perspective, the environment has been rapidly changing as well, following the changes of customer needs and the innovation of industrial technologies. In order to stay competitive and continue to grow in this quickly changing business world, companies are trying to introduce and apply new and innovative management techniques.

Organizations began to realize that it is not enough to improve efficiencies within the organization, but that their whole supply chain has to be made competitive

S.-J. Hong (✉)
Kedge Business School, 680 cours de la Liberation, 3405 Talence Cedex, Bordeaux, France
e-mail: seockjin.hong@kedgebs.com

V. Kachitvichyanukul et al. (eds.), *Toward Sustainable Operations of Supply Chain and Logistics Systems*, EcoProduction,
DOI 10.1007/978-3-319-19006-8_2

(Childerhouse and Towill 2002; Li et al. 2006). Supply chain management (SCM) has been defined to explicitly recognize the strategic nature of coordination between trading partners, and its purpose is to improve the performance of an individual organization and the whole supply chain. The organizations measure and evaluate the operational performance to compare it with other supply chains (Childerhouse and Towill 2002; Feldmann and Muller 2003; Li et al. 2006). Therefore, each company needs to set its own performance metrics, and each metric requires data collection. SCM performance metrics can be differently created according to the company's own needs and requirements.

Operational innovation has been central to some of the greatest success stories in recent business history, including Wal-Mart, Toyota, and Dell (Hammer 2004). Operation innovation can give high revenue growth, lower inventory using cross-docking and responsive purchase and distributed goods, lower prices, and high profits, but operational performance is difficult to realize (Hammer 2004). To improve the operational performance, a firm must to apply supply chain practices (Rice and Hoppe 2001; Hammer 2004; Li et al. 2006), change business culture (Hammer 2004), and introduce six sigma and lean. Supply chain practice could improve cash flows and reduce the cash-to-cash cycle time (C2C), which would help free up working capital to be invested in other products, better processes, and better financial performance (Rice and Hoppe 2001)—specifically growth, profitability, and capital utilization. Some studies have been done on the relationship between liquidity management and operating performance of a company. The concept of C2C leads to the premise that a reduction in the cash conversion cycle time will lead to a financial and operational improvement; however, the C2C concept assumes that a shortening of cycle time can be achieved without increasing costs and decreasing sales (Soenen 1993). However, to reduce C2C means reducing the terms of credit for the receiver and delaying payment to suppliers. This assumption would lead to a reduction of the product's attractiveness from the customer's perspective and an increase in the cost of goods supplied.

A number of researchers have analyzed the industry using C2C, such as shrimp suppliers in Thailand to a major retailer in the USA (Banomyong 2005), the relationship between C2C and return on asset (ROA), return on equity (ROE) in Japan and Taiwan (Wang 2002), and the relationship between corporate profitability and working capital in Greece (Lazaridis and Tryfonidis 2006). C2C is also used to measure the impact on profitability using ROA and ROE for manufacturing companies in Pakistan (Anser and Malik 2013), comparing C2C of merchandising and manufacturing industries in Turkey (Uyar 2009) and in Korea (Lee et al. 2010).

C2C is one of the liquidity ratios which measure a company's ability to meet cash needs (Fraser and Ormiston 2007). Traditional measures of corporate liquidity such as the current ratio, the quick ratio, and even net working capital are static in terms of what cash resources are ready for use at a given moment in time to satisfy the current obligations (Wang 2002). From this point of view, investors should focus their concern to cash flows from mobilizing inventory and receivable investments within the normal firm's operations. The operating cash flows are sensitive to declining sales and earnings (Richards and Laughlin 1980).

The previous research showed the results that reducing the C2C improved operating performance of a firm (Wang 2002); statistical significance between profitability and the C2C (Lazaridis and Tryfonidis 2006); and a strong negative relationship between variables of working capital management and profitability (Raheman and Nasr 2007). It suggested that companies could increase profits by correctly managing the C2C and keeping the components of C2C to an optimum level.

Some researchers tried to find the different C2C characteristics industry by industry. Retail and wholesale industry was found to have had a shorter C2C than manufacturing industry. The textile industry had the longest C2C, which might imply liquidity problems (Uyar 2009). Moreover, the findings indicated a significant negative correlation between the length of C2C and the firm's size in terms of net sales and total assets, and between the length of C2C and profitability. It concluded that smaller companies have longer C2Cs, and the longer the C2C, the less profitable they are (Raheman and Nasr 2007; Uyar 2009).

This paper focuses on C2C and four other finance performance metrics: revenue change rate (%), profit margin (%, net and gross), ROA (%), and current ratio. With the data from the selected industries, a comparison of the C2C of each industry has been conducted. This was then the correlation analysis of the C2C and finance performance metrics to see whether there is a positive or negative relationship between them.

2 C2C and SCM Performance

One important factor in business is ongoing performance measurement. Measuring supply chain performance is not a new practice. The problem is not a lack of possible metrics, but an overwhelming abundance of choices (Hofman 2004). A survey found that organizations that are best at SCM hold a 40–65 % advantage in their C2C over average organizations, while the top organizations carry 50–85 % less inventory than competitors (Sheridan 1998). C2C is measured using accounts receivable, inventory, and accounts payable data. C2C is defined as the sum of the days sales outstanding (DSO), plus the days of inventory (DOI), minus the days payable outstanding (DPO), that is C2C = DSO + DOI − DPO. DSO typically grows as revenue increases, and it shrinks as revenue decreases (Wojcik 2009). On the other hand, it is important to pay the suppliers earlier to be more profitable, as it might bring discount terms from the suppliers for early payment, which decreases the total costs.

The goal of many companies has been to shorten C2C, but in this paper, the relationship between C2C and financial performances is tested industry by industry (Fawcett et al. 2007). The C2C is a critical performance measure and was also selected as the measure which has the greatest impact on supply chain practice as it showed the direct financial benefits of SCM (Banomyong 2005; Christopher and Gattorna 2005; Farris et al. 2005; Fawcett et al. 2007).

Gartner has measured the top 25 firms in terms of supply chain performance since 2003 using peer opinion, return on assets, inventory turns, and revenue growth. The ranking gives which firms demonstrate operational and innovation excellence based on health of their value chains. To measure the overall operational excellence, Gartner identified key metrics with perfect order rates and total supply chain costs. Gartner also recommended the more detailed functional metrics that must be aligned with end-to-end supply chain goals to achieve the overall goal of operational excellence. C2C is a metric to assess the perfect order and SCM cost. The C2C metric is an important measure as it bridges across inbound material activities with suppliers, through manufacturing operations and outbound sales activities with customers (Farris and Hutchison 2002). C2C offers increased visibility of more decision variables; increases optimization of decisions in the supply chain; reduces suboptimization of financial decision within firms; and aids supplier decision making by eliminating the uncertainty of customer actions (Farris et al. 2005).

Gartner's measurement is based on Hofman's suggestions of the hierarchy of supply chain metrics (Gartner 2013; Hofman 2004). Hofman clarifies three levels of metrics for measuring supply chain performance. The top tier includes demand forecast, perfect order, and SCM cost, mid-level includes C2C to access cash flow and customer versus supplier balance, while bottom level to comprehend operational effectiveness includes supplier quality and raw material inventory purchasing costs. Hofman's top tier focuses on demand visibility based on push strategies such as demand visibility. C2C is not the sole metric to cover the end to end of supply chain, but it gives a certain diagnostic view based on inventory, accounts receivable, and accounts payable.

3 Research Methodology

The purpose of this paper is to see whether or not the C2C metrics could be used to measure SCM performance of a company. The second objective is to know the differences in C2C industry by industry and to know whether there is a positive or negative correlation between C2C and financial performance in eleven industries. These factors are closely studied to see how they affect the performance metrics, either positively or negatively, and the differences among industries. To see the tendencies of different industries, eleven industries are selected: the apparel industry (8 companies with 46 data from 2008 to 2014); the airline industry (27 companies with 138 data from 2007 to 2013); integrators (4 companies with 21 data from 2008 to 2013); the automobile industry (12 companies with 59 data from 2006 to 2013); the electronics industry (17 companies with 77 data from 2009 to 2013); the retail industry (8 companies with 40 data from 2006 to 2014); the logistics industry (9 companies with 45 data from 2008 to 2013); consumer packaged goods (CPG) industry (8 companies with 32 data from 2010 to 2013); the beverage and food industry (5 companies with 20 data from 2010 to 2013); the equipment and machine industry (4 companies with 12 data from 2010 to 2013); and finally the restaurant industry (2 companies with 8 data from 2010 to 2013). The data used in this paper were predominantly collected from www.investing.businessweek.com operated by Bloomberg, finance.yahoo.com, investing.money.msn.com, and companies' annual report. The total number of companies is 103 and 509 sample data from 2006 to 2014.

3.1 Data Collection and Descriptive Statistics

Tables 1, 2, 3, 4, 5, 6, 7, 8, 9, 10, 11, 12, and 13 give the descriptive statistics of the variables including weighted revenue change (%), weighted gross profit margin (GPM) (%), net profit margin (%), return on assets (%), current ratio, C2C, DSO, days in inventory (DII), and DPO. Table 1 presents the total descriptive statistics of all surveyed companies and shows moderate fluctuations on revenue change (the coefficient of variable (CV) is 1.43). But it is highest among the variables from −31.16 to 53.42 % with a weighted average change of 8.71 %. The variation of (GPM) (0.82) is less than net profit margin (1.21). In this context, presumably, we could assume that the cost of goods sold is less varied than expenses even revenue change is high. Table 2 shows average financial and operational performance including current ratio (CR) of all surveyed industries. Table 3 presents integrators including DHL, FedEx, TNT, and UPS, while the average C2C varies from 2.25 to 35.62 days with an average of 22.03 days and median of 26.79 days. DII (2.74 days) and DPO (26.07 days) show the lowest levels, followed by the retail industry (5.43 days) for DII and the restaurant industry (26.23 days) for DPO.

Table 4 shows the descriptive statistics of the selected apparel industry companies. The (GPM) is the biggest with 48.38 %, followed by the restaurant industry with 48.20 %. Inventory level is also the highest number of days with 127.27 that varied from 63.56 to 338.75 days, followed by the food and beverage industry with 11.39 days. Table 5 shows the electronics industry as having very high revenue

Table 1 Descriptive statistics of total surveyed, 2006–2014

	N	Min	Max	GM	Median	CV
Revenue change (%)	391	−31.16	53.42	8.71	5.93	1.43
Gross profit margin (%)	495	1.08	111.28	23.51	22.50	0.82
Net profit margin (%)	495	−6.90	34.20	6.47	4.27	1.21
ROA (%)	495	−6.64	27.46	5.57	5.19	0.82
Current ratio	495	0.33	4.78	1.31	1.15	0.52
C2C	495	2.25	35.62	22.03	26.79	0.55
DSO	495	0.05	150.34	34.69	28.28	0.71
DII	495	0.01	334.32	44.81	33.75	1.17
DPO	495	2.53	168.30	55.73	45.71	0.62

Table 2 Average of financial variables industry by industry

	REV.	GPM	NPM	ROA	CR	C2C	DII
Integrator	3.0	14.6	2.5	3.7	1.45	21.45	2.80
Apparel	10.8	58.1	11.8	13.0	2.19	84.06	128.65
Electronics	2.9	34.4	8.8	7.3	1.8	23.18	39.76
Automobile	11.1	18.1	4.7	3.8	1.22	29.70	46.83
Retail	12.2	20.0	2.4	4.5	1.02	−3.15	45.53
Airlines	16.0	17.3	3.5	1.4	0.91	−2.17	11.49
Logistics	2.6	14.9	2.5	4.9	1.24	15.46	5.44
CPG	3.2	35.8	13.1	11.0	1.23	26.63	73.75
Restaurant	7.9	47.4	12.9	12.7	1.21	24.24	35.14
Food and beverage	4.2	21.7	15.7	8.8	0.96	46.27	112.64
Equipment	6.4	35.2	7.5	6.7	1.88	92.21	101.60

Table 3 Descriptive statistics of integrators companies, 2008–2013

DPW, FedEx, TNT, UPS							
	N	Min	Max	GM	Median	CV	GM (1)
Revenue change (%)	16	−15.13	13.61	2.35	2.78	3.41	2.56
Gross profit margin (%)	21	6.95	25.52	15.99	13.77	0.46	16.26
Net profit margin (%)	21	−3.75	7.89	2.22	2.19	1.44	2.37
ROA (%)	21	−5.74	12.07	3.32	3.03	1.37	3.55
Current ratio	21	0.66	2.13	1.45	1.56	0.29	1.46
C2C	21	2.25	35.62	22.43	26.47	0.50	21.93
DSO	21	34.87	56.58	45.81	43.26	0.14	45.38
DII	21	0.60	5.47	2.69	2.35	0.63	2.78
DPO	21	15.57	47.32	26.07	21.40	0.38	26.23

Table 4 Descriptive statistics of apparel industry companies, 2008–2014

Benetton, Inditex, H & M, GAP, Ralph Lauren, LVMH, Adidas, Nike							
	N	Min	Max	GM	Median	CV	GM(1)
Revenue change (%)	28	−2.63	25.65	8.62	8.50	0.83	9.27
Gross profit margin (%)	36	7.56	65.80	48.38	57.09	0.28	45.90
Net profit margin (%)	36	3.53	17.22	10.26	10.29	0.36	10.87
ROA (%)	36	2.36	31.57	13.39	13.80	0.57	11.71
Current ratio	36	1.21	3.47	2.18	1.98	0.31	1.91
C2C	36	−42.17	252.10	80.85	85.73	0.98	88.82
DSO	36	3.86	56.94	24.09	25.78	0.69	40.10
DII	36	63.56	338.75	127.21	93.41	0.65	120.65
DPO	36	19.22	140.96	71.37	45.42	0.59	71.75

Table 5 Descriptive statistics of electronics industry, 2009–2013

Sony, Toshiba, Apple, Dell, RIMM, Samsung, LG, Nokia, and 9 others							
	N	Min	Max	GM	Median	CV	GM(1)
Revenue change (%)	60	−57.88	65.96	3.89	1.22	5.14	2.93
Gross profit margin (%)	77	10.78	80.16	33.72	32.79	0.52	34.37
Net profit margin (%)	77	−10.51	33.10	8.73	5.37	1.26	8.79
ROA (%)	77	−11.63	26.49	7.21	6.83	1.14	6.81
Current ratio	77	0.83	9.37	1.71	1.44	0.64	1.80
C2C	77	−78.54	133.33	21.38	21.79	2.34	23.18
DSO	77	18.10	122.68	53.91	49.76	0.45	53.87
DII	77	3.29	90.62	39.66	37.97	0.56	39.76
DPO	77	17.53	144.87	72.19	63.29	0.46	70.45

Table 6 Descriptive statistics of automobile industry, 2006–2013

Renault, Hyundai, Peugeot, Toyota, Honda, Nissan, GM, and 5 others							
	N	Min	Max	GM	Median	CV	GM(1)
Revenue change (%)	47	−26.76	25.59	4.95	5.67	2.24	11.07
GPM (%)	59	1.53	27.50	18.38	18.44	0.28	18.38
Net profit margin (%)	59	−9.27	14.83	4.15	4.55	1.04	4.72
ROA (%)	59	−7.73	12.87	3.29	3.07	1.12	3.79
Current ratio	59	0.50	1.81	1.17	1.16	0.25	1.18
C2C	59	−18.73	173.12	28.65	23.72	1.35	25.31
DSO	59	10.12	182.73	23.03	21.55	1.44	20.09
DII	59	19.00	85.86	47.45	49.74	0.31	43.68
DPO	59	18.10	85.82	51.80	47.06	0.34	46.39

change fluctuations with CV 5.14 from minimum level −57.88 % (Nokia in 2013) to 65.96 % (Apple in 2011). Table 6 displays data of the automobile industry. The industry's DSO varied greatly compared to other industry from 10.12 days (Renault in 2012) to 182.73 days (Nissan in 2013). Table 7 presents 8 retail industry companies from 2006 to 2014. Retailers' C2C fluctuates from −49.70 (Amazon in 2010) to 54.07 (Home Depot in 2011) with a high value of CV (−3.03).

Table 8 shows the descriptive statistics of airline industry companies. The net profit margin, ROA, and (CR) are the lowest with 1.34, 0.10, and 0.90 %. But the C2C level is also the shortest with −6.55 days. The level of fluctuation of ROA is 148.36 from −27.48 (United Airlines in 2008) to 20.17 (Delta in 2013). It is also important to note that 10 of 27 airlines have been recording negative average weighted ROA through recent 7–4 years. Even though airlines have the best weighted average C2C, their financial performance is the worst.

Table 9 shows the logistics industry, in which the revenue change fluctuates greatly with CV 7.35 from minimum level −84.77 % (Li and Fung in 2010 and

Table 7 Descriptive statistics of retail industry companies, 2006–2014

Carrefour, Wal-Mart, Costco, Metro, Tesco, Target, Home Depot, Amazon							
	N	Min	Max	GM	Median	CV	GM(1)
Revenue change (%)	36	−7.46	40.56	3.68	4.16	2.59	2.59
Gross profit margin (%)	45	7.07	34.57	20.30	21.25	0.37	20.99
Net profit margin (%)	45	−0.15	6.17	2.31	1.88	0.75	2.58
ROA (%)	45	−0.33	11.04	4.32	4.50	0.71	4.72
Current ratio	45	0.65	1.71	0.95	0.88	0.30	1.03
C2C	45	−49.70	54.07	−9.70	−14.77	−3.03	−0.04
DSO	45	0.47	38.90	13.95	5.57	0.86	9.50
DII	45	19.08	86.77	46.66	44.85	0.35	46.97
DPO	45	31.92	110.64	70.31	52.62	0.42	56.51

Table 8 Descriptive statistics of airline industry companies, 2007–2013

Companies name refer to Table 1							
	N	Min	Max	GM	Median	CV	GM(1)
Revenue change (%)	111	−41.05	84.87	8.84	8.41	2.27	15.96
GPM (%)	138	−11.32	56.47	19.70	20.09	1.27	19.65
Net profit margin (%)	138	−39.31	162.15	1.34	1.65	55.21	3.49
ROA (%)	138	−27.48	20.17	0.10	1.07	148.36	1.42
Current ratio	138	0.19	2.28	0.90	0.81	0.53	0.80
C2C	138	−131.23	82.85	−6.55	−2.93	−5.41	−2.17
DSO	138	4.19	99.59	21.92	21.04	0.59	22.90
DII	138	0.21	55.35	11.39	9.94	0.84	11.07
DPO	138	1.80	214.85	39.64	28.43	0.92	36.09

Table 9 Descriptive statistics of logistics companies, 2008–2013

Con-Way, Ryder Sys., ND, Yamato Holdings, Li & Fung, and 4 others							
	N	Min	Max	GM	Median	CV	GM(1)
Revenue change (%)	16	−84.77	25.98	−2.4	2.65	−7.35	2.6
Gross profit margin (%)	22	3.82	45.48	17.2	18.15	0.55	16.6
Net profit margin (%)	22	−2.52	5.23	2.3	2.72	0.63	2.5
ROA (%)	22	−3.72	21.18	4.3	2.97	1.27	4.9
Current ratio	22	0.14	1.87	1.21	1.24	0.32	1.24
C2C	22	5.88	28.81	15.81	16.26	0.34	15.46
DSO	22	36.53	76.89	51.65	51.27	0.21	51.65
DII	22	0.11	22.18	5.43	1.79	1.02	4.23
DPO	22	21.32	58.64	39.70	42.74	0.28	40.43

−18.53 % for Ryders in 2009) to 25.98 % (Norbert Dentressangle in 2011). Tables 10 and 11 show data of the CPG and restaurant industries. The industries' parameters are stable within CV = 1 except C2C 1.51 (CPG) and 1.28 (restaurant). Table 12 shows the descriptive data of the beverage and food industry with revenue change 1.88 from −4.82 % (Nestlé in 2011) to 32.53 % (Coca-Cola in 2011) and C2C 2.58 from −107.42 days (Anheuser Busch in 2013) to 322.18 days (Diageo in 2013) in terms of CV. Table 13 presents four companies in the equipment and machine industry. Its revenue changes vary from −15.51 % (Caterpillar in 2013) to 41.21 % (Caterpillar in 2011) with a CV of 1.69.

Companies' data were taken 2–7 years from their financial statements weighted more on recent year. When the author takes four years data for a company, the weighted data was calculated by years of the first year (y_i) by 40 % plus the second year (y_{i-1}) by 30 % plus the third year (y_{i-2}) by 15 % plus the fourth year (y_{i-3}) by 5 % (F2).

Table 10 Descriptive statistics of CPG companies, 2010–2013

Unilever, P&G, J&J, Kimberly-Clark, Colgate-Palmolive, and 2 others							
	N	Min	Max	GM	Median	CV	GM(1)
Revenue change (%)	24	−2.98	12.21	3.94	3.39	0.95	3.24
Gross profit margin (%)	32	12.42	114.01	43.2	42.49	0.75	43.3
Net profit margin (%)	32	7.41	21.65	13.2	14.31	0.30	13.1
ROA (%)	32	6.28	19.72	11.1	10.37	0.31	11.0
Current ratio	32	0.46	2.38	1.24	1.14	0.46	1.11
C2C	32	−146.86	97.85	30.93	29.63	1.51	26.63
DSO	32	19.86	61.40	41.53	42.25	0.28	41.86
DII	32	52.74	128.70	74.07	67.08	0.27	73.75
DPO	32	38.21	261.16	84.87	76.15	0.47	88.24

Table 11 Descriptive statistics of restaurant companies, 2010–2013

Starbucks, Mcdonald's							
	N	Min	Max	GM	Median	CV	GM(1)
Revenue change (%)	6	1.95	13.67	8.4	10.62	0.62	7.9
Gross profit margin (%)	8	38.89	58.75	48.20	48.16	0.20	47.9
Net profit margin (%)	8	0.06	20.55	13.6	15.24	0.55	12.9
ROA (%)	8	0.07	16.92	13.8	15.46	0.41	12.7
Current ratio	8	0.24	1.90	1.34	1.47	0.39	1.21
C2C	8	−4.02	66.29	22.46	15.51	1.28	24.24
DSO	8	10.32	18.21	15.09	15.45	0.21	15.33
DII	8	2.61	77.95	33.60	23.84	1.02	35.14
DPO	8	21.50	40.10	26.23	24.37	0.23	26.23

Table 12 Descriptive statistics of beverage and food industry, 2010–2013

Coca-Cola, Pepsico, Nestle, Anheuser-Busch, Diageo							
	N	Min	Max	GM	Median	CV	GM(1)
Revenue change (%)	15	−4.82	32.53	4.84	2.72	1.88	4.18
Gross profit margin (%)	20	9.65	64.64	25.1	17.29	0.73	25.1
Net profit margin (%)	20	9.42	38.89	16.5	15.83	0.47	15.7
ROA (%)	20	3.52	30.66	9.5	8.69	0.59	8.8
Current ratio	20	0.34	1.76	0.97	0.98	0.43	0.96
C2C	20	−107.42	322.18	52.38	19.83	2.58	46.27
DSO	20	24.14	65.66	39.13	36.96	0.30	39.16
DII	20	39.83	346.80	111.39	63.06	0.99	112.64
DPO	20	37.72	189.58	98.14	91.70	0.49	105.54

Table 13 Descriptive statistics of equip. and machine ind., 2010–2013

Caterpillar, Cummins, Deere & Co, Komatsu							
	N	Min	Max	GM	Median	CV	GM(1)
Revenue change (%)	12	−15.51	41.21	10.4	8.53	1.69	6.4
Gross profit margin (%)	16	11.28	72.87	41.1	46.72	0.54	40.2
Net profit margin (%)	16	0.67	10.24	7.4	8.31	0.34	7.5
ROA (%)	16	0.50	15.84	6.7	6.01	0.59	6.7
Current ratio	16	1.34	2.57	1.85	1.87	0.21	1.88
C2C	16	−19.51	225.46	90.08	96.49	0.85	92.21
DSO	16	36.75	117.52	66.50	55.19	0.42	65.20
DII	16	58.06	167.90	99.84	94.35	0.36	101.60
DPO	16	38.19	133.16	76.26	68.48	0.44	74.59

$$P_i = [(0.5y_i) + (0.3y_{i-1}) + (0.2y_{i-2})] \quad \text{(F1)}$$

$$P_i = [(0.4y_i) + (0.3y_{i-1}) + (0.15y_{i-2}) + (0.05y_{i-3})] \quad \text{(F2)}$$

$$P_i = [(0.4y_i) + (0.3y_{i-1}) + (0.15y_{i-2}) + (0.1y_{i-3}) + (0.05y_{i-4})] \quad \text{(F3)}$$

$$P_i = [(0.4y_i) + (0.3y_{i-1}) + (0.15y_{i-2}) + (0.075y_{i-3}) + (0.04y_{i-4}) + (0.035y_{i-5})] \quad \text{(F4)}$$

$$P_i = [(0.4y_i) + (0.3y_{i-1}) + (0.15y_{i-2}) + (0.075y_{i-3}) + (0.04y_{i-4}) + (0.025y_{i-5}) + (0.01y_{i-6})] \quad \text{(F5)}$$

$$P_i = [(0.4y_i) + (0.3y_{i-1}) + (0.15y_{i-2}) + (0.07y_{i-3}) + (0.04y_{i-4}) + (0.275y_{i-5}) + (0.0075y_{i-6}) + (0.005y_{i-7})] \quad \text{(F6)}$$

P_i Weighted financial or operational performance variables.

3.2 Variables and Hypothesis

SCM Performance: The short-term objectives of SCM are primarily to increase productivity and reduce inventory and cycle time, while the long-term objectives are to increase market share and profit for all members of the supply chain (Tan et al. 1998). Some of the inventory strategies can be used for improving the variable of the inventory days of supply, real-time inventory tracking for example, using advanced technology such as RFID provides real-time inventory information throughout the supply chain, CPFR (Collaborative Planning, Forecasting, and Replenishment), and synchronizing supply/demand planning (Farris et al. 2005). Excellent supply chain performance leads to strong financial performance such as earnings per share, ROA, and net profit margin (Hofman 2004).

Hypothesis 1: The organizations with a low level of C2C will have a high level of financial performance (gross, net profit rate, and revenue growth).

Organizational performance refers to how well an organization achieves its market-oriented goals as well as its financial goals (Yamin et al. 1999). Financial metrics have served as a tool for comparing organizations and evaluating an organization's behavior over time (Holmberg 2000). A number of prior studies have measured organizational performance using both financial and market criteria, including return on investment (ROI), market share, profit margin on sales, the growth of sales, the growth of market share, and overall competitive position (Vickery et al. 1999; Stock et al. 2000). Gross profit and net profit rate are used to as profitability ratios for comparing a company's management with industry competitors (Fraser and Ormiston 2007). Gross profit rate is measured as one of the performance measures to see at first how efficiently a company is managed. Generally, as a company grows, gross profit grows as well. However, it does not mean that a company which has higher gross profits is more efficient than one which has smaller gross profits. Net profit rate is also another performance measure. The objective of generating high net profit is the one which shows the company's competitive power. Therefore, in order to take a company's performance and profitability into consideration, net profit rate is also reviewed as sometimes a company shows a positive operating profit rate with a negative net profit rate. It also indicates whether the business has enough sales volume in order to cover the fixed costs and still generate an acceptable profit.

ROA is measured as another indicator of a company's profitability. ROA shows how efficiently a company manages its assets to generate profits, or how effectively a company allocates its resources, debt, and equity, to convert them into profits.

Hypothesis 2: The organizations with a low level of C2C will have a high level of liquidity (CR) and ROA.

Hypothesis 3: The companies within the same industry category will have homogeneity characteristics and heterogeneity characteristics in comparison with the other industries.

Although organizations have understood the importance of implementing good SCM, they often do not know exactly how to realize the goal, due to a lack of understanding of what constitutes a comprehensive set of SCM practices and strategy. By matching SCM metrics and financial performance, this study provides SCM managers with a useful tool for evaluating their SCM activities.

4 Results and Analysis

C2C and financial performance (H1): The results of the analysis in Tables 14 and 15 show that C2C has a relationship with gross profit, not with revenue growth rate nor net profit margin. The relationship between C2C and financial performance, such as revenue growth rate and net profit margin, is not statistically significant, but significant with the GPM. But the higher GPM will make longer the C2C (0.00109). It means the bigger the C2C, the bigger the profits. C2C presents the length of the payment of purchases to the supplier and remittance from the customers. Previous literature has concluded that companies can increase their profitability by shortening the C2C (Lazaridis and Tryfonidis 2006; Raheman and Nasr 2007; Shin and Soenen 1998). This study shows the negative relationship between C2C and revenue growth, as well as net profit margin, but this was not significant. So, the author found it difficult to prove the hypothesis. Gartner has published the top 25 companies in terms of supply chain using peer opinion, ROA, inventory turns, and revenue growth (Gartner 2013). Table 16 shows the differences between top 25 companies and the other unranked companies. No significant difference was found for C2C and revenue growth rate between the top 25 and other unranked companies.

C2C with liquidity and ROA (H2): Tables 17 and 18 showed the relationship of C2C with liquidity and ROA. Current ratio (CR) is used to measure a company's ability to meet payment obligations pay in the short term using current assets over

Table 14 Pearson's coefficient of correlation

	GPM	NPM	C2C	Rev. growth
Gross profit margin (GPM)	1	0.525**	0.313**	−0.046
Net profit margin (NPM)		1	0.127	−0.008
C2C			1	−0.162
Revenue growth rate				1

*$p < 0.05$

Table 15 Regression analysis with C2C

Independent variables	Rev. growth rate	GPM	NPM
Coefficient	−0.00036	0.00109	−0.00017
Constant	0.101	0.263	0.064
Sig.	0.101	0.001	0.200
R^2	0.026	0.098	0.016

Table 16 Independent *t*-test between SCM top 25 and other companies

Independent variables	*N*	Average	Sig.
Gross profit margin	78	26.66 %	0.052
	25	36.13 %	
Return on asset	78	3.91 %	0.000***
	25	11.42 %	
Inventory turns	74	71.77	0.016**
	25	16.15	
Net profit margin	78	4.84 %	0.000***
	25	12.68 %	
Current ratio	78	1.18	0.016**
	25	1.70	
Revenue growth rate	78	9.37 %	0.996
	25	9.36 %	
C2C	78	21.22	0.629
	25	27.46	

$^{***}p < 0.01$, $^{**}p < 0.05$

Table 17 Pearson's coefficient of correlation

	ROA	C2C	Current ratio
ROA	1	0.151	0.455***
C2C		1	0.343***
Current ratio			1

$^{***}p < 0.01$

Table 18 Regression analysis with C2C

Independent variables	ROA	Current ratio
Coefficient	−0.00016	0.00422
Constant	0.541	1.212
Sig.	0.127	0.000***
R^2	0.013	0.118

$^{***}p < 0.01$

current liabilities. The results of study showed the significant relationship between C2C and CR, but not between C2C and ROA, profitability on assets.

Comparing C2C industry by industry (H3): Table 19 presented the result of ANOVA (analysis of variance) among industries with financial performance along with C2C and Scheffe test comparing industry by industry. According to the ANOVA, each industry has its own characteristics of financial performance and C2C. Table 2 shows the differences. The retail and airline industries have negative C2Cs with the highest revenue growth rate compared to the other 9 industries; the equipment and apparel industries have longer C2Cs with higher average revenue growth rates. There exist differences between industries based on the Scheffe test. The GPM of apparel industry is much higher than that of integrator, automobile, retail, airline, and logistics industries.

Table 19 ANOVA and Scheffe test for industry by industry

	ANOVA	Scheffe test			
		Ind. A	Ind. B	Diff. (A − B) (%)	Sig.
Gross profit margin	0.000	Integrator	Apparel	−42.73	0.041
		Apparel	Automobile	40.40	0.002
		Apparel	Retail	36.93	0.024
		Apparel	Airline	38.88	0.000
		Apparel	Logistics	41.88	0.003
Return on asset	0.000	Apparel	Airline	11.73	0.002
		Airline	CPG	−9.57	0.020
Net profit margin	0.001	No significant differences			
Current ratio	0.000	Apparel	Airline	1.27	0.008
		Electronics	Airline	0.88	0.017
Revenue growth rate	0.020	No significant differences			
C2C	0.011	No significant differences			

5 Conclusion

The result of this study is that it is very difficult to prove the relationship between C2C and financial performance. The C2C has homogeneity characteristics within industry and different characteristic compared to other industries, but no significances statistically when pairwise compared between industries (Table 19). Shortening the payments to supplier creates liquidity pressures to other companies of the supply chain. Within the supply chain, a leading player, likely located downstream, could take initiative to shorten C2C significantly (Losbichler et al. 2008). A strong player in the supply chain could finance weak suppliers and customers. If the strong player tries to reduce C2C with a value chain perspective instead of a company and focus more on reducing inventory in place of shortening payable and receivable, the supply chain could achieve a durable competitive advantage. This study focuses only on company-wide performance rather than on supply chain performance of companies. Even with a focus on the company, shorten C2C would not contribute to financial nor operational performance.

Acknowledgments This article is made possible through the support of Mr. Varun Banthia to collect data.

References

Anser R, Malik QA (2013) Cash conversion cycle and firms' profitability: a study of listed manufacturing companies of Pakistan. IOSR J Bus Manag 8(1):83–87

Bhagwat R, Sharma MK (2006) Performance measurement of supply chain management: a balanced scorecard approach. Comput Ind Eng 53:43–62

Banomyong R (2005) Measuring the cash conversion cycle in an international supply chain. In: Proceeding of annual Logistics Research Network (LRN) conference
Childerhouse P, Towill DR (2002) Simplified material flow holds the key to supply chain integration. Omega Int J Manag Sci 31(1):17–27
Christopher M, Gattorna J (2005) Supply chain cost management and value based pricing. Ind Mark Manag 34:115–121
Farris MT II, Hutchison PD (2002) Cash-to-cash: the new supply chain management metric. Int J Phys Distrib Logistics Manag 32(4):288–298
Farris MT II, Hutchison PD, Hasty RW (2005) Using cash-to-cash to benchmark: service industry performance. J Appl Bus Res 21(2):113–124
Fawcett SE, Ellram LM, Ogden JA (2007) Supply chain management: from vision to implementation. Pearson Education Inc., Upper Saddle River
Feldmann M, Muller S (2003) An incentive scheme for true information providing in supply chains. Omega Int J Manag Sci 31(2):63–73
Fraser LM, Ormiston A (2007) Understanding financial statements. Pearson Education Inc, Upper Saddle River
Gartner (2013) The Gartner supply chain top 25 for 2013. http://www.gartner.com/imagesrv/summits/docs/na/supply-chain/Gartner-2013-SupplyChain-Top25.pdf
Gunasekaran A, Patel C, McGaughey RE (2004) A framework for supply chain performance measurement. Int J Prod Econ 87:333–347
Hammer M (2004) Deep change: how operational innovation can transform your company. Harv Bus Rev Logistics Manag 82(4):84–93
Hofman D (2004) The hierarchy of supply chain metrics: diagnosing your supply chain health. AMR Res rep. http://www.concise-management.ch/medialibrary/site1/AMR_Research_REPORT_16962_The_Hierarchy_of_Supply_Chain_Metrics.pdf
Holmberg S (2000) A systems perspective on supply chain measurements. Int J Phys Distrib Logistics Manag 30(10):847–868
Kim D, Kwon OK, Baik I (2003) An empirical study on the supply chain asset performance of Korean companies. IE Interfaces 16(2):167–173
Kim J, Kim D (2004) An exploratory study on developing SCM performance measurements in small and medium enterprises. J Inf Syst 13(2):195–223
Lazaridis L, Tryfonidis D (2006) The relationship between working capital management and profitability of listed companies in the Athen stock exchange. J Financ Manag Anal 19(1):1–12
Lee HJ, Song SH, Hong SJ (2010) Linking supply chain management and financial performances: evidence from Korean companies. Korea Logistics Rev 20(4):119–143
Li S, Ragu-Nathan B, Ragu-Nathan TS, Subba Rao S (2006) The impact of supply chain management practices on competitive advantage and organizational performance. Omega Int J Manag Sci 34:107–124
Losbichler H, Mahmoodi F, Rothboeck M (2008) Creating greater shareholder value from supply chain initiatives. Supply Chain Forum Int J 9(1):82–91
Raheman A, Nasr M (2007) Working capital management and profitability: case of Pakistani firms. Int Rev Bus Res Pap 3(1):279–300
Rice JB, Hoppe RM (2001) Supply chain vs. supply chain: the hype & the reality. Supply Chain Manag Rev 5(5):46–54
Richards VD, Laughlin EJ (1980) A cash conversion cycle approach to liquidity analysis. Financ Manag 9:32–38
Shen T (2005) Linking supply chain practices to operational and financial performance. Supply chain 2020 project working paper
Sheridan JH (1998) The supply chain paradox. Ind Week 247(3):20–29
Shin H, Soenen L (1998) Efficiency of working capital and corporate profitability. J Fin Pract Educ 8:37–45
Soenen LA (1993) Cash conversion cycle and corporate profitability. J Cash Manag 13(4):53–58
Stock GN, Greis NP, Kasarda JD (2000) Enterprise logistics and supply chain structure: the role of the fit. J Oper Manag 18(5):531–547

Supply-Chain Council (2008) Supply chain operations reference model version 9.0. Supply Chain Council, Inc.
Tan KC, Kannan VR, Hanfield RB (1998) Supply chain management: supplier performance and firm performance. J Purchasing Mater Manag 34(3):2–9
Uyar A (2009) The relationship of cash conversion cycle with firm size and profitability: an empirical investigation in Turkey. Int Res J Financ Econ 24:186–193
Vickery S, Calantone R, Droge C (1999) Supply chain flexibility: an empirical study. J Supply Chain Manag 35(3):16–24
Wang YJ (2002) Liquidity management, operating performance, and corporate value: evidence from Japan and Taiwan. J Multinatl Financ Manag 12:159–169
Wojcik TC (2009) Simple metrics for your supply chain: how to define the first key measurements: encore consulting white paper. http://www.encore-c.com/spotlight/Simple_Metrics_for_Your_Supply_Chain.pdf
Yamin S, Gunasekruan A, Mavondo FT (1999) Relationship between generic strategy, competitive advantage and firm performance: an empirical analysis. Technovation 19(8): 507–518

Key Performance Indicator Framework for Measuring Healthcare Logistics in ASEAN

Soriya Hoeur and Duangpun Kritchanchai

Abstract Performance measurement or benchmarking in healthcare logistics requires right components and meaningful key performance indicators (KPIs), but having them is very challenging since it involves factors both outer and inner dependence among those factors. This chapter presents a model, based on analytic network process (ANP) model, for sorting and prioritizing the meaningful and a sufficient number of components with individually unique KPIs in healthcare logistics. They are used to benchmark the healthcare logistics performance in different countries in ASEAN such as Singapore, Malaysia, Thailand, Myanmar, and Lao. Those components and their KPIs are identified based on the literature review, and checked and weighted by experts. ANP model is used to prioritize each logistics component and its KPIs. Super Decisions software is used to do all related computations in ANP model such as supermatrix (limit matrix) to get synthesized priorities. The research result offers a good and implementable healthcare logistics performance measurement framework from which managers or researchers in healthcare industry can use to carry out the healthcare logistics performance benchmarking within particular organizations in the same or different countries.

Keywords Key performance indicator · Logistics component · Healthcare logistics

1 Introduction

Logistics activities, which have been practiced since pyramid in ancient Egypt (DHL-logbook 2008), now are becoming very important to healthcare industry due to complexities deriving from higher standard requirements such as lead time, strict corporate governance requirements, rigorous regulatory environments, local

S. Hoeur (✉) · D. Kritchanchai
Mahidol University, 73170 Salaya, Nakhon Pathom, Thailand
e-mail: hoeur.soriya@gmail.com

V. Kachitvichyanukul et al. (eds.), *Toward Sustainable Operations of Supply Chain and Logistics Systems*, EcoProduction,
DOI 10.1007/978-3-319-19006-8_3

healthcare systems, and pharmacovigilance (or drug safety). Moreover, healthcare market is growing increasingly. Although it accounts for only 17 % of the size of the consumer goods in Asia Pacific (without Japan), it boasted an impressive average market expansion service penetration rate of 45 % in 2013 (DKSH 2014). Therefore, it is beneficial to have good logistics practices for healthcare industry.

Recently, there are extensive literatures reviewing on the performance measurement and improvement technique in healthcare logistics. Dembińska-Cyran (2005) suggested that to reduce the cost associated with supply chain process, there is a need of better logistics management. He studied on four major activities in hospital logistics: inventory management activities, transportation management activities, production activities, and distribution activities. Kelle et al. (2012) discussed pharmacy supply chain and managerial practice of the hospital focusing merely on inventory management and its key performance indicators, namely the expected number of daily refills, the service level, and storage space utilization as the trade-off for the tactical implementation. Jørgensen et al. (2012) presented a framework measuring the performance of the overall flow and individual processes of healthcare logistics flow across the countries, Denmark and Japan. The measurement is focusing on four parameters, namely logistics, technology, structure, and procedure. Kim (2005) proposed the supply chain management (SCM) system focusing on inventory control optimization based on three components, namely pharmaceutical companies, a wholesaler, and hospital itself. Grigoroudis et al. (2012) used UTASTAR method based on balance scorecard approach to develop a strategic performance measurement system measuring on four perspectives, namely financial, customer, internal business, and learning and growth perspective.

Recently, multi-criteria decision-making methods play a very important role in making important decisions, which cannot be determined straightforwardly, and application of analytic network process (ANP) has been seen in many extensive researches in various industries. ANP can be used in both qualitative and quantitative research for screening the best alternative (company, location, or system)-based criteria influencing on each other (Çelebi et al. 2010; Cheng et al. 2005; Percin 2008; Jharkharia and Shankar 2007). Additionally, ANP was also used to give the weight for performance indicators or factors in performance measurement (Kayakutlu and Buyukozkan 2011; Yüksel and Dagdeviren 2007; Van Horenbeek and Pintelon 2014; Görener 2012).

As a conclusion, it can be summarized that there are available researches about conceptual frameworks and discussion on healthcare supply chain management or healthcare logistics performance measurement; however, there is a lack of methodological approach of developing key performance indicator framework under the constraint to the relationship between those components. The objective of this study, therefore, is to propose the key performance indicator framework scoping out the healthcare logistics components from supplier to point of care (pharmacy storeroom). In developing the framework, ANP model is used to determine the relationship between the components. Since the logistics components, best practices, or activities do influence each other, the link between the components is necessary to develop the framework reflexing the real world of logistics.

After the introduction, the remainder of this paper is organized as follows. Section 2 of this paper describes the developed key performance indicator framework. An overview and application of the ANP methodology are presented in detail in Sects. 3 and 4. Finally, a discussion and a conclusion are given in Sects. 5 and 6.

2 Literature Review

The literature review mainly aims to identify the criteria which need to be considered in healthcare logistics. Moreover, logistics components and their KPIs have been also captured. The outcome of literature review, together with the inputs from academia, has been used to set up the healthcare logistics KPI framework. In addition to this, an ANP-based model has been developed for the final KPIs prioritizing.

2.1 Healthcare Logistics

Traditionally, the healthcare industry has identified itself differently from other businesses in terms of operations. It fundamentally had unstable productions or operations schedules that is so hard to be controlled and projected (Gary Jarrett 1998). However, healthcare industry, to fulfill the market demand nowadays, has been learning from the best practices of manufacturing or retail industry, which has more experiences in adopting advanced logistics techniques enabled by technology. The best practices such as e-commerce, performance management, scanning technology, and data standardization normally are complemented by scientific and mathematical model, so that the performance of the organization can be highly achieved. Therefore, it is believed that healthcare industry can benefit applying the best practices from manufacturing or retail industry. However, it is required a specific KPIs reflecting the healthcare logistics components or practices due to the different nature or context between the industries (Pohl et al. 2012; Cho et al. 2012; Everard and CPM 2000). This reflexes that the business process in healthcare industry is not quite different from the one in manufacturing industry.

Healthcare logistics is a kind of series of network or system (of supplier, manufacturer, wholesaler, and hospital) performing different activities, so that various products and services of the healthcare business can be served to customers over time and places (Gary Jarrett 1998; Kim 2005; Grigoroudis et al. 2012; Kumar et al. 2008; Chandra and Kachhal 2004). Activities (purchasing, information technology, inventory control of stock and supply, transportation of patients, medical products delivery, warehousing, etc.) should aim for right item, place, quantity, costs, and information (Dembińska-Cyran 2005; Bose 2003; Coyle et al. 2003). The activities are apparently different, but they strongly influence each other. As a matter of the fact, purchasing is different from warehousing or even inventory management. However, to have an effective and efficient healthcare logistics, these activities should be connected (Chandra and Kachhal 2004).

2.2 Performance Measurement

Performance measurement is good for visualizing a particular status. It monitors performance of particular activity, visuals the current state of organizational behaviors, and helps the organization reach its potential success and achieving strategic goals (Behn 2003; Fawcett and Cooper 1998). Moreover, it is believed that performance measurement models the behaviors of both managers and their staff, so that they can improve the business process, archive democratization, and transparency, and make the most of limited resources (Delorme and Chatelain 2011). Performance measurement system requires identification of indicators. The indicator translates activity, situation, or behavior and provides information statistically and logically in both quantitative and qualitative mean.

2.3 Criteria for Developing KPIs Framework

To measure the performance of healthcare logistics aiming for efficiency and effectiveness, logistics components and their key performance indicators have been identified. From the literature review and discussion with experts and academia, the criteria related to logistics components and their KPIs are presented as following. These criteria form the basis for the development of framework and an ANP model.

Commodity-centralized purchasing and supply: It is an activity related to purchasing of supply management and distribution functions between the supplier and the purchasing team of an organization (hospital) (Pohl et al. 2012).

- Delivery reliability: It refers to the capacity to deliver the products to the customer on promising criteria (time, quantity, location, etc.) (Çelebi et al. 2010; Stock et al. 1998; Supply Chain Council 2012).
- Quick response: It refers to the capability to provide the services or products to meet a particular customer delivery requirement. Quick respond promotes speed advantages in the supply chain (Çelebi et al. 2010; Gunasekaran et al. 2004).
- E-procurement: It involves the use of internet technology to access online cataloges, conduct data interchange, and price comparison, perform product evaluations, etc. More importantly, it is used to exchange of data to facilitate the financing and payment aspect of the business transaction (Pohl et al. 2012; Truman 2011; Allsop et al. 2010).

Warehousing: Warehouse refers to a large building for storing goods which is cataloged, shipped, and received. In warehouse, there are performances such as managing the storage and preservation of goods or materials until they are dispatch to the consumers.

- Space utilization: It is about how efficient the resource (space of the warehouse) is utilized. Good space utilization promotes productivity, operating ratio of actual to planned working house (Kelle et al. 2012; Van Horenbeek and Pintelon 2014; Gu et al. 2010).

- Order sorting: It is associated with picking area, storage area, replenishment of the picking, and a sorter (Bozarth and Vilarinho 2010).
- Receiving completeness: It refers to the activities needed to be done after receiving goods. They are unloading the product of the transport carrier, updating inventory record, inspecting to find whether there is any quality and quantity inconsistency (Supply Chain Council 2012).
- Cross-docking: It is the process that goods is kept in shipping docks for being distributed directly to the customers. Cross-docking commonly consists of trucks and dock doors on inbound and outbound side at the cross-docking distribution terminal, and it requires only a small warehouse space (discuss with expert).

Inventory management: It involves processes that identify physical inventory monitoring, inventory valuation, visibility, report actual and projected inventory status, and other functions associated with tracking (Abbasi 2011).

- Inventory visibility: It allows the organization to access and share the information related to the inventory status, so that the organization can have more confident in making decision in ordering and demand management (Zhang et al. 2011; Caridi et al. 2014).
- Inventory availability: Inventory availability refers to on-hand inventory available to service current requirements. It is associated with inventory level, safety stock, and stock out (Zinn et al. 2002; Kumar and Rahman 2014).
- Inventory accuracy: It refers to accurate statistics and status of the inventory. Inventory inaccuracy happens due to the inability of capturing transaction error, scanning errors, shrinkage errors, supply errors, thieves, and so on (Kumar and Rahman 2014)

Transportation and distribution: Transportation and distribution is the link allowing goods flow from origin point to consumption point, so buyers and sellers are always connected. However, in this paper, it aims access the performance of inventory flow from warehouse (seller) pharmacy storeroom (buyer) (Coyle et al. 2003).

- Delivery perfect condition: It refers to the percentage of delivery in undamaged state that meets specification, has the correct configuration, and faultlessly installed (Supply Chain Council 2012).
- Order delivery in full: It is the percentage of orders which all of the items are received by customer in the quantities committed (Supply Chain Council 2012).
- Order delivery to customer commit date: It is all about the order received on times as defined by the customer, and the delivery is made to the correct location and customer entity (Supply Chain Council 2012).
- Urgent delivery: Urgent delivery is what the hospital cannot avoid. Thus, urgent delivery in healthcare logistics refers to the ability to cope with such unscheduled and un-planned delivery (discuss with experts).

Information and technology management: Technology today is an important component for logistics. It serves as the enabler of logistics operation improvement (Jørgensen et al.; Kim 2005)

- Each of use and usefulness: It refers to the favorite level of the staff using technology in healthcare logistics (discuss with experts).
- Product identification: It refers to the unique identity of the medical products. It normally refers to the product identification carriers (barcode, RFID, QR code) whether they are the standard (belong to GS1) or just the identification of the organization itself (Ochiai 2010).
- Accurate and reliable tracking: Tracking helps locate products in the case of a recall without employees going into each and every ward, or hospital pharmacy or operating suite and physically checking whether the product is there (Halstead 2010).
- Information availability: It refers to the information technology capability in display the information in full visibility down to the products movement across the logistics components (Truman 2011; Allsop et al. 2010).
- Information accuracy: It refers to the reliable level of the information displayed by the technology (Truman 2011; Allsop et al. 2010).

Total logistics costs: It refers the total costs associated with logistics components from the point of origin to the point of care (discuss with experts).

- Transportation costs: The expenditure related to products movement from the organization to the customer such as manhour and loading and unloading facilities (Coyle et al. 2003; Behn 2003).
- Warehousing costs: Warehousing costs is the costs associated with warehouse as the building such as maintaining the warehouse (discuss with experts).
- Inventory holding costs: The expense associated with holding the goods in the storage that could be in warehouse, shipping containers, trailers, or railcar (Taylor 2007).
- Administrative costs: It refers to shipper-related and administration cost. Shipper-related costs refer to expense-related to shipper-related functions performed by the shippers. Administration costs refer to the expense associated with cooperated management and support staff providing logistics support such as supply chain planning and analyzing staff. Computer hardware and software are also included in administration costs (Taylor 2007).

3 Research Methodology: The Analytic Network Process

A suitable methodology that can decode the high-level and complex relationship of the model in Fig. 1, in order determine the importance of each logistics component, is a critical issue. This methodology should be able to use qualitative and intangible factors to achieve goal. In this case, healthcare logistics KPIs framework is

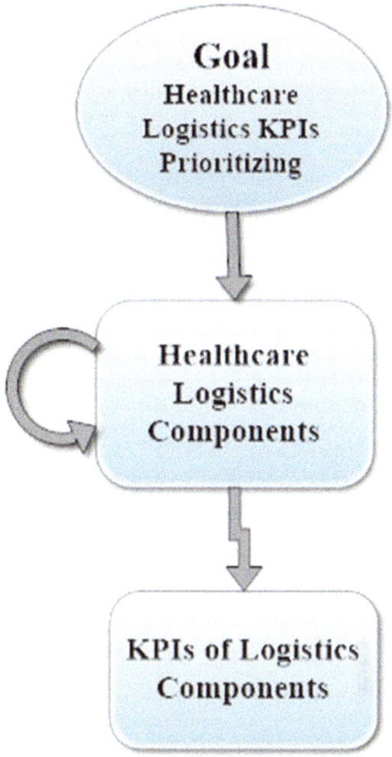

Fig. 1 ANP conceptual framework

developed using the analytic network process, which allows measurement of inner-dependency among the logistics components. At the same, the AHP is used in order to determine the factor weights of the dependency or independency, and the influence of logistics components on their KPIs and their importance.

The initial multi-criteria decision making with a general theory of measurement known as analytic hierarchy process (AHP) can be used to establish in both physical and social domain (Saaty and Vargas 2013). Whether there is certainty or not, AHP is designed to cope with the intuitive, the rational, and the irrational when multi-objective, multi-criterion, and multi-actor decision are made for any number of alternatives (Bose 2003). Under no circumstances, can all problems be structured hierarchically. Some problems involve the interaction and dependency between higher level elements on the lower level or within the same level elements. It is why AHP was extended; it is called ANP (Saaty and Vargas 2013; Yüksel and Dagdeviren 2007; Van Horenbeek and Pintelon 2014).

4 Application of ANP Methodology

The application of ANP has been implemented in extensive literature in different purposes and different industries with different steps (Van Horenbeek and Pintelon 2014; Jharkharia and Shankar 2007). In this research, the ANP is composed of five steps.

4.1 Step 1: Model Construction and Problem Formulation

In this paper, the ANP model has been developed based on the literature review and discussion with academia and experts. The discussion with academia and experts helps us verify the model in the extensive researches, maps the literature with our research problem, and finally classifies the various logistics components and the KPIs as shown in Fig. 2. The logistics components were contracted in purpose of healthcare logistics effectiveness and efficiency. The literature on logistics accords a very important relationship and impacts of the components on each other. The academia and experts also supported the idea of having relationship between the components. In addition, there are performance indicators reflecting the performance of the components. Since the logistics components were set up to have influence on each other, and they also do in the real world, there is no need such an influence between the KPIs of each logistics component.

4.2 Step 2: Pairwise Comparison

In this step, pairwise comparison between the goal of the model and logistics components, logistics components with logistics components themselves, and logistics components with their individually KPIs is made. Comparison in ANP, as in AHP, uses ration scale of absolute number from number 1 (equal important) to

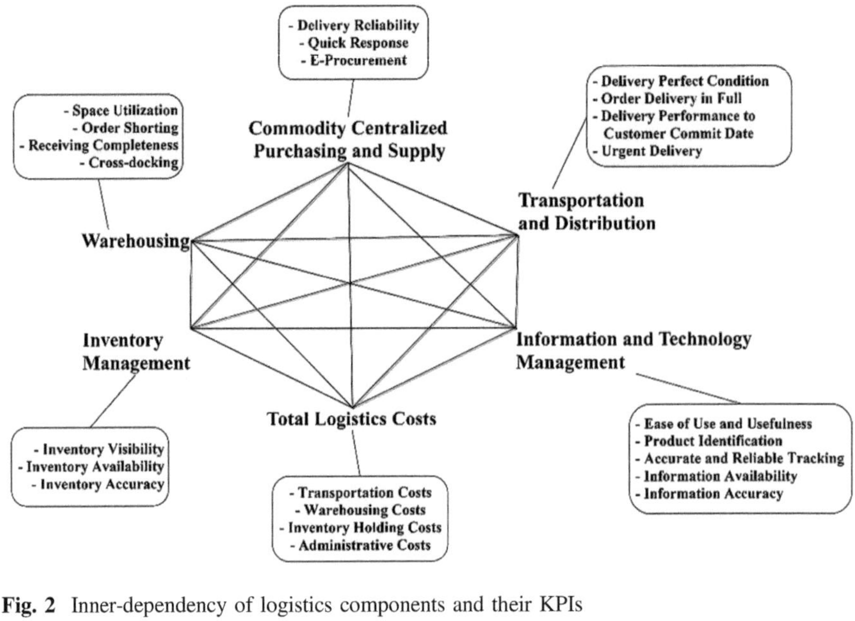

Fig. 2 Inner-dependency of logistics components and their KPIs

number 9 (extreme strong). However, the situation in real world is abstract and complex, so the comparison allows further refinement such as 1.1, 1.2, …, 1.8, and 1.9 (Saaty 2004; Saaty and Vargas 2013). Since pairwise comparison is mainly associated with qualitative and judgment, there is always biased attitude of the decision maker involved. To reduce such bias, there are several ways such as consensus, vote or compromise, geometric mean of the individual's judgments, and a separate model (Jharkharia and Shankar 2007; Dyer and Forman 1992; Forman and Peniwati 1998). In this paper, geometric mean of the individual's judgments is used (Forman and Peniwati 1998; Escobar et al. 2004).

$$W_i = \left(\prod_{j=1}^{n} a_j\right)^{\frac{1}{n}}; \quad i = 1, 2, 3, \ldots, n;$$

W_i total pairwise compairson of two elements,
a_j pairwise comparison value given by experts, and
n number of experts.

4.3 Step 3: Increasing Consistency Ratio

Naturally, consistency is very important to everything. Things will not be ordered if they have no consistency. In our ANP model, the consistency is essential for value of the elements in our framework given by the experts based on their experiences and expertise. However, due to the complexity of the decision problem, there is always inconsistency among the elements. In real practice, it is impossible to obtain a perfect consistency. AHP, therefore, allows a certain level of inconsistency generated from the pairwise comparison, and the inconsistency must be less than 10 %. If the inconsistency is greater than 10 %, it is recommended to discuss with the experts again. In our study, three are some pairwise comparisons which have inconsistency over 10 %. From the inconsistency report generated by the software, Fig. 3, it is quite easy to modify the value of the pairwise comparison among element by keeping the value remained in the same interest of the experts. The experts also supported the idea. Thus, the way to make such a modification is to change only the number in the same color, blue or red.

4.4 Step 4: Supermatrix Formation

The supermatrix shows priority which each element influences the others in the network. If an element has no such influence, the priority is equal to zero. In our study, the framework goal and logistics components are designed as hierarchy, logistics

Inconsistency Report

Rank	Row	Col	Current Val	Best Val	Old Inconsist.	New Inconsist	% Improvement
1.	4. Transportation and Di	6. Total Logi:	2.809999	3.049578	0.177034	0.043978	75.16 %
2.	1. Commodity Centralize	6. Total Logi:	1.080000	4.835583	0.177034	0.089088	49.68 %
3.	3. Inventory Managemer	4. Transporta	5.550006	1.425598	0.177034	0.142676	19.41 %
4.	1. Commodity Centralize	4. Transporta	5.369992	2.238466	0.177034	0.157113	11.25 %
5.	3. Inventory Managemer	6. Total Logi:	1.659999	2.429922	0.177034	0.166723	5.82 %
6.	1. Commodity Centralize	3. Inventory I	1.649999	1.059070	0.177034	0.168759	4.67 %
7.	1. Commodity Centralize	2. Warehous	4.060007	2.848189	0.177034	0.169858	4.05 %
8.	2. Warehousing	4. Transporta	1.490000	1.387602	0.177034	0.174383	1.50 %
9.	2. Warehousing	3. Inventory I	2.564103	2.620353	0.177034	0.177119	-0.05 %
10.	2. Warehousing	6. Total Logi:	1.150000	1.425794	0.177034	0.179310	-1.29 %

Fig. 3 Inconsistency report

components are designed as network, and the KPIs are designed as the subnet of the components. The unweighted supermatrix is therefore the same as weighted supermatrix. The weighted supermatrix shows how much an element of a cluster individually influences another element in the same cluster or the other cluster. On the other hand, limit matrix provides a stable priority of each element in the network as shown in Fig. 4. Warehousing influences the network totally 0.122928.

Since the system of our framework composed of six logistics components connected to each other and 6 subnets under each logistics components, there are seven limit matrices can be derived. Once is the limit matrix for the logistics components, and the other six are belong to the subnet each located under each logistics component.

Super Decisions Main Window: All.sdmod: Limit Matrix

Cluster	Node Labels	Logistics Components: 1. Commodity Centralized Purchasing and Supply	2. Warehousing	3. Inventory Management	4. Transportation and Distribution	5. Information and Technology Management	6. Total Logistics Cost	Prioritizing Logistics Components: Goal Node
Logistics Components	1. Commodity Centralized Purchasing and Supply	0.207851	0.207851	0.207851	0.207851	0.207851	0.207851	0.207851
	2. Warehousing	0.122928	0.122928	0.122928	0.122928	0.122928	0.122928	0.122928
	3. Inventory Management	0.249893	0.249893	0.249893	0.249893	0.249893	0.249893	0.249893
	4. Transportation and Distribution	0.078820	0.078820	0.078820	0.078820	0.078820	0.078820	0.078820
	5. Information and Technology Management	0.242606	0.242606	0.242606	0.242606	0.242606	0.242606	0.242606
	6. Total Logistics Cost	0.097902	0.097902	0.097902	0.097902	0.097902	0.097902	0.097902
Prioritizing Logistics Components	Goal Node	0.000000	0.000000	0.000000	0.000000	0.000000	0.000000	0.000000

Done

Fig. 4 Limit matrix of the logistics components

4.5 Step 5: Overall Priority of the Framework

To see how much each component and KPI have the influence in the system or framework, the overall priorities are calculated by multiplying the limit matrix of an individual component with its limit matrix of subnets.

5 Discussion

A methodological approach to develop a healthcare logistics KPI framework is presented. The framework provides the implication of a complex multi-criteria decision-making problem mainly related to qualitative judgments, which are necessary to evaluate different logistics components and their KPIs. The proposed framework illustrates the behavioral healthcare logistics, and dependency and independency of the components and KPIs. This is a good input for decision makers in healthcare industry and academic researchers. With the framework, decision makers can see the overview of the components' performance priorities from suppliers to point of care. In addition to the components priorities, the decision makers are also able to get the overview of performances under each component.

Based on framework derived from ANP result, the most critical component is inventory management followed by information and technology management, commodity-centralized purchasing and supply, warehousing, total logistics costs, and transportation and distribution accordingly (Fig. 5). Actually, in supply chain or logistics management, all the components are important because they link and support each other. However, if there is a question of priority, answering that all components are important is not acceptable. Based on the framework, inventory management, information and technology management, and commodity-centralized purchasing and supply are the components, which affect the overall performances of healthcare logistics the most. It is pretty logic for commodity-centralized purchasing and supply and inventory management because there must be providers, goods or services, and customers (outside our scope) to do business. However, the framework gave the priority to information and technology management rather than transportation and distribution. The reason is that nowadays, it is not hard to have transportation means but to control them. Moreover, technology is distinguishable enabler for the logistics management of any businesses. This result was supported by the experts since it really reflects the current practices in the real world of healthcare logistics. In addition to the components priority, there are also sets of KPIs which explain how those components could be improved. It also statistically provides how each KPI contributes to its component. For example, in order to improve commodity-centralized purchasing and supply, the improvement should be focusing more on delivery reliability and e-procurement rather than quick response.

Logisics Components: CR Goal: 9.19%		Group Priority	KPIs: CR		KPIs Priority within the Best Practice via ANP	Overal Priority of KPIs
Commodity Centralized Purchasing and Supply	2.25%	0.208	**Delivery Reliability (DR)**	4.11%	**0.339**	**0.070**
			Quick Response (QR)		0.248	0.052
			E-Procurement (EP)		**0.413**	**0.086**
Warehousing	2.23%	0.123	Space utilization (SU)	9.80%	0.190	0.023
			Order Shorting (OS)		**0.397**	**0.049**
			Receiving Completeness (RC)		**0.332**	**0.041**
			Cross-docking		0.081	0.010
Inventory Management	4.36%	**0.250**	Inventory Visibility (IV)	9.30%	0.293	0.073
			Inventory Availability (IAv)		**0.506**	**0.126**
			Inventory Accuracy (IAc)		0.201	0.050
Transportation and Distribution	5.71%	0.079	*Delivery Perfect Condition (DPC)*	1.34%	0.169	0.013
			Order Delivery in Full (ODF)		0.130	0.010
			Delivery Performance to Customer Commit Date (DPC2D)		**0.337**	**0.027**
			Urgent Delivery (UD)		**0.364**	**0.029**
Information and Technology Management	3.84%	0.243	*Ease of Use and the Usefullness (EU2)*	9.59%	0.062	0.015
			Product Identification (PI)		**0.296**	**0.072**
			Accurate and Reliable Tracking (ART)		**0.260**	**0.063**
			Information Availability (InAv)		0.182	0.044
			Information Accuracy (InAc)		**0.201**	**0.049**
Total Logisics Costs	9.67%	0.098	*Transportation Costs (TC)*	3.04%	0.096	0.009
			Warehousing Costs (WC)		**0.189**	**0.018**
			Inventory Holding Costs (IHC)		**0.609**	**0.060**
			Administrative Costs (AC)		0.106	0.010

Fig. 5 Overall priorities of elements in the system

Overall, the framework of this study provides quantitative influence of the logistics components and KPIs in healthcare context. It is a good input for making the strategic plan to improve healthcare logistics targeting effectiveness and efficiency.

6 Conclusion

This research mainly lies in the development of a comprehensive methodology for prioritizing the healthcare logistics components and their KPIs. The study addresses the healthcare logistics components and their individual KPIs. Moreover, it shows the relationship among the components, and their influence on KPIs. Because of such interactions and influences, ANP has become the right model for the research. ANP model can present the components, KPIs, and their relationship within and across levels. Moreover, ANP has also made them ranked in quantitatively.

The proposed framework has a few limitations as well. The result of overall priorities of each logistics component and KPI depends on the response of the experts and academia. Although number of the response provided by the academia and experts, the bias cannot be completely removed. In addition to bias, inconsistency is also possible. Thus, it is why the inconsistency has to be less than 10 %. Moreover, the operation inside the hospital and the patients were not included in the research. Hence, the future research should take these two components into account.

References

Abbasi M (2011) Storage, warehousing, and inventory management. Logistics Oper Manage: Concepts Models 181

Allsop N, Halstead G, Keyes A, Larmour I (2010) eCommerce within the hospital pharmaceutical supply chain lays foundation for improved patient safety. 2009/2010 GS1 Healthcare Reference Book

Behn RD (2003) Why measure performance? Different purposes require different measures. Public Adm Rev 63(5):586–606

Bose R (2003) Knowledge management-enabled health care management systems: capabilities, infrastructure, and decision-support. Expert Syst Appl 24(1):59–71

Bozarth C, Vilarinho P (2010) Analyzing the impact of space utilization and production planning on plant space requirements—a case study and methodology. Int J Ind EngTheory Appl Pract 13(1):81–89

Caridi M, Moretto A, Perego A, Tumino A (2014) The benefits of supply chain visibility: a value assessment model. Int J Prod Econ 151:1–19

Çelebi D, Bayraktar D, Bingöl L (2010) Analytical Network Process for logistics management: a case study in a small electronic appliances manufacturer. Comput Ind Eng 58(3):432–441

Chandra C, Kachhal SK (2004) Managing health care supply chain: trends, issues, and solutions from a logistics perspective. Paper presented at the proceedings of the sixteenth annual society of health systems management engineering forum, Feb 2004

Cheng EW, Li H, Yu L (2005) The analytic network process (ANP) approach to location selection: a shopping mall illustration. Constr Innov 5(2):83–97

Cho DW, Lee YH, Ahn SH, Hwang MK (2012) A framework for measuring the performance of service supply chain management. Comput Ind Eng 62(3):801–818

Coyle JJ, Bardi EJ, Langley CJ (2003) The management of business logistics: a supply chain perspective. South-Western/Thomson Learning, Mason

Delorme P, Chatelain O (2011) Policy steering—the role and use of performance measurement indicators. Aid Delivery Methods Programme

Dembińska-Cyran I (2005) Internal and external supply chain of hospital. LogForum 1(1):5

DHL-logbook (2008) The great logistics success story. Discover Logistics

DKSH Holding Ltd., Roland Berger AG Strategy Consultants, Zurich (2014) Digitization and the market expansion services industry: driving omni-chanel growth. Fourth Global Market Expansion Services Report

Dyer RF, Forman EH (1992) Group decision support with the analytic hierarchy process. Decis Support Syst 8(2):99–124

Escobar MT, Aguarón J, Moreno-Jiménez JM (2004) A note on AHP group consistency for the row geometric mean priorization procedure. Eur J Oper Res 153(2):318–322

Everard LJ, CPM C (2000) Blueprint for an efficient health care supply chain: White Paper. Medical Distribution Solutions, Norcross, Georgia

Fawcett SE, Cooper MB (1998) Logistics performance measurement and customer success. Ind Mark Manage 27(4):341–357

Forman E, Peniwati K (1998) Aggregating individual judgments and priorities with the analytic hierarchy process. Eur J Oper Res 108(1):165–169

Gary Jarrett P (1998) Logistics in the health care industry. Int J Phys Distrib Logistics Manage 28 (9/10):741–772

Görener A (2012) Comparing AHP and ANP: an application of strategic decisions making in a manufacturing company. Int J Bus Social Sci 3(11):194–208

Grigoroudis E, Orfanoudaki E, Zopounidis C (2012) Strategic performance measurement in a healthcare organisation: a multiple criteria approach based on balanced scorecard. Omega 40 (1):104–119

Gu J, Goetschalckx M, McGinnis LF (2010) Research on warehouse design and performance evaluation: aA comprehensive review. Eur J Oper Res 203(3):539–549

Gunasekaran A, Patel C, McGaughey RE (2004) A framework for supply chain performance measurement. Int J Prod Econ 87(3):333–347
Halstead G (2010) Building traceability in Australian healthcare. 2009/2010 GS1 Healthcare Reference Book
Jharkharia S, Shankar R (2007) Selection of logistics service provider: an analytic network process (ANP) approach. Omega 35(3):274–289
Jørgensen PMT, Jacobsen P, Itoh K (2012) Assessing technology in hospital logistical settings: comparing Danish and Japanese healthcare. Paper presented at the 13th international CINet conference
Kayakutlu G, Buyukozkan G (2011) Assessing performance factors for a 3PL in a value chain. Int J Prod Econ 131(2):441–452
Kelle P, Woosley J, Schneider H (2012) Pharmaceutical supply chain specifics and inventory solutions for a hospital case. Oper Res Health Care 1(2):54–63
Kim D (2005) An integrated supply chain management system: a case study in healthcare sector E-commerce and web technologies. Springer, Berlin, pp 218–227
Kumar A, Ozdamar L, Ning Zhang C (2008) Supply chain redesign in the healthcare industry of Singapore. Supply Chain Manage: Int J 13(2):95–103
Kumar A, Rahman S (2014) RFID-enabled process reengineering of closed-loop supply chains in the healthcare industry of Singapore. J Clean Prod 85:382–394
Ochiai C (2010) RFID and barcode based management of surgical instruments in a theatre sterile supply unit. 2009/2010 GS1 Healthcare Reference Book
Percin S (2008) Using the ANP approach in selecting and benchmarking ERP systems. Benchmarking: Int J 15(5):630–649
Pohl EA, Rossetti M, Varghese V, Nachtmann H (2012) Healthcare vs retail gap analysis: measuring the gap between healthcare and retail in supply chain best practice. Center for Innovation in Healthcare Logistics
Saaty TL (2004) Decision making—the analytic hierarchy and network processes (AHP/ANP). J Syst Sci Syst Eng 13(1):1–35
Saaty TL, Vargas LG (2013) Decision making with the analytic network process, economics, political, social and technological applications with benefits, opportunities, costs, and risks. Springer, Berlin
Stock GN, Greis NP, Kasarda JD (1998) Logistics, strategy and structure: a conceptual framework. Int J Oper Prod Manage 18(1):37–52
Supply Chain Council (2012) SCOR: supply chain operations reference model, Revision 11.0
Taylor GD (2007) Logistics engineering handbook. CRC Press, New York
Truman T (2011) Health procurement leader turns to Australia's National Product Catalogue to improve tendering. 2010/2011 GS1 Healthcare Reference Book
Van Horenbeek A, Pintelon L (2014) Development of a maintenance performance measurement framework—using the analytic network process (ANP) for maintenance performance indicator selection. Omega 42(1):33–46
Yüksel İ, Dagdeviren M (2007) Using the analytic network process (ANP) in a SWOT analysis—a case study for a textile firm. Inf Sci 177(16):3364–3382
Zhang AN, Goh M, Meng F (2011) Conceptual modelling for supply chain inventory visibility. Int J Prod Econ 133(2):578–585
Zinn W, Mentzer JT, Croxton KL (2002) Customer-based measures of inventory availability. J Bus Logistics 23(2):19–43

Supply Chain Modelling Under Uncertainty: A Supplier's Perspective

Shruti Maheshwari and Pramod Kumar Jain

Abstract One of the main challenges in supply chain management is to manage uncertainties within its environment, which yields increased total operational costs. Hence, there is a need to consider all the aspects that are responsible for uncertainties in supply chain management environment. There has been increasing dependence on supplier's side that triggers companies more prone to risk and uncertainties. This chapter considers all the uncertainty factors from supplier's perspective in supply chain management. The major factors are production, transportation, inventory, penalty and discounts. Thus, a supply chain model is established that explicitly incorporates all the factors responsible for the uncertainty from supplier's side in supply chain environment. The problem objective is to minimize overall cost considering demand satisfaction and profit. The chapter also highlights the root causes of uncertainty and various modelling approaches used in supply chain under uncertainty.

Keywords Supply chain management · Uncertainty · Supplier · Modelling

1 Introduction

Supply chain plays the most vital role in managing businesses in today's competitive world. Supply chain management has become a hot topic to research various related topics and has seen tremendous growth during the last two decades. Since the 1980s, interest in the concept of supply chain management has steadily increased due to the fact that companies observed huge benefits through collaborative relationships within and beyond their own organizations. Also, since 1990s,

S. Maheshwari · P.K. Jain (✉)
Indian Institute of Technology, Roorkee, India
e-mail: pkjain123@gmail.com

S. Maheshwari
e-mail: gudzieshruti@gmail.com

V. Kachitvichyanukul et al. (eds.), *Toward Sustainable Operations of Supply Chain and Logistics Systems*, EcoProduction,
DOI 10.1007/978-3-319-19006-8_4

companies have been forced to improve efficiency in many aspects in this competitive business world. And, also at the same time, the uncertainties in the business environment have increased which made companies to consume more resources to offset their adverse effect on demand and supply and for better sustainability in the competitive environment.

Modelling of uncertainties in supply chains has become important with the proliferation of product varieties and the increased volatility of the global market places of the contemporary business environment. For instance, product range and product style in a fashion industry are renewed constantly; while in the basic apparel industry, long production and distribution lead times still prevail. Managing business environment dynamics in an uncertain and competitive market is a challenging task and forced many organizations to reform their operational processes.

Nowadays, there is an emerging business trend to focus on outsourcing, reduction of the supplier base, long-term supplier relationships, reduced inventory and shorter lead times. Due to these new thrust activities, the vulnerability to risks increased in the supply chains and delineated the function of business units. Hence, the supplier selection has become an important decision to be taken at strategic level and to mitigate the risks by planning for the various uncertainties. The dependency on supplier has become the main consequence of the selection. This increasing dependency on suppliers is more prone to uncertain events; thus, the supply chain risk management (SCRM) has become necessary part of the supplier selection.

Many articles have been published on SCRM in last two decades. Several authors have worked on quantitative models for managing supply chain risk (e.g. Tang 2006; Goh et al. 2007). Various SCRM strategies were examined with case studies from the industries. Also Mark Goh et al. presented a stochastic model and designed an algorithm for profit maximization and risk minimization in a multi-stage global supply chain. Foroughi et al. (2006) explained the outcome of risk assessment from supplier side. You et al. (2009) proposed a two-stage stochastic linear programming approach that considers transportation modes, times of shipment, customer service level, etc.

Few authors have also worked on supplier selection problem, pricing and inventory decision, distributor location, relationship between buyer–supplier partnership quality, etc. They have worked on supply chain performance in the presence of uncertain environment (e.g. Srinivasan et al. 2011; Huang and Huang 2012; Sadigh et al. 2013, etc.). This paper aims to incorporate various aspects that are responsible for the uncertainty from supplier's side and then to model the same in order to minimize the cost.

2 Supply Chain Risk Management

Tang (2006) defined supply chain management as "the management of material, information and financial flows through a network of organizations (i.e. suppliers, manufacturers, logistics providers, wholesalers/distributors, retailers) that aims to produce and deliver products or services for the consumers. It includes the coordination and collaboration of processes and activities across different functions such as marketing, sales, production, product design, procurement, logistics, finance, and information technology within the network of organizations". Norman and Jansson (2004) defined "SCRM is to [collaborate] with partners in a supply chain apply risk management process tools to deal with risks and uncertainties caused by, or impacting on, logistics related activities or resources". Based on the definitions of supply chain management and SCRM, it appears that one can address the issue of SCRM along two dimensions:

1. Supply chain risks—operational risks or disruption risks.
2. Mitigation approach—supply management, demand management, product management or information management (Tang 2006).

The first dimension addresses the risk level of certain events. Operational risks are mainly mentioned to the inherent uncertainties such as uncertain customer demand, uncertain supply and uncertain cost. Disruption risks are referred to the major disruptions caused by natural and man-made disasters such as earthquakes, floods, hurricanes, terrorist attacks, or economic crises such as currency evaluation or strikes. In most cases, disruption risks have much greater impact than that of the operational risks.

Types of Supplier-Side Risks

Supply chain faces numerous supplier-side vulnerabilities from many factors in everyday performance of the suppliers. Broadly, the various types of risks are as follows:

Material flow risks: Many examples of natural disasters such as labour strikes and fires have brought an end to supply of material. In supply chain material flow, demand fluctuations and supply disruptions are two types of primary uncertainties which involve issues as single sourcing risk, sourcing flexibility risk, supplier selection/outsourcing, supply product monitoring/quality and supply capacity.

Financial risks: It takes into account foreign exchange, currency risk, and tariffs and taxes as well as product price, markups and rebates. It also includes the inability to settle payments and improper investment. The common risks are exchange rate risk, price and cost risk, financial strength of supply chain partners and financial handling/practice.

Integrity risks: Supply chain integrity has become an important necessity for brand owners across the industries. It is essential to keep supply chain from being compromised or interrupted and to ensure its integrity. Supply chain integrity protects the brand, minimizes costs due to loss or damage and enables to provide a quality product to the end consumer. These issues go beyond fraud alone to include risks associated with regulatory compliance, conflicts of interest, brand and reputation.

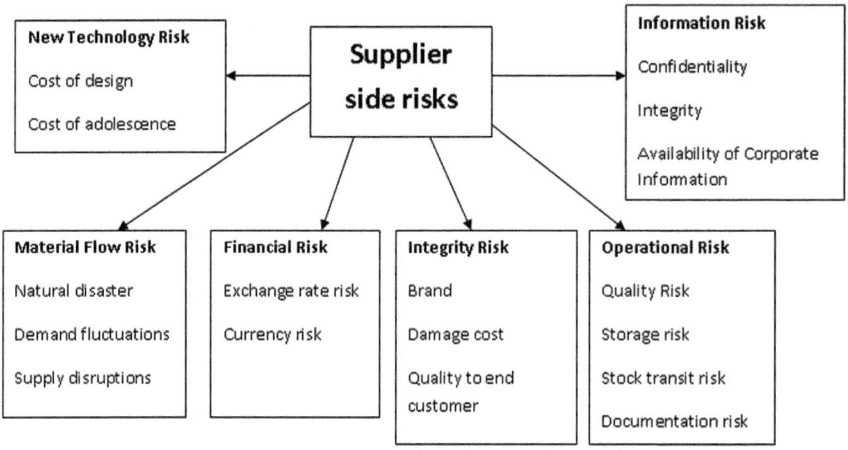

Fig. 1 Types of suppliers' side risk

Operational risks: Many times the risks are associated with the execution of business functions of a company. It includes risks of tangible and intangible assets. These risks address not only cost, efficiency, and contracting issues but also business disruption risk and misalignment of supply chains.

Information risk: Supply chain is one of the most collaborative environment in an organization; thus, it inherently poses greater risks to the confidentiality, integrity, and availability of corporate information. They should consider the accuracy, timeliness and relevance of data shared among parties, information system security and disruption, intellectual property and information outsourcing risk.

New technology risks: Technology risks emerge as smart phones, tablets, social media, cloud computing and new types of technology continue to develop (Fig. 1).

3 Approaches to Managing Risks

White (1995) suggests that the most approaches tend to follow the generic process despite of number of different risk management systems have been put forward. This consists of three critical stages:

1. Risk identification: Its purpose being to determine all risk factors that are likely to occur on a project.
2. Risk analysis: Its purpose being to understand the likelihood and extent of the most significant risks.

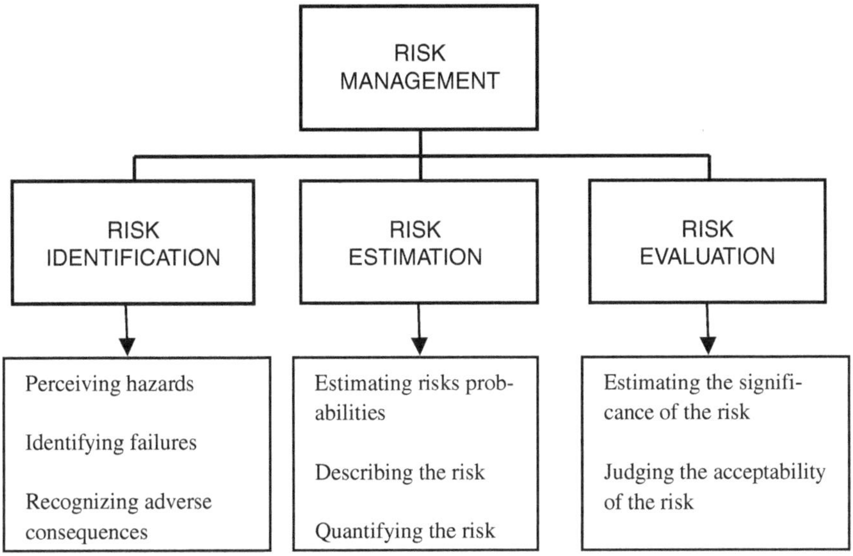

Fig. 2 Risk management stages (Khan and Burnes 2007)

3. Risk evaluation: Its purpose being to decide on the most appropriate management response for each risk/combination of risks and which party is most appropriate to manage each of the risks identified.

Further details in each stage are shown in Fig. 2.

3.1 *Robust Strategies for Mitigating Operational and Disruption Risks*

1. **Robust supply management strategies**: This is the most common strategy focusing on extending the supplier's network by increasing the number of suppliers. This helps in reducing supply chain risks through proper monitoring of exchange rate fluctuations, major operational disruptions, supply disruptions, etc. In addition, having multiple suppliers in multiple countries can make a supply chain more resilient during a major disruption.In many instances, the customer may not have the luxury to shift production among different suppliers due to the limited number of quality suppliers available. Hence in order to cultivate additional suppliers, it is recommended to design effective supply contracts that could serve as robust strategies to make a supply chain more efficient and resilient.
2. **Robust demand management strategies**: There are two robust demand management strategies. First, there is shifting demand across products. This strategy

can make a supply chain more efficient and resilient as it has the capability to shift demand across products by two mechanisms: product substitution and product bundling. In addition, a responsive pricing strategy could improve supply chain resiliency as well.

Besides these strategies, the other strategy that would help in enhancing supply chain efficiency and resiliency is the demand postponement strategy. It can be a robust demand management strategy. Under the demand postponement strategy, a manufacturer may offer price discounts to some retailers to accept late shipments. By having the capability to shift some of the demands to a later period, it would certainly help a firm to manage both operational risks and disruption risks.

3. **Robust product management strategies**: Among the product management strategies, the postponement strategy is a robust strategy for enhancing the efficiency and the resiliency of a supply chain when facing uncertain demands for different products. When selling products online, e-trailers can change their product assortments dynamically according to the supply and demand of different products
4. **Robust information management strategies**: strategies based on information sharing, VMI or collaborative forecasting and replenishment planning would increase "supply chain visibility" in the sense that the upstream partners have access to information regarding the demand and inventory position at downstream stages. As supply chain visibility improves, each supply chain partner can generate more accurate forecast of future demands and better coordination (Tang 2006).

3.2 Supply Chain Risk Mitigation through Collaboration

When two or more companies adopt long-term perspective and work together to construct unique value that none of the partner can attain alone, then such an environment is referred to as supply chain collaboration. Due to intensified competition, individual companies have found it difficult to compete alone but need to align their supply chain partners to achieve collaborative advantage. In a collaborative ethnicity, supply chain partners work together and communicate openly. They share information to improve the supply chain visibility which reduces uncertainty; they also share knowledge and expertise in all joint efforts such as joint problem-solving and new products development to smooth the operations and enhance the competitiveness.

In supplier collaboration, the buying company is involved directly with the processes and activities of its suppliers. To ensure the quality of supplied items, the buying company may help suppliers to implement quality management programme in their facilities. They can visit the suppliers' premises and provide training to their employees or even locate their own employees at suppliers' bases.

To reduce the damage caused by the capacity constraints of the suppliers, buying companies can assist by upgrading suppliers' technical capabilities and fostering continuous improvement programmes. They can also invite the suppliers to their plant to see how their items are used and include suppliers into their new product development processes, which enables suppliers to have a better understanding of manufacturing and thus better coordinate operations. As a result, suppliers' capability and performance is improved, operations of the two companies are better coordinated, the continuity of supply is ensured and supply risk is reduced.

4 Research Methodology

Owing to the investigative nature of this study, the case study design was considered to be an appropriate methodology. A qualitative case study captures the reality of a given situation in significant features and is particularly useful when a natural setting or a focus on contemporary events is necessitated. Inevitably, this approach has been criticized for having a limited capacity for scientific generalization. However, it can be argued that the objective of qualitative research is to refer back to a theory or application rather than to draw inferences about some larger population. Furthermore, as aimed to develop a systemic perspective on the flexibility strategies adopted in different situations, cases with differing backgrounds and configurations were needed to observe the differences in the phenomena under study and their relevance.

4.1 Statement of the Problem

The supply chain studied is made of supplier and customer as shown in Fig. 3. The main objective is to find optimum production lot and transportation quantity from supplier's perspective so that the total cost of supply chain is minimized. This problem-solving can help the supply chain to improve their efficiency by predicting demand and transportation cost in different situations. In the studied supply chain, as of the technological availability and hygienic standards that need to be followed, each product can just be produced in their respective plant and, hence, the assignment is not in the scope of this paper and, hence, the same is not considered in this problem. Demands will be fulfilled from the finished products that will be stored in warehouses. Although market demands are uncertain, it will be calculated and production forecasts will be done which are the Poisson processes and certain demand will be fixed at the beginning of each production period. Unsatisfied demand in each period is lost. Demands come as a Poisson process with different independent rates and the lead time for the production is exponentially distributed with parameter μ (>0).

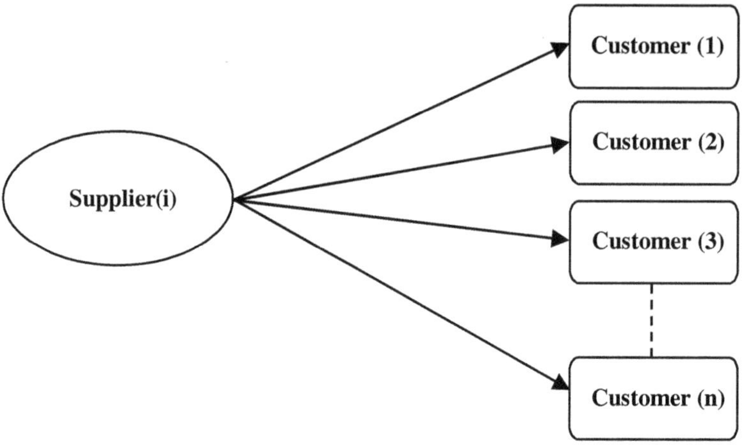

Fig. 3 Supply chain configuration

4.2 Problem Formulation

This section is dedicated to mathematical formulation of model. The problem will be solved using particle swarm optimization (PSO) algorithm.

In the present research, mainly the following types of costs are taken into the account:

1. **Total Variable Cost**: It is the multiplications of the variable production cost per unit (C_{ij}) and production volume (x_{ij}), where the production amount will be randomly generated using PSO technique.

$$\text{Total Variable Cost} = C_{ij}x_{ij}$$

   This represents total variable cost of all the products to be supplied from pre-defined set of suppliers to pre-defined set of customers.
2. **Total Shipment Cost**: It is assumed that transportation cost follows bivariate exponential distribution which depends upon demand (u) and capacity (v) as

$$\mathrm{P}\{\mathrm{X} > u, \mathrm{Y} > v\} = \mathrm{F}(u, v) = \exp\{-\lambda_1 u - \lambda_2 v - \lambda_{12}\max(u, v)\}, \quad \text{for} \quad u, v \geq 0$$

   It depends on the number of suppliers, number of customers, units to be shipped from supplier (i) to customer (j), and cost of transportation/unit. It is the product of average shipment cost per unit (t_{ij}) and total shipment lot to be supplied (S_{ij}).
   Total transportation cost = TSC
   Subjected to $S_{ij} \leq d_{ij}$
   Also, R = demand/shipment size = $\frac{d_{ij}}{S_{ij}}$

Case 1: If the shipment size is FULL
R = integer
then, TSC $= t_{ij}S_{ij}N$, where, $R = N$

Case 2: If the shipment size is PARTIAL
R = fraction
then, TSC = $t_{ij}s_{ij}N_0 + \frac{t_{ij}s_{ij}(R-N_0)}{N_1-R}$
where

N_0 = just previous integer to R
N_1 = just adjacent integer to R

3. **Inventory Holding Cost**: Inventory carrying cost influences many decisions in the strategic, analytic, and operations level of the business. Holding costs are a major component of supply chain management, since businesses must determine how much of a product to keep in stock. Mathematically, it is the multiplication of average inventory holding cost per item (h_{ij}) and the inventory of finished products (I_{ij}).
Inventory holding cost = $h_{ij}I_{ij}$
Subjected to
$$\text{if } d_{ij} - x_{ij} < 0, \quad \text{then} \quad I_{ij} = x_{ij} - S_{ij}$$
$$d_{ij} - x_{ij} = 0 \quad \text{then} \quad I_{ij} = 0$$

4. **Penalty Cost**: Penalty cost occurs when lead time is greater than the maximum value of supply lead time, i.e. (LT > LT*)

where

LT = lead time

LT* = the maximum value of the supply lead time not causing a stock out event

$$\text{Hence, } I_{ij}^{\Delta} = d_{ij} \int_{LT*}^{\propto} (LT - LT*) \cdot \text{pdf(LT)} \times \text{dLT}$$

It is the multiplication of the penalty cost per unit due to shortage of the inventory (δ_{ij}) and amount of the lost inventory (I_{ij}^{Δ}).

Penalty cost = $\delta_{ij}I_{ij}^{\Delta}$

Subjected to $d_{ij} - x_{ij} > 0$, then $I_{ij}^{\Delta} = d_{ij} - x_{ij}$

Penalty cost has been used to incorporate the uncertainty in lead time in the present model.

The authors have formulated the problem as a mixed integer programming model as follows: firstly defining the parameters and variables:

Parameters

C_{ij} variable production cost per unit for customer j from supplier i
t_{ij} transportation cost per unit from supplier i to customer j
h_{ij} inventory holding cost per unit from supplier i to customer j
δ_{ij} penalty cost due to shortage of inventory from supplier i to customer j
A_j minimum amount of product required for the customer j to purchase from the supplier i
M_i maximum quantity of a supplier i can deliver to the customer j

Variables

x_{ij} amount of production to be produced by supplier i
S_{ij} amount of production to be shipped from supplier i to customer j
I_{ij} inventory of the finished products from supplier i to customer j
I_{ij}^{Δ} lost inventory of the finished products from supplier i to customer j
d_{ij} demand of an item
i number of suppliers
j number of customers

The model is then minimized to

$$\sum_{i=1}^{m}\sum_{j=1}^{m}\left(C_{ij}x_{ij} + \mathrm{TSC} + h_{ij}I_{ij} + \delta_{ij}I_{ij}^{\Delta}\right) \tag{1}$$

subject to

$$\sum_{i=1}^{m} x_{ij} \geq A_{ij} \tag{2}$$

$$\sum_{j=1}^{n} x_{ij} \geq M_i \tag{3}$$

$$I_{ij} = x_{ij} - \sum_{j=1}^{n} S_{ij} \tag{4}$$

$$S_{ij} \leq d_{ij} \tag{5}$$

$$\text{Also, if } d_{ij} - x_{ij} < 0, \quad \text{then} \quad I_{ij}^{\Delta} = \frac{d_{ij} - x_{ij}}{2} \tag{6}$$

$$d_{ij} - x_{ij} > 0, \quad \text{then} \quad I_{ij}^{\Delta} = d_{ij} - x_{ij} \tag{7}$$

$$d_{ij} - x_{ij} = 0 \quad \text{then} \quad I_{ij} = 0 \tag{8}$$

$$x_{ij} \geq 0 \tag{9}$$

$$I_{ij} \geq 0 \tag{10}$$

$$S_{ij} \geq 0 \tag{11}$$

$$d_{ij} \geq 0 \tag{12}$$

where in all above equations: $i = 1, \ldots, m,\ j = 1, \ldots, n.$

In this formulation, the objective function (1) seeks to minimize the total production costs, transportation costs and inventory costs. Constraint (2) represents that production amount should be greater than equal to the minimum amount customer required. Similarly, constraint (3) embodies the maximum amount a supplier can deliver to the customer. Constraint (4) signifies the relationship between the inventories built up after the shipment of the product. Also, constraint (5) exemplifies the relationship between demand and supply.

Constraint (6)–(8) represents the different situation of demand of the customer and production amount supplier can make available. In constraint (6), inventory is built up if production amount exceeds the demand and lost inventory situation is represented by constraint (7). Lost inventory situation leads to penalty which shows uncertainty in lead time. There is no inventory in case if demand is equal to the production amount which is embodied by constraint (8). Constraints (9)–(12) stipulate the non-negativity and discrete requirements for the variables.

Since it was very difficult to solve the model using the latest state-of-the-art integer programming techniques, even after fixing the parameters, hence, using PSO algorithm is used to solve the problem.

5 Particle Swarm Optimization

PSO is a swarm-intelligence-based, approximate, nondeterministic optimization technique. It is a robust stochastic technique based on the movement and intelligence of swarms. It applies the concept of social interaction to problem-solving (people.scs.carleton.ca).

The algorithm of PSO emulates from behaviour of animals societies that do not have any leader in their group or swarm, such as bird flocking and fish schooling. Typically, a flock of swarm that have no leaders will find food by random, follow one of the members of the group that has the closest position with a food source (potential solution). The flocks, through communication among members who already have a better situation, achieve their best condition simultaneously. Swarm

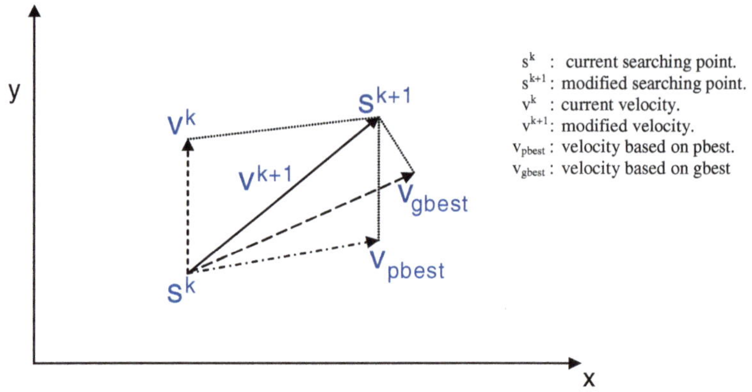

Fig. 4 Concept of modification of a searching point by PSO (people.scs.carleton.ca)

which has a superior condition will update it to its flocks and the others will move simultaneously to that place. This would happen repeatedly until the best conditions or a food source is discovered. PSO consists of a swarm of particles, where particle represents a potential solution (Rini et al. 2011).

In PSO, each particle keeps track of its coordinates in the solution space which are associated with the best solution (fitness) that has achieved so far by that particle. This value is called personal best, *pbest*. Another best value that is tracked by the PSO is the best value obtained so far by any particle in the neighbourhood of that particle. This value is called *gbest*. The basic concept of PSO lies in accelerating each particle towards its *pbest* and the *gbest* locations, with a random weighted acceleration at each time step as shown in Fig. 4.

Steps used to solve the problem using algorithm are as follows:

(i) Evaluate fitness of each particle,
(ii) Update individual and global bests and
(iii) Update velocity and position of each particle.

These steps are iterated until the objective of the problem (i.e. minimize) is met.

Mathematically, particles position can be modified by the following equation:

$$\mathbf{V}_i^{k+1} = w\mathbf{V}_i^k + \mathbf{c}_1\mathbf{rand}_1(\ldots) \times \left(pbest_i - \mathbf{s}_i^k\right) + \mathbf{c}_2\mathbf{rand}_2(\ldots) \times \left(gbest - \mathbf{s}_i^k\right)$$

where

V_i^k	velocity of agent *i* at iteration *k*,
W	weighting function,
c_j	weighting factor,
rand:	uniformly distributed random number between 0 and 1,
s_i^k	current position of agent i at iteration k,

pbest$_i$ *pbest* of agent I,
gbest *gbest* of the group[people.scs.carleton.ca].

During each iteration of the algorithm, each solution is calculated by the objective function to determine the fitness of the problem. In comparison with the other algorithms, this method is very simple, easily completed and it needs fewer parameters, which made it fully developed. Also, the impact of the solution is small as compared to other optimization techniques.

6 Computational Experiments

The PSO algorithm was used in MATLAB platform. The experiments were carried out with presently taking one supplier and one customer, i.e. $i = 1$ and $j = 1$. The input of variable cost per unit (C_{ij}), transportation cost per unit (t_{ij}), inventory holding cost per unit (h_{ij}) and penalty cost (δ_{ij}) was taken from a case study (Teimoury et al. 2010). The number of iterations made was 30. Once the values are entered for different products, the minimum cost was calculated by PSO algorithm in MATLAB and the graph for each product was plotted in MATLAB.

As gained the information from the case study, many different detergent products were produced by a chemical manufacturing company. Among the varied range of products, few numbers of products were used in this case study. Planning horizon in a year consists of 12 production periods. To avoid complicated calculations and a large numbers of iterations in computational algorithm and as all products are transported by truck, the demand has been changed to the truck scale and the capacity is taken randomly 640 fixed for all the products, and hence, transportation cost per unit is fixed to be 30 for the domestic demand and 50 for

Table 1 Input parameters and global best (*gbest*) of a chemical company for 12 products

Product	C_{ij}	t_{ij}	h_{ij}	δ_{ij}	*gbest*
1	43.776	30	0.288	81.408	7380.7
2	61.286	30	0.288	113.971	9132.3
3	39.398	30	0.288	73.267	6943.0
4	26.266	30	0.256	48.845	5629.1
5	55.158	30	0.288	102.574	8518.9
6	13.133	30	0.224	24.422	4315.1
7	43.776	50	0.288	89.549	9380.6
8	61.286	50	0.288	125.368	11,132.2
9	39.398	50	0.288	80.594	8942.9
10	26.266	50	0.256	53.729	7630.4
11	55.158	50	0.288	112.831	10,519.4
12	13.133	50	0.224	26.865	6315.1

international demand, and all cost are scaled to one million. Input parameters for the case study of a chemical company are given in Table 1 (Teimoury et al. 2010).

Using these values in Table 1 in MATLAB, column (5) gives the minimum cost for every 12 products using PSO algorithm and the graph is plotted for each value.

The Fig. 5 represents the graphs of *gbest* (cost in millions) for 40 different iterations.

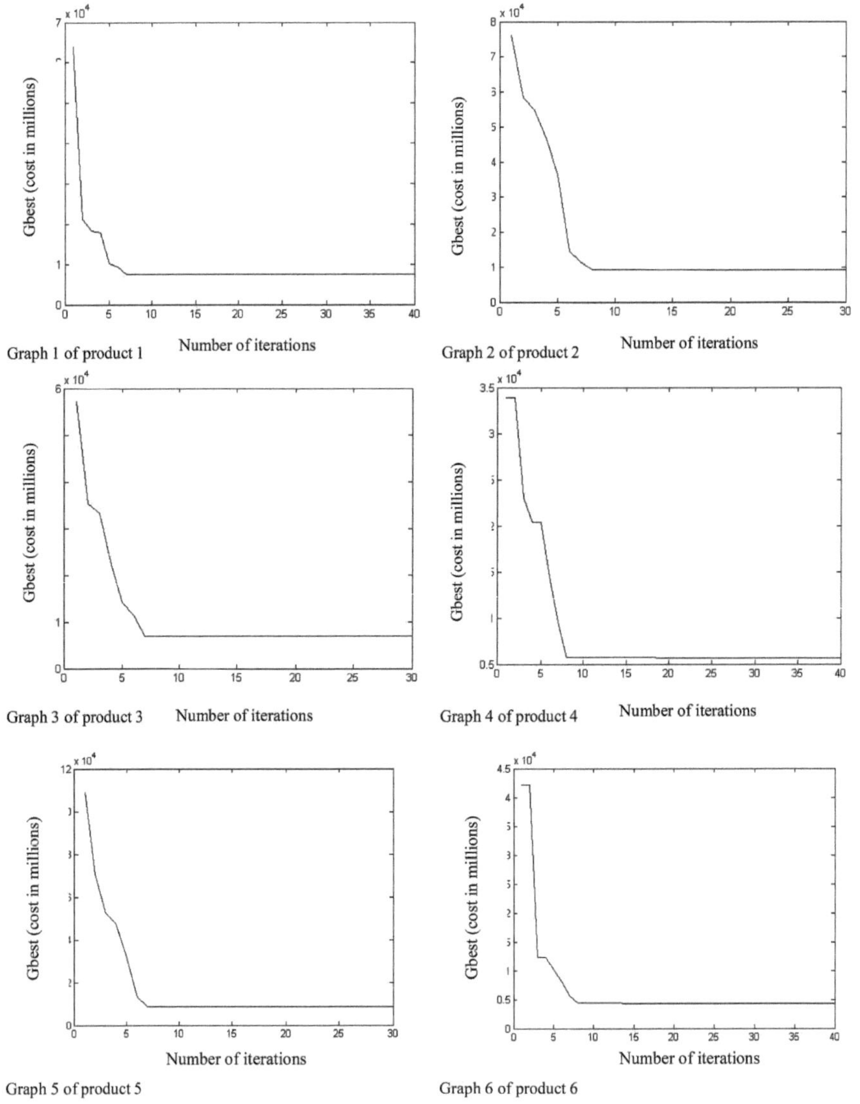

Fig. 5 Graphs of *gbest* of 12 products in Table 1

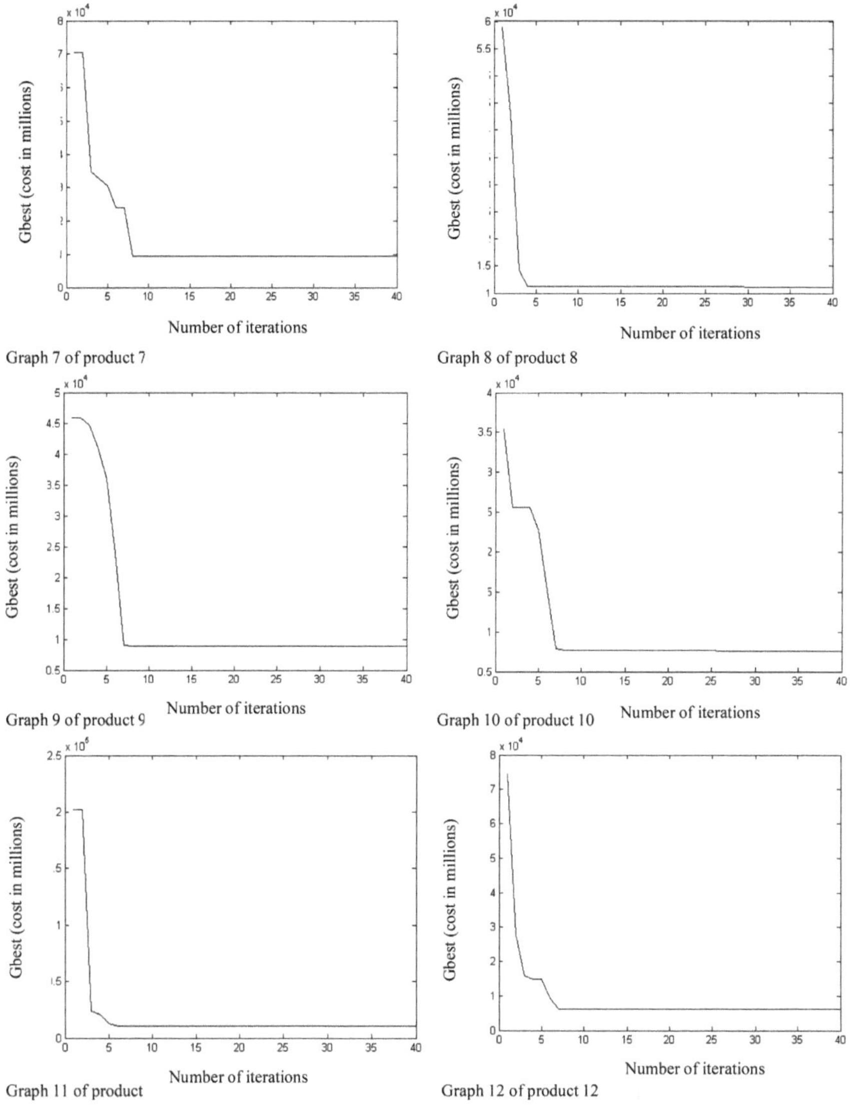

Fig. 5 (continued)

7 Conclusion and Future Scope

The real-time process in chemical industry of supply chain is studied over here. The main objective was to minimize cost using PSO algorithm which is successfully done under real demand situations. The proposed model is a dynamic model using mainly three varieties of costs, viz. variable cost, transportation cost and inventory holding cost. To make the model more applicable, the model was designed under uncertain demand situations, despite of common approaches such as robust optimization, stochastic programming, games theory, linear programming and parametric programming. It is the first time that anyone has used PSO algorithm to solve a supply chain problem under uncertainty. PSO has made computational time very less and can be effectively used in real-world problems.

The problem will be further carried to more uncertain situation considering penalty and bulk discounts situations in future. Also, situation of full loaded or partially loaded truck will be considered using the same PSO algorithm. And, an attempt will be made to solve the problem with multiple numbers of suppliers, multiple numbers of customers and multiple numbers of products simultaneously. This will make the algorithm more near to real-world problems and efficiently reduce the cost of the supply chain under uncertainty from supplier's side.

References

Foroughi A, Albin M, Kocakulah M (2006) Perspectives on global supply chain supply-side risk management. In: Portland international center for management of engineering and technology (pp 2732–2740)

Goh M, Lim JYS, Meng F (2007) A stochastic model for risk management in global supply chain networks. Eur J Oper Res 182:164–173

Huang Y, Huang GQ (2012) Integrated supplier selection, pricing and inventory decisions in a multi-level supply chain. Decis Making Supply Chain Integr Decis Eng 1:47–62

Khan O, Burnes B (2007) Risk and supply chain management: creating a research agenda. Int J Logistics Manage 18(2):197–216

Norrman A, Jansson U (2004) Ericsson's proactive supply chain risk management approach after a serious sub-supplier accident. Int J Physical Distribution Logistics Manage 34(5):434–456

Rini DP, Shamsuddin SM, Yuhaniz SS (2011) Particle swarm optimization: technique, system and challenges. Int J Comput Appl 14(1):19–27

Sadigh AN, Fallah H, Nahavandi N (2013) A multi-objective supply chain model integrated with location of distribution centers and supplier selection decisions. Int J Adv Manuf Technol 69:225–235

Srinivasan M, Mukherjee D, Gaur AS (2011) Buyer-supplier partnership quality and supply chain performance: moderating role of risks, and environmental uncertainty. Eur Manag J 29: 260–267

Tang CS (2006) Perspectives in supply chain risk management. Int J Prod Econ 103:451–488

Teimoury E, Modarresc M, Ghasemzadeh F, Fathi M (2010) A queueing approach to production-inventory planning for supply chain with uncertain demands: case study of PAKSHOO chemicals company. J Manuf Syst 29:55–62

White D (1995) Application of system thinking to risk management: a review of the literature, management decision. 33(10):35–45

You F, Wassick JM, Grossmann IE (2009) Risk management for a global supply chain planning under uncertainty: models and algorithms. Am Inst Chem Eng J 55(4):931–946. http://people.scs.carleton.ca/~arpwhite/courses/95590Y/lectures/pso.pdf (approached 11 March 2015)

A Study on the Changing Structure of Retail Logistics

Masafumi Nakamura and Kuninori Suzuki

Abstract The purpose of this chapter is to analyze the changing structure of retail logistics in Japan through the comparison with that from the UK. The grocery retail supply chain in Japan and UK is quite advanced, and the focus is on the retailers of the grocery. This chapter consists of four sections. In the first section, the research on the theory of retail logistics and marketing is presented. In the second section, the logistical transformation at Tesco is explained. Tesco is a biggest supermarket in the UK and had changed the logistics strategy and adopted a strategy to sell private brand goods (PB). In the third section, the logistics and marketing strategy are described about Japanese retailer, Daiei, Seven-Eleven, and AEON. In the fourth section, a description is given on how the retailer of the grocery has the power over the manufacturers by using IT and selling PB. The conclusions of the study are given in the last section.

Keywords Retail logistics · Convenience store · Supermarket · Efficient customer response (ECR) · Private brands

1 Introduction

Michael Porter analyzed the successful companies and provided three generic strategies: cost leadership, differentiation, and focus. The focus strategy has two variants: cost focus and differentiation focus (Porter 1985).

Cost leadership is the strategy that the company wins in the market by appealing to price-sensitive customers. Michael Porter defined the two types of competitive

M. Nakamura
JR Freight Research Center, Tokyo, Japan
e-mail: m-nakamura@jrf-rc.co.pl

K. Suzuki (✉)
Nihon University, Tokyo, Japan
e-mail: qwl02764@nifty.ne.jp

V. Kachitvichyanukul et al. (eds.), *Toward Sustainable Operations of Supply Chain and Logistics Systems*, EcoProduction,
DOI 10.1007/978-3-319-19006-8_5

Fig. 1 Three generic strategies. *Source* Porter (1985)

		Competitive Advantage	
		Lower Cost	Differentiation
Competitive Scope	Broad Target	1.Cost Leadership	2.Differentiation
	Narrow Target	3A. Cost Focus	3B. Differentiation Focus

advantages that an organization can achieve relative to its rivals. The competitive advantages are lower cost and differentiation (Porter 1985). In the theory of Michael Porter, the logistics is the main activity in the value chain.

According to Michael Porter, we can assume that successful retailers have the competitive advantage in the strategy of business and logistics (Fig. 1).

Galbraith stated that a large buyer has more power to obtain lower prices than his rival.

Galbraith explained that the retailer gets price concessions from the food manufacturer in concentrated downstream markets. He named it the countervailing power.

He regarded A&P Company and cornflakes in USA as a typical example the use of countervailing power (Levi 1954). When the rate of the concentration in the food market increases, large food retailer gets to extend price concession from the manufacturers. The countervailing power of downstream company enables low price. The company with high market share can take the cost leadership strategy.

Ogawa (2001) analyzed about demand pull supply chain in Japan. He pointed out that the demand pull supply chain changed the distribution system.

Sparks (2010) explained about Tesco's transformation. Tesco and other big supermarket got the cost leadership in the UK. We described about Tesco's transformation in Sect. 2.

Yahagi (2014) indicated that three big alterations occurred in Japanese retail industry. Three changes were the paradigm shift of the retail business, brand competition of NB versus PB, and sale competition in the retail industry.

Previous studies have shown that private brand goods (PB) strategy is popular among the food retailers and they use IT for the improvement of their business.

Tesco in the UK and Japanese retailer, such as AEON and Seven-Eleven, took PB strategy. PB may give the countervailing power to the retailers.

Tesco, AEON, and Seven-Eleven had good IT systems. The advance of IT enables that retailer to control the manufacturers by consumer data. We have the hypnosis that PB and IT give a good effect to the retailer by the countervailing power and the competitive advantage. In this report, we analyzed the impacts of IT and PB for retail industry.

2 Tesco's Logistical Transformation

According to Euro trend 2001 of JETO, half of the retail trade is food retail trade in the UK. The ratio of the British food retail trade is high.

In this section, we describe the logistical transformation by five phases. Tesco is No. 1 supermarket in the UK, and it has good logistical transformation.

2.1 First Phase, Direct to Store Delivery (DSD)

In the 1970s, Tesco grew in the UK retail industry, selling the cheap and simple goods. Tesco's motto was "pile high, sell it cheap," because consumer needed goods that were the completive price.

At that time, manufacturers did directly to store delivery (DSD). In the UK, Tesco had 800 stores and were needed to deliver for numbers of small stores.

There was a problem in DSD. The manufacturer delivered the goods that did not necessarily match the customers' needs to the store. This is because Tesco could not decide the price and range. Moreover, manufacturer could not deliver with full load vehicle of product by the DSD system. So empty shelve emerged in the some store. Also, the trucks had to wait in front of the store for delivery. As the volume of goods increased, the DSD system did not work well. At the same time, the demands of consumers began to change and it is required for Tesco to revise its concept "pile high, sell it cheap." DSD system inevitably had to change to the new system (Table 1).

2.2 Second Phase, Centralization

In the 1980s, Tesco changed the strategy for the warehouse and transportation to be sharp in the quality and adopted centralization strategy in place of DSD. Centralization involved building new distribution centers and extending to distribution facilities. Tesco invested regional distribution center (RDC).

Table 1 Tesco's supply chain transformation (1). *Source* Sparks (2010)

Phase	Key issues
Direct to store delivery, in the 1970s	Supplier organized and controlled no retailer range, price, control
Centralization, in the 1980s	New facilities and approaches; manufacturers' and suppliers' products to distribution center
Composite, in the 1980s and 1990s	Multi-temperature regional centers and vehicles sharing of stock holding increased frequency of delivery to stores

Under the new strategy, manufacturers had to deliver the goods to RDC. Inventory had been shifted from store to RDC. This change enabled the head office of Tesco to control the distribution and the store, taking initiative away from manufacturers. Tesco enabled to reduce the lead time of delivery and to minimize its inventory.

2.3 Third Phase, Composite Distribution

In 1989, Tesco had 42 depots at which only a single temperature. These depots were handled single temperature and single product. In the 1980s, some disadvantages of the centralization appeared. The frequent delivery from single product depot to store led to empty running of truck, thereby making the process expensive.

Tesco realized the needs of multi-temperature type of warehouse and vehicles.

Tesco made a new strategy "*Composite distribution of warehouses and vehicles.*"

The composite distribution of warehouses was controlled by three types of temperature (chilled, fresh, and frozen). Eventually, Tesco established 10 composite distribution centers. The merits of composite distribution were as follows.

First: reduction of stock at the stores, Second: reduction in wastage, and Third: reduction of costs through less congestion at the store.

Composite distribution led to reduction of stock holding. The daily deliveries of composite product to all stores reduced stock at the stores.

2.4 Fourth Phase, Vertical Collaboration

Tesco thought that the ordering and replenishment were important for its business, so it shared the information about customer needs with their supplier. Tesco also started to improve the goods depending on the needs of customer. Tesco managed ordering and replenishment by itself.

Scanning the goods at store made it possible to order and replenish the product efficiently.

Thus, Tesco's IT network was eventually upgraded as efficient consumer resource (ECR). ECR is a joint trade system that responds to the consumer demand in the SCM.

Besides, Tesco introduced IT system that fostered collaborative planning, forecasting, and replenishment (CPFR) in 1999. Sharing of information using CPFR led to reduce the response time. Suppliers obtained Tesco customer needs and led to produce the appropriate goods. Thus, the use of information became more important in logistics. Tesco introduced a loyalty card and got the consumer needs from its loyalty card. Loyalty card fostered up store loyalty.

Tesco tried to sell the differentiated goods by PB. The distributor's margin was reduced by direct dealings with a manufacturer. The manufacturer could have the merit stability of sales volume by PB. Tesco and other big supermarkets got the cost leadership. The market share of big supermarkets (Tesco, Asda, and Siansbury, etc.) as well as the profit rate of those became high.

2.5 Fifth Phase, Managing Complexity

Tesco aimed the inbound supply chain to more efficient and the low cost by the consumer need data. Tesco thought that the change of the primary distribution (manufacturer to distribution center) enabled the supply chain better.

Tesco began to focus the primary distribution, and Tesco controlled directly the primary distribution by selecting logistics company which could reduce cost and give the good supply chain (Table 2).

Tesco took new strategy "*Green and sustainably initiative, reverse logistics, collaborative transport.*" Tesco secured the efficiently increasing transportation to RDC from a manufacturer. The modal shift is one of the results of green and sustainably initiative strategy, because the truck transport increases congestion and reduces speeds of delivery.

On the other hand, freight railways have superiority in long-distance transportation. As the freight train can transport heavy goods, its cost is cheaper than truck in the long distance.

Tesco started to use Stobart railway for a daily delivery train in 2006. Stobart collected Tesco goods from the Midlands, and the train leaves from the Stobart depot at Daventry International Rail Freight Terminal (Midlands) at Rugby for the Grangemouth Rail Terminal in Scotland.

The train then returned to Rugby. The daily shipping amount of each direction was 90 % filled with Tesco and rest was Coca Cola shipping the equivalent of 26 Lorries.

Tesco chose the railway for primary distribution because of its economic benefit and contribution to reduce CO_2.

Table 2 Tesco's supply chain transformation (2). *Source* Sparks (2010)

Phase	Key issues
Vertical collaboration and lean supply chains (in the 2000s)	Primary distribution focus, flow principles and lean principles, stockless distribution centers, sales-based ordering-driven system continuous replenishment customer focused from Clubcard data
Managing complexity (in the 2000s)	Internationalization of supply and operation, multi-format store development with new sizes, green and sustainably initiative reverse logistics, collaborative transport Internet-based retailing, 24/7 operation

According to Freight Best Practice case study 1094 (Department for Transport 2009), the Tesco train link between Daventry and Grangemouth saved 3.18 million road miles per year and reduces CO_2 emission 2424 tons per year.

3 Japanese Retail Transformation

3.1 Revolution Retailer

Daiei was very famous as a revolution retailer in the 1980s. Daiei's President Nakauchi aimed that Daiei became the manufacturer who had no factory to sell PB. Daiei tried to wrest pricing power from manufacturers due to the power of price determination. When Daiei ordered PB to manufacturer, Daiei had superior position to manufacturer. As a result, Daiei succeeded to build the power concentration mechanism by single-item enormous volume purchase.

In the 1980s, many big manufacturers started to produce the goods for Daiei. Daiei produced many PB, for example, food, detergent, and videotape.

However, the Great Hanshin/Awaji Earthquake occurred at the Daiei's main market area in 1995 and the business depression hit the Japanese economy in 1998.

Daiei tried to survive by discounting its goods. Daiei set a new policy "*every day, low price, low cost operation.*" But Daiei's sales kept on getting worse. Consumers chose convenience store which they could buy the cheap and convenient goods at the store.

Moreover, convenience store introduced ECR and CPFR like Tesco and created cooperative relationship with manufacturer by consumer data. They offered products that satisfied the consumer needs. Convenience store had grown; meanwhile, the sales of supermarket had come to decline.

In the 2000s, big supermarkets (Daiei, Yaohan, Seiyu, etc.) run out. Daiei was under AEON in 2013. AEON merged not only Daiei but also Yaohan and Maikal and became No. 1 supermarket in Japan (Fig. 2).

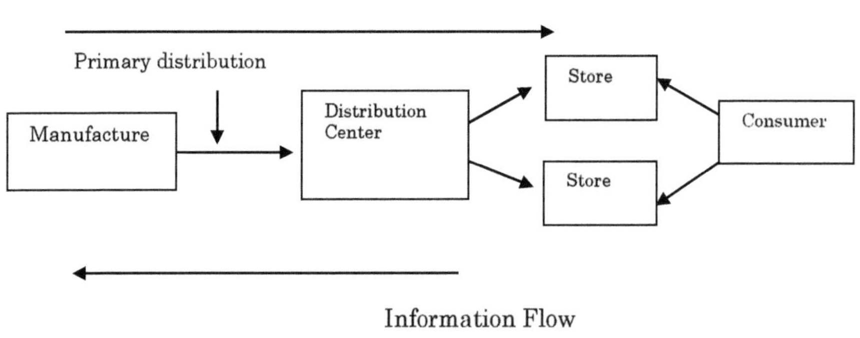

Fig. 2 SCM of the supermarket

3.2 Convenience Store Power

When the convenience store was opened for the first time in 1974 by Seven-Eleven, many trouble happened. For example, the delivery truck visited every 14 min to the store and store clerk dealt with not only customer but also goods for 16 h per day, because the manufacturers delivered their goods to the store by each truck.

Even worse, store clerk had to make order to the manufacturers 100 times by the phone for a day. These were non-efficiency work in the store. It became a big problem to eliminate inefficient works and reduce the number of deliver trucks.

To deal with this, Seven-Eleven changed a transport company to a new wholesale company that the manufacturers invested in. This change gave good effects on delivery to the store. For example, each beer maker (4 major beer makers) delivered its products by each truck to the store at first. After introducing the new system, the truck of a wholesale company delivered the products of all makers to the store. This appropriate response improved the inefficiency of truck delivery, and the store management became more efficient.

As Fig. 3 shows, the number of delivery truck was 70 per day in 1974. In the meantime, the number was 12 in 1990. This cooperative delivery system also led to the cut down of the cost.

Seven-Eleven realized the cost saving by cooperative distribution. Figure 4 shows the Seven-Eleven cooperative distribution (joint distribution) system. In this system, each manufacturer delivers to the exclusive distribution center under appropriate temperature and the goods are sent to the store by cooperative distribution. The system enabled Seven-Eleven to ensure better quality of goods and efficient delivery.

A business model of Seven-Eleven shows a good example of efficient delivery system.

Factors behind the Seven-Eleven success are cooperative distribution, inventory control, and prediction of demand by IT.

Fig. 3 The number of the delivery truck each store

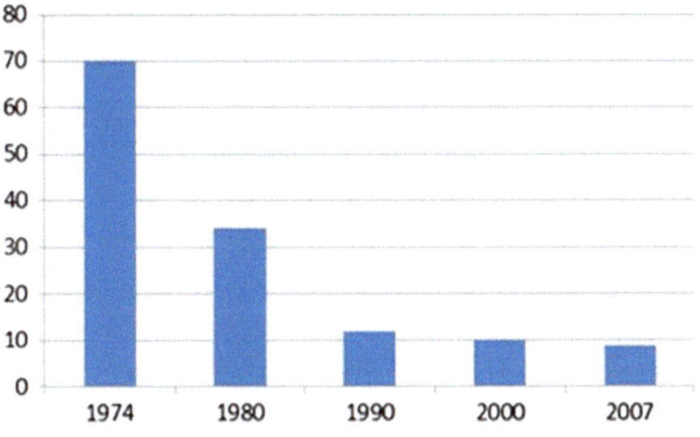

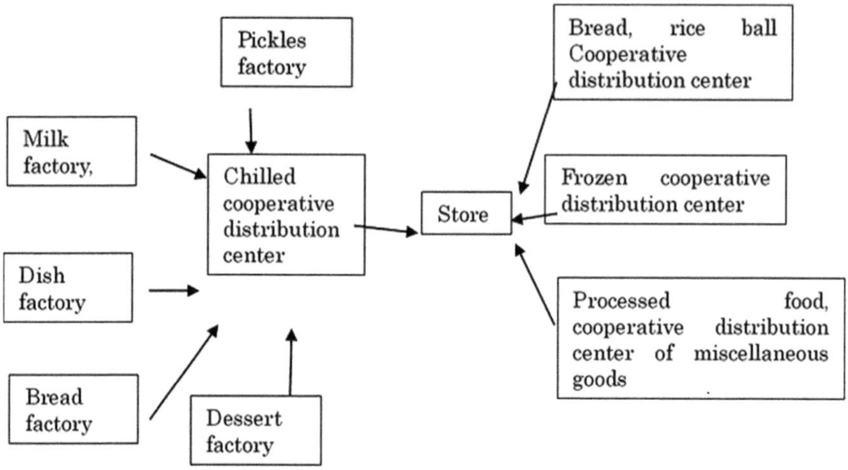

Fig. 4 Seven-Eleven cooperative distribution system

3.3 *The Effect of IT to the Retail Business*

Seven-Eleven introduced a POS (point of sales) register and barcode reader in 1976. These new IT systems made it possible to decrease order mistakes and reduce the cost. The head office of Seven-Eleven sent the information about the weekly sales of every single item and shortage of goods to stores.

In addition, Seven-Eleven connected the POS system to supplier in 1982. Due to this, the accurate order and appropriate delivery were implemented. Seven-Eleven could grasp customer needs and order popular goods to manufacturers by the use of the IT system. It had grown rapidly since the 1980s in Japan.

IT system of Seven-Eleven was one of the systems of ECR. A manufacturer knew easily the customer needs from a convenience store by ECR and supplied goods to the store. The convenience stores had a superiority to the manufacturer.

3.4 *The Impact of PB for the Retail Business*

The new trend occurred in the retail market by private label brand goods (PB). AEON has begun "*Top value*" as PB since the 1990s. PB has given competitive advantage to retailer by the reduction of lead time and cost. The number of PB item orders made by convenience stores and supermarkets to manufacturers has been increasing. PB items match consumer demand and raise the market share rapidly. The penetration of PB enables a retailer to establish customer loyalty. On the other hand, manufacturer can supply popular goods by the production of PB goods and also has a chance to plan and produce new goods.

AEON and Seven-Eleven are now big PB retailers in Japan. AEON has 5500 PB items, and its sales amount to 460 billion yen. Seven-Eleven has 1300 goods of PB, and its sales amount is 320 billion

Famous Japanese food manufacturers, such as Ajinomoto, House Foods, Nipponham, and Suntory, produce PB goods for AEON and Seven-Eleven. These two retailers are good customers for such PB manufacturers. According to the survey conducted by Japan Finance Corporation about the possibility of making PB goods for food manufacturers, 49 % of those surveyed would like to produce PB goods.

The rate of PB in the UK and Germany is beyond 20 %, but Japanese PB rate remains only 5 %. It is expected that PB rate of Japan will rise in the future.

4 The Business Feature of Tesco and Seven-Eleven

We had the hypnosis that PB and IT took effect to the retailer by the countervailing power and the competitive advantage.

We checked our hypnosis by the 4P classification.
McCarthy proposed 4P classification to analyzed company business.
4P are product, price, promotion, and place (Distribution).

4.1 The Features of Tesco Business

We analyzed Tesco business by the classification of 4P as follows:

1. Product
   Tesco shared the information about customer needs with their supplier, and Tesco improved the goods to sell by ECR and CPFR. And Tesco tried to sell the differentiated goods by PB. The distributor's margin was reduced by direct dealings with a manufacturer.
2. Price
   PB was sold at a reasonable price.
3. Place (Distribution)
   Tesco used composite distribution facilities (the multi-temperature type of warehouse and the vehicles). The daily deliveries of composite product to all stores reduced stock at the stores.
   Tesco controlled directly the primary distribution (manufacturer to distribution center) by selecting logistics company which could reduce cost and give the good supply chain. Tesco chose the railway for primary distribution because of its economic benefit and contribution to reduce CO_2.

4. Promotion
   Tesco introduced a loyalty card and got the information of the consumer needs from its loyalty card. Loyalty card fostered the store loyalty.

4.2 The Features of Seven-Eleven Business

We analyzed Seven-Eleven business by the classification of 4P as follows:

1. Product
   Small quantity matches consumer needs. It is transported to a small store efficiently by cooperative distribution. Seven-Eleven managed the inventory control and prediction of demand by IT. Popular goods were predicted by IT.
2. Price
   PB (reasonable price). The management that makes wasteful stock decrease by IT. Inventory control and prediction of demand.
3. Place (Distribution)
   Seven-Eleven had changed the delivery truck company to new wholesale company. This system change improves delivery, and the store management had become efficient. Seven-Eleven did cooperative distribution (joint distribution) system. Seven-Eleven used the exclusive distribution center with multiple temperatures.
4. Promotion
   Dominatrix strategy and franchise strategy (Fig. 5).

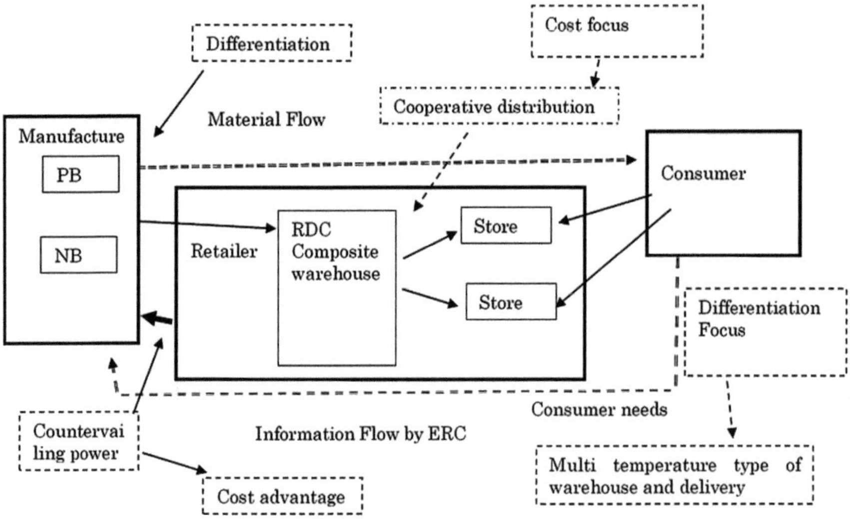

Fig. 5 Competitive advantage of food retailer

5 Conclusion

5.1 The Impact of PB and IT

The number of PB item orders made by Tesco and Seven-Eleven to manufacturers has been increasing. PB items match consumer demand, and the market share of Seven-Eleven and Tesco has grown rapidly. The penetration of PB items enables a retailer to establish customer loyalty. Tesco and Seven-Eleven got cost leadership and differentiation by PB and the IT.

5.2 The Impact of Cooperative Distribution and Multi-Temperature Type of Warehouse

We described Tesco's logistical transformation in Sect. 2 and logistics and marketing strategy about Japanese retailer in Sect. 3. We had some idea of focus strategy from these sections. Porter stated that the focus strategy has two variants: cost focus and differentiation focus. We think Seven-Eleven and Tesco did cooperative distribution and kept multi-temperature type of warehouse and delivery as a focus strategy in the grocery retail industry.

The cooperative distribution reduced the logistics cost.

The multi-temperature type of warehouse and delivery enables to hold the quality of goods. If the grocery retailer aimed to achieve cost leadership in the retail industry, they had to win a sale competition in the retail industry. Logistic strategy supported a cost strategy and a differentiation strategy.

The cooperative distribution, multi-temperature distribution as well as IT and PB are regarded as a successful factor of Seven-Eleven and Tesco.

5.3 The Change of Retail Logistics Structure

Now, it is clear that a logistics strategy as well as product and price strategies is important for a grocery retailer to gain competitive advantage in the industry. The improvement of the transportation quality is also a key to win in the sales competition.

Based on the analyses performed above, we have drawn the following conclusion.

1. Grocery retailers gain the power to manufacturers by using the IT and selling PB products.

PB and IT took effect to the retailer by the countervailing power and the competitive advantage.

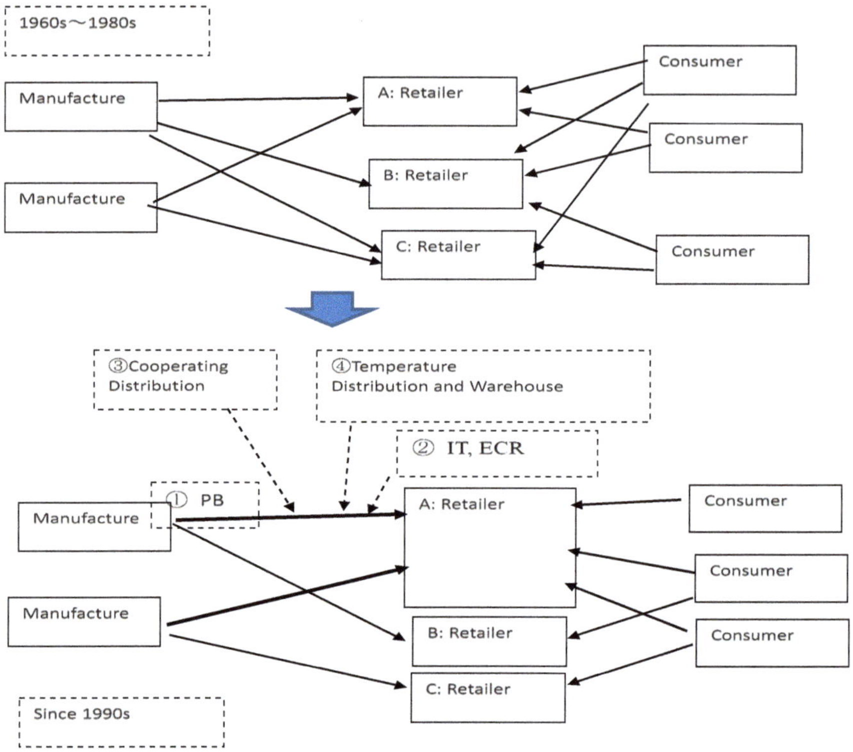

Fig. 6 The change of retail logistics structure

2. Big grocery retailers have the competitive advantage to its rivals by the use of cooperative and temperature-controlled logistics.

Composite distribution led to reduction of stock holding. The daily deliveries of composite product to all stores reduced stock at the stores.

Big grocery retailer with countervailing power offered cooperative and temperature-controlled logistics to manufacturers.

For example, Tesco chose the railway for primary distribution because of its economic benefit in the UK. There is also a similar trial in Japan. AEON began a railway study with PB manufacturer, and AEON is taking the lead and composing an exclusive train of PB product. Retailer has become the channel captain of SCM in Japan. The changing of retail logistics has occurred in Japan (Fig. 6).

References

Department for Transport (2009) Low carbon transport: a green future, p 80
Levi EH (1954) Symposium review: Galbraith's 'concept of countervailing power' and Lilienthal's 'big business'. University of Chicago Law School Chicago Unbound, Chicago, pp 176–178
Ogawa S (2001) Demand chain management, HITOTSUBASHI business review, pp 45–57
Porter ME (1985) Competitive advantage. Free press, New York, pp 11–16
Sparks L (2010) Supply chain management and retailing. Int J 4:4–12
Yahagi T (2014) Dual brand strategy NB and/or PB, pp 2–18

Investigating the End-Customers' Acceptance of the Virtual Supply Chain: The Case of Grocery Retailers in Egypt

Sama Gad, Sara El-Zarka and Mohammed Abdel Qader

Abstract One of the emerging concepts that gained popularity in many countries is virtual supply chain (VSC). Since the end-customer is a key player in the success of integrating the VSC in any business, this chapter focuses on examining the end-customers' acceptance of using technology through the technology acceptance model (TAM) to investigate the potential application of VSC in the grocery retailers' sector in Egypt. A survey-based questionnaire was used to assess the Egyptian customers' acceptance of virtual shopping. The questionnaire consisted of 44 questions that focused on the five variables of TAM: perceived usefulness, perceived ease of use, attitude toward using the virtual shopping, intention to use virtual shopping, and the actual use of the virtual shopping. The structural equation model (SEM) was applied to test the research hypotheses. The data analysis revealed a strong relationship between the five TAM variables which concludes the acceptance of the Egyptian customers' for VSC. The data only focused on customers in the city of Alexandria which might limit the generalizability of the research results.

Keywords Virtual supply chain (VSC) · Virtual shopping · Technology acceptance model (TAM) · Egypt · Grocery

S. Gad (✉) · S. El-Zarka
College of International Transport & Logistics, Arab Academy for Science, Technology and Maritime Transport, Alexandria, Egypt
e-mail: Sama.hossam@aast.edu

S. El-Zarka
e-mail: sara_elzarka@aast.edu

M.A. Qader
Dar AL Uloom University, Riyadh, Saudi Arabia
e-mail: makader@dau.edu.sa

V. Kachitvichyanukul et al. (eds.), *Toward Sustainable Operations of Supply Chain and Logistics Systems*, EcoProduction,
DOI 10.1007/978-3-319-19006-8_6

1 Introduction

World business nowadays witnesses drastic changes and improvements toward adopting new technologies that benefit both; the businesses and the customers which in turn will boost the business's turnover and consequently improve the country's economy and encourage the world's trade. This research is concerned with several key-enabled technologies and management systems supporting the application of the virtual supply chain (VSC) concept within the enterprise and its acceptance by the customers. The research investigates the effectiveness of the high-technology route in helping customers feel informed about the products and their acceptance and familiarity with using the advanced technologies represented in using their smart mobile devices and Internet in the process of purchasing their groceries. This will be measured by applying the technology acceptance model (TAM) to assess the degree to which the Egyptian customers are familiar and prepared to use the virtual shopping. This research motivates customers and grocery retail stores to adopt new technologies such as the VSC supported by the provision of a detailed analysis of the virtual shopping dimensions, as a new way of shopping while commuting by using mobile phones to scan the codes of the products to be purchased from smart posters connected with the Internet, placed in major locations in the city, e.g., train or metro stations, shopping malls, and gas stations.

This research investigates the effectiveness of the high-technology route in helping customers feel informed about products. The acceptance and familiarity of the Egyptian customers with the advanced technologies represented in their smart mobile devices and Internet and using it in buying their groceries will be measured by applying the TAM model.

This research paper starts with a research background that reviews the literature related to the TAM and its applicability in the Egyptian market supported by statistical reports that focused on the Egyptian market. Based on the literature, the research model, variables, and hypotheses are then formulated, and consequently tested by the survey described in the research methodology. The collected data are then measured and analyzed to present the research findings and to provide the conclusions and the recommendations for further research.

2 Literature Review

According to the analysis of previous research literature of the VSC concept and in coherence with researches that analyze the Egyptian culture and grocery sector, this research provides a link between both and a solid analysis of its applicability in the Egyptian market.

2.1 Virtual Supply Chain

A virtually enabled supply chain network can be described as a series of value-added processes owned by one or more enterprises, starting with material/information suppliers and ending with consumers.

Kisperska-Moroń (2010) proposed another definition which is an organization that is subject to constant changes, demonstrating a specific potential when required, and overcoming time and space barriers. VSCs could be described by such attributes such as temporary character, focus on customers, geographical dispersion, and intensive support of IT systems, network structure, and an extensive use of key competencies of their members.

In virtual environment systems, integration means to integrate disparate data sources, to present real-time content and information to all virtual network participants and finally to create an environment where every participant and every activity in the process is integrated. Virtual environment involves limited human interaction (seamless integration) as manual processes are too slow and error-prone to support the new business model. The goal of a structured data exchange is to move data from one organization to another with near-zero tolerance for the loss of the actual content or meaning of it (Manthou et al. 2004).

The key elements that form the virtual network in terms of supply chain management are suppliers, who actually perform, provide/sell, and deliver the product/service in demand; intermediaries, who own the supplier/customer relationship and knowledge; and customers, who directly purchase and acquire the product/service. Their roles tend to change as a supplier or an intermediary in a given business relationship may play the role of a customer in another one. Not all suppliers and business partners are alike. Some suppliers provide materials that are used in production, while others provide indirect goods (Manthou et al. 2004).

Supply chain integration in a virtual environment considered the process by which suppliers, partners, and customers within a shared market space collaboratively plan, implement, and manage the flow of information, services, and products along the supply chain in a way that improves business operations in terms of speed, agility, real-time control, or customer response (Liu et al. 2008).

Virtualization as indicated by Zarour et al. (2005) implies the use of IT and communication technology by organizations in managing their interactions and key business operations with customers, suppliers, and employees.

From other researchers point of view that a number of technology advocates have claimed that virtual reality (and other advanced technology) will encourage playful interaction with the product and thereby make online shopping a more engaging and enjoyable experience (Manthou et al. 2004).

Applied to the online shopping context, the theory predicts that shoppers will have positive attitudes toward stores that are easy to use and useful and that positive attitudes are, in turn, associated with ongoing intention to visit the online shop. Simulation technologies are, by their nature, likely to be more complex and harder to use than catalogs and static Web pages (Manthou et al. 2004).

In particular, Johnston et al. (2006) have demonstrated in his research which contributes to the e-commerce literature by increasing the understanding of how customer interfaces affect perceptions, and the relationships between system design and different types of informativeness, and between informativeness and system acceptance.

These technologies represent fundamentally different means of providing online vicarious experience. As anticipated by Johnston et al. (2006) that they will also prove problematic in some ways, but others may compensate for some of the problems found in virtual personalized experience systems.

2.2 Implementing a Virtual Supply Chain

Virtualization of supply chain systems involves both IT and organizational changes Liu et al. (2008). In the VSC, hierarchical virtualization of a network provides a flexible, granular, protection resource. Using the Hardware Abstract Layer (HAL) concept, network services are handled by a number of virtual servers (Dawson 2008). In the event that a physical server hosting the virtual servers fails, another physical server already operating on the network with the virtual systems can stand in as an alternative physical host.

One of the key aspects related to the VSC that make the SC processes easier and faster that has been addressed by Lorchirachoonku and Mo (2010) is the quick response (QR) code and he stated that it was developed by a Japanese corporation Denso-Wave in 1994 that has been widely used in Japan.

QR code can be quickly decoded on virtually any computing device that connects to the Internet. At the moment, apart from normal desktop and laptop computers, most mobile phones have built-in cameras that can read the QR code, decode it with free software inside the phone, and then either link them to a Web site, or store the entire business information somewhere in a server (Lorchirachoonku and Mo 2010). Typical mobile platforms such as Apple iPhone and Nokia phones have been proved to work seamlessly with this code. Hence, to virtualize the identification system using QR code, a mobile phone with camera scans and decodes the image which can then link to URLs or a global information server is needed.

2.3 The Retailing Sector in Egypt

Recently, the growing youth population who are educated, technologically savvy and accepting a more westernized life style lead to a growth in the Egyptian retail market which is an evidence that the changes occurring in Egypt not only economic (Ramzy et al. 2011). Indications in Egypt show the modernization of food retailing being accepted, which may be related to the age of the customers (Ramzy et al. 2011).

Goldman et al. (2002) mentioned that large supermarkets were accepted mainly by the wealthy category of customers in developing countries. The time-starved upper classes prefer the one-stop shopping that larger retail formats provide. In the Egyptian retailing market, low-income residents visit the larger food stores to benefit from the store's economies of scale which provides them lower prices of goods and services (Goldman et al. 2002).

Ramzy et al. (2011) mentioned that the consumer survey in the year 2010 that reveals Egyptian retail purchasing habits recorded that Egypt is ranked the 13th on the most attractive retail market worldwide and that it is one of the promising and fast expanding markets in its region. Thus, the economy is lucrative for foreign investment. Although the trend is spreading, the variation among emerging market countries of the split of market share between small and big box retailers is large. In addition, Egypt is currently experiencing the growth of large supermarkets and the loss of domination by small retailers (Ramzy et al. 2011). Some of the Egyptians are still culturally loyal and bond to small independent retailers due to the ease of transactions. This loyalty is not as strong as in the past due to the lower prices offered by larger food retailers.

Low-income consumers represent the majority of purchasing power in the market in Egypt, like most of the other developing countries. Food retailers like Carrefour started to capture market share by attracting the attention of lower-income customers. A positive relationship between retail concentration and price has been found in several studies (Ramzy et al. 2011).

Education is needed to teach consumers modern marketing trends. Most low-income Egyptians do not differentiate between branded and non-branded packaged items, especially when it comes to food and electronics. The difference between the prices of branded versus non-branded products still not well understood in the Egyptian culture. In emerging markets, customers do not realize the full benefits of products; some even do not know how to access different providers' offerings. Also store employees suffer from the same lack of knowledge (D'Andrea et al. 2010).

2.4 Retailing Strategies

There are different strategies being applied in the food sector, where the emphasis is on price reduction in order to gain market share or market entry. Many low-income customers in Egypt understand neither value-based quality nor the concept of paying more for customer service. Large retailers are able to offer assortment and variety with lower prices, but often at the expense of services and convenient locations. In developed nations, store quality is defined by location, price assortment, fast checkout, friendly and courteous service, weekly specials, and pleasant shopping environments (Goldman et al. 2002). These concepts have not reached their potential when it comes to the Egyptian retail landscape.

Bellona et al. (2010) ensures that the strategies in the retail grocery industry are as diverse as its players. Well-known grocery retail store pursues an enhanced differentiation strategy that has redefined the experience of grocery shopping for consumers willing to trade price with unique service. While these companies pursue an array of distinct strategies, all seek to strengthen existing structures, create new value such as convenience, low cost, and quality, and create better methods to capture value. The design of these strategies highlighted as a way of relating service design and profits and used in assessing the industry over the next 3–5 years.

2.5 The Grocery Sector Trends and Obstacles in Egypt

Design quality provided by large grocery retail stores will be informed by some apparent trends in the market during the next 3–5 years and beyond. According to Bellona et al. (2010), these trends include the following: (1) healthier eating habits, and a more sustainable lifestyle, (2) an increased passion for food due to growing media attention, and (3) a shift from grocery shopping as a necessity toward grocery shopping as a user experience. These trends will lead to customers' satisfaction in terms of searching for new food experiences demanding variety and more unusual goods in the product line. A larger segment of the market might also be willing to pay more for enhanced customer experience in terms of service and shopping atmosphere.

Ramzy et al. (2011) mentioned some obstacles to the modernization of retailing in Egypt: (1) New retail establishments are opening in crowded areas with no parking facilities, (2) Out-of-stock situations are experienced daily in many stores, (3) Returning merchandise is neither a pleasant nor an easy task for customers, (4) Telephone services numbers are not usually working and if occasionally customers are often placed on hold for extended periods, and (5) Consumers often receive conflicting messages from retailers, resulting in frustration.

2.6 The Grocery Retail Sector Growth

As the Egyptian youth buying power increase, this leads to increase in the expected growth in the e-commerce business, as well as an increase in the customer's knowledge which will lead to increase in the demand for service quality. Increasing access to the Internet enables Egyptians to customize orders and obtain products which were previously inaccessible. A movement toward dual household income is also increasing Egyptians buying power. Consequently, middle-income Egyptian families have more discretionary income than the past (Ramzy et al. 2011).

In 2008 and 2009, the consumer food retail sector has witnessed a huge growth despite the global downturn in 2009, in addition to the presence and opening of new international and domestic retail companies (Maldonado and Mansour 2009). Recently, Egyptian consumers face changes in their lifestyle as consumers became more westernized in terms of their work and their eating habits, taking in consideration the increasing percentage of women workforce. All these elements contributed to the rise in food services and retail sales.

A massive reduction in the catering business in Egypt recorded by about 15–20 % due to the global financial crisis; then, the increasing number of the international chains in Egypt with variety of products and the increase in the level of income and brand advertising led to a growth in the retail sales by the same portion (Maldonado and Mansour 2009).

Rapid development in the food retail sector continues in Egypt. More hypermarkets and supermarkets are spreading in the two largest cities in Egypt, Cairo and Alexandria, as the Egyptian population is exceeding the 80 million citizens based on the recent World Bank report in 2012 http://data.worldbank.org/indicator/SP.POP.TOTL, in addition to spreading in other governorates of Egypt, primarily the Red Sea resort area (Maldonado and Mansour 2009).

Ramzy et al. (2011) compared the Egyptian retail sector to the American and European retail sector and mentioned that small grocers and independent stores dominated the grocery market in the past and they were the primary source of products. However, today the large store format dominates the USA and Europe as supermarkets and hypermarkets take the majority of market share.

According to Egypt Retail Report (2010), the sales of the Egyptian retail industry is expected to grow in the coming years from EGP 147.80 billion (US $27.22 billion) in 2010 to EGP 208.37 billion (US$38.37 billion) by 2014.

In 2010, Egypt's GDP was US$226 billion. GDP per capita is predicted to increase by 63.9 % to reach US$4.463 between the years 2010 and 2014, which will lead to fluctuation in demand for the luxurious products and necessitates more modern retail technology. Attraction of the international brands will be witnessed for the Egyptian retailing market in a result of that shift in demand and income (Ramzy et al. 2011).

2.7 Grocery Stores Classification

Maldonado and Mansour (2009) highlighted the introduction of hypermarkets with more than 5000 m^2, supermarkets with more than 450 m^2, and mini markets with more than 150 m^2 is helping to reshape the retail industry in Egypt in addition to the shopping habits of customers. They also clarified that one of the major obstacles to expansion is the limited availability of shelf space. Larger land space is needed for hypermarkets to fulfill sufficient retail space for customers and larger storage area and more parking spots which is not available in central Cairo. However, sufficient land spaces are available in other cities such as Alexandria and other suburban areas which are rapidly developing.

Ramzy et al. (2011) also mentioned in their research that the grocery chains are also experiencing success in Egypt. For example, Carrefour, Spinneys and the locally owned Hyper One are opening stores (Egypt: Retail sector to reach US $25.33 billion, 2011) and also they highlighted that there is a potential for more retail expansion since the market is undersupplied (Economic Update Egypt: The rise of retail Egypt, 2011).

Maldonado and Mansour (2009) also analyzed the Egyptian retail industry, by focusing on Carrefour France which recently operates six stores and has a future plan to expand in the market and open eighteen new stores during the next five years including in their plan another sixteen small express stores. The researchers included in their research other chain stores and referred to Metro, a local supermarket chain that has 39 branches and Khair Zaman operates 21 branches. Also they provided a detailed focus on Carrefour the well-known international hypermarket as follows:

"The French chain Carrefour has revolutionized the retail industry in Egypt by introducing a new format for hypermarket shopping. Carrefour opened six hypermarkets: four in Cairo and two in Alexandria. The first Carrefour opened in December 2002, covering 28,000 m^2 of retail space. Two of them opened in 2003, covering 32,000 m^2, one opened in 2008, and two opened in 2009. Carrefour is now reporting an average of 17,000 visitors per day on weekdays and 40,000 visitors on the weekends; an average of 7500 transactions a day during week days and 13,000 daily on weekends. Their focus is on middle-income families and they rely primarily on local goods."

2.8 Technology Acceptance Model

According to Liu et al. (2008), TAM appears to be the most widely accepted theory among information systems research for studying users' system acceptance behavior. Many researchers relied on TAM where its functionality as a basic model was integrated with different theories and tested in order to predict the behavior of different users.

As indicated by Schepers and Wetzels (2007), TAM was inspired by the theory of reasoned action (TRA) of Fishbein and Ajzen, which assured that both the attitude toward a specific technology has an impact on the behavioral intention of people, which in turn affects how those people perform the action toward that technology.

To date, TAM has been the only model which has mostly captured the attention of the information systems community. Thus, it is essential for anyone willing to study user acceptance of a specific technology to have an understanding of the TAM (Chuttur 2009).

Chuttur (2009) mentioned that most of the studies carried out in this research area failed to produce reliable measures that could explain system acceptance or rejection. Chuttur also indicated in his research the TAM as a system that is used as a response that can be explained or predicted by user motivation, which, consequently is directly influenced by an external motivation consisting of the actual system's features and capabilities.

Schepers and Wetzels (2007) defined TAM in their research, as one of the best known models that explains the intention to use a technology. It assumes that the perceived usefulness (PU) and the perceived ease of use (PEOU) are central in influencing a person's attitude and behavioral intention toward using it.

Schepers and Wetzels (2007) based on their review for several research papers provided a quantitative and qualitative meta-analysis of the TAM's variables. The quantitative analysis only focused on three relationships: (1) perceived usefulness and perceived ease of use, (2) perceived usefulness and technology acceptance (i.e., named in other researches the actual use of technology), and the last one (3) perceived ease of use and technology acceptance (actual use). The researchers highlighted and focused on some other variables that may affect the TAM, one of these variables was the culture.

The TAM identifies perceived ease of use and perceived usefulness as key independent variables. Perceived usefulness is also influenced by perceived ease of use (Dishawa and Strong 1999).

Dishawa and Strong (1999) mentioned that TAM is an excellent model by itself, as it is internally sound and based directly on well-tested attitude/behavior models. Tests of TAM have consistently produced similar results. TAM, however, does have some weaknesses. While TAM provides excellent explanation of intention to use, it is much weaker for actual use. TAM's two independent variables, namely perceived usefulness and perceived ease of use, are still important contributors to explaining utilization (Fig. 1).

A weakness of TAM for understanding IT utilization was provided by Dishawa and Strong (1999) which is its lack of task focus. IT is a tool by which users accomplish organizational tasks. The lack of task characteristics in evaluating IT and its acceptance, use, and performance contributes to the mixed results in IT evaluations. While TAM's usefulness concept implicitly includes task, more explicit addition of task characteristics may provide a better model of IT utilization.

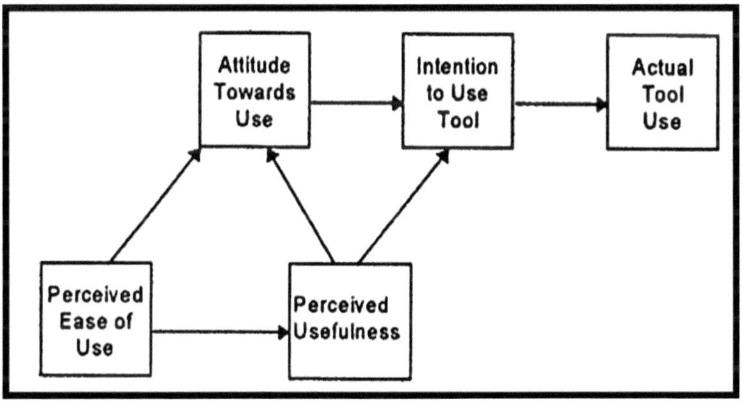

Fig. 1 Technology acceptance model (TAM). *Source* Dishawa and Strong (1999)

Users' motivation can be explained by three factors: perceived ease of use, perceived usefulness, and attitude toward using the system. He hypothesized that the attitude of a user toward a system was a major determinant of whether the user will actually use or reject the system (Chuttur 2009).

Chuttur (2009) relied on several other related studies to identify only two distinct beliefs, perceived usefulness and perceived ease of use, that were sufficient enough to predict the attitude of a user toward the use of a system. Related studies on perceived usefulness and perceived ease of use highlighted the importance of these two variables in predicting a person's behavior.

Chuttur (2009) provided a definition to "perceived usefulness" as the degree to which an individual believes that using a particular system would enhance his or her job performance, where "Perceived ease of use" is the degree to which an individual believes that using a particular system would be free of physical and mental effort.

Schepers and Wetzels (2007) mentioned that attitude can be defined as the degree to which a person has a favorable or unfavorable evaluation or appraisal of the behavior.

3 Research Model and Hypotheses

The typical TAM shows the relation between the five variables: perceived usefulness, perceived ease of use, attitude toward using the virtual shopping, intention to use virtual shopping, and the actual use of the virtual shopping. The effect of each variable on the other is highlighted by using arrows as clarified in the following figure with indicating the hypotheses number on the model (Fig. 2).

H1 Perceived ease of use is positively associated with perceived usefulness.

H2 Perceived ease of use is positively associated with attitude to the virtual shopping.

H3 Perceived usefulness is positively associated with attitude to the virtual shopping.

H4 User's attitude to the virtual shopping is positively associated with intention to use the virtual shopping.

Fig. 2 Research model

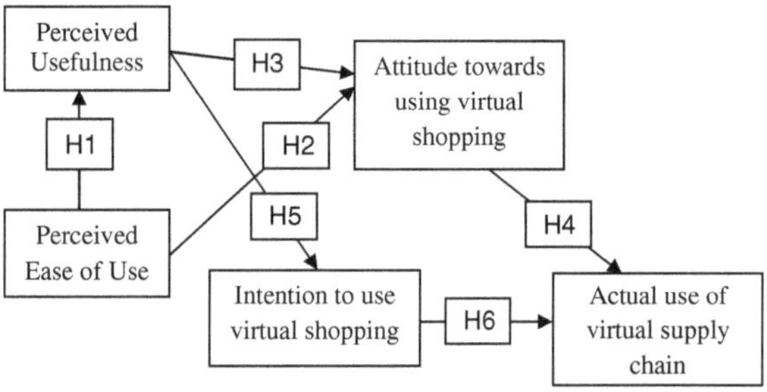

H5 User's perceived usefulness is positively associated with intention to use the virtual shopping.

H6 If users have intention to use the VSC, then the virtual shopping will be actually used.

4 Research Methodology

This research is empirical by obtaining data from individuals and industry experts. The research is based on two phases: the first phase describes the adoption of VSC by the grocery retail stores in Alexandria and the second phase examines the Egyptian customer's willingness to accept the virtual idea based on the application of TAM a questionnaire targeting potential Egyptian customers.

A printed questionnaire is used to study the degree of preparedness of the Egyptian customer to accept the virtual shopping concept. The questionnaire is used to empirically test the TAM's hypotheses and the data collected were quantitatively analyzed on a statistical basis. All the variables of the TAM reviewed in the literature review part have been included in the questionnaire.

The questionnaire was distributed in Arabic as it is the native language of the Egyptian customers. Simple random sampling was used to select respondents and the total number of respondents was 312 full responses of 400 distributed which represent a response rate of 76 %. A pilot test was conducted in order to validate the instrument in terms of language and structure.

A personalized cover letter was provided at the beginning of the questionnaire highlighting the importance of the study. The questionnaire consists of forty four questions divided into four groups. The statements of the questionnaire are related to variables under study (perceived ease of use = 5 statements, perceived usefulness = 5 statements, behavioral intention = 4 statements, attitude = 4 statements, and actual use = 3 statements).

A lack of multivariate normality is particularly troublesome, because it substantially inflates the chi-square statistic and provides parameter estimates with too much statistical power (Abdel Hamid 2013). AMOS provides some additional goodness-of-fit indices researchers can use it to resolve the problem of chi-square biasness to large sample size and normality issue.

The most common measure of model goodness-of-fit is chi-square/degrees-of-freedom (CMIN/DF). The value of (CMIN/DF) ratio should not exceed 5. AMOS provides some additional goodness-of-fit indices researchers can use it to resolve the problem of chi-square biasness to large sample size and normality issue (Abdel Hamid 2013).

5 Findings and Analysis

The research model is measured and analyzed by using structural equation model (SEM) by using AMOS18 and SPSS. The analysis of the TAM evaluates the extent to which the Egyptian customer will accept the new technology, i.e., the virtual shopping. The analysis revealed a significant relation between the TAM variables (perceived ease of use, perceived usefulness, attitude toward use, intention to use, and actual use) (Fig. 3).

PEOU = perceived ease of use, PU = perceived usefulness of the virtual shopping, Int. = behavioral intention toward the virtual shopping, Att. = attitude toward the virtual shopping, and AU = actual use of the virtual shopping.

In order to determine the goodness-of-model fit, some measures need to be observed and highlighted; the chi-square statistic and the model fit that has the CFI, TLI, and RMSEA. As noted in the following Table 1, the chi-square equals to 386.060 which represents a good fit of the model and considered a significant value. P indicates significant regression path when ($P < 0.001$), also the path is considered as significant when the C.R. is at 0.5 level or higher.

As the research sample size equals to 312, AMOS software automatically sums responses to keep the research results accuracy high and dependable, so the distinct sample responses according to AMOS calculation equal to 231 and the estimated parameters equal to 48 resulting from the total of variances and the regression weights of the unlabeled parameters that are neither fixed nor labeled. Such parameters are free to take on any value.

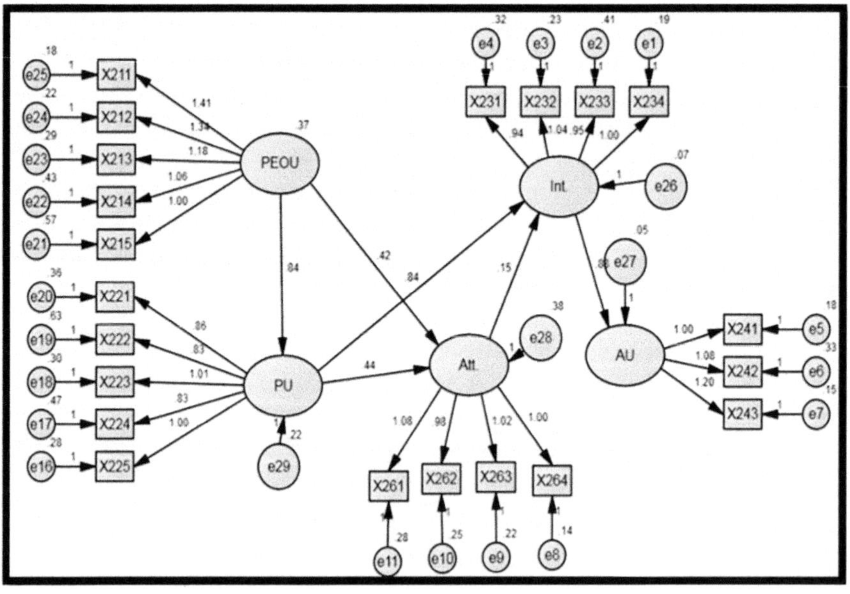

Fig. 3 Technology acceptance model (measurement model)

Table 1 Computation of degrees of freedom

Number of distinct sample moments	231
Number of distinct parameters to be estimated	48
Degrees of freedom (231 − 48)	183
Chi-square	368.060
Probability level (*P*)	0.000
Sample size	312

Also other indicators that measure the goodness-of-model fit is the CMIN/df = 2.011, CFI = 0.961, TLI = 0.955, and RMSEA = 0.057. These values demonstrate that the hypothesized model has a good fit with the observed data, supported by the chi-square, indicates that the probability level is significant as shown in Table 1.

The proposed model of the study contains the following hypotheses:

H1: Perceived ease of use is positively associated with perceived usefulness. (Supported)

The findings of testing H1 revealed that there is a highly significant relation between perceived ease of use and perceived usefulness, as the Egyptian customers perceives the ease of using the virtual shopping related to its usefulness. This relationship yields ($P < 0.001$, C.R. = 9.562).

H2: Perceived ease of use is positively associated with attitude to the virtual shopping. (Supported)

The relation between the perceived ease of use of the virtual shopping is positively associated with attitude of the Egyptian customers toward using this technology, and its significance achieves ($P < 0.001$, C.R. = 3.657).

H3: Perceived usefulness is positively associated with attitude to the virtual shopping. (Supported)

Testing the relation between the two variables; perceived usefulness of trying the virtual shopping and the attitude toward using it achieved an acceptable significance *P* value ($P < 0.001$, C.R. = 4.320).

H4: User's attitude to the smart mobile phones is positively associated with intention to use the virtual shopping. (Supported)

H4 examines that if there is a positive relation between the attitude of the Egyptian customers and their intention to use their smart mobile phones or devices in trying the virtual shopping, the result of the hypothesis also yields a high significant *P* value ($P < 0.001$, C.R. = 3.421).

H5: User's perceived usefulness is positively associated with intention to use the virtual shopping. (Supported)

The relation between the Egyptian customer's perspective of finding the virtual shopping useful in their life and their intention toward using it records in testing the

Table 2 Regression weights and hypotheses testing sample size (n = 312)

#	Path direction			Standardized regression weights	Sig.	C.R
1	Perceived usefulness	←	Perceived ease of use	0.842	***	9.562
2	Attitude	←	Perceived ease of use	0.419	***	3.657
3	Attitude	←	Perceived usefulness	0.440	***	4.320
4	Behavioral intention	←	Perceived usefulness	0.838	***	12.436
5	Behavioral intention	←	Attitude	0.147	***	3.421
6	Actual use	←	Behavioral intention	0.881	***	17.225

***p < .001

model that there is a strong relationship between the two variables represented in a high significance P value ($P < 0.001$, C.R. = 12.436).

H6: If users have intention to use the VSC, then the virtual shopping will be actually used. (Supported)

Testing H6 examines the effect of the Egyptian customer's intention to use the virtual shopping and actually using this technology. The relation between the two variable yields the highest significance level of P value ($P < 0.001$, C.R. = 17.225).

Table 2 shows the summary of the hypotheses testing.

6 Conclusions

In summary of research results, this research provided a detailed analysis about the development of the VSC concept in integration with the analysis of the Egyptian populations and their willingness to adopt new technology and deal with it on a daily basis that may enhance their lifestyle in terms of saving time and effort.

The analysis of the research survey results shows that the Egyptian population is on a satisfactory level of involvement in using smart mobile devices and technology. Thus, the application of the VSC, translated to the virtual shopping to customers, will not be complicated.

Applying the VSC in the Egyptian grocery retail market would enhance the business and boost this sector's outcome, which in turn help in improving the country's economy.

7 Recommendations

The high demand from the Egyptian customers on the large retail stores represents a real opportunity for retailers, promising profit margins, and competitive advantage in the Egyptian retail sector and the global market as well.

Based on the questionnaire analysis of this research, the sample chosen encourages the idea based on their familiarity with using smart mobile phones and devices, so a user-friendly application uploaded to their devices would be beneficial to encourage those who find the virtual shopping complicated.

International retail investment in Egypt must be developed in the coming period. Good partnership must be developed between the international retailers and the local suppliers and the other members and intermediaries in the distribution network, in addition to understanding the local culture and the consumer's preferences of the new environment.

References

Abdel Hamid, Mohamed Abdel Kader, (2013). Investigating the factors that influence continue receive mobile SMS advertisement: an empirical study applied in Egypt. Ph.D. University of Lincoln

Bellona D, Breivik K, Cannon C, Yap M (2010) Industry analysis: retail grocery industry analysis, value creation and best practices

Chuttur MY (2009) Overview of the technology acceptance model: origins

Dawson P (2008) IT virtualization—virtual tales and real life stories. Eng Technol 3(3):56–58

Dishawa MT, Strong DM (1999) Extending the technology acceptance model with task-technology fit constructs. Inf Manag 36(1):9–21

D'Andrea G, Marcotte D, Morrison GD (2010) Let emerging market customers be your teachers, Harvard Business Review. pp 2301–2304

Goldman A, Ramaswami S, Krider RE (2002) Barriers to the advancement of modern food retail formats: theory and measurement. J Retail 78:281–295

Kisperska-Moroń D (2010) Evolution of competencies of logistics and supply chain managers. LogForum. [online] Available at: http://www.logforum.net/vol6/issue3/no3. Accessed 9 Apr 2012

Liu SH, Liao HL, Peng CJ (2008) Applying the technology acceptance model and follow theory to online E-learning users' acceptance behavior. Inf Syst 6(2)

Lorchirachoonkul W, Mo JPT (2010) RFID implementation with virtual infrastructures. Bus Process Manag J 16(6):917–931

Maldonado J, Mansour S (2009) Egypt retail food sector. Global agricultural information network report

Manthou V, Vlachopoulou M, Folinas D (2004) Virtual e-Chain (VeC) model for supply chain collaboration. Int J Inf Manag 87:241–250. Available through: ScienceDirect database. Accessed 17 Mar 2011

Ramzy O, James RO (2011) The Egyptian retailing industry. J Int Bus Eco 11(3):1–2. [online] Available at: http://www.hu.edu.eg/sites/default/files/files/Egyptian%20Retail%20Industry.pdf. Accessed 4 Mar 2014

Johnston RB, Smith SP, Howard S (2006) Virtual models in online shopping: do they help or hinder customers. [online] Available at: http://is2.lse.ac.uk/asp/aspecis/20060120.pdf. Accessed 10 Apr 2012

Schepers J, Wetzels M (2007) A meta-analysis of the technology acceptance model: investigating subjective norm and moderation effects. Elsevier. [Online] Available at: www.sciencedirect.com. Accessed 29 Dec 2012

Zarour N, Boufaida M, Seinturier L, Estraillier P (2005) Supporting virtual enterprise systems using agent coordination. Knowl Inf Syst 8(3):330–349

Part II
Sustainable Logistics

The Effect of Institutional Pressures and Top Managers' Posture on Green Supply Chain Management

Yi-Chun Huang and Min-Li Yang

Abstract This chapter applies stewardship theory to explain the effects of top managers' postures toward green supply chain management (GSCM). It combines institutional theory and stewardship theory to examine the relationship among institutional pressure, top managers' posture, and GSCM. The method of questionnaire investigation is adopted for electrical and electronic industries in Taiwan. A total of 1000 questionnaires was mailed, and the 180 valid questionnaires were returned. There are some main findings. First, institutional pressures have significant positive effect on the GSCM. Second, institutional pressures have significant positive effect on top managers' posture. Third, top managers' posture has a significant positive effect on GSCM. Finally, institutional pressures through top managers' posture have indirect effect on GSCM.

Keywords Green supply chain management (GSCM) · Institutional pressure · Top managers' posture

1 Introduction

The dramatic rise of the climate agenda has had a significant effort on the environmental policy area. Therefore, today as never before people are concerned with the environment and climate change (Walker et al. 2008). Moreover, with the rise of international environmental regulations such as the Montreal Agreement, the Basel Treaty, the Kyoto Protocol, the Waste Electrical and Electronic Equipment (WEEE) Directive, the Restriction of Hazardous Substances (RoHS) Directive, and Eco-design for Energy using products (EuP) Directive, increasingly, companies are

Y.-C. Huang (✉) · M.-L. Yang
National Kaohsiung University of Applied Sciences, Kaohsiung, Taiwan
e-mail: peterhun@cc.kuas.edu.tw

M.-L. Yang
e-mail: minly@kuas.edu.tw

V. Kachitvichyanukul et al. (eds.), *Toward Sustainable Operations of Supply Chain and Logistics Systems*, EcoProduction,
DOI 10.1007/978-3-319-19006-8_7

recognizing that environmental management is a key strategic issue with the potential for a lasting impact on organizational performance (Diabat and Govindan 2011; Huang et al. 2009; Huang and Wu 2010). In the field of business and management, there is an increasing onus on the role of organisations in society (Ambec and Lanoie 2008; Huang et al. 2012), and their responsibilities to minimize impacts upon the environment (Hart 1995; Christmann 2000). Firms are facing increasing pressure to become responsible and greener. Companies should rethink the way they plan to do business in the future to stay profitable (Kumar et al. 2012).

How does a company respond to the green issues challenge? Huang et al. (2012) argued that when organizations deliberate as to how to respond to the green challenge, they should consider the importance of taking more direct green's action; therefore, green supply chain management (GSCM) plays a critical role (Azevedo et al. 2011; Kumar et al. 2012; Rao and Holt 2005; Walker et al. 2008; Zhu et al. 2011). GSCM is an approach that aims at the overall optimization of material and information flows along the value chain. Development of environmentally friendly processes, products, and services requires a unified effort by all members of the supply chain (SC) to avoid sub-optimization at the partner level. GSCM implementation could include reducing packaging and waste, assessing vendors on their environmental performance, developing more eco-friendly products, and reducing carbon emissions associated with transport of goods (Walker et al. 2008). The implementation of GSCM is expected to result in improved environmental performance as measured by reductions in air emissions, effluent waste, solid waste, and the consumption of toxic materials (Zhu and Sarkis 2004). There is concern, however, whether such environmental sustainability efforts will ultimately translate into improved market share and profitability (Kumar et al. 2012). Ultimately, manufacturing managers are responsible for the performance of the organizations for which they work.

GSCM has emerged as an organizational philosophy which helps organizations and their partners to achieve corporate profit and market-share objectives by reducing environmental risk and impacts while improving ecological efficiency (Azevedo 2011; Rao and Holt 2005; Zhu et al. 2008). There are some works to investigate the factors that drive organization to implement GSCM (Diabat and Govindan 2011; Walker et al. 2008; Zhu and Sarkis 2006; Zhu et al. 2011). Still some questions need to clarify? Why should companies care? Why do organizations that face same institutional pressures to adopt different practices of GSCM? In order to investigate the questions further, this study use stewardship theory to explain the effects of top managers' postures toward GSCM. Therefore, this study draws both institutional theory and stewardship theory to explore the relationship among institutional pressure, top managers' posture and GSCM. GSCM has emerged as an organizational philosophy which helps organizations and their partners to achieve corporate profit and market-share objectives by reducing environmental risk and impacts while improving ecological efficiency (Azevedo 2011; Rao and Holt 2005; Zhu et al. 2008). There are some works to investigate the factors that drive organization to implement GSCM (Diabat and Govindan 2011; Walker et al. 2008; Zhu and Sarkis 2006; Zhu et al. 2011). Still some questions need to clarify? Why should companies care? Why do organizations that face same institutional pressures to adopt different practices of GSCM? In order to investigate the questions further, this study use stewardship theory to explain the effects of top managers' postures toward GSCM. Therefore, this study draws both institutional theory and stewardship theory to explore the relationship among institutional pressure, top managers' posture, and GSCM.

2 Theoretical Background and Research Hypotheses

2.1 Previous Literature

A number of authors have referred to the green supply chain (GSC) over the past decade due to emerging environmental management topics. The growth in this GSC literature extends back to the early 1990s with the advent of corporate environmental management, environmentally conscious manufacturing strategy, and supply chain management literature (Zhu and Sarkis 2005).

A number of authors have studied GSCM, which is one of the more popular emerging corporate environmental management topics that have arisen over the past decade. For example, Bowen et al. (2001a) analyzed implementation patterns of green supply and found three types of green supply, including product-based green supply and advanced green supply. Sarkis (2003) presented a strategic decision framework to evaluate GSC alternatives using an analytical network process. Holt and Ghobadian (2009) examined the level and nature of greening the SC in the UK manufacturing sector. Hu and Hsu (2010) explore factors that are critical for implementing GSCM in the Taiwanese electrical and electronics industries relative to European Union directives, and extract 20 critical factors along four dimensions (supplier management, product recycling, organization involvement, and life cycle management). Zhu et al. (2008) evaluated GSCM practices by relating GSCM to closing the SC loop for four Chinese industries (power generation, chemical/petroleum, electrical/electronic, and automobile). They concluded that adoption of GSCM implementation in different industrial contexts is not uniform across the four industries.

In addition, the work explores the driving forces behind environmental behavior, the specific management practices that result, and the relationships between them. Bowen et al. (2001b) analyzed the relationship between supply management capabilities and green supply practices and identified internal drivers for implementing green supply policies (strategic purchasing and supply, corporate environmental proactivity, and supply management capabilities). Trowbridge (2001) distinguished between internal and external drivers for the implementation of GSCM at a chip manufacturer. Internal drivers include the willingness to improve risk management due to potential interruptions in the SC, and the collaboration with suppliers to find alternative materials and equipment that minimize environmental impacts. External drivers include customers, investors, and non-governmental organizations. Lee (2008) identified the main drivers for companies to participate in GSCM implementation as buyer influence, government involvement, and GSC readiness. Walker et al. (2008) reviewed the literature and identified the factors that drive or hinder organizations to implement GSCM initiatives; these includes internal drivers, such as organizational factors, and external drivers such as regulation, customers, competitors, society, and suppliers. Based on interviews conducted at seven different private and public sector organizations, they further identified the internal barriers such as cost and lack of legitimacy, as well as external barriers such as regulation, poor supplier commitment, and industry specific barriers. Diabat and Govindan (2011) used an interpretive structural modeling (ISM) framework to find the drivers affecting the implementation of GSCM.

Rao and Holt (2005) observed that greening different phases of the SC leads to an integrated GSC, which in turn leads to competitiveness and better economical and operational performance. Azevedo et al. (2011) investigated and tested the relationships between green practices of supply chain management and SC performance from five case studies taken from the Portuguese automotive SC. They found that green practices have positive effects on quality, customer satisfaction, and efficiency, but have negative effects on SC performance. Kumar et al. (2012) explores companies can use to understand and improve SC sustainability practices. The results demonstrate that eliminating waste throughout the SC will make the SC more profitable. Green Jr. et al. (2012) found that implementation of GSCM practices such as green purchasing, cooperation with customers, eco-design, and investment recovery will lead to improved environmental and economic performance.

2.2 Institutional Theory and GSCM Implementation

2.2.1 Institutional Theory

Institutional theory (Hirsch 1975; Parsons 1956; Thompson 1967; Weber 1930) investigates how external forces influence an organization. DiMaggio and Powell (1983) argues that managerial decisions are strongly influenced by three institutional mechanisms—coercive, mimetic, and normative isomorphism—that create and diffuse a common set of values, norms, and rules to produce similar practices and structures across organizations that share a common organizational field. Jennings and Zandbergen (1995) were the first to apply institutional theory to explain firms' adoption of environmental management practices. Green institutional theory offers that institutional forces—coercive, mimetic, and normative—impact how firms address green issues (Jennings and Zandbergen 1995). Recently, critics of existing institutional research (Hirsch 1975; Hirsch and Lounsbury 1997; Hoffman 2001) have argued that institutional literature (Scott 2014) places too much emphasis on the homogeneity of organizational populations and not the processes that may or may not create this outcome.

An organizational field is defined as 'those organizational that constitute a recognized area of institutional life: key suppliers, resource and product consumers, regulatory agencies and other organizations that produce similar services or products (DiMaggio and Powell 1983, p. 148). Therefore, we propose that all institutional pressures (i.e., coercive, mimetic, and normative) have the capacity to influence an organization's responsiveness to GSCM.

2.2.2 Institutional Pressures on GSCM Implementation

Vachon and Klassen (2006) mentioned the GSCM practices and defined that as management practices that in essence organizations can choose to directly devote or invest their own resources in order to improve environment of SC members. Rao and Holt (2005) classified the GSCM practice as internal logistics, internal SC, external logistics, and reverse logistics. Zhu and Sarkis (2007) note that the GSCM practices include five parts which are internal environmental management, green procurement, collaboration with customers, investments in waste recycling, and eco-design.

In the issues of institutional pressures and environments, Jennings and Zandbergen (1995) firstly cite the institutional theory to interpret enterprises' adoption of environmental management practices and that green institutional theory contains coercive, mimetic, and normative institutional power and will influence enterprises to resolve green issues. Zhu and Sarkis (2007) explore that Chinese manufacturers are driven to implement GSCM because of institutional pressures, while Hoffman (2001) believes that governments, regulators, competitors and consumers, community and environmental interest groups, and industry associations can directly affect plant's environmental practices. Delmas and Toffel (2004) propose the interested parties will take coercive power and norms to put pressure on companies, and make them adopt environmental management practices. Clemens and Douglas' (2006) study shows that institutional pressures will affect the green voluntary environmental action. Consumer pressure spurs organizations into the implementation and response to environmental protection to enhance performance (Kagan et al. 2003). Ye et al. (2013) indicate that coercive pressure, normative pressure, and mimetic pressure are three essential pressures, which are the key environmental factors that reverse logistics practices. In conclusion, this study suggests that the institutional pressure which enterprises are facing will affect the GSCM implementation. Consequently, the following hypothesis is proposed:

H_1: The institutional pressures have direct positive influence on the enterprises' GSCM implementation.

2.3 *Top Managers' Posture*

2.3.1 Stewardship Theory

Stewardship theory is used as the theoretical driver of this research. Stewardship theory is a complementary theory to agency theory, as managers find that they identify with their organization and do not instinctively act in an opportunistic way (Davis et al. 1997). Along with psychological factors centered on the individual, stewardship theory defines situational aspects of management philosophy and organizational culture. In the realm, managers are more concerned about the well-being of the organization and their identity is tied to that of organization (Craig and Dibrell 2006).

After weighing the individual needs and organizational objectives, the executives believe that to meet organizational goals through hard work for the organization will also maximize personal effectiveness. Stewardship behaviors are collectivism and the pursuit of organizational interests. Hernandez (2012) defines stewardship behaviors of top managers as on the basis that personal interests are closely related to interests of the organization, key management personnel naturally do everything they can to protect the interests of the organization and to maximize the benefits of the organization.

In this research, we are concerned with using the stewardship theoretical framework to understand top managers' posture that potentially is influenced institutional pressures and affects GSCM implementation.

2.3.2 Institutional Pressures on Top Managers' Posture

From the point of view of institutional theory, the organization will be definitely affected by external pressures. The organization must try to respond to these pressures, otherwise they will endanger the survival of the organization (Selznick 1949). These pressures exerted on the organization will also affect the organizational practices and structures (Scott 1995). Oliver (1997) also considers the organization in the face of the institutional pressure will produce the following actions: Organizational choices will be limited by outside institutional pressures; organizational survival depends on whether the organization can respond and meet the requirements of the outside world and seeks legitimacy of the organization itself; and the organization will be driven by outside interests.

On the relationship between institutional pressures and stewardship behaviors, Sharma (2000) states that top executives' cognition on the external environment will affect the actions adopted and the environmental strategy selected by the organization. Zhu and Cote (2004) advocate that the driving forces for the aspects of green trade barriers on exports, investment in overseas downstream manufacturers, consumer pressure, environmental care, etc., also make top executives aware of the importance of environmental issues, and start thinking about the GSCM practices. Seuring (2004) considers that GSCM is a continuing need for cost effectiveness and long-term sustainable development in the growth process of a company, and when companies are forced to be competitive in the image and reputation of environmental protection, the top executives' concept of environmental consciousness will be the key issue in the growth process. Government regulatory power and other interested parties make the executive cognizes that whether the company's products are in compliance with environmental laws and regulations, and this will affect the implementation of the company's environmental protection policies (Ye et al. 2013). Zhu and Sarkis (2006) state that governmental regulations and corporate environmental goals are two major business pressures, and companies cannot ignore these issues and must be integrated them into business execution planning. To sum up, this study suggests that the institutional pressure which enterprises are facing will affect the Stewardship behaviors of top managers, and further proposes the hypothesis:

H2: The institutional pressures have direct positive influence on top managers' posture toward GSCM.

2.3.3 Top Managers' Posture on GSCM Implementation

Some research works have emphasized that top management plays a critical role in establishing an organization's norms and expectations (Chatterjee et al. 2002). For instance, top managers' posture has a positive implementation of product recovery (Ye et al. 2013). On the relationship between stewardship behaviors of top managers and GSCM, Banerjee et al. (2003) mention that the key management personnel's commitment is the strong internal force to promote corporate environmental theory of top executives. Sharma (2000) also states that top executives' cognition on the external environment will affect the actions adopted and the environmental strategy selected by the organization. Steger (1996) suggests that the GSC needs the value chain which takes the environmental protection as the basis as well as creates and learns organizational culture in the company's management system.

In addition, Handfield et al. (2005) believe that the company's top executives should realize the importance of environmental issues in the SC. Trowbridge (2001) also points out that the strong support of top executives on GSCM practices and risk management will be an internal driver that can enhance cooperation mechanism among the various units. From the foregoing, this study considers that the cognition and commitment of stewardship behaviors of top managers on environmental management will affect GSCM implementation, and the hypothesis is proposed:

H3: The top managers' postures toward GSCM have direct positive influence on GSCM implementation.

2.4 Top Managers' Postures Mediating Effect

On the relationship among institutional pressures, top managers' posture towards GSCM, and GSCM implementation, when the top executives respond to the institutional environment's concerns or pressures for the purpose of environmental protection, the top executives need to understand this issue and implications of the issue on organizations and in what ways they should respond to this issue (Jennings and Zanbergen 1995). When companies face the issues of environmental protection, top executives would respond to the market at the time in order to remain competitive, especially frontline staff within the enterprise often takes the lead in grasping customer needs and market information (Ye et al. 2013). Many top executives of Taiwan's manufacturing industry believe that the enterprise can amplify its market share when taking green products as production orientation, which is also in line with international environmental regulations. To explore markets overseas for the products, this will be of great benefit (Huang and Wu 2010).

Today's environmental regulations have become increasingly stringent, especially in the face of the EU environmental directives, the electronic products of Taiwan's electrical and electronic industry exported to the EU are forbidden from containing toxic substances, the directives are of high standards of environmental protection, such external regulatory pressures and the driving pressure from relevant interested parties also make top executives aware of the importance of environmental issues, and start thinking about the implementation of GSCM (Zhu and Cote 2004). The enterprises are driven to implement GSCM when executives perceive external environmental pressures, and it makes products comply with environmental laws and regulations and also can enhance the company's image and a competitive advantage (Zhu et al. 2007). In summary, this study suggests that the institutional pressure which enterprises are facing will affect the enterprises' GSCM implementation through top managers' posture. Consequently, further puts forward the hypotheses:

H4: Top managers' postures toward GSCM have indirect influence on the relationship between institutional pressures and GSCM implementation.

3 Research Methods

3.1 Operational Definition

3.1.1 Institutional Pressures

The current study defines institutional pressure as the capacity to influence an organization's responsiveness to GSCM, including coercive, mimetic, and normative pressures (DiMaggio and Powell 1983). This study adopts what Delmas and

Toffel (2004), Khanna and Speir (2007) propose that the interested parties use normative and coercive power to bring pressure on the enterprise, including pressures from the government, consumers, interest group pressures, competitors, and industry associations.

3.1.2 Top Managers' Posture

Referring to Hernandez's (2012) definition, this study defines stewardship behaviors of top managers as on account of personal cognition and affection for the organization, top managers naturally do everything they can to protect the long-term interests of the organization, and to maximize the benefits of the organization. According to this view, this study suggests that corporate managers perceive the threat of external pressure and will conduct the GSCM implementation on account of the manager's commitment to the organization and the environment. Following the recommendation suggested by Sharma (2000), Banerjee et al. (2003), we divide top managers' posture into two constructs as cognitive mechanism and emotional mechanism.

3.1.3 GSCM Implementation

This study adopts Srivastava's (2007) definition of GSCM implementation which is adding environmental concepts into the supply chain management including product design, material purchase, manufacturing processes, ultimate consumer, and product recycling. Following Zhu and Sarkis (2007) research, the current study divides GSCM practices into five constructs as internal environmental management, green procurement, collaboration with customers, investment in waste recycling, and ecological design.

3.2 Sampling and Data Collection

This study selected the members of 2011–2012 of Taiwan Electrical and Electronic Manufacturers' Association as the population totaling 3604 members, and then, randomly selected 1000 member companies out of the companies with a capital of more than NT $10 million taken as a sampling standard. Questionnaires were sent out by postal mail.

Among 180 valid questionnaires, currently the first generation businesses occupy the majority of the respondents, accounting for 78.3 %, followed by the second generation businesses, accounting for 14.4 %; enterprises with 200–1000 employees make up the majority of all, accounting for 52.2 %; followed by the ones with less than 200 employees, accounting for 37.8 %.

4 Analysis and Results

4.1 Reliability and Discriminant Validity

SPSS 19.0 and AMOS 18.0 statistical software packages were undertaken to assess the relations of constructs on the questionnaires. Statistical methods are used primarily as follows: reliability and validity analysis, confirmatory factor analysis, and structural equation.

This study assessed construct reliability by calculating Cronbach's alpha coefficients. Based on Cuieford (1965), Cronbach's α value higher than 0.7 indicates high reliability, the value between 0.7 and 0.35 represents acceptable reliability, and that below 0.35 means unacceptable reliability. The results of our study demonstrate that all variables' Cronbach's α values exceed 0.7 which represents measuring scales have high reliability and internal consistency.

In addition, according to the standard of convergent validity assessed by Fornell and Larcker (1981), all standardized factor loadings should be greater than 0.5 and should be above 0.6 in terms of composite reliability, as shown in Table 1 that all constructs' indicator-factor loadings are above 0.5, all composite reliabilities (CR) are above 0.7, average variance extracted (AVE) is above 0.5. Measurement scales see Appendix A.

Discriminant validity was assessed by comparing (a) the correlations between a given construct and all other constructs with (b) the AVE for the focal construct (Hair et al. 2006). Table 2 shows the correlation matrix for the constructs; the diagonal elements have been replaced by the square root of the AVE construct. The constructs demonstrate adequate discriminant validity because these diagonal elements were greater than the off-diagonal elements in the corresponding rows and columns.

4.2 Linear Structural Equations

Linear structural model analysis of this study uses AMOS18.0 as an analytical tool and refers to Bagozzi and Yi's (1988) recommended fit indices and judgment criteria, and results of the initial model analysis are shown in Table 3. In the current study, only GFI and AGFI are close to recommended value, but Hair et al. (2006) suggest that as for GFI, AGFI, and NFI value, the closer to one, the better it is; however, there is no standard absolute value to determine the fitness between observational data and models, while Baumgartner and Homburg's (1996) study between 1977 and 1994 analyzes 184 literatures with SEM and indicates GFI; AGFI of which the literature rates below the recommended values (greater than 0.9), respectively, is 24 and 48 % that are still within the acceptable range. Moreover, according to Table 4, estimating significant degree of parameters, reliability of all indices, and potential variables within the measurement mode, *t*-value

Table 1 Measurement model and confirmatory factor analysis

Variable	Construct	Factor loading	CR	AVE	Cronbach's α
GSCM implementation	Internal environmental management (IEM)	IEM1 0.576	0.877	0.622	0.871
		IEM2 0.823			
		IEM3 0.899			
		IEM4 0.838			
		IEM5 0.640			
	Green procurement (GP)	GP1 0.765	0.937	0.849	0.932
		GP2 0.837			
		GP3 0.853			
		GP4 0.948			
		GP5 0.914			
	Collaboration with customers (CC)	CC1 0.937	0.945	0.815	0.941
		CC2 0.980			
		CC3 0.957			
		CC4 0.710			
	Investment in waste recycling (IWR)	IWR1 0.670	0.712	0.556	0.692
		IWR2 0.754			
		IWR3 0.653			
	Ecological design (ED)	ED1 0.874	0.956	0.878	0.953
		ED2 0.957			
		ED3 0.653			
	Governmental pressure (GMP)	GMP1 0.969	0.853	0.612	0.852
		GMP2 0.984			
		GMP3 0.519			
		GMP4 0.521			
	Customers pressure (CMP)	CMP1 0.952	0.840	0.643	0.821
		CMP2 0.618			
		CMP3 0.802			

(continued)

Table 1 (continued)

Variable	Construct	Factor loading	CR	AVE	Cronbach's α
Institutional pressures	Interest group pressures (IGP)	IGP1 0.852	0.867	0.621	0.829
		IGP2 0.706			
		IGP3 0.802			
		IGP4 0.788			
	Competitors pressures (CTP)	CTP1 0.884	0.879	0.652	0.875
		CTP2 0.895			
		CTP3 0.799			
		CTP4 0.613			
	Industrial pressures (IP)	IP1 0.600	0.868	0.705	0.814
		IP2 0.991			
		IP3 0.863			
Top managers' posture	Cognitive mechanism (CM)	CM1 0.821	0.762	0.520	0.953
		CM2 0.717			
		CM3 0.611			
	Emotional mechanism (EM)	EM1 0.828	0.721	0.512	0.961
		EM2 0.679			
		EM3 0.539			

is higher than the absolute value 1.645. Overall, the models and observations represent an acceptable fitness.

4.3 Analysis of Mediation Effects

According to Baron and Kenny's (1986) analysis method of the medication variables, model (1), model (2), and model (3) are significant; In model (4), the effect of institutional pressure on GSCM implementation is significant (the path coefficients is 0.771 and *t*-value is 13.286); the effect of stewardship behaviors of top managers on GSCM implementation is significant (the path coefficient is 0.509 and *t*-value is 9.869). According to the criterion, stewardship behaviors of top managers have partial mediation effects on institutional pressure and GSCM implementation.

According to the constructs' analysis results of the linear structural relations model, hypotheses of institutional pressure, top managers' posture, and GSCM implementation were supported. Figure 1 and Table 4 shows the results.

Table 2 Descriptive statistics and correlations

	(1)	(2)	(3)	(4)	(5)	(6)	(7)	(8)	(9)	(10)	(11)	(12)
(1) IEM	0.80											
(2) GP	0.70**	0.92										
(3) CC	0.71**	0.70**	0.90									
(4) IWR	0.44**	0.48**	0.53**	0.74								
(5) ED	0.61**	0.71**	0.75**	0.49**	0.94							
(6) GMP	0.20*	0.25**	0.16*	0.14*	0.25**	0.78						
(7) CMP	0.30**	0.35**	0.37**	0.16*	0.31**	0.03	0.80					
(8) IGP	0.59**	0.59**	0.61**	0.45**	0.60**	0.08	0.23**	0.78				
(9) CTP	0.61**	0.55**	0.74**	0.42**	0.73**	0.23**	0.32**	0.63**	0.76			
(10) IP	0.39**	0.37**	0.38**	0.26**	0.42**	0.19*	0.11*	0.38**	0.45**	0.84		
(11) CM	0.38**	0.44**	0.46**	0.36**	0.50**	0.19*	0.20**	0.41**	0.46**	0.30**	0.72	
(12) EM	0.48**	0.62**	0.61**	0.41*	0.67**	0.26**	0.24**	0.50**	0.47**	0.33*	0.62**	0.71

Note $n = 180$, $^{*}p < 0.05$, $^{**}p < 0.01$, values on the diagonal are the average variance extracted

IEM Internal environmental management; *GP* Green procurement; *CC* Collaboration with customers; *IWR* Investment in waste recycling; *ED* Ecological design; *GMP* Governmental pressure; *CMP* Customers pressure; *IGP* Interest group pressures; *CTP* Competitors pressures; *IP* Industrial pressures; *CM* Cognitive mechanism; *EM* Emotional mechanism

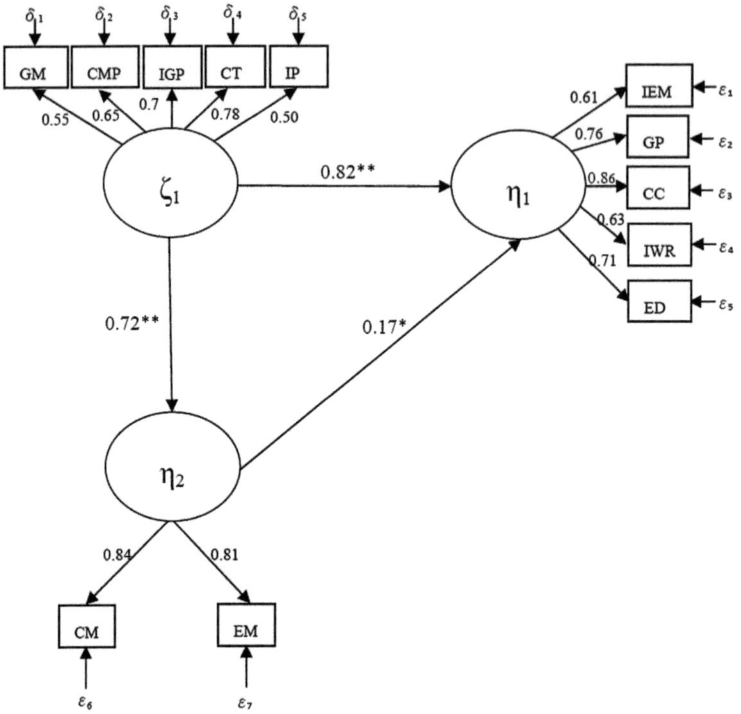

Fig. 1 Overall fit of the proposed model and path analysis of the latent constructs. *Note* $n = 180$ observations, $*p < 0.05$, $**p < 0.01$. ζ_1 Institutional pressure; *GMP* governmental pressure; *CMP* customers pressure; *IGP* inter-est group pressures; *CTP* competitors pressures; *IP* industrial pressures. η_1 GSCM practices; *IEM* internal environmental management; *GP* green procurement; *CC* collaboration with customers; *IWR* investment in waste recycling; *ED* ecological design. η_2 Top managers' posture; *CM* cognitive mechanism; *EM* emotional mechanism

Table 3 The fit of measurement model

Fit indices	Judgment criteria	The results of this research	Fit or not
Chi-squre		208	
DF		86	
Chi-squre/DF	< 3	2.41	Yes
GFI	> 0.9	0.86	Closed
AGFI	> 0.9	0.81	Closed
NFI	> 0.9	0.90	Yes
CFI	> 0.9	0.93	Yes
RMR	< 0.05	0.01	Yes
RMSEA	< 0.1	0.08	Yes

Table 4 The results of path coefficient and hypotheses

Path	Result	*t*-value	Support
Institutional pressure → GSCM implementation	0.82	2.759**	Yes
Institutional pressure → Top managers' posture	0.72	2.802**	Yes
Top managers' posture → GSCM implementation	0.17	1.970*	Yes
Institutional pressures → Top managers' posture → GSCM implementation	0.52	9.869**	Yes

Note $n = 180$, $*p < 0.05$, $**p < 0.01$

5 Conclusion and Suggestions

5.1 Conclusions

The results show that the higher the degree of institutional pressure that companies face governmental pressure, customer pressure, the pressure from interest groups, pressure from competitors, and industrial pressure, the more it helps companies promote GSCM implementation on internal environmental management, green procurement, collaboration with customers, investment in waste recycling, and eco-design. This result is in line with Clemens and Douglas (2006), and Zhu and Sarkis (2007), which propose that owing to the impact from the institutional pressure on organizations, companies will go beyond the statute's limitations, and further perform the environmental management measures. Institutional pressure is also the key factor driving the middle managers to support GSCM implementation. Therefore, this study suggests that the institutional pressure has a strong influence on the GSCM implementation.

The finding suggests that that the higher the degree of institutional pressure that companies face governmental pressure, customer pressure, the pressure from interest groups, pressure from competitors, and industrial pressure is, the more the top managers' awareness of the institutional pressure will develop with positive point of view. This is consistent with Banerjee et al. (2003), Sharma (2000), and Zhu and Sarkis (2006), which indicates that facing the pressure exerted by the institutional environment will affect the enterprises executives' cognition and emotion on the institutional environment, thus leading to an environmental commitment.

The results show that executives' positive cognition and emotion on environmental issues will help companies promote GSCM implementation in aspects of internal environmental management, green procurement, collaboration with customers, investments in waste recycling, and eco-design. This means that executives are an important key role in the implementation of GSCM implementation. This is consistent with the findings of previous studies (Banerjee et al. 2003; Handfield et al. 2005; and Trowbridge 2001), which suggest that when the decision-makers pay attention to environmental protection, their cognition and emotion on environment would enable enterprises to implement GSCM implementation.

It can be seen from the results that the higher the degree that companies face institutional pressures, the executives' cognition and emotion on environment will be more positive, and they will more actively enable enterprises to implement GSCM implementation, which means that executives' behaviors under institutional pressure will affect the corporate different options on environmental strategies. Therefore, this study supports Sharma's (2000) findings that executives' cognition on external environment will affect the actions adopted and environmental strategy selected by the organizations, which explains why organizations adopt different environmental management implementation.

5.2 Management Implications

How business managers effectively improve the GSCM implementation is an important issue. According to the empirical results, we provide with several suggestions as follows:

The rise of global environmental awareness coupled with various regulatory pressures from WEEE, RoHS, and EuP, etc., enterprises should actively promote GSCM implementation, because when companies are internally working on GSCM implementation, it improves the enterprises' environmental performance, economic performance, and operating performance, ultimately leading to sustainable development. The institutional pressure is an important factor in promoting GSCM implementation for enterprises. Companies must understand that the pressures exerted by various demands including governmental pressure, customer pressure and pressure from interest groups, pressure from competitors, and industrial pressure and enterprises will not enhance the competitive advantage in the trend of green issues unless they carry on communication and concern.

Top managers' posture is referring to the higher the degree that executives' cognition and emotion on environmental issues is more positive, the more it helps companies promote GSCM implementation, such as internal environmental management, green procurement, collaboration with customers, investment in waste recycling, and eco-design. This illustrates the importance of implementation of GSCM implementation conducted by top managers. Institutional pressure will implement GSCM implementation through executives' cognition and emotion on environment. Executives should focus on external environmental information with an optimistic attitude. When an enterprise promotes GSCM, executives must recognize the effect of GSCM on the enterprise and commit to implementing GSCM.

5.3 Research Contributions

5.3.1 In Academic Fields

Previous institutional theory describes the institutional pressures faced by companies will lead to inter-enterprise performance of similar organizational implementation. However, recent studies have shown that companies will adopt different management practices when they face the same institutional pressure. This study cites a green institutional theory to explore the impact of institutional pressure on GSCM implementation. In the GSCM implementation issue, the researches on how enterprises transform the impact on performance of GSCM implementation when they face institutional pressures are rare, and this study thus combines institutional theory and stewardship theory to explore the impact of these two theories on GSCM implementation; secondly, steward behaviors were assessed a mediation variables to explore how the institutional pressure indirectly affect the GSCM implementation by stewardship behaviors of top managers. In the previous studies, stewardship theory mostly applied in the relationship between the board and the shareholders, and explored executives' psychological level. This study initially attempted to apply the stewardship theory to environmental issues and found that top managers' behaviors would affect GSCM implementation.

5.3.2 In Practices Field

Taiwan is an export-oriented country in trade, while the EU's environmental regulations formed trade barriers. The current study suggests that companies should be aware of the importance of environmental management, especially of RoHS and WEEE environmental regulations, and the results of this study confirm that the institutional pressure is an important factor affecting companies' promotion of GSC. Taiwan's electrical and electronic industries are affected by the international regulations; corporate executives must make a response; the empirical results of this study demonstrate that when external pressure is exerted on the enterprise, it affects the executives' environmental cognition and emotional commitment, and make environmental action in line with regulations and interested parties' expectation. The results of this study confirm that enterprises' promotion of GSCM implementation is not only affected by the institutional pressure, but also must cooperate with corporate executives' support and top managers' positive thinking, and this will affect the corporate to more actively promote GSCM. The study findings can provide to the top managers of Taiwan electrical and electronic companies for reference when they implement GSCM implementation.

5.4 Limitations and Future Directions

A limitation of this research is its restricted generalizability. The results of this study apply to the electrical and electronic industries in Taiwan, may not apply to other industries. This study is aimed at Taiwan's electrical and electronic industries, and targeting companies with capital amounted to 10 million or more, the results of this study are only suitable for partial objects. This study is only aimed at Taiwan's electrical and electronic industries, future research could explore other industries about the relevance of institutional pressure, stewardship behaviors, and performance. Taiwan manufacturers currently set up factories in China, and it is possible to study the performance of GSC implementation of Taiwanese manufacturers in China and makes a comparison between the two sides in the future. The companies of electrical and electronic industry with capital of 10 million or more are targeted, and it may focus on the companies with ISO14001 certification as survey objects in the follow-up research.

Appendix A: Measurement scales

Green supply chain management
Internal environmental management (IEM) (Zhu et al. 2007)
Please indicate the extent to which you perceive that your company is implementing each of the following. (1 = not considering it; 7 = very successfully)

- Support for GSCM from mid-level managers (IEM 1)
- Cross-functional cooperation for environmental improvements (IEM 2)
- Total quality environmental management (IEM 3)
- Environmental compliance and auditing programs (IEM 4)
- Environmental Management Systems (IEM 5)

Green purchasing (GP) (Zhu et al. 2007)
Please indicate the extent to which you perceive that your company is implementing each of the following. (1 = not considering it; 7 = very successfully).

- Providing design specification to suppliers that include environmental requirements for purchased item (GP1)
- Cooperation with suppliers for environmental objectives (GP2)
- Environmental audit for suppliers' internal management (GP3)
- Suppliers' ISO14000 certification (GP4)
- Second-tier supplier environmentally friendly practice evaluation (GP5)

Cooperation with customers (CC)(Zhu et al. 2007)
Please indicate the extent to which you perceive that your company is implementing each of the following. (1= not considering it; 7 = very successfully).

- Cooperation with customers for eco-design (CC1)
- Cooperation with customers for cleaner production (CC2)
- Cooperation with customers for green packaging (CC3)
- Cooperation with customers for using less energy during product transportation (CC4)

Investment in waste recycling (IWR) (Zhu et al. 2007)
Please indicate the extent to which you perceive that your company is implementing each of the following during the past year. (1 = not considering it; 7 = very successfully).

- Investment recovery (sale) of excess inventories/materials (IWR1)
- Sale of scrap and used materials (IWR2)
- Sale of excess capital equipment (IWR3)

Eco-design (ED) (Zhu et al. 2007)
Please indicate the extent to which you perceive that your company is implementing each of the following. (1 = not considering it; 7 = very successfully).

- Design of products for reduced consumption of materials/energy (ED1)
- Design of products for reuse, recycle, recovery of materials, component parts (ED2)
- Design of products to avoid or reduce use of hazardous products and/or their manufacturing process (ED3)

Institutional pressures
Governmental Pressure (GMP) (Delmas and Toffel 2004)
Please evaluate the degree of importance for each pressure on your company implementing green supply chain management (1 = very unimportant; 7 = very important).

- Complying with current government environmental regulations (GMP1)
- Taking environmentally friendly actions to reduce regulatory inspections and make it easier to get environmental permits (GMP2)
- Being better prepared for meeting anticipated environmental regulations (GMP3)
- Preempting future environmental regulations by voluntarily reducing regulated pollution beyond compliance levels (GMP4)

Customers Pressure (CMP) (Delmas and Toffel 2004)
Please evaluate the degree of importance for each pressure on your company implementing green supply chain management (1 = very unimportant; 7 = very important).

- Customer desire for environmentally friendly products and services (CMP1)
- Customer willingness to pay higher prices for environmentally friendly products/services (CMP2)

- Ability to earn public recognition and customer goodwill with environmentally friendly actions (CMP3)

Interest Group Pressures (IGP) (Delmas and Toffel 2004)
Please evaluate the degree of importance for each pressure on your company implementing green supply chain management (1 = very unimportant; 7 = very important).

- Preventing boycotts or other adverse actions by environmental interest groups (IGP1)
- Promoting an environmentally friendly image to environmental interest groups (IGP2)
- Being environmentally responsible attracts quality employees and reduces employee turnover (IGP3)
- Being environmentally responsible improves employee morale, motivation and productivity (IGP4)

Competitors Pressures (CTP) (Delmas and Toffel 2004)
Please evaluate the degree of importance for each pressure on your company implementing green supply chain management (1 = very unimportant; 7 = very important).

- Investing in cleaner products and services differentiates our products or our facility with competitors (CTP1)
- Improving environmental performance to help us differentiate with competitors (CPT2)
- Environmentally friendly actions result in product or process innovations to help us differentiate with competitors (CPT3)
- Environmentally friendly actions can reduce costs to help us differentiate with competitors (CPT4)

Stewardship Behaviors
Cognitive mechanism (CM)
Please indicate the degree to which you agree with the following statements regarding top manager's cognitive posture in your company implementing green supply chain management (1 = very strongly disagree; 7 = very strongly agree).

- Top managers are likely to gain rather than lose by actions to green supply chain management (CM1)
- Any actions that I may take for green supply chain management are encouraged by top manager (CM2)
- Top managers have the technical knowledge to reduce the environmental impact of company operations (CM3)

Emotional mechanism (EM)
Please indicate the degree to which you agree with the following statements regarding top manager's emotional posture in your company implementing green supply chain management (1 = very strongly disagree; 7 = very strongly agree).

- The top managers in our firm commit to implement green supply chain management (EM1)
- Our firm's green supply chain management receive full support from our top managers (EM2)
- Our firm's green supply chain management is driven by the top managers (EM3)

References

Ambec S, Lanoie P (2008) Does it pay to be green? A systematic overview. Acad Manag Perspect 22(4):5–62

Azevedo SG, Carvalho H, Machado VC (2011) The influence of green practices on supply chain performance: a case study approach. Transp Res Part E 47:850–871

Bagozzi R, Yi Y (1988) On the evaluation of structural equation models. J Acad Mark Sci 16 (1):74–94

Banerjee SB, Iyer ES, Kashyap RK (2003) Corporate environment antecedents and influence of industry type. J Mark 67(2):106–122

Baron RM, Kenny DA (1986) The moderator-mediator variable distinction in social psychological research. J Pers Soc Psychol 51(6):1173–1182

Baumgartner H, Homburg C (1996) Applications of structural equation modeling in marketing and consumer research: a review. Int J Res Mark 13(2):139–161

Bowen FE, Cousine PD, Lamming RC, Faruk AC (2001a) Explaining the gap between the theory and practice of green supply. Greener Manag Int 35:41–59

Bowen FE, Cousins PD, Lamming RC, Faruk AC (2001b) The role of supply management capabilities in green supply. Prod Oper Manag 10(2):174–189

Chatterjee D, Grewal R, Sambamurthy V (2002) Shaping up for e-commerce: institutional enablers of the organizational assimilation of web technologies. MIS Q 26(2):65–89

Christmann P (2000) Effects of 'best practices' of environmental management on cost advantage: the role of complementary assets. Acad Manag J 43:663–680

Clemens B, Douglas T (2006) Dose coercion drive firm to adopt voluntary green initiatives? J Bus Res 59:483–491

Craig J, Dibrell C (2006) The natural environment, innovation, and firm performance: a comparative study. Fam Bus Rev 19(4):275–288

Cuieford JP (1965) Fundamental statistics in psychology and education. McGraw-Hill, New York

Davis JH, Schoorman FD, Donaldson L (1997) Toward a stewardship theory of management. Acad Manag Rev 22:20–47

Delmas M, Toffel MW (2004) Stakeholders and environment management practices: an institutional framework. Bus Strategy Environ 13:209–222

Diabat A, Govindan K (2011) An analysis of the drivers affecting the implementation of green supply chain management. Resour Conserv Recycl 55(6):659–667

DiMaggio PJ, Powell WW (1983) The iron cage revisited: institutional isomorphism and collective rationality in organizational fields. Am Sociol Rev 48(2):147–160

Fornell C, Larcker D (1981) Evaluating structural equation models with unobservable variables and measurement error. J Mark Res 18:39–50

Green KW Jr, Zelbst PJ, Meacham J, Bhadauria VS (2012) Green supply chain management practices: impact on performance. Supply Chain Manag Int J 17(3):290–305

Hair JF, Black WC, Babin BJ, Anderson RE, Tatham RL (2006) Multivariate data analysis, 6th edn. Prentice-Hall, New Jersey

Handfield R, Sroufe R, Walton S (2005) Integrating environmental management and supply chain strategies. Bus Strategy Environ 14:1–19

Hart SL (1995) A natural-resource-based view of the firm. Acad Manag Rev 20(4):986–1014
Hernandez M (2012) Toward an understanding of the psychology of stewardship. Acad Manag Rev 37:172–193
Hirsch PM (1975) Organizational effectiveness and the institutional environment. Adm Sci Q 20:327–344
Hirsch PM, Lounsbury M (1997) Putting the organization back into organization theory". J Manag Inq 6(1):79–88
Hoffman AJ (2001) Linking organizational and field-level analyses. Organ Environ 14:133–156
Holt D, Ghobadian A (2009) An empirical study of green supply chain management practices amongst UK manufacturers. J Manuf Technol Manag 20(7):933–956
Hu HA, Hsu CW (2010) Critical factors for implementing green supply chain management practice: an empirical study of electrical and electronics industries in Taiwan. Manag Res Rev 33:586–608
Huang YC, Wu YCJ (2010) The effects of organizational factors on green new product success. Manag Decis 48:1539–1567
Huang YC, Ding HB, Kao MR (2009) Salient stakeholder voices: family business and green innovation adoption. J Manag Organ 5(3):309–326
Huang YC, Wu YC, Shams YCR (2012) Reverse logistics activities, the task environment and performance. Prod Plann Control 23(10–11):851–863
Jennings JA, Zandbergen PA (1995) Ecologically sustainable organization: an institional approach. Acad Manag J 20(4):1015–1053
Kagan RA, Gunningham N, Thornton D (2003) Explaining corporate environment performance. Law Soc Rev 37:51–90
Khanna M, Speir C (2007) Motivations for proactive environmental and innovation pollution control. Am Agric Econ Assoc Annu Meet 1–35
Kumar S, Teichman S, Timpernagel T (2012) A green supply chain is a requirement for profitability. Int J Prod Res 50(5):1278–1296
Lee SY (2008) Drivers for the participation of small and medium-sized suppliers in green supply chain initiatives. Supply Chain Manag Int J 13(3):185–198
Oliver C (1997) Sustainable competitive advantage: combining institutional and resource-based views. Strategy Manag J 18(9):697–713
Parsons T (1956) Suggestions for a sociological approach to the theory of organization. Adm Sci Q 1:225–239
Rao P, Holt D (2005) Do green supply chains lead to competitiveness and economic performance? Int J Oper Prod Manag 25:898–916
Sarkis J (2003) A strategic decision framework for green supply chain management. J Clean Prod 11:397–409
Scott WR (1995) Institutions and organizations. Sage Publications Inc, Thousand Oaks
Scott WR (2014) Institutions and organizations: ideas, interests, and identities, 4th edn. SAGE Publications Inc, Thousand Oaks
Selznick P (1949) Leadership in administration: a study in the sociological interpretation. Hamper and Row, New York
Seuring S (2004) Industrial ecology, life cycles, supply chains: differences and interrelations. Bus Strategy Environ 13:306–319
Sharma S (2000) Managerial interpretations and organization context as predictors of corporate choice of environment strategy. Acad Manag J 26:159–180
Srivastava SK (2007) Green supply chain management: a state-of-the-art literature review. Int J Manag Rev 9(1):53–80
Steger U (1996) Management issues in closing the loop. Bus Strategy Environ 5:252–268
Thompson JD (1967) Organizations in action. McGraw Hill, New York, p 7
Trowbridge P (2001) A case study of green supply chain management at advanced micro devices. Greener Manag Int 35:121–135
Vachon S, Klassen RD (2006) Extending green practices across the supply chain: the impact of upstream and downstream integration. Int J Oper Prod Manag 26:795–821

Walker H, Di Sisto L, McBain D (2008) Drivers and barriers to environmental supply chain management practices: lessons from the public and private sectors. J Purch Supply Manag 14:69–85
Weber M (1930) The protestant work ethic and the spirit of capitalism: an outline of interpretive sociology. Allen and Erwin, New York
Ye F, Zhao X, Prahinski C, Li Y (2013) The impact of institutional pressures, top managers' posture and reverse logistics on performance—evidence from China. Int J Prod Econ 143:132–143
Zhu QH, Cote R (2004) Integrating green supply chain management into an embryonic eco-industrial development: a case study of the Guitang group. J Clean Prod 12(8–10):1025–1035
Zhu Q, Sarkis J (2004) Relationship between operational practices and performance among early adopters of green supply chain management practices in Chinese manufacturing enterprises. J Oper Manag 22:265–289
Zhu Q, Sarkis J (2006) An intersectoral comparison of green supply chain management in China: drivers and practices. J Clean Prod 14:472–486
Zhu Q, Sarkis J (2007) The moderating effects of institutional pressures on emergent green supply chain practices and performance. Int J Prod Res 45:333–355
Zhu Q, Sarkis J, Geng Y (2005) Green supply chain management in China: pressures, practices and performance. Int J Oper Prod Manag 25:449–468
Zhu Q, Sarkis J, Lai KH (2007) Green supply chain management: pressures, practices and performance within the Chinese automobile industry. J Clean Prod 15:1041–1052
Zhu Q, Sarkis J, Cordeiro JJ, Lai KH (2008) Firm-level correlates of emergent green supply chain management practices in the Chinese context. Omega 36(4):577–591
Zhu Q, Geng Y, Sarkis J, Lai KH (2011) Evaluating green supply chain management among Chinese manufacturers from the ecological modernization perspective. Transp Res Part E 47:808–821

Deployment of Sustainable Logistics Optimization Incorporated with Modal Shift and Emission Trading on Carbon Dioxide

Yoshiaki Shimizu, Tatsuhiko Sakaguchi and Hiroki Shimada

Abstract This chapter discusses a green logistics optimization problem associated with production method of manufacturers and green attitude of consumers. A novel hierarchical method is developed for a three-echelon logistics network to optimize the production methods with different structures regarding cost and emission of carbon dioxide (CO_2) at production sites (PSs), the available collection center, the paths between members of the logistics network, and circular routes over consumers. First, the problem is aimed at minimizing either the total cost or CO_2 emission through controlling prone and aversion behaviors on sustainability of each member. Then, as a promising glue to integrate these individual problems and to evaluate them on the same basis, an economic mechanism known as the emission trading rate on CO_2 is introduced. Furthermore, to discuss the sustainability in a broader logistics system, the modal shift in transportation is noted. To show the significance of the proposed approach, a case study is provided to explore some prospects for green logistics in whole society based on the computational results.

Keywords Sustainability · Green logistics optimization · Hybrid meta-heuristic method · Emission trading · Modal shift

Y. Shimizu (✉) · T. Sakaguchi · H. Shimada
Department of Mechanical Engineering, Toyohashi University of Technology, Hibarigaoka 1-1,Tenpaku, Toyohashi 441-8580, Japan
e-mail: shimizu@me.tut.ac.jp

T. Sakaguchi
e-mail: sakaguchi@me.tut.ac.jp

H. Shimada
e-mail: shimada@ise.me.tut.ac.jp

V. Kachitvichyanukul et al. (eds.), *Toward Sustainable Operations of Supply Chain and Logistics Systems*, EcoProduction,
DOI 10.1007/978-3-319-19006-8_8

1 Introduction

Under a growing importance of environmental issues, many interests have been paid to the studies generally termed reverse, closed, green or sustainable logistics (Fiksel 2010; Nikolopoulou and Ierapetritou 2012; Srivastava 2007). In fact, we concerned with a reverse logistics optimization in a practical manner (Shimizu and Sakaguchi 2013). Recently, we have noticed the deployment of green logistics incorporated with co-existence/co-operation of manufacturing and consumption is a new paradigm realizing an infrastructure aligned with sustainable development. Then, we deployed a model for three-echelon logistics network by introducing an idea associated with production methods at manufacturers and consumption behaviors of consumers (Shimizu 2014).

This study tries to extend such model to discuss on the sustainability in a more manageable way. Actually, we apply a novel hierarchical method that enables us to optimize the production methods (CLs) with different structures regarding cost and CO_2 emission at production sites (PSs), the available collection centers (CCs), the paths from PSs to CCs, and the circular routes traveling client retailers (REs) from every CC. In this development, we particularly note that transportation cost and CO_2 emission depend not only on the distance traveled but also on the loading weight[1]. It is practical, therefore, to consider these two factors in parallel or to adopt a bilinear model of distance and load (Weber) or nonlinear one (generalized Weber). Moreover, it had better to consider the fixed operational cost of vehicle for practical accounting. To the best of our knowledge, however, these ideas have been never considered elsewhere associated with vehicle routing problem (VRP) except for our studies (Shimizu 2011a, b).

The original problem aimed at minimizing either total cost or CO_2 emission through controlling prone and aversion behaviors on sustainability of each logistics member. As a promising glue to integrate these problems solved individually and to evaluate the decision on the same basis, i.e., cost, we introduce an economic mechanism known as the emission trading for CO_2. Moreover, to discuss the sustainability from a broader concern, we note a transportation system known as the modal shift. Through similarly extending the solution procedure developed before, we show the significance of the proposed approach. For this purpose, we provide a case study with a few plausible scenarios and explore some prospects for green logistics in the global society based on the computational results.

The rest of the manuscript is organized as follows. Section 2 describes a mathematical formulation of the problem. In Sect. 3, we explain the proposed hybrid method to solve the problem practically. Discussing some key issues through a case study in Sect. 4, we will give a conclusion in Sect. 5.

[1]This value should be counted including unladen weight of vehicle.

2 Problem Formulation

Taking a global logistics network composed of production site (PS), collection center (CC), and customer or retailer (RE) as shown in Fig. 1, we try to decide the available PSs and their production methods (CLs), paths from PSs to CCs, and circular routes from every CC over its client REs. The goal is to provide a practical solution method that enables us to minimize either total cost or CO_2 emission. Then, it will be extended so that we can discuss on trade-off to realize co-operative and competitive green logistics. In this study, such problem is formulated as the following mixed-integer programming problem under mild assumptions, e.g., round-trip transportation between PS and CC, circular transportation over REs, certainly defined system parameters, etc.

Then, we define an objective function of the first problem $f_1(\boldsymbol{X})$ called pure cost problem as follows.

$$\begin{aligned} f_1(X) = &\sum_{i\in I}\sum_{j\in J} T_c L_{ij}(f2_{ij} + 2w_0 t_i) + \sum_{i\in I}\sum_{i'\in I'} \mathrm{Hp}_{ii'} f1_{ii'} + \sum_{i\in I}\mathrm{Sh1}_i\left(\sum_{j\in J} f2_{ij}\right) + \sum_{j\in J}\mathrm{Sh2}_j\left(\sum_{i\in I} f2_{ij}\right) \\ &+ \sum_{v\in V}\sum_{p\in P}\sum_{p'\in P} c_v d_{pp'}(g_{pp'v} + w_v) z_{pp'v} + \sum_{v\in V} F_v y_v + \sum_{i\in I}\sum_{i'\in I} \mathrm{Fc}_{ii'} x_{ii'} \end{aligned} \tag{1}$$

where $\boldsymbol{X}$ denotes a vector standing for all decision variables shown below. Each term in the right-hand side of equation describes the round-trip transportation cost between PS and CC, production cost at PS, shipping cost at PS and CC, circular transportation cost covering every RE, and fixed charge for production and vehicle operation, respectively. Just replacing the coefficients of the cost factors in $f_1(\boldsymbol{X})$ with those of CO_2 emission, we can give the second objective function $f_2(\boldsymbol{X})$ to evaluate emission of carbon dioxide. Moreover, introducing another

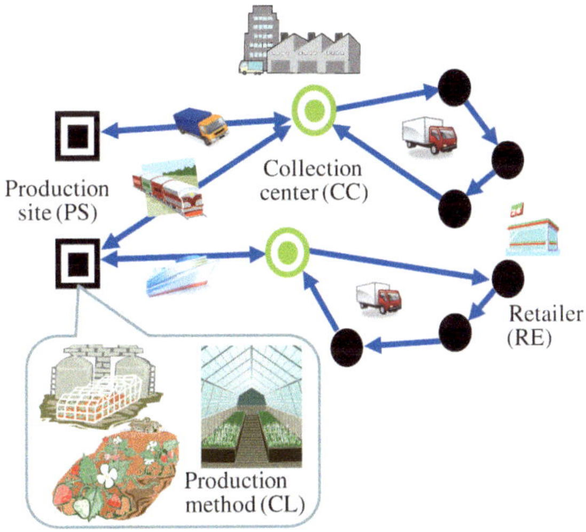

Fig. 1 Global logistics network under consideration

coefficients λ associated with the emission trading rate on CO_2, we can transform the amount of CO_2 emission into the cost (surrogate cost). Finally, we come up to the following mixed-integer programming problem under consideration.

(p.1) Minimize $f_1(\boldsymbol{X}) + \lambda f_2(\boldsymbol{X})$ subject to

$$\sum_{p \in P} z_{kpv} \leq 1, \quad \forall k \in K;\ \forall v \in V \tag{2}$$

$$\sum_{p' \in P} z_{pp'v} - \sum_{p' \in P} z_{p'pv} = 0, \quad \forall p \in P;\ \forall v \in V \tag{3}$$

$$\sum_{j' \in J} z_{jj'v} = 0, \quad \forall j \in J,\ \forall v \in V \tag{4}$$

$$g_{pp'v} \leq W_v z_{pp'v}, \quad \forall p \in P;\ \forall p' \in P;\ \forall v \in V \tag{5}$$

$$\sum_{p \in P} \sum_{p' \in P} z_{pp'v} \leq M y_v, \quad \forall v \in V \tag{6}$$

$$\sum_{j \in J} \sum_{k \in K} z_{jkv} = y_v, \quad \forall v \in V \tag{7}$$

$$\sum_{j \in J} \sum_{k \in K} z_{kjv} = y_v, \quad \forall v \in V \tag{8}$$

$$\sum_{k \in K} g_{kjv} = 0, \quad \forall j \in J;\ \forall v \in V \tag{9}$$

$$\sum_{v \in V} \sum_{p \in P} g_{pkv} - \sum_{v \in V} \sum_{p \in P} g_{kpv} = D_k, \quad \forall k \in K \tag{10}$$

$$\sum_{p \in P} \left(g_{pkv} - D_k z_{pkv} \right) = \sum_{p \in P} g_{kpv}, \quad \forall k \in K,\ \forall v \in V \tag{11}$$

$$\sum_{p \in \Omega} \sum_{p' \in \Omega} z_{pp'v} \leq |\Omega| - 1, \quad \forall \Omega \subseteq P \backslash \{1\},\ |\Omega| \geq 2,\ \forall v \in V \tag{12}$$

$$\sum_{j \in J} f2_{ij} = \sum_{i' \in I'} f1_{ii'}, \quad \forall i \in I \tag{13}$$

$$\sum_{i \in I} f2_{ij} = \sum_{v \in V} \sum_{k \in K} g_{jkv}, \quad \forall j \in J \tag{14}$$

$$\sum_{i \in I} f2_{ij} \leq U_j, \quad \forall j \in J \tag{15}$$

$$f1_{ii'} \leq S_{ii'}^{\max} x_{ii'}, \quad \forall i \in I, \forall i' \in I' \tag{16}$$

$$P_i^{\min} t_i \leq \sum_{j \in J} f2_{ij} \leq P_i^{\max} t_i, \quad \forall i \in I \tag{17}$$

$$x_{ii'} \in \{0,1\}, \ \forall i \in I, \forall i' \in I'; \ y_v \in \{0,1\}, \quad \forall v \in V; \ t_i \in \{0,1\}, \ \forall i \in I;$$
$$z_{pp'v} \in \{0,1\}, \ \forall p \in P; \forall p' \in P, \forall v \in V$$
$$f2_{ij} \geq 0, \ \forall i \in I, \forall j \in J; \ g_{pp'v} \geq 0, \ \forall p \in P, \forall p' \in P, \forall v \in V$$

Variables

$f1_{ii'}$	amount of production at PS i by production method i'
$f2_{ij}$	shipping amount from PS i to CC j
$g_{pp'v}(t)$	load of vehicle v on the path from $p \in P$ to $p' \in P$
$t_i = 1$	if PS i is selected; otherwise 0
$x_{ii'}$	production method i' is selected at PS i; otherwise 0
$y_v(t) = 1$	if vehicle v is used; otherwise 0
$z_{pp'v}(t) = 1$	if vehicle v travels on the path from $p \in P$ to $p' \in P$; otherwise 0

Parameters

c_v	transportation cost per unit load per unit distance of vehicle v
D_k	demand of retailer k
$d_{pp'}$	path distance between $p, p' \in P$
$Fc_{ii'}$	fixed charge to take production method i' at PS i
F_v	fixed charge to operate vehicle v
$Hp_{ii'}$	production cost per unit load at PS i for production method i'
L_{ij}	distance between PS i and CC j
M	auxiliary constant (Large integer number)
$P_i^{\max}$	maximum available amount at PS i
$P_i^{\min}$	minimum available amount at PS i
$S_{ii'}^{\max}$	maximum production at PS i for production method i'
$\mathrm{Sh1}_i$	shipping cost per unit load at PS i
$\mathrm{Sh2}_j$	shipping cost per unit load at CC j
T_c	transportation cost per unit load from PS to CC
U_j	maximum capacity at CC j
w_0, w_v	unladen weight of vehicle used at PS and CC, respectively
W_v	maximum capacity of vehicle v
λ	emission trading rate on CO_2

Index set

I: PS; I': CL; J: CC; K: RE

The objective function of (p. 1) is composed of the pure cost and the surrogate one mentioned already. When we are interested only in the pure cost minimization, $f_1(\boldsymbol{X})$ becomes the objective function, while $\lambda f_2(\boldsymbol{X})$ when interested in the surrogate one. On the other hand, constraints mean that Eq. (2): vehicles cannot visit a customer twice; Eq. (3): vehicles entering a certain RE must leave it; Eq. (4): no travel between CCs; Eq. (5): upper bound on loading capacity for vehicle; Eq. (6): each vehicle must travel once on a certain path; Eqs. (7) and (8): each vehicle leaves only one CC and returns there; Eq. (9): vehicles return to the CC with empty; Eq. (10): customer demand is satisfied; Eq. (11): sum of inlet must be greater than that of outlets by the demand; Eq. (12): sub-tour elimination; Eqs. (13) and (14): material balances at PS and CC, respectively; Eq. (15): holding capacity at CC is upper bounded; Eq. (16): production is upper bounded at PS per method; and Eq. (17): total production at PS must be within the prescribed range. Moreover, either binary or non-negative integer condition is put on the decision variables in terms of their properties.

From this problem, we can decide the available PSs and their production methods, paths from PS to CC, and circular routes touring every client REs from each CC under the relevant constraints. Due to the high NP-hardness, however, it is almost impossible to solve this problem with practical size by any currently available commercial software. Against this, our hybrid method (Wada and Shimizu 2006) can successfully solve several variants, including complicated situations resulting from a variety of real-life conditions. This method is not only practical and powerful but also flexible and adaptive to various applications. So, we intended to extend such idea to the present problem.

3 Multi-level Method Incorporating Multi-depot Vehicle Routing Problem and Modified Tabu Search

To solve the above problem practically, we have introduced a few ideas and integrated them into the framework of our conventional method. The first one is how to choose the available PSs and their production methods and to decide the paths from PSs to REs via CCs. Here, every path from each CC to REs determines the client REs for each CC. This is equivalently to solve a customer assignment problem. For the pure cost problem, it is described below by a linear programming problem (LP).

$$\text{(p.2)}\ \min \sum_{i\in I}\sum_{j\in J} T_c L_{ij} f2_{ij} + \sum_{i\in I}\sum_{j\in J} c_v d_{ij'} f2_{ij} + \sum_{i\in I}\sum_{i'\in I'} \mathrm{Hp}_{ii'} f1_{ii'} + \sum_{i\in I} \mathrm{Sh1}_i\left(\sum_{j\in J} f2_{ij}\right) + \sum_{j\in J} \mathrm{Sh2}_j\left(\sum_{i\in I} f2_{ij}\right)$$

subject to

$$f1_{ii'} \le S_{ii'}^{\max}, \quad \forall i \in I, \forall i' \in I' \tag{18}$$

$$P_i^{\min} \le \sum_{j \in J} f2_{ij} \le P_i^{\max}, \quad \forall i \in I \tag{19}$$

$$\sum_{i \in I} f2_{ij} \le U_j, \quad \forall j \in J \tag{20}$$

$$\sum_{j \in J} f2_{jk} = D_k, \quad \forall k \in K \tag{21}$$

$$f1_{ii'} \ge 0, \ \forall i \in I, \forall i' \in I', \quad f2_{jk} \ge 0, \ \forall j \in J, \forall j \in J$$

Solving the above LP, we can decide the amount of each production method at PS associated with the available paths between PSs and CCs and the client REs of each CC. The assignment by this approach is truly suitable compared with the conventional ones relying on a certain geometric reasoning such as Voronoi diagram (Man et al. 2012), cluster divisions (Şakir and Küçükdeniz 2009), polar angles between the depot and the customers (Gillett and Miller 1974), etc. These just claim the rationality qualitatively and neglect every condition appeared in the formulation of (p. 1). That is, they never consider the capacity constraints, the handling cost, and the practical transportation cost accounting in the objective function. Against this, since the auxiliary problem (p. 2) might partly reflect some conditions of (p. 1), we can assert its rationality even quantitatively.

Secondly, instead of solving (p. 2) directly, we transform it into the equivalent minimum cost flow (MCF) problem by introducing some virtual nodes and edges to the physical structure so that we can reflect some constraints of the present formulation. This graph structure and its label information are shown in Fig. 2 and Table 1, respectively. Then, we applied a graph algorithm of MCF problem like RELAX4 (URL1)[2] to solve this problem extremely fast by incorporating its sensitivity analysis. After all, we can speed up the outer-loop search where we need to solve the similar LP repeatedly. This graph is still valid for the problems with other objective functions, i.e., the surrogate cost or the augmented one. It is necessary only to replace the label information regarding the costs with those of the respective problem. Apparently, for every node of CL and PS without inflow in the MCF graph, CL will not be used and PS not be opened, respectively.

Thirdly, letting REs thus allocated as the clients for each CC, we move on the inner-loop search that refers to individual VRP per CC. It should be noted here the Weber basis cost accounting is essential for this VRP to hold consistency when evaluating the transportation cost from PS to CC and that of VRP. Though there are several methods to provide a good approximated solution of VRP like saving method (Clarke and Wright 1964), none of the conventional methods adopt the

[2]URL1, http://mit.edu/dimitrib/www/home.html, Massachusetts Institute of Technology, Lab. for information and decision systems. (Accessed 31 May 2013).

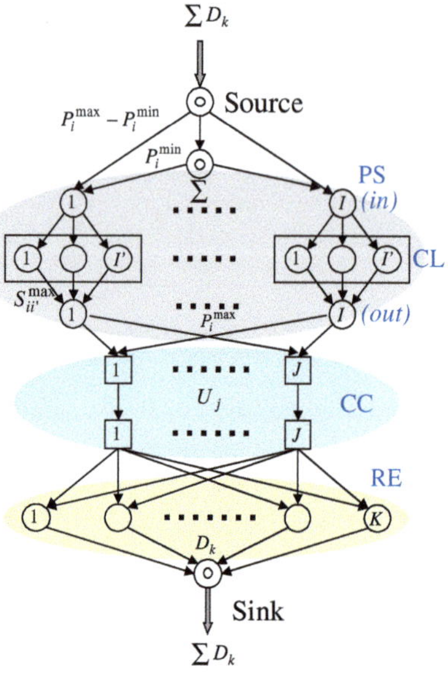

Fig. 2 MCF graph structure

Table 1 Labeling on the edge for MCF graph

Edge (from to)	Cost	Capacity
Source—$\sum$ (dummy)	$-M$	$\sum_{i\in I} P_i^{min}$
Source—PS(in) i	0	$P_i^{max} - P_i^{min}$
$\sum$—PS(in) i	0	P_i^{min}
PS(in) i—CL ii'	$Hp_{ii'}$	$S_{ii'}^{max}$
CL ii'—PS(out) i	0	$S_{ii'}^{max}$
PS(out) i—CC j	$T_cLij + Ship1_i$	P_i^{max}
Between double nodes of CC j	$Ship2_j$	U_j
CC j—RE k	$c_v d_{jk}$	D_j
RE k—Sink	0	D_k

Weber basis. Against this, our Weber basis saving method is conveniently applied to derive an initial solution for this aim. Its algorithm is outlined as follows.

Step 1: Create a round-trip route from every CC to each RE. Then, compute the Weber basis saving value by $s_{ij} = (d_{0j} - d_{0i} - d_{ij})D_j + (d_{0j} + d_{i0} - d_{ij})w_v$ (refer to Fig. 3). Sort the savings in descending order and store them in the saving list.

Step 2: Starting from the top of the saving list, determine whether there exist two routes, one containing $(i, 0)$ and the other containing $(0, j)$. If so, merge the route as long as the combined demand is less than the vehicle capacity and the saving is greater than $-F_v/c_v$. Within this range, we can save the additional expense due to another fixed cost of vehicle.

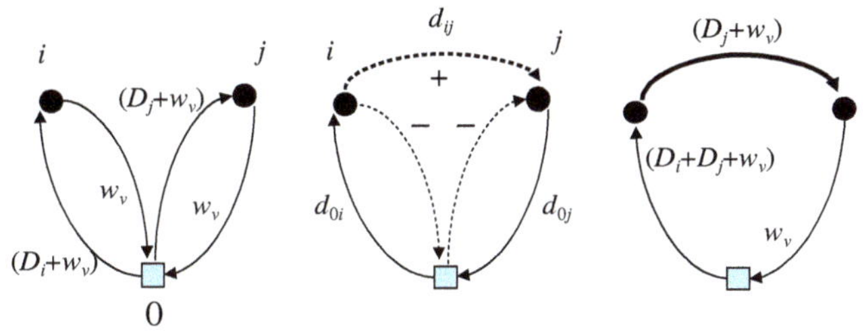

Fig. 3 Scheme to compute Weber basis saving value

Since the Weber basis saving method derives only an approximated solution, we attempt to improve it through the modified tabu search. The original tabu search (Glover 1989) is a simple but powerful heuristic method that refers to a local search with a certain memory structure. In the present local search, we randomly generate a neighbor solution by either of insert, swap, or 2-opt operations for the cases between the routes (extra-loop) and within the route (intra-loop). These operations of the inner-loop are shown in Fig. 4. To avoid trapping into a local minimum, our modified method allows even a degraded neighbor solution to be a new tentative solution as long as it would be feasible and not be involved in the tabu lists. Such decision is made in terms of the probability p whose distribution obeys the following Maxwell–Boltzmann function used in simulated annealing (Kirkpatrick et al. 1983).

$$p = \begin{cases} 1 & \text{if } \Delta e \leq 0 \\ \exp(-\Delta e/T) & \text{if } \Delta e < \varepsilon\ . \\ 0 & \text{if } \Delta e > \varepsilon \end{cases} \tag{22}$$

where Δe denotes the difference of objective function from the tentative best value, and ε, a small positive number. Moreover, T is the temperature that will decrease along with the iteration k geometrically, i.e., $T^k = \sigma T^{k-1}, \sigma < 1$.

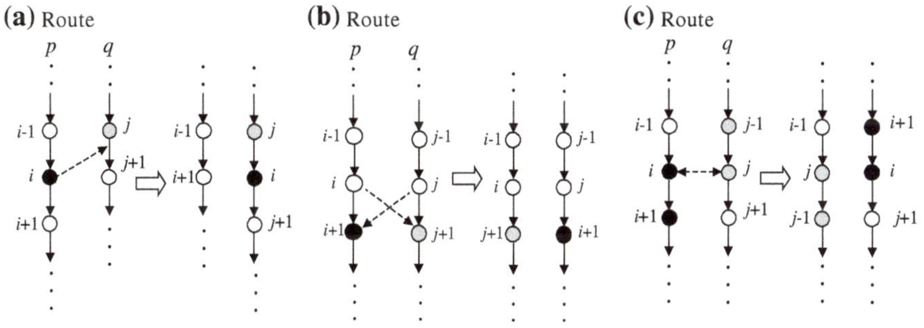

Fig. 4 Scheme to generate neighbor solution in inner-loop search. **a** Insertion, **b** exchange, and **c** 2-opt

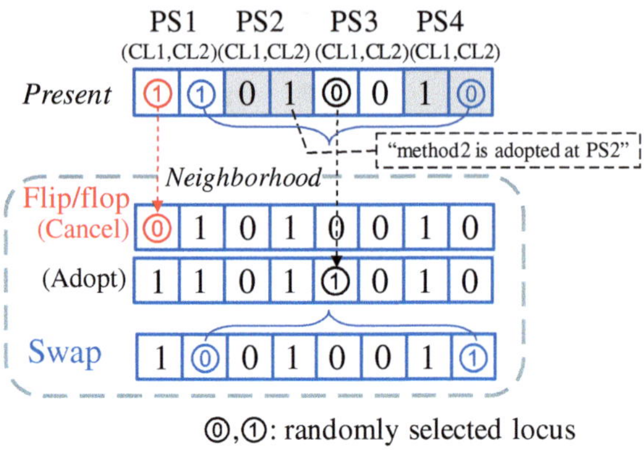

Fig. 5 Scheme to generate neighbor solution in outer-loop search (|I| = 4, |I′| = 2)

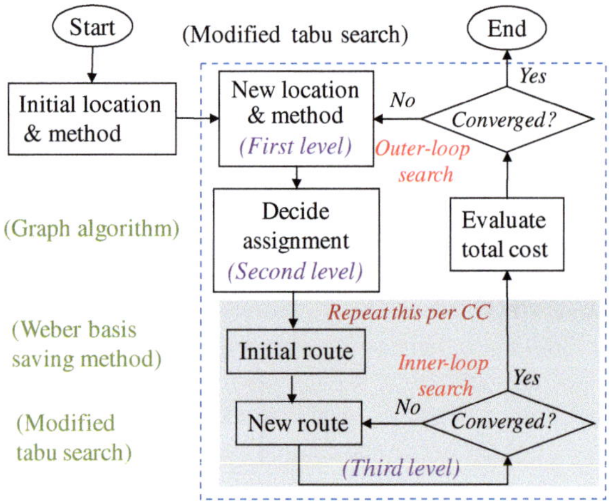

Fig. 6 Outline of the algorithm

Thus, derived result is accumulated over the network and the resulting total evaluation is fed back to update the first-level problem. Fourthly, such outer-loop search is also preceded by the modified tabu search mentioned above. This time, however, its neighbor solution is randomly generated by either of flip/flop or swap operation for the element(s) of binary sequence that represents the available PSs and CLs as illustrated in Fig. 5. In a summary, the outline of this algorithm is shown in Fig. 6.

4 Numerical Experiments and Discussion

In our previous study (Shimizu 2014), we examined the solution ability of our multi-level method through a few benchmark problems with different problem sizes, i.e., {|I|, |I′|, |J|, |K|}. Even for the larger size problems, we ascertained

Table 2 Supposed scenarios on transportation module

Production site	PS1; Kumamoto (far from Nagoya)	PS2; Gifu (very near)	PS3; Tahara (near)
Scenario 1	15 ton truck	4 ton truck	10 ton truck
Scenario 2	Rail + 4 ton truck (multi-modal)	4 ton truck	10 ton truck
Scenario 3	Ferry + 4 ton truck (multi-modal)	4 ton truck	10 ton truck

well-approximated solutions are obtained within an acceptable computation time. We also carried out a case study incorporated with the Google map API to retrieve the real distance data from address of the locations and draw the resulting routes on the map.

Adding some necessary information for the present concern, we have preceded further investigations using the same case study that concerns grocery logistics supplying tomato in Nagoya city in Japan (Nakajima et al. 2011). First, we outline the description of the model under consideration below:

- 3 production sites located in Kumamoto area (PS1; far from Nagoya), Gifu (PS2; very near) and Tahara (PS3; near)
- 3 collection centers located in Nagoya (CC1, CC2, CC3)
- 17 super markets located in Nagoya as retailers (RE1–RE17)
- 2 production methods at production sites (CL1 = ground, CL2 = in house)
- 3 scenarios on transportation modules listed in Table 2
- Distances between the members of logistics are retrieved in terms of Google map API.
- Various parameters listed in Tables 3, 4, and 5 are collected from literatures (URL2–URL7)[3] as much as possible except for someone assumed adequately.

We summarize the results of the pure cost and the surrogate cost (CO_2 emission) minimization problems under the prescribed scenarios in Tables 6 and 7, respectively. The secondary objective function is evaluated by the optimal solution of each problem, i.e., CO_2 emission in Table 6 and total const in Table 7, respectively.

[3]URL2, http://www.e-stat.go.jp/SG1/estat/List.do?lid=000001061833, Statics of production states for vegetables and fruits, Ministry of Agriculture, Forestry and Fisheries (2007)

URL3, http://www.e-stat.go.jp/SG1/estat/List.do?lid=000001112277, Report on cost accounting for food logistics, Ministry of Agriculture, Forestry and Fisheries (2011)

URL4, http://www.enecho.meti.go.jp/policy/images/060518pamph.pdf, Guideline to compute CO_2 emission in logistics, Ministry of Economy, Trade and Industry Minister of Land, Infrastructure, Transport and Tourism

URL5, http://www.ajinomoto.com/jp/activity/environment/pdf/2010/lcco2.pdf, Database of CO_2 emission rate on food materials, Environmental report of Ajinomoto Inc. (2006)

URL6, http://www.shijou.metro.tokyo.jp/gyosei/04/04.html, White Paper on the environment of Tokyo Metropolitan Central Wholesale Market (2008)

URL7, http://www.city.osaka.lg.jp/port/cmsfiles/contents/0000002/2591/12.pdf, Report on the fair of business transportation in Osaka city

Table 3 Parameter setting at production site

Production site		PS1		PS2		PS3	
Shipping amount (kg)	Upper	65,300		73,007		86,700	
	Lower	32,700		36,900		43,400	
Shipping cost (¥/t)		24,150		29,000		31,000	
Emission from shipping (kg-CO_2/t)		20		40		30	
Production method		Ground	House	Ground	House	Ground	House
Production cost (¥/t)	Operating	65,370	103,890	63,000	12,3000	58,000	98,000
	Fixed charge	45,950	103,890	63,000	77,240	58,000	98,000
Available production (kg)		4000	1500	4500	1800	4000	3500
Emission by production (kg-CO_2/t)		354	640	572	872	790	408

Table 4 Parameter setting at market

	Available capacity (kg)	Shipping cost (¥/t)	Emission (kg-CO_2/t)
CC 1	7370	6400	30
CC 2	6530	5120	24
CC 3	8670	6800	20

Table 5 Parameter setting for vehicle

Module	Truck payload			Rail	Ferry
	4 ton	10 ton	15 ton		
Trans. cost unit (¥/t • km)	35.7	21.3	16.4	10.4	11.6
Fixed charge (¥)	12,000	18,000	22,000	22,000	17,000
CO_2 emission unit (kg-CO_2/t • km)	0.177	0.132	0.11	0.022	0.04

Middle part of the tables shows the trading scheme between the production sites and collection centers, i.e., the resulting path from PS to CC and production method at PS for shipping product. Moreover, bottom part of the tables describes the results for VRP for each CC. As supposed easily, we obtained the different decisions depending on the adopted objective function and scenarios. Here, we know the effect of the modal shift to rail or ferry transportation both regarding the cost and CO_2 emission. Under the present parameter setting, scenario 2 seems to be the best result for this logistic system, i.e., gained the minimum values at both objective functions among the scenarios.

In particular, we can observe the trade-off relation between the results of cost and CO_2 emission minimization problems under every scenario. So it is meaningful to consider these two objectives at the same time to lead a decision from more total viewpoint. Actually, as mentioned already, we integrated these two objectives through the emission trading rate and solved the aimed problem (p. 1). We show the result under every scenario in Figs. 7, 8, and 9, respectively. Apparently, the

Table 6 Results of pure cost minimization problem

Scenario #		1 (only truck)	2 (multi-modal: rail)	3 (multi-modal: ferry)
Objective value	**Total cost**	**1,298,496.1** (¥)	**1,290,265.46** (¥)	**1,292,969.87** (¥)
	CO_2 emission	7377.621 (kg-CO_2)	7232.351 (kg-CO_2)	7305.981 (kg-CO_2)
Path from PS to CC & (method)		(PS1, ground) → (CC1)[a]	(PS1, ground) → (CC2)	(PS1, ground) → (CC1)
		(PS2, ground) → (CC1, CC2)	(PS2, ground) → (CC1, CC2)	(PS2, ground) → (CC2, CC1)
		(PS3, ground) → (CC2)	(PS3, ground) → (CC2)	(PS3, ground) → (CC2)
Route over REs	From CC1	V1: 14 → 5→2 → 1→11[b]	V1: 14 → 5→2 → 1→11	V1: 14 → 5→2 → 1→11
		V2: 8 → 17 → 16 → 7→12 → 13	V2: 8 → 17 → 16 → 7→12 → 13	V2: 8 → 17 → 16 → 7→12 → 13
	From CC2	V1: 15 → 3→10 → 9→4 → 6→11	V1: 15 → 3→10 → 9→4 → 6→11	V1: 15 → 3→10 → 9→4 → 6→11
	From CC3	N/A[c]	N/A	N/A

[a]Product produced at PS1 by ground is shipped to collection center CC1
[b]Vehicle V1 departing from CC1 visits customer 14 first and move on the rests in turn
[c]No shipping from CC3 in this scenario

Table 7 Results of CO_2 emission minimization problem

Scenario #		1 (only truck)	2 (multi modal: rail)	3 (multi modal: ferry)
Objective value	**CO_2 emission**	**5602.88** (kg-CO_2)	**5321.76** (kg-CO_2)	**5463.76** (kg-CO_2)
	Total cost	1,528,255 (¥)	1,513,349 (¥)	1,517,790 (¥)
Path from PS to CC and (method)		(PS1, ground) → (CC3)	(PS1, ground) → (CC3)	(PS1,Ground) → (CC3)
		(PS2, ground) → (CC1, CC3)	(PS2, ground) → (CC1, CC3)	(PS2, ground) → (CC1, CC3)
		(PS3, house) → (CC3)	(PS3, house) → (CC3)	(PS3, house) → (CC3)
Route over REs	From CC1	V1: 3 → 10 → 15	V1: 11 → 4 → 9 → 10 → 15	V1: 4 → 9 → 10 → 15
	From CC2	N/A	N/A	N/A
	From CC3	V1: 17 → 8 → 4 → 9 → 6 → 15 → 11	V1: 5 → 6 →15 → 3 →10 → 14	V1: 8 → 17 → 7 → 13 → 12 → 16
		V2: 5 → 2→1	V2: 2 → 1	V2: 2 → 1
		V3: 16 → 7 → 12 → 13	V3: 16 → 17 → 8	V3: 5 → 14 → 15 → 6

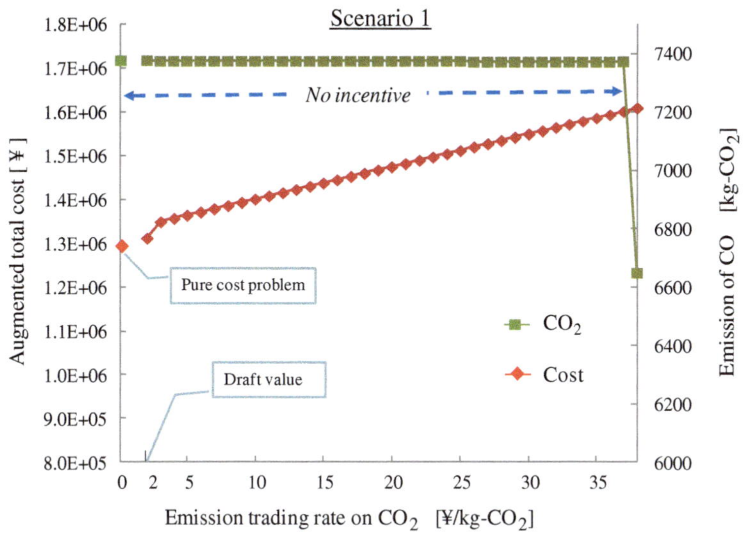

Fig. 7 Trend of two factors in objective function with emission trading rate under scenario 1

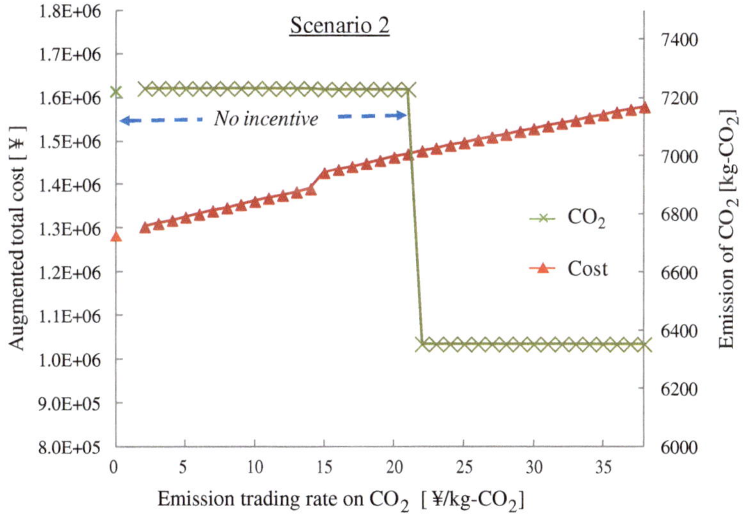

Fig. 8 Trend of two factors in objective function with emission trading rate under scenario 2

augmented total cost expands along with the increase in emission trading rate in every case. On the other hand, CO_2 emission is kept constant by a certain point where we have the same optimal solution as that of the pure cost minimization problem. Against this, when the emission trading rate for each scenario becomes greater than 37, 22, and 28 (¥/kg-CO_2), respectively, we know it possible to

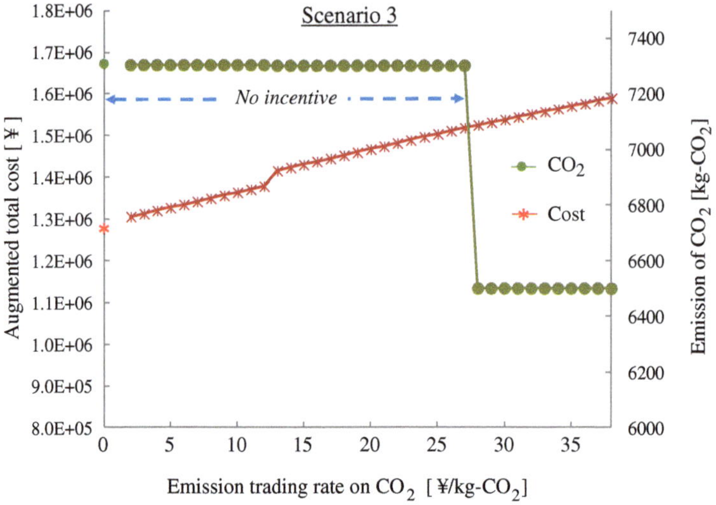

Fig. 9 Trend of two factors in objective function with emission trading rate under scenario 3

decrease CO_2 emission. Thereat, the optimal solutions are changed so as to increase the available amount of product at site PS1 by ground made.

In fact, however, those emission trading rates are too high compared with the value that is presently viewed plausible in Japan, say 2 (¥/kg-CO_2) (Kakumoto and Irohara 2010). We could discuss this issue more generally and certainly if the proposed idea is applied in terms of more reliable parameters and deployed in more global and total logistics systems. However, this study, as it is, has a great significance to show the possibility of reducing the CO_2 emission by controlling the emission trading rate. To facilitate such movement in advance, earnest efforts that make turn consumption behavior to the eco-efficient products become essential. The proposed approach may contribute to provide some successful scenarios for such co-existence and/or co-operation among the logistics members.

5 Conclusion

Provision for essential infrastructure aligned with sustainable development is urging us to realize green logistics incorporating co-existence of manufacturers and consumers. This paper has concerned the green logistics optimization involving a decision-making on production methods. Such idea has not concerned elsewhere so far. Actually, taking a multi-layer logistics network, we have developed a novel approach to optimize the augmented cost comprised the pure economic cost and the surrogate one referring to CO_2 emission. Through a case study under some plausible scenarios associated with the modal shift, we have shown the significance of the proposed approach and explored some prospects for green logistics in the global society.

Future studies should be devoted to establish a total decision support system associated with a generic multi-objective optimization. Various real-world applications are also meaningful to accumulate the wisdom toward the sustainable development.

Acknowledgments This research was partially supported by the Ministry of Education, Science, Sports and Culture, Grant-in-Aid, 25350442, Japan.

References

Clarke G, Wright J (1964) Scheduling of vehicles from a central depot to a number of delivery points. Oper Res 12(4):568–581

Fiksel J (2010) Evaluating supply chain sustainability. CEP 5:28–36

Gillett BE, Miller LR (1974) A heuristic algorithm for the vehicle dispatch problem. Oper Res 22:340–349

Glover F (1989) Tabu search-part I. ORSA J Comput 1(3):190–206

Kakumoto K, Irohara T (2010) Optimization and sensitivity analysis of an international transportation scheduling problem incorporating CO_2 emission levels. J Jpn Ind Manage Assoc 61(2):46–53 (in Japanese)

Kirkpatrick S, Gelatt CD, Vecch MP (1983) Optimization by simulated annealing. Science 220:671–680

Man Q, Mingyi D, Liu Y (2012) Application of Voronoi diagrams and multiangle measurable image in the urban POI location and site generation. In: Proceedings of information science and control engineering 2012 (ICISCE 2012), Shenzhen, pp 1–5

Nakajima H, Ohno T, Ikemori F, Takagi K (2011) Estimate by LCA technique of CO_2 exhaust according to vegetables and fruits consumption of Nagoya City. Annual report of Nagoya City Environmental Science Research Institute, vol 41, pp 15–19 (in Japanese)

Nikolopoulou A, Ierapetritou MG (2012) Optimal design of sustainable chemical processes and supply chains: a review. Comput Chem Eng 44(9):94–103

Şakir E, Küçükdeniz T (2009) A fuzzy clustering-based hybrid method for a multi-facility location problem. J Intell Manuf 20(2):259–260

Shimizu Y (2011a) Advanced saving method to evaluate economic concern. Trans Inst Syst Control Inf Eng 24(2):39–41 (in Japanese)

Shimizu Y (2011b) A meta-heuristic approach for variants of VRP in terms of generalized saving method. Trans Inst Syst Control Inf Eng 24(12):287–295 (in Japanese)

Shimizu Y (2014) Incorporating green into production and consumption behaviours toward sustainable logistics optimization in process systems. In: 24nd European symposium on computer aided process engineering, Budapest, Hungary, 15–18 June 2014

Shimizu Y, Sakaguchi T (2013) Generalized vehicle routing problem for reverse logistics aiming at low carbon transportation. Ind Eng Manage Syst 12(2):161–170

Srivastava S (2007) Green supply-chain management: a state-of-art literature review. Int J Manage Rev 9:53–80

Wada T, Shimizu Y (2006) A hybrid metaheuristic approach for optimal design of total supply chain network. Trans Inst Syst Control Inf Eng 19(2):69–77 (in Japanese)

Sustainability Classification for SMEs—A Guidance of Sustainability Assessment with the Use of Averaged Traits Quality Method

Monika Kosacka, Rafał Mierzwiak and Paulina Golinska-Dawson

Abstract This chapter presents a method for aggregation of different indicators, which are relevant for sustainability assessment. The originality of the procedure relies on the use of Averaging Quality Rating method that was used on the level of indicators analysis for making them comparable despite different measurement method (qualitative or quantitative). It is suitable for small- and medium-sized companies (SMEs) where the information system is rather undeveloped.

Keywords Sustainability assessment · Indicators · Averaged traits quality method · Relativization

1 Introduction

The competitive environment forces companies to improve their performance systematically. With the increasing importance of sustainability, companies aim to apply the sustainability policy, what can be defined as the better utilization of resources without negative influence on the natural environment and the surrounding communities.

In order to better identify the current situation and to find the improvement potential, companies should build own sustainability assessment system which aims to help the managers in decision making by facilitating the control and improvement of different aspects of the business operations. The sustainability measures should include the following dimensions: economic, ecological, and social.

M. Kosacka (✉) · R. Mierzwiak · P. Golinska-Dawson
Poznan University of Technology, Strzelecka 11, 60-965 Poznan, Poland
e-mail: monika.kosacka@put.poznan.pl

R. Mierzwiak
e-mail: rafal.mierzwiak@gmail.com

P. Golinska-Dawson
e-mail: paulina.golinska@put.poznan.pl

V. Kachitvichyanukul et al. (eds.), *Toward Sustainable Operations of Supply Chain and Logistics Systems*, EcoProduction,
DOI 10.1007/978-3-319-19006-8_9

Determination of customized Sustainability Indicators System (SIS) is the first step on the road leading to higher sustainability class. Each company is different and should establish its own SIS, which will be helpful in decision making toward more sustainable societies and should be simple, relevant, intentional, data accessible, and customized.

In order to become more sustainable, small- and medium-sized companies (SMEs) are facing problems of the limited resources (human and financial), what makes impossible to implement complex performance measurement systems. SMEs need some guidelines for decision making in the context of better sustainability class.

This paper presents the concept of decision-making procedures for assessing the sustainability with the use of universal Sustainability Classification System (SCS) for SMEs. In this paper, there is no imposed universal indicators system providing appropriate sustainable assessment for each SME. There is presented a guidance for classification of SMEs from the sustainability perspective which will be suitable regardless of approved indicators system.

The originality of the procedure relies on the use of Averaging Quality Rating method that was used on the level of indicators analysis for making them comparable despite different measurement method (qualitative or quantitative). There are often many problems with establishing only measureable indicators particularly for sustainability assessment in SMEs. High variety of measurement's units and the combination of quantitative criteria to be evaluated by descriptive, makes impossible to compare the sustainability criteria and determine the strength of their impact on the overall assessment of the companies. Averaging Quality Rating method was used for the estimation of quality index of sustainability which is used to sustainability assessment in SMEs.

2 Sustainability Assessment for SMEs—State of the Art

Sustainable development is equal to the development of people, economy, and environment (WCED 1987).

It can be observed growing attention given to sustainable development, which encourages companies to integrate sustainable issues into their activities. In the past, companies presented an approach where the environment and sustainable development were the problems causing costs and risk factors. Today, "the sustain" state is seen as an opportunity and a source of efficiency improvement or growth. For example, if energy is saved, both the production costs and the negative outputs like waste are reduced.

Popularity of the sustainability issues results in many papers referred to that topic (e.g., it can be obtained from 1,040,000 results with the "sustainable development" issue and 1,110,000 results of "sustainability" with the use of Google Scholar) (https://scholar.google.pl).

According to the global trend, companies are aiming to achieve the "sustain" state due to a higher value for companies as well as the increase of their competitiveness particularly for small and medium enterprises (SME) (Henriques and Catarino 2014).

In order to ensure SMEs' long-term survival in global markets, companies must pursue sustainability as a strategic goal. A prerequisite for this is the ability to perform a sustainability assessment of products and processes (Rosen and Kishawy 2012; Garetti and Taisch 2012). Such an assessment includes a sustainability accounting and an impact analysis, too. Sustainability accounting refers to measurements of resource utilization, waste generation and pollution emission from all activities in manufacturing. The impact analysis applies the impact of chosen measurements on the people, on the environment, and on the economy (Feng et al. 2010).

The need for assessment was recognized more than forty years ago. At the beginning, the focus was on environmental impacts only. With the expansion of sustainability definition to include social and economic goals, assessment began to include the three pillars of sustainability (Pope et al. 2004). Since that, there appeared many assessment methods in the literature (e.g., Devuyst and Hens 2001; Verheem 2002; Buselich 2004; Ness et al. 2007).

It was assumed that sustainability becomes the key for the success. In the consequence, there is a strong desire for the comprehensive assessment of sustainability level (changes in economic, environmental, and social conditions). The information about actual "state of the sustainability" in a company is desirable data possible to achieve with the appropriate system of sustainability indexes (SIS).

Sustainability assessments required answer for two questions:

- What is the object of the measurement?—determining what should be measured. The answer needs understanding the origins, fundamentals, and principles of sustainability.
- How to measure sustainability? Measurements can range from objective and quantitative to more subjective or qualitative metrics. There is established the SI adopted to the company.

The significance of indicators is high—they allow to assess the current position and determine the direction of further improvement actions. In order to better identify the current situation in a company and find the optimization potential, there is a requirement of a system of performance measures which helps the managers to follow up, coordinate, control, and improve different aspects of the organizational activities (Kollberg et al. 2005).

Indicators are built by the values (a measurement is related to the object of a care) and indicators create values (care about everything what is measured (Sustainable Measures 2010).

The concept of sustainability is pointless without the parameterization—"It has become merely a marketing slogan" (Darton 2005).

The sustainability measures should allow to assess the company performance in the three dimensions as proposed by Brundtland Commission: economic, ecological, and social (WCED 1987).

In the literature, there are plenty of propositions on indicators and methods for evaluating the performance of sustainable development (Rinne et al. 2013; Li et al. 2009; Singh et al. 2009; Böhringer and Jochem 2007; Hopwood et al. 2005). Böhringer and Jochem (2007) and Pintér et al. (2012) reviewed the consistency and meaningfulness of various indicators and proposed many principles and theoretical frameworks about the assessment of sustainable development. Despite the fact that there are many possibilities of expressing SI, there are problems with relating them to the SME. In the literature can be found indexes referred to different global issues such as:

- Human Development Index (HDI) (UNDP 2005);
- Ecological Footprint (EF) (Monfreda et al. 2004);
- Environmental Performance Index (EPI) (Esty et al. 2006);

Appropriate sustainability assessment for SMEs with the use of presented aggregated indexes is impossible.

Except of aggregated sustainability indicators, there are available indexes dedicated for each pillar of sustainability:

- Environmental—e.g., energy consumption, water consumption, waste generation, and GHG emissions.
- Economic (cost perspective) with the most popular method in the literature quoted of assessing the economic aspects of sustainable development—life cycle cost (LCC).
- Social—including the following issues: health and safe, human rights, employment, living conditions, crime, corruption, etc. (Fatimah 2013; Schau et al. 2011; UN 2008).

Most of the companies in UE are SMEs.[1] They have limited human and financial resources that enables them the implementation of complex performance measurements systems. It is crucial to ensure an effective decision support model for a goal-oriented analysis and implementation of appropriate measures for increasing sustainability in SMEs with the use of appropriate SIS. In the majority of the cases, there is lack of SIS dedicated for SME. There can be found some examples of SI prepared for chosen company (e.g., system of indicators for remanufacturing companies presented in Golinska et al. 2014).

3 Procedure for Sustainability Classification for SMEs

In order to classify companies into one of three classification groups of the sustainability (S) averaged traits, quality method was used, which allow for the carrying out of the classification process in a quantifiable way (Fig. 1).

[1]A total of 99.8 % of all companies in Poland are SMEs. Structure of Polish companies is very similar to the structure in UE (PARP).

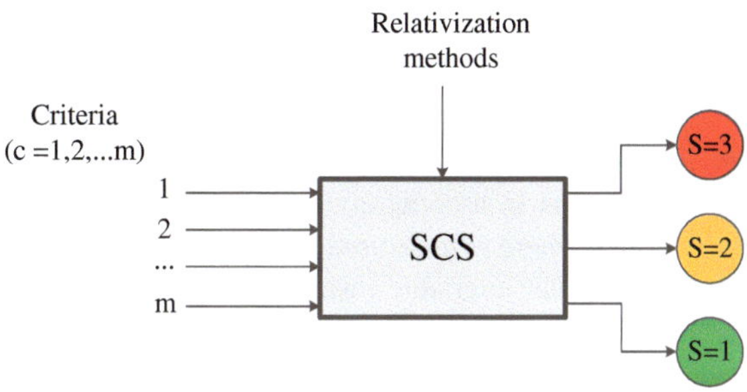

Fig. 1 The SCS process (own elaboration)

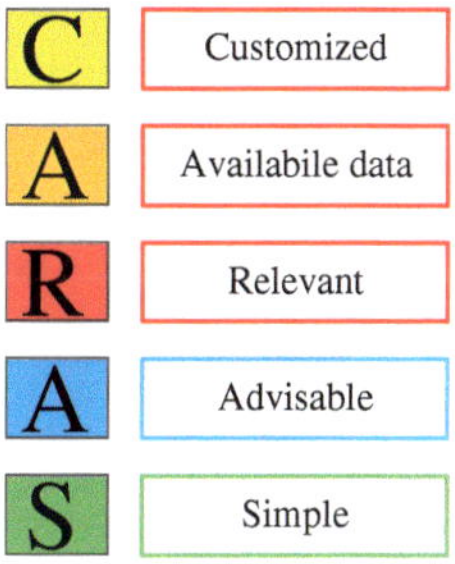

Fig. 2 CARAS—requirements of sustainability indicator for SMEs (own elaboration based on Feng and Joung 2009; UN 2008)

SCS is a procedure for transforming input data—criteria of the sustainability assessment in output—and a class of the sustainability (S).

The procedure of classification can be presented in a few steps, which are presented below.

Step 1: Defining of assessment criteria (measures) There are defined set criteria c = 1, 2, …, m. It is recommended to create a SIS for analyzed company that portrays sustainability in an analyzed SME. It is a crucial stage.

The criteria selection process is a holistic approach based on a definition of the fundamental objectives of sustainable development and features of good SIS (CARAS) (Fig. 2).

Features of good sustainability indicator are as follows:

- **Customized**: Indicator should be suitable for the company, including specific features of SME as well as industry. Consequently, it should take into account the goals from the sustainability perspective and resources used to meet the objective.
- **Available data**: Indicator should be based on data that are readily available; otherwise, indicator has no value for SME, where there is no time and money for the development of a data collection system for indicator. It should not be too difficult or too expensive to collect the information.

- **Relevant**: An indicator must fit the purpose for measuring. It must have a direct relationship to the structure, process, or outcome that it is measuring what is important from the sustainable perspective.
- **Advisable**: The indicator is established to present the actual situation of analyzed areas in a company with the improvement potential in the sustainability context. There is measured that what should be improved to be more sustain—to utilize better available resources—and to be more competitive. The indicator is required to make some decisions improving actual situation.
- **Simple**: The indicator should be simple and logical, in the construction and interpretation, that even a non-expert is able to understand and interpret it for future decision making. If indicator is simple, two different people measure the same indicator using the same tool and will get the same result. Simple indicators are understandable.

  Each requirement is equally important.

Step 2: SIS analysis and relativization SIS does not require measurability, which is often hard to achieve in the context of measuring SME, particularly in the sustainability context. To measure sustainability there are often use some qualitative indicators. The sustainability assessment of SME is similar to the process of requirement definition of products which can be expressed as physical quantities of different units, as well as various characteristics expressed in words (Kolman 2013).

It is known that the quality is not directly measurable, but its level is possible to determine with the use of means of numerical indicators expressing the states of studied criteria. The method of quality evaluation aims to present the quantitative level of quality on the basis of analysis of the state of quality criteria. Value and unit of measure represents the absolute state of any quantity (Kolman 2013).

In practice, when there is a need to compare the different features of the object, the possibility of two different metrics requires their standardization. Hence, in the method of valuation, there is a rule of relative state consideration.

There are often many problems with establishing only measureable indicators particularly for sustainability assessment in SMEs. High variety of measurement's units and the combination of quantitative criteria to be evaluated by descriptive, what makes impossible to compare the sustainability criteria and determine the strength of their impact on the overall assessment of the companies in the sustainable use of resources. Consequently, there was a need for relativization, transforming the absolute state of relative state (for measurable factors), or selecting the proper relative state (from the universal scale of relative states)—for unmeasured factors. Thanks relativizing all the analyzed criteria are expressed in unitless terms, in the range of between 0 and 1 (Dudek-Burlikowska 2006). In the result, comparing SI will be possible because of representation in the form of relative states, but it requires universal scale which will be a reference for the analyzed states, which was presented in Fig. 3.

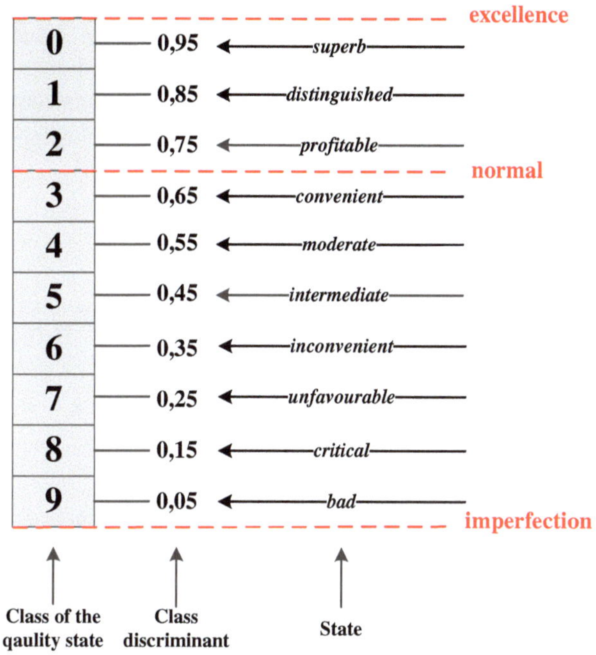

Fig. 3 Universal scale of quality relative states (adapted from Dudek-Burlikowska 2006)

Table 1 Set of SI features required during relativization (own work based on Kolman 2013)

No.	Aspect of SI	SI realization	
1	Measurability	Measurable	Immeasurable
2	Direction of the evaluative function	Maximum	Minimum
3	Range of variability of the criteria	Full	Acceptable by the tolerance
4	Requirements for detailed analysis	YES	NO
5	Conditions of the evaluation	YES	NO

Universal scale of quality relative states becomes a sample for relative state interpretation. There are 10 classes of quality states which are described in words and by numbers.

Quality class in the general meaning is a set of information about the qualitative properties of the analyzed object that facilitates the identification of requirements and identification of the achieved results. It was assumed that the class "0" corresponds to the highest level of quality and class "9" the lowest level of quality (Dudek-Burlikowska 2006).

From the perspective of the relativization, there should be taken into account the following set of criteria features (Table 1).

Step 3 In this step, the validity of the classification is determined by various criteria, η_c, where $0 < \eta_c \leq 1$. The closer the value of η_c to 1 means the more important the criterion is. In order to determine η_c, the third-quarter preferential Thurstone's analysis can be a detailed explanation of this procedure which is in basic old paper from 1927 year (Thurstone 1927). Other method is analytical hierarchy process (AHP) well presented, e.g., by Saaty (2008).

Step 4 In this step, it is necessary to designate x_c, which is the real value of the c-th criterion for the company classified in accordance with the accepted operational definition of each of the criteria.

Step 5 Calculating the value of relative state—S_c on the basis of x_c and method of relativization.

Step 6 Calculating the W_{xc}—discriminate of criteria according to the formula (1):

$$W_{xc} = \eta_c * s_c \tag{1}$$

Step 7 Calculating aggregate indicator (S), which can be calculated on the basis of each pillar of sustainability as well as summarized indicator (see formula 2):

$$S_i = \frac{\sum W_{xc}}{\sum \eta_j} \tag{2}$$

where

- For i = so—assessment of the social aspect;
- For i = ec—assessment of the ecological aspect;
- For i = en—assessment of the economic aspect;
- For i = s—summarized sustainability assessment.

After comparing the result of the assessment, there can be concluded the SCS according to the following scale (Fig. 4).

In SCS, there were distinguished 3 classes (referring to the traffic light), characterized in the Table 2. It is noteworthy that the scale can be developed by using

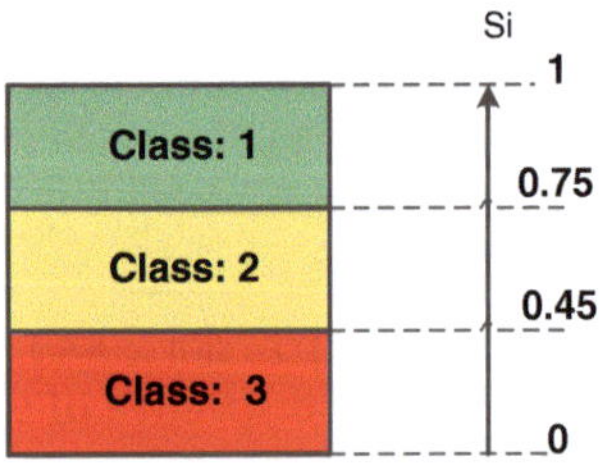

Fig. 4 Scale for SCS (own work on the basis of Dudek-Burlikowska 2006)

Table 2 Characteristic of sustainability classes (own work on the basis of Dudek-Burlikowska 2006)

FEATURE	**SUSTAINABILITY CLASS**		
	1	2	3
Interval of S_i	(1-0,75>	(0,75-0,45>	(0,45;0)
General assessment	Favorable	Normal	Negative
States from the Universal scale of quality relative states	superb; distinguished; profitable.	convenient; moderate; intermediate.	inconvenient; unfavorable; critical; bad.
Class signature			

the fuzzy sets theory or gray systems theory, which will be the direction for future researches (experts' researches are required). Intervals' limits were arbitrarily defined.

Each of the sustainability class is assigned depending on the value of the aggregate index—S_i on the analyzed level (there are social, economic, and ecologic platform as well as complex sustainability). Intervals were determined arbitrarily by the authors of the paper based on a universal scale of unity relative states, taking into account the semantic definitions of the state term:

- For negative terms—CLASS 3 (red);
- For states with moderate preferred semantic overtones—CLASS 2 (orange); and
- For benefit terms—CLASS 1 (green).

The application of the proposed algorithm in the decision making for the remanufacturing process sustainability assessment is presented in Fig. 5.

In Fig. 5, the algorithm facilitates the decision-making process and helps the company to find out what is the aggregated sustainability class as well as at the level of each sustainability dimension, at the each criteria level ending that helps in decision-making process for sustainability improvement actions.

The algorithm was validated based on the real-life data coming from the group of small-size companies. The algorithm can be used in companies which on regular basis collect the minimum scope of data regarding their operations (production volume, lead times, resource utilization, etc.). By the definition of the SIS are taken into consideration the limitations of SMEs which have limited resources and usually do not apply sophisticated IT systems, for example, MRP/ERP.

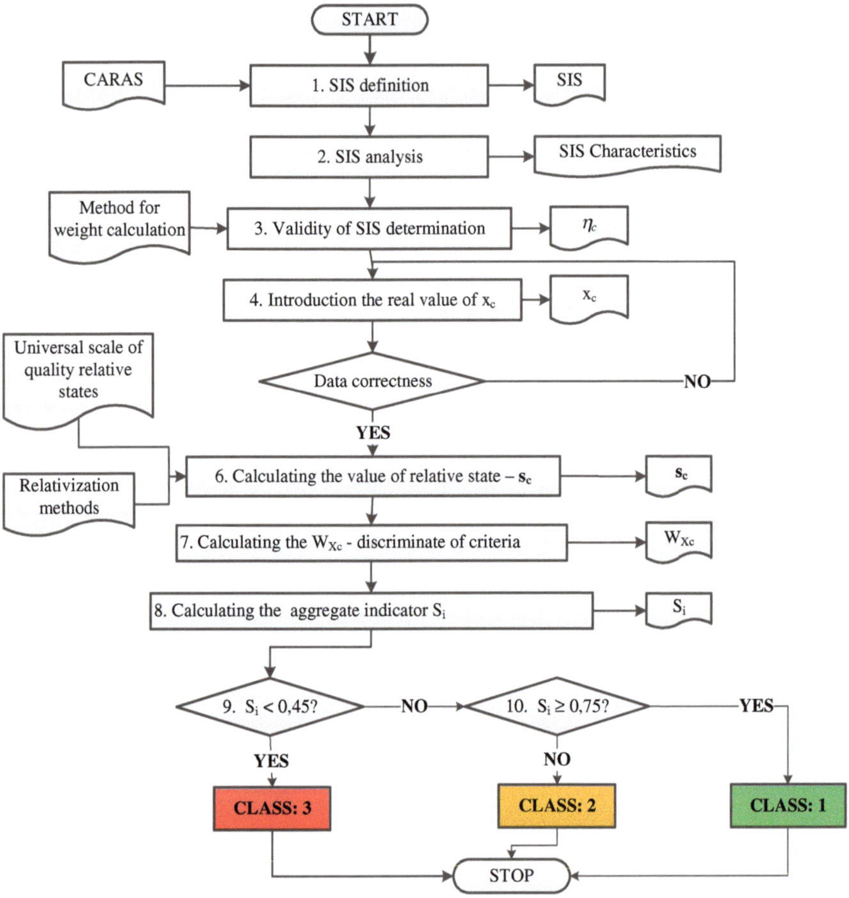

Fig. 5 The algorithm of SCS

4 Conclusions

The presented method allows the aggregation of qualitative and quantitative indicators values. It helps to classify and compare companies with different approaches to the measurement of indicators which are important for the assessment of the sustainability case of companies. This method of aggregation is especially suitable for SMEs which were due to the limited resources and relatively undeveloped information structure and it is difficult to obtain relevant indicators values. This method allows to include in the assessment not only the collected numerical data but also expert knowledge of management staff. The further research will include more extended method testing in SMEs from different industries.

References

Böhringer C, Jochem PE (2007) Measuring the immeasurable—a survey of sustainability indices. Ecol Econ 63(1):1–8

Buselich K (2004) An outline of current thinking on sustainability assessment—a background paper prepared for the Western Australian state sustainability assessment. Murdoch University, Western Australia

Darton R (2005) Measuring our future—the role of sustainability metrics. In: The 11th Hartley lecture delivered at the royal society on 10th November 2005. Available via http://www.eng.ox.ac.uk/chemeng/people/darton/Hartleylecture.pdf. Accessed 10 July 2014

Devuyst D, Hens N (2001) How green is the city? sustainability assessment and the management of urban environments. Columbia University Presss, New York

Dudek-Burlikowska M (2006) Quality research methods as a factor of improvement of preproduction sphere. J Achiev Mater Manuf Eng 18(1–2):435–438

Esty DC, Levy MA, Srebotnjak T, de Sherbinin A, Kim CH, Anderson B (2006) Pilot 2006 environmental performance index. Yale Center for Environmental Law & Policy, New Haven

Fatimah YA, Biswas W, Mazhar I, Islam MN (2013) Sustainable manufacturing for Indonesian small-and medium-sized enterprises (SMEs): the case of remanufactured alternators. J Remanuf 3(1):1–11

Feng SC, Joung CB (2009) An overview of a proposed measurement infrastructure for sustainable manufacturing. In: Proceedings of the 7th global conference on sustainable manufacturing. Available 2009

Feng SC, Joung CB, Li G (2010) Development overview of sustainable manufacturing metrics. In: Proceedings of the 17th CIRP international conference on life cycle engineering, pp 6–12

Garetti M, Taisch M (2012) Sustainable manufacturing: trends and research challenges. Prod Plann Control 23(2–3):83–104

Golinska P, Kosacka M, Mierzwiak R, Werner-Lewandowska K (2014) Grey decision making as a tool for the classification of the sustainability level of remanufacturing companies. J Clean Prod

Henriques J, Catarino J (2014) Sustainable value and cleaner production—research and application in 19 Portuguese SME. J Clean Prod

Hopwood B, Mellor M, O'Brien G (2005) Sustainable development: mapping different approaches. Sustain Dev 13(1):38–52. https://scholar.google.pl/. Accessed 15 Jan 2015

Kollberg B, Elg M, Lindmark J (2005) Design and implementation of a performance measurement system in Swedish health care services: a multiple case study of 6 development teams. Qual Manag Healthc 14(2):95–111

Kolman R (2013) Different varieties of quality and their practical use. Publisher Placet, Warsaw (in Polish)

Li F, Liu X, Hu D, Wang R, Yang W, Li D, Zhao D (2009) Measurement indicators and an evaluation approach for assessing urban sustainable development: a case study for China's Jining city. Landsc Urban Plan 90(3):134–142

Monfreda C, Wackernagel M, Deumling D (2004) Establishing national natural capital accounts based on detailed ecological footprint and biological capacity assessments. Land Use Policy 21 (3):231–246

Ness B, Urbel-Piirsalu E, Anderberg S, Olsson L (2007) Categorising tools for sustainability assessment. Ecol Econ 60(3):498–508

Pintér L, Hardi P, Martinuzzi A, Hall J (2012) Bellagio STAMP: principles for sustainability assessment and measurement. Ecol Indic 17:20–28

Pope J, Annandale D, Morrison-Saunders A (2004) Conceptualising sustainability assessment. Environ Impact Assess Rev 24(6):595–616

Rinne J, Lyytimäki J, Kautto P (2013) From sustainability to well-being: lessons learned from the use of sustainable development indicators at national and EU level. Ecol Indic 35:35–42

Rosen MA, Kishawy HA (2012) Sustainable manufacturing and design: concepts, practices and needs. Sustainability 4(2):154–174

Saaty TL (2008) Decision making with the analytic hierarchy process. Int J Serv Sci 1(1):83–98

Schau EM, Traverso M, Lehmann A, Finkbeiner M (2011) Life cycle costing in sustainability assessment—a case study of remanufactured alternators. Sustainability 3(11):2268–2288

Singh RK, Murty HR, Gupta SK, Dikshit AK (2009) An overview of sustainability assessment methodologies. Ecol Indic 9(2):189–212

Sustainable Measures (2010) Characteristics of effective indicators. Available via http://www.sustainablemeasures.com/Indicators/Characteristics.html. Accessed 14 July 2014

Thurstone L (1927) The method of paired comparisons for social values. J Abnorm Soc Psychol 21:384–400

UNDP—United Nations Development Programme (2005) Human development report 2005. Oxford University Press, Oxford

United Nations (UN) (2008) Guidance on corporate responsibility indicators in annual reports. United Nations conference on trade and development, New York and Geneva. Available via http://unctad.org/en/docs/iteteb20076_en.pdf. Accessed 20 July 2014

Verheem R (2002) Recommendations for sustainability assessment in the Netherlands. In: Proceedings of the 22nd annual conference of the international association for impact assessment

World Commission on Environment and Development (WCED) (1987) Our common future. Oxford University Press, Oxford, pp. XV, 400

Decision-Making Criteria for Sustainable Remanufacturing

Passaporn Kanchanasri, Seung Ki Moon and Gary Ka Lai Ng

Abstract Due to significant economic and environmental benefits, remanufacturing has been deployed and enhanced in various market sectors. Determining product parts whether they should be reused, repaired, or disposed is an important task in the remanufacturing. There are a variety of criteria such as cost optimization, minimizing disposed items, and so on. Many researches considered cases with one or two criteria while it is possible to have more factors that should be considered in an industrial case. In this chapter, highlighting on sustainability, available decision-making criteria are reviewed and categorized into groups, based on the objective in the literature, if they are economical or environmental focused. Additionally, the decision-making criteria are classified if they are related to products, processes, and people in the system. Decision makers can select the criteria based on the sections which they concentrate on. The decision criteria are covered by various aspects to support decision makers to set product recovery objectives easily and quickly. Criteria can be conflicted with each other to be minimizing cost and maximizing product quality, simultaneously. This chapter investigates conflicting criteria to assist decision makers in selecting a decision method to determine product recovery option. In future work, the decision-making criteria will be identified to develop a framework by integrating an intelligent decision-making agent system.

Keywords Product recovery · Reverse logistics · Decision making

P. Kanchanasri · S.K. Moon (✉)
School of Mechanical and Aerospace Engineering,
Nanyang Technological University, Singapore 639798, Singapore
e-mail: skmoon@ntu.edu.sg

P. Kanchanasri
e-mail: pass0001@e.ntu.edu.sg

G.K.L. Ng
Advanced Remanufacturing and Technology Centre,
Singapore 637143, Singapore
e-mail: gary-ng@artc.a-star.edu.sg

V. Kachitvichyanukul et al. (eds.), *Toward Sustainable Operations of Supply Chain and Logistics Systems*, EcoProduction,
DOI 10.1007/978-3-319-19006-8_10

1 Introduction

Moving forward to a sustainable development requires a proper decision making. Product recovery is one way to contribute to sustainability. It leads to the salvaging of material and energy of a product at its end of life. People increasingly take a consideration of reducing waste (Kim et al. 2013; European Commission 2014). Instead of disposing entire end-of-life products, we can reduce waste and new manufactured product volume by considering other recovery alternatives such as reuse, remanufacturing, and so on. Generally, a product is manufactured and then used until its end-of-life phase when the product seems probably unable to function well anymore. The product recovery can yield benefits to many parties. Original equipment manufacturers (OEMs) can reduce cost (Toffel 2004). When production cost decreases, a product price could be reduced. As a result, end users can buy a product at lower price. In addition, the product recovery affects positively our environment and society. It can help reduce natural resource consumptions. Pollutant emissions or negative environmental impacts are able to be reduced also. Benefits to human, society, and environment cause the product recovery to get attention by stakeholders and researchers.

Remanufacturing is an industrial process in which worn-out products are restored to like-new product. It acts on end-of-life products to recover manufacturing standards and specifications (ARTC 2013). The remanufacturing is one key of a sustainable development. It can lead to economic, environmental, and societal sustainability (Matsumoto et al. 2013). It can save cost and resources while reducing waste (Lund 1985).

There are several procedures in a remanufacturing process. A discarded product is disassembled into parts. The product is reassembled from reusable parts to produce a unit fully equivalent in product quality, durability, performance, and expected lifetime to the original new product (Lund 1984). It is inspected, reassembled, and tested, respectively.

Inspecting an end-of-life product and selecting a proper recovery option are crucial. There are various product end-of-life options (Thierry et al. 1995; Krikke et al. 1998; Rose et al. 2002; Parlikad et al. 2003; Jun et al. 2007). A product can be determined to be reused if it can be used further with no or minimal treatments such as cleaning. If the product can be used after getting a treatment such as damage fixing, it can be determined to be repaired. If the product is determined that it cannot be used, it will be disposed. Others options could be refurbishing, remanufacturing, cannibalization, recycling with disassembly, recycling without disassembly, reconditioning, etc.

Product recovery option selection is challenging. Various criteria can be involved in decision making. Basically, decision criteria are selected relying on a specific business objective in each industry case. Economic factors are commonly found in many studies. To gain the highest revenue or minimizing cost, product recovery option is selected based on cost factors. Review of literature is presented in Sect. 2. Apart from the cost, there are other factors involving in product recovery.

There are concerns about sustainability and social impacts (Miemczyk 2008). It leads to a challenge in finding out a way to balance economic and non-economic factors. This research aims to help decision makers consider decision factors to provide a product recovery decision support for long-term benefits. Decision criteria for product recovery option selection in economic and societal aspects were presented in Sect. 3. They are classified and sorted covering economic and environmental factors in a decision criteria factors list. The decision criteria from this study can be further utilized. Closing remarks and future work are described in Sect. 4 finally.

2 Literature Review

There are various factors or criteria to decide how to manage an end-of-life product. Law clearly assigns an explicit responsibility to producers or OEMs for their products. For example, car manufacturers were obliged to take their products back and recycle them at 85 % by 2005 and at 95 % by 2015, respectively. Electronic product producers had to do so as well (Feldmann et al. 1999). Apart from the law, others factors were stated in studies as follows. O'Shea et al. (1998) considered reliability, flexibility, and environmental aspects, adding destructive operations, costs specification, and integration in the disassembly planning system. Goggin et al. (2000) focused on material cost, inventory material cost, fixed cost associated with process allocated to products or options undergoing the process according to their volume and a standard absorption rate, variable cost associated with the processing of a certain volume of products or options, and inventory carrying cost. Lee et al. (2001) determined the optimal end-of-life options by considering cost and environmental impact. Santochi et al. (2002), in order to decide the appropriate strategy, considered influencing factors which were age of products, general conditions, service status, innovations, market demands, wear damage, reusability, material, toxicity, and value. List of criteria presented in Bufardi (2003) work included logistics cost (collection and transport), disassembly cost, product value (what is gained from incineration recycling, landfilling, etc.), product cost, number of employee necessary to perform relevant operations, exposure to hazardous materials, CO_2/SO_2 emissions, and energy consumption. Staikos et al. (2007) considered three main factors in their decision-making model for waste management in footwear industry. First factors were economic factors which were cost and benefit. Second factors were environmental factors (global warming potential, ozone depletion, acidification, eco-toxicity, human toxicity, hazardous waste). Third factors were technical factors (technical feasibility, compliance with legislation, market pressures, public opinion). Gomes et al. (2008) considered investments, operating costs, disposal and treatment costs, CO_2 emissions, corporate image, and benefits to select an optimal alternative for plastic waste. According to Xanthopoulos et al. (2009) study, financial, environmental, legislative, and technological criteria were considered for an optimal recovery operation.

Wadhwa et al. (2009) considered time, material cost, labor, overheads, and administrative. Environmental impacts which were resource consumption, resource conservation, waste release, and waste impact were also taken into consideration. Moreover, they also included market factors (demand, supply), quality factors both technical-related and operational-related, and a legislative impact. Dhouib (2014) selected a market factor, profits, jobs created, and environmental impact to be criteria for reverse manufacturing alternative selection. Ghazilla et al. (2014) considered return product volume, demand, maintenance plan, engineering design, ease of disassembly, and ability to disassemble a product. From literature, various decision criteria can be involved in product recovery. Observably, they are related to products, processes, and people. Cost minimizing and engineering quality are the criteria that are commonly focused in many researches. In this study, pushing forward to sustainable development, environmental factors are highlighted in addition to economic factors to prompt decision makers to carefully make a decision. It becomes a challenge for the decision makers to consider both economic and non-economic aspects which can conflict with each other. Decision criteria for product recovery decision making are presented in the following section.

3 Decision Criteria for Product Recovery Option Selection

This section presents decision criteria for product recovery option selection including economic and non-economic factors. They cover business aspects, work safety, impacts to products, processes and people, environment, etc. The criteria of environmental factors are useful to make a product recovery decision and consider long term benefits to end users and manufacturers. Decision makers can consider economic and environmental decision criteria. Furthermore, the decision criteria include product-, process-, and people-related factors. The decision factors for product recovery option selection are presented in Table 1. In this section, each criterion is classified by two dimensions which are as follows:

- Category Dimension I indicates that a criterion is environment-oriented or economy-oriented.
- Category Dimension II indicates that a criterion is related to products, processes, or people.

Economic factors are basically focused. One is a return product volume. Decision makers should concern about handling uncertainty of return product volume. A product which has regularly high return product volume can be inspected and considered to be reused or remanufactured. A selected recovery option and volume can depend on market demand as well. For example, if market or customers accept and require the remanufactured product, remanufacturing will be added as a recovery option. On the other hand, a decision maker might consider

Table 1 Decision factors for product recovery option selection

Decision criteria/factors	Category dimension I (environment-oriented or economy-oriented)	Category dimension II (Product/process or people-related)
Return product volume	Economy-oriented	Product
Demand/market need	Economy-oriented	Product
Product reliability	Economy-oriented	Product
Product/part before recovery quality	Economy-oriented	Product
Product/part after recovery quality	Economy-oriented	Product
Product design (ease of disassembly)	Economy-oriented	Product
Inventory cost	Economy-oriented	Product
Material cost	Economy-oriented	Product
Landfill cost	Economy-oriented	Process
Logistic cost—location, transportation, collecting	Economy-oriented	Process
Machine cost including maintenance	Economy-oriented	Process
Processing cost by machine—disassembly, repair/restoration, remanufacture, disposal, etc.	Economy-oriented	Process
Processing time by machine—disassembly, repair/restoration, remanufacture, disposal, etc.	Economy-oriented	Process
Machine capability/failure	Economy-oriented	Process
Labor—human factors, skill	Economy-oriented	People
Labor cost/processing cost by people—disassembly, repair/restoration, remanufacture, disposal, etc.	Economy-oriented	People
Processing time by people—disassembly, repair/restoration, remanufacture, disposal, etc.	Economy-oriented	People
Legislation	Environment-oriented	Process
Gas emission—CO_2, SO_2, etc.	Environment-oriented	Process
Pollutant or toxic emission	Environment-oriented	Process
Heat emission	Environment-oriented	Process
Product useful life time	Economy-oriented/ environment-oriented	Product
Natural resource consumption—material, fuel, energy, etc.	Economy-oriented/ environment-oriented	Product
User preference	Economy-oriented/ environment-oriented	Product, process, people

other options instead. Other is product reliability. Failure frequency of recovering product can help determine whether a product should be recovered or not. Additionally, a condition or quality of the product is a common factor in product recovery. If the product has a huge damage or is not good enough to recover, it will be disposed. Furthermore, an ease of product disassembling can be taken into consideration. Disassembly is a process in which a product is separated into its components and/or subassemblies (Bogue 2007). Easy-to-disassembling designed product takes shorter time in processing. It is attractive for remanufacturing. According to disassembly experiments, disassembly processing time could greatly affect cost also (Ghazilla et al. 2014). Other factor is a cost relating with product recovery process such as inventory cost, material cost, landfill cost, logistic cost, machine cost, disassembly cost, and recycling cost. A disposal cost is one influential factor. The effect of disposal costs on a disassembly-for-recycling process is shown by Dini et al. (2001). High disposal costs can cause recycling to be more attractive. Commonly, product recovery option is selected based on cost optimization or profit maximizing. Other factors which are related to process can be processing time, machine capability, number of equipment, number of machines, machine failure, dependent machines failure, and process bottleneck. Decision makers can consider human factors also. Human capabilities such as working memory, eyesight, and different skill can affect product recovery effectiveness. Number of required manpower and effective labor working hour can be additionally considered if further needed in product recovery. Moving to non-economic-focused factors, many factors highlight environmental and societal benefits. Obviously, legislation pays attention to save the earth. Law in different countries probably assigns different responsibilities in product recovery. Gas, heat, and pollutant emission can be limited in order to not harm an environment and human. These can affect the number of products to be recovered. The following factors are focused in both economy and environment. Product useful lifetime is one factor that the decision makers can consider. A product which has short and long useful lifetime can have different strategies in managing its end of life. Additionally, according to Ondemir et al. (2014), a useful lifetime can be considered as a measure of quality. Furthermore, natural resource consumption including energy, electricity, and material can be considered. In a case of limited resources, options such as reuse and remanufacture should be included. The natural resource consumption is related to economic aspect as well. The less natural resource is consumed, the less money is spent. Also less harm is occurred in our environment. Last but not the least, user preferences can be taken into consideration. Product recovery option selection can be varied by user type (Kiritsis et al. 2003). OEMs may prefer to remanufacture the product while recycling companies may prefer to recycle the product. The decision makers should know the user preferences (if any) and select the product recovery options to finally satisfy the end user.

In case the decision makers consider more than one factor in product recovery, some factors can conflict with each other when setting decision objective such as minimizing recovery cost while expecting high recovered product after quality, and minimizing recovery cost while maximizing number of recovered product (demand).

4 Closing Remarks and Future Work

This chapter presents decision factors for selecting product recovery options. Decision makers can consider the factors to make a proper product recovery decision. There are a variety of decision criteria for product recovery in remanufacturing. Each criterion is related to products, processes, or people that should be considered. Various available criteria focus on economic factors. To increase concern about sustainability, environmental factors are added in decision criteria for product recovery. Therefore, the decision makers can select a proper method (conflicting or non-conflicting factors decision-making method) to solve the problem efficiently.

The decision criteria will be identified and configurable in the developed intelligent multi-criteria decision making for automated product recovery option selection. It will support conflicting criteria such as natural resource consumption, product quality, and cost. The decision makers can consider criteria covered by environmental and economic aspects. They can observe the developed model and analyze long-term benefits while increasing sustainability in not only remanufacturing but society.

Acknowledgments This research was partially supported by Advanced Remanufacturing and Technology Centre (ATRC), Singapore and an AcRF Tier 1 grant (RG 94/13) from Ministry of Education, Singapore.

References

ARTC (2013) Manufacturing and Remanufacturing from http://www.a-star.edu.sg/artc/About-ARTC/Manufacturing-Remanufacturing.aspx

Bogue R (2007) Design for disassembly: a critical twenty-first century discipline. Assem Autom 27(4):285–289

Bufardi A, Sakara D, Gheorghe R, Kiritsis D, Xirouchakis P (2003) Multiple criteria decision aid for selecting the best product end of life scenario. Int J Comput Integr Manuf 16(7–8):526–534

Dhouib D (2014) An extension of MACBETH method for a fuzzy environment to analyze alternatives in reverse logistics for automobile tire wastes. Omega 42(1):25–32

Dini G, Failli F, Santochi M (2001) A disassembly planning software system for the optimization of recycling processes. Prod Plann Control 12(1):2–12

European Commission (2014) Communication from the commission to the European parliament, the council, the European economic and social committee and the committee of the regions—towards a circular economy: a zero waste programme for Europe

Feldmann K, Trautner S, Meedt O (1999) Innovative disassembly strategies based on flexible partial destructive tools. Annu Rev Control 23:159–164

Ghazilla RAR, Taha Z, Yusoff S, Rashid SHA, Sakundarini N (2014) Development of decision support system for fastener selection in product recovery oriented design. Int J Adv Manuf Technol 70(5–8):1403–1413

Goggin K, Browne J (2000) The resource recovery level decision for end-of-life products. Prod Plann Control 11(7):628–640

Gomes CFS, Nunes KR, Xavier LH, Cardoso R, Valle R (2008) Multicriteria decision making applied to waste recycling in Brazil. Omega 36(3):395–404
Jun HB, Cusin M, Kiritsis D, Xirouchakis P (2007) A multi-objective evolutionary algorithm for EOL product recovery optimization: turbocharger case study. Int J Prod Res 45(18–19): 4573–4594
Kim M, Jang YC, Lee S (2013) Application of Delphi-AHP methods to select the priorities of WEEE for recycling in a waste management decision-making tool. J Environ Manag 128: 941–948
Kiritsis D, Bufardi A, Xirouchakis, P (2003) Multi-criteria decision aid for product end of life options selection. In: IEEE international symposium on electronics and the environment. IEEE, New York, pp 48–53
Krikke HR, Van Harten A, Schuur PC (1998) On a medium term product recovery and disposal strategy for durable assembly products. Int J Prod Res 36(1):111–140
Lee SG, Lye SW, Khoo MK (2001) A multi-objective methodology for evaluating product end-of-life options and disassembly. Int J Adv Manuf Technol 18(2):148–156
Lund RT (1984) Remanufacturing. Technol Rev 87(2):19–23, 28
Lund RT (1985) Remanufacturing: the experience of the United States and implications for developing countries, Technical paper No. 31. World Bank, New York
Matsumoto M, Ijomah W (2013) Remanufacturing. In: Kauffman J, Lee K-M (eds) Handbook of sustainable engineering. Springer, Netherlands, pp 389–408
Miemczyk J (2008) An exploration of institutional constraints on developing end-of-life product recovery capabilities. Int J Prod Econ 115(2):272–282
O'Shea B, Grewal SS, Kaebernick H (1998) State of the art literature survey on disassembly planning. Concurr Eng 6(4):345–357
Ondemir O, Gupta SM (2014) Quality management in product recovery using the internet of things: an optimization approach. Comput Ind 65(3):491–504
Parlikad AK, McFarlane D, Fleisch E, Gross S (2003) The role of product identity in end-of-life decision making. In: Auto ID Centre White Paper CAM-AUTOID-WH017
Rose CM, Ishii K, Stevels A (2002) Influencing design to improve product end-of-life stage. Res Eng Des 13(2):83–93
Santochi M, Dini G, Failli F (2002) Computer aided disassembly planning: state of the art and perspectives. CIRP Ann Manuf Technol 51(2):507–529
Staikos T, Rahimifard S (2007) A decision-making model for waste management in the footwear industry. Int J Prod Res 45(18–19):4403–4422
Thierry M, Salomon M, Van Nunen J, Van Wassenhove L (1995) Strategic issues in product recovery management. Calif Manag Rev 37(2):114–135
Toffel MW (2004) Strategic management of product recovery. Calif Manag Rev 46(2):120–141
Wadhwa S, Madaan J, Chan FTS (2009) Flexible decision modeling of reverse logistics system: a value adding MCDM approach for alternative selection. Robot Comput Integr Manuf 25 (2):460–469
Xanthopoulos A, Iakovou E (2009) On the optimal design of the disassembly and recovery processes. Waste Manag 29(5):1702–1711

A Framework for Sustainable Food Supply Chain: Reflections from the Indian Dairy Producers

Gyan Prakash

Abstract This chapter presents a framework for sustainable operations of food supply chains. A case study approach is used. The data were collected using exploratory interviews and through a structured questionnaire. The sustainable food supply chain involves operational levers of robustness, transparency, traceability and information flow and performance measure for monitoring and control of day-to-day operations for efficiency, flexibility, responsiveness and product quality. New product development, research and development, productivity improvement, entrepreneurial orientation, quality control and conducive policy support across key stages of production, procurement, processing, distribution and consumption of the dairy supply chain emerge as key elements of the sustainable framework.

Keywords Public policy · Sustainability · Dairy supply chain

1 Introduction

India is the world's largest producer of milk, second largest producer of fruits and vegetables and third largest producer of fish. With a large agriculture sector, abundant livestock and cost competitiveness, India is fast emerging as a sourcing hub of processed food. India produced 250 million tonnes of food grains in the financial year 2012. India's comparative advantage lies in its favourable climate, geographic location, large agriculture sector and livestock base, long coastline and inland water resources and closeness with key export destinations such as Middle East and Southeast Asia (India Brand Equity Foundation (IBEF) 2012; Economic Survey 2013). Food processing involves value addition to farm produces and includes processes such as grading, sorting and packaging which enhances shelf life of the produce. Due to natural and perishable nature of food, the food processing sector has

G. Prakash (✉)
Indian Institute of Information Technology and Management Gwalior, Gwalior, India
e-mail: gyanprakasha@yahoo.com

V. Kachitvichyanukul et al. (eds.), *Toward Sustainable Operations of Supply Chain and Logistics Systems*, EcoProduction,
DOI 10.1007/978-3-319-19006-8_11

significant quality, safety and performance implications. The concern is more so in the case of developing countries such as India which is one of the agri-based economies where agriculture accounts for 19 % of gross domestic product (GDP) but employs over 60 % of its population (IBEF 2010). In these countries, there are wastages of food grains, vegetables, fruits, dairy products and other food items due to lack of processing capacity, infrastructure facilities, storage facilities and other related constraints (Reardon et al. 2001; Dharni and Sharma 2008; Ruteri and Qi Xu 2009) which necessitates the investigation of underlying supply chain. In case of developed countries, supply chain of food products has received a great deal of attention due to issues related to public health. In near future, the design and operation of food supply chains will be subject to more stringent regulations and closer monitoring. As a result, the traditional food supply chain practices and corresponding performance measures would be subject to revision and change (Ahumada and Villalabos 2009; Bigliardi and Bottani 2010). The dairy segment occupies an important position in the agriculture economy of India, as milk is the second largest agriculture commodity contributing to the gross national product (GNP) next only to rice. The strength of the Indian dairy sector lies in the fact that in spite of limited investment, it has shown consistent and sustainable growth (Venugopal 2008; Patil et al. 2009). About 70 % of the milk production is by small and marginal farmers having less than five milch animals (Chand et al. 2010). The author's purpose is to study the existing dairy supply chain in Madhya Pradesh (MP) state of India and to identify issues contributing towards sustainability. It assesses the dairy supply chain configuration, information flow and traceability, robustness and performance measurement. Central findings are that dairy supply chains are inherently complex and achievement of sustainable operations requires coordinated efforts. It is inferred that sustainability in the Indian dairy context is defined in terms of achievement of broad social needs and appropriateness of technologies. Broad social needs underscore livelihood of small producers and associated socio-economic goals. Appropriateness of technologies is defined in terms of low-cost solutions for processing of milk.

2 Review of Literature

2.1 Supply Chain Management and Food Industry

The food industry is categorized by different segments such as fresh food industry, processed food industry and livestock food industry. Each segment has its unique characteristics and associated complexities; therefore, each segment needs different supply chain strategies for its procurement and sourcing, inventory management, warehouse management, packaging and labelling system, and distribution management, thus providing uniqueness to underlying food supply chains (Georgiadis et al. 2005).

2.2 Partnering in Food Supply Chains

A successful supply chain requires coordination and cooperation between its components (Hobbs and Young 2000). Absence of coordination results in inefficient supply and dissatisfied customers (Chung-Chi and Cheng-Han 2008). Uncoordinated information from downstream to upstream stages results in wastages and losses for most of the food processors. The distorted information implies that the processors work on unreliable amplified demand data which have serious cost implications (Ouyang and Daganzo 2008). Food products have limited shelf life, and it is not easy to recover any material whenever expiry date is due (Minegish and Thiel 2000). Losses can be minimized through coordination between partners within the supply chain including customers by forming alliances or sharing information and knowledge to create a collaborative competitive and cost-effective supply chain (Wee and Yang 2004; Ketikidis et al. 2008). Collaboration appears where working and operating alone is not sufficient to resolve common problems (Wagner et al. 2002; Matopoulos et al. 2007). Supply chain collaboration involves design and governance of supply chain activities and the establishment and maintenance of supply chain relationships (Matopoulos et al. 2007).

2.3 Transparency in Food Supply Chains

Transparency of a supply chain is the extent to which all its stakeholders have a shared understanding of and access to the product-related information, without loss, noise, delay and distortion (Hofstede et al. 2005; Deimel et al. 2008). Transparency in the food supply chain is essential to guarantee food quality. Food supply chains plan to achieve transparency with respect to multitude of food properties and are linked through governance mechanisms that are supported by information systems and aim to achieve pre-defined production standards, specified in quality and safety standards (Trienekens et al. 2012).

2.4 Traceability in Food Supply Chains

Food traceability not only helps in managing food quality and safety risks but also promotes development of effective food supply chain management (Manzini and Accorsi 2013). Two main types of traceability technologies and devices are identification tags (i.e. barcode, label, RFID tag) and data loggers (also called as black boxes) (Abad et al. 2009). Identification tags identify a product with a specific code denoting its lot number, shelf life, manufacturer's name, etc. Data loggers aim to trace and record environmental conditions and profiles on-the-move experiences of a product across supply chain processes. Food preservation and deterioration depend on intrinsic and extrinsic factors such as storage temperature, concentration of oxygen, relative humidity, solar radiation, acidity, microbial growth, endogenous enzyme activities, etc. (Alasalvar et al. 2001; Zhang et al. 2009). Thus, the purpose of traceability devices and systems is to preserve specificities of food products for their safety and hygiene requirements as per the food standards (Sarc et al. 2010).

2.5 Performance Measurement and Sustainability

In SCM theory, robustness and vulnerability are perceived as opposite though not mature concepts (Wagner and Bode 2006). Disruption can have direct effect on the organizations' ability to distribute products into a market and provide critical services to customers (Bhamra et al. 2011). Vulnerability refers to the capacity of the system to preserve its structure (Gallopin 2006) or as the extent to which a system is susceptible to the effects of change (Bhamra et al. 2011). Resilience represents sum total of vulnerability of a system as well as its adaptive capacity (Dalziell and McManus 2004). Robustness is mainly considered as the ability of the system to continue to function in the event of a disturbance (Dong 2006). Supply chain robustness is a desired property that is reflected in supply chain performance (Vlajic et al. 2012). In case of food supply chains, the vulnerability is high due to inherent factors such as seasonality, perishability and variability.

Utilization of resources in terms of cost and profit, social–economic welfare and use of resources to address broad social goals impacts efficiency (Lai et al. 2002). Use of environmental technologies in operational decisions such as equipment, methods, practices and delivering systems can reduce negative impact on the environment (Angell and Klassen 1999). Improved farming techniques can alter associated cost structure which impacts sustainability (Maloni and Brown 2006). Inventory of farm produce as well as work in process food items is linked with storage costs, stock turns, wastes (Bourlakis et al. 2014) and energy-linked cold storage cost and in turn impacts sustainability (Coley et al. 2009). Process improvements through innovative dairy technologies may reduce waste which in turn results in efficiency gains (Sarkis 2001), positive economic outcome (Handfield et al. 1997) and improved environmental sustainability (Gerbens-Leenes et al. 2003). Location and volume-based flexibility of dairy processes help in matching demand with supply (Shepherd and Gunter 2006). Responsiveness in terms of lead time, point of sale time and delivery schedule impacts sustainability. Therefore, partners need to take shared responsibility, absence of which may lead to search for better upstream or downstream partner which in turn impacts sustainability. Transportation of agricultural produce and processed foods is linked with variable cost of transportation mode, and fuel consumption (food miles), therefore, affects cost and sustainability (Maloni and Brown 2006). Use of green logistical options (Llbery and Maye 2005) may reduce carbon emissions, thereby improving economic and environmental outcomes (Gerbens-Leenes et al. 2003).

Balanced score card (BSC) assimilates perspectives of finance, customer, internal business processes and learning and growth (Kaplan and Norton 1992) and presents a holistic performance measure framework. Bigliardi and Bottani (2010) have used it for performance measurement of food supply chain.

3 Methodology

This paper follows a case study approach which is an inquiry of a real-life phenomenon having blurred boundaries (Yin 1994) and involving cycles of description, explanation and testing (Meredith 1998). This method is also used to serve the purpose of exploring, describing and explaining empirical setting (Yin 1994). Barratt et al. (2011) has defined qualitative case study as an empirical real-world setting to investigate a focussed phenomenon. This approach has appealed the researchers for integration of operation management with other functional areas of the supply chain (Pagell 2004). This paper primarily used elaboration approach of case research (Ketokivi and Choi 2014). The underlying reason is that the Indian dairy supply chain context is not known well enough to obtain sufficiently detailed premises that could be used in conjunction with the general theory to deduce precise testable hypotheses. The unit of analysis is the dairy supply chains in MP state of India. Unit of analysis is critical for relating the case to pertinent body of knowledge (Dube and Pare 2003) and helps in defining boundaries of a theory which in turn sets the limitations in applying the theory. Review of literature has been undertaken to develop a holistic view on the supply chain issues in FPI sector. Data have been collected from different types of dairy supply chains in and around Gwalior region of MP state of India. Exploratory interviews have been performed to understand the structure and functioning of the dairy supply chain at stages of milk collection, sourcing, processing and distribution and their associated impact on sustainability. Table 1 shows description of exploratory interview and elaborates on data sources and associated issues of discussion. Data have been collected during January–October 2014. The items of measurement have been taken from the literature.

4 Findings and Discussion

4.1 The Indian Dairy Supply Chains

India has a unique pattern of production, processing and marketing of milk, which is not comparable with any large milk-producing country (Ministry of Food Processing Industries (MoFPI) 2009). Indian dairy supply chains are differentiated based on the processing firm which is the main integrator and differentiator of the supply chain. Indian dairy supply chains involve dairy cooperative society

Table 1 Description of exploratory interviews

S. no	Data sources	Issues of discussion
1	*District government-level officials:* The chief medical and health officer (CMHO), food inspector and superintendent of police Gwalior	About existing policy framework for administration of dairy products
2	*Officials of organized sector dairy operators: Gwalior Sahkari Dugdh Sangh, Gwalior,* Sterling Agro Industries, VRS Food Limited, S.M. Milkose, Reliance Fresh and Cadbury, India	Issues involved in production, collection, sourcing, processing and distribution stages of dairy supply chain
3	*Unorganized sector dairy operators:* Reputed small dairy owners and halwaiis from the region.	
4	*Academicians*: Food science researchers from Agriculture University Gwalior, Supply chain researchers from Indian Institute of Technology Kanpur, and ABV-Indian Institute of Information Technology and Management Gwalior	About emerging technologies for production and processing of milk and dairy products. Issues of supply chain management and dairy supply chain management

(DCS) supply chain; large private dairy processing firm supply chain; and small dairy and halwaiis (local-level small dairy operators) supply chain. The processing plants owned by the dairy cooperatives manufacture wide range of Western and the traditional Indian dairy products. Some of these products are skimmed milk powder (SMP), table butter, cheese, packet milk, ultra-high-temperature (UHT) milk, flavoured milk, khoya,[1] chhena,[2] gulabjamun,[3] burfi[4], peda[5], shrikhand,[6] rasogolla,[7] rajbhog,[8] rasmalai[9], etc. Figure 1 shows typical route of a procurement cycle of a dairy cooperative Gwalior Sahkari Dugdh Sangh (GSDS). This shows collection of milk from various village-level societies and its movement through bulk milk

[1]*khoya*: it is solid concentrated remains obtained by continuously steering boiling milk.

[2]*Chhena*: it is obtained by curding milk that separates chhena (cottage cheese) from whey.

[3]*Gulabjamun*: it is spherical or cylindrical in shape, brown in colour and soaked in sugar syrup.

[4]*Burfi:* it is prepared with kyoya and is cream in colour with firm granular body.

[5]*Peda*: it is a khoya-based product, flattened circular in shape and prepared by mixing *khoya* and sugar.

[6]*Shrikhand*: it is made from concentrated yogurt (*dahi)* with a sweet and sour taste and it is semi-soft.

[7]*Rasogolla*: it is prepared using soft *chhena in the* shape of small balls and dipped in sugar syrup.

[8]*Rajbhog*: a large yellowish and somewhat less soft variety of rasogolla is the *rajbhog*.

[9]*Rasmalai*: it is prepared by kneading *chhena* with wheat flour dipped in thickened milk.

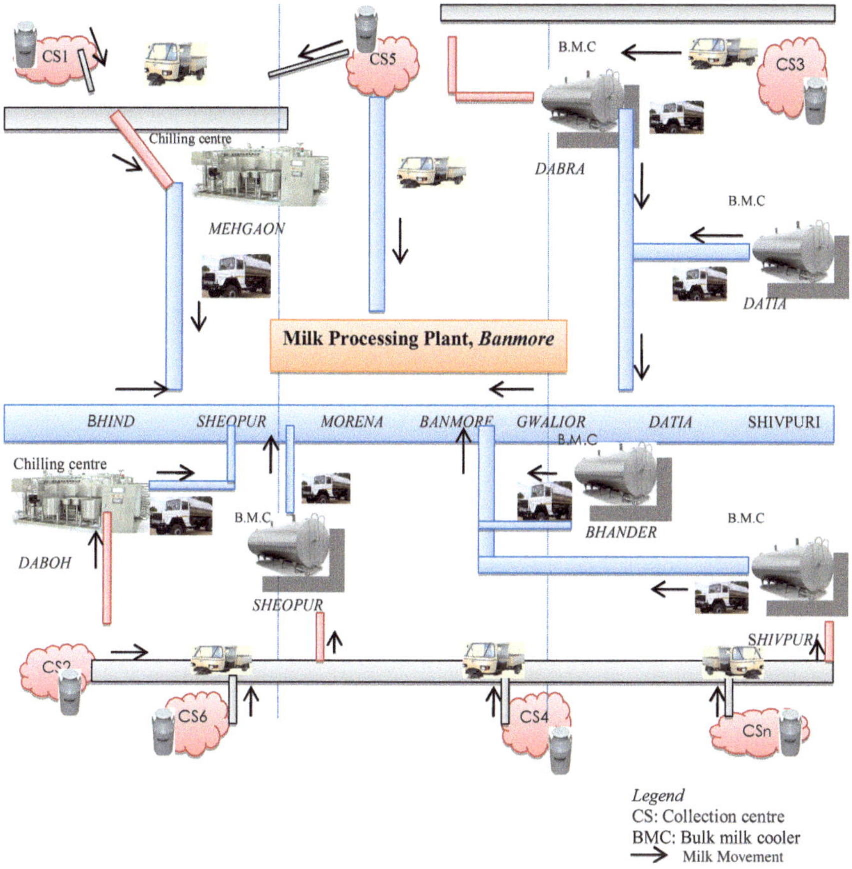

Fig. 1 Typical route of a milk procurement cycle of organized dairy supply chain of GSDS

coolers (B.M.C) and chilling centres to processing plant. Figure 2 depicts various issues involved in various stages of milk procurement cycle.

The sourcing stage faces various challenges such as variable supply of milk from farmers, their lack of commitment for regular milk supply and losses in a form of leakage during transportation. The measurement of fat and solid not fat (SNF) is still manual which results in time delays and poor utilization of man hours. Further, transportation management of vehicles does not use ICT-based track and trace systems. This results in lack of alignment between arrival of a truck and a production schedule. The main differentiating factor of organized and unorganized dairy sector is in terms of the difference in investments by them for quality, shelf life, processing equipment and compliance to the standard. The unorganized sector basically comprised of small dairies and halwaiis (GoI 2005). As per the Indian Food Safety Standard (FSS 2011), dairy operators are required to register and

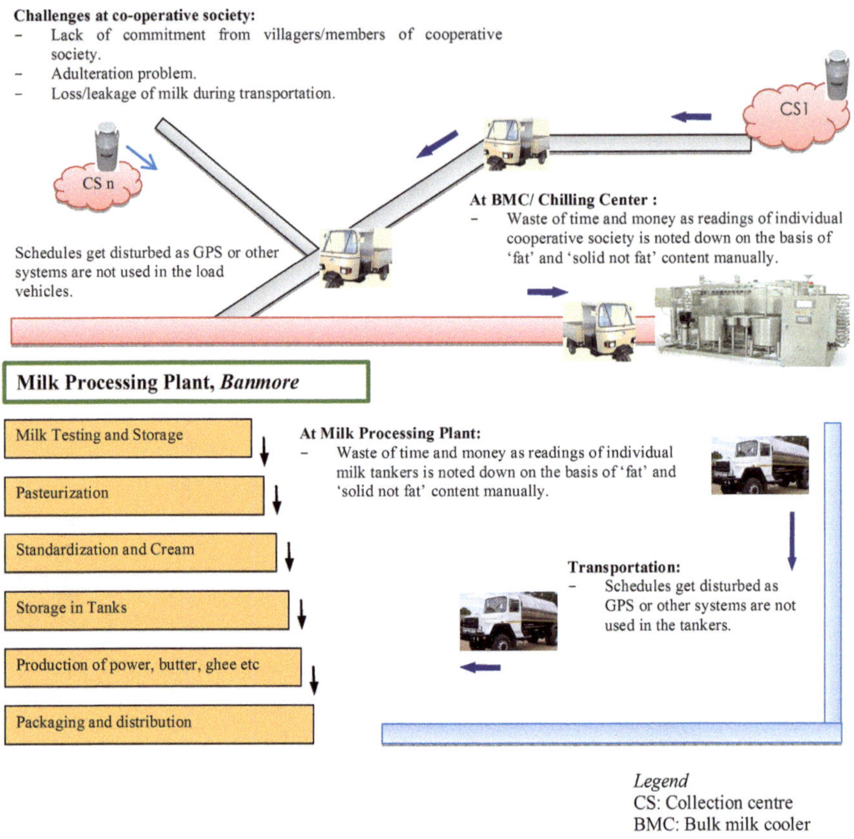

Fig. 2 Issues in milk procurement cycle of GSDS

comply with the quality, safety and hygiene standards. Various issues across these three types of supply chains are elaborated in Table 2.

Table 3 proposes various modes of linkages for collaborative alliances. These competing supply chains may collaborate with each other and co-create unique values for participant actors. However, the level of collaboration is very poor. In this, co-creation value is a bidirectional flow between the involved actors. Organized actors own higher installed capacity for milk processing which is currently underutilized. Dairy cooperatives and private dairy operators may collaborate for the development of sourcing of milk, development of transport network and utilization of each other's distribution channel. These organized actors may involve small dairy actors either by developing their milk processing capacity or converting them into an efficient sourcing hub.

Table 2 Issues across three types of dairy supply chains

S. no	Issues	DCS	Private	Halwaiis
1	Philosophy	Cooperative	Market opportunity	Market opportunity
2	Primary objective	Social	Market oriented	Sustenance
3	Business model	Make to stock	Make to stock	Make to stock, Build to order
4	Stages	Many	Many	Single or two
5	Market coverage	Whole market	Whole market	Local market
6	Structure	Organized	Semi(Organized)	Unorganized
7	Technology	Extensive use	Extensive use	Minimal or no use
8	Cold chain	Use cold chains	Use cold chains	Minimal or none
9	Partnerships	Formal partnerships	Formal partnerships	No formal partnership
10	Relationship	Collaborative, win-win	Arms length, control type	Opportunistic
11	Trust	High	Moderate	Moderate
12	Governance	Formal	Formal	Situational
13	Quality	High-quality control	Acceptable quality control	Poor-quality control
14	People, task and responsibilities	Presence of silos, poor accountability	Feeling of ownership and accountability	Owner's traits are the reflection of firm

Table 3 Various modes of linkages

S.no	Linkage	Description	Linkage state
1	Dairy cooperative society and large private actors	Sharing across all stages	Not present
2	Dairy cooperative society and small operators	Supply of milk by small operators, Processing know-how transfer to small operators	To some extent To some extent
3	Large private actors and small operators	Processing know-how transfer to small operators. Small operators as sourcing hub for organized actors	Not present To some extent

4.2 The Challenge of Food Processing

The food processing supply chain comprises of input suppliers, farmers, processors, logistic providers and consumers. Seasonality of production, perishability and variability of the raw material and products affect its performance parameters such as quality, delivery and price. These cause hindrances, which discourage the processing units to reach the optimum size and achieve economies of scale. Also, lack of consistent quality of raw materials hinders small-scale units to build brand equity

Table 4 Overview of the product and process characteristics and their impact

Product and process characteristics	Impact on operations
Overall	
–Shelf-life constraints for milk, intermediates and finished products and changes in product quality level while progressing the supply chain—recycling of materials required	–Timing constraints –Information requirements –Return flows
Growers/Producers	
–Long production times (producing new or additional products takes time) –Seasonality in production –Variability of quality and quantity of supply	–Responsiveness –Flexibility in process and planning
Food processing industry	
–High volume, low variety production systems –Highly sophisticated capital-intensive machinery leading to the need to maintain capacity utilization –Variable process yield in quantity and quality due to biological variations, seasonality, weather, pests, other biological hazards –A possible necessity to wait for the results of quality tests –Alternative installations, alternative recipes, product-dependent cleaning and processing times, carry-over of raw materials between successive product lots –Storage buffer capacity is restricted, when material, intermediates or finished –Products can only be kept in special tanks or containers –Necessity to value all parts because of the complementary nature of agricultural inputs Necessity for lot traceability of work in process due to quality and environmental requirements and product responsibility	–Importance of production planning and scheduling focusing on high capacity utilization –Flexibility of recipes –Timing constraints, ICT possibility to confine products –Flexible production Planning that can handle this complexity –Need for configurations that facilitate tracking and tracing
Auctions/Wholesalers/Retailers	
–Variability of quality and quantity of supply of farm-based inputs –Seasonal supply of products requires global (year-round) sourcing –Requirements for conditioned transportation and storage means	–Pricing issues –Timing constraints –Need for conditioning –Pre-information on quality status of products

(Adapted from: van der Vorst et al., 2007)

for themselves in international and domestic markets (Morgan 2007). The supply chain comprises functions, processes and associated flows and is geared towards value-added activities (Zhou and Benton 2007). A list of specific process and product characteristics of food supply chain networks (FSCNs) is summarized in Table 4. Each characteristic has an impact on the way the operational processes are organized. It is vital for industrial producers to contract suppliers to guarantee the supply of raw materials with the right volume, right quantity, right quality, at the right place and at the right time.

4.3 Traceability and Transparency in the Food Supply Chains

Tracking is the ability to follow the downstream path of a product along the supply chain (Dabbene et al. 2014), and traceability refers to access of product-related records in the upstream stages of supply chain (Bechini et al. 2008). Traceability involves ability to trace history in terms of recorded identification (International Standard Organization (ISO): 8402) or movement of items (Food and Agriculture Organization (FAO)/World Health Organization (WHO): 1997) and intends to capture data about each constituent item used (European Union 2002) or other measures of records under consideration (ISO 9000, ISO 22005). Traceability systems are formulated (ISO 22000: 2005) on the principles and design guidelines (ISO 22005:2007) for product identification (ISO 9001:2008) through data encoding on RFID devices and interoperable bar-coded devices (ISO/International Electro-technical Commission (IEC) 15962, 24,791, 15,459, 15,418 and 15,434). Aung and Chang (2014) define traceability in terms of what, how, where, why and when aspects of underlying product along a supply chain. In European Union (EU), the emphasis is on food hygiene (EU 2004) which is operationalized through European Food Safety Authority (EFSA) and Rapid Alert System for Food and Feed (RASFF). In the USA (US), regulation for complete traceability is enacted in 2005, whereas in India, Food Safety Standard (FSS 2011) was formulated to cover all types of dairy operators. However, in India, the traceability of food products is limited to track back lot number of processed product only. Traceability mechanisms are required in case of detection of unsafe food in the market or in situations of an outbreak of food-borne disease. Traceability helps in identifying root cause of the problem, taking timely corrective actions and development of robust system.

4.4 Complexity in Dairy Supply Chains

Milk being a natural product has in-build variability. Quality and composition of milk is highly depended on various external factors. The dairy supply chain like other food supply chain is a directed network of business processes with precedence relationships. Quality and safety of dairy products along a supply chain stage depend upon product and process quality at the preceding stage. The final acceptance of the product is by the consumers, which depends on the combination of three factors quality, price and safety. The dairy and food supply chain possesses

Table 5 Dairy product and process attributes

Types of special characteristics	Description
Product intrinsic attributes	1. Perishability and rapid deterioration in quality of milk and milk products with time and temperature. 2. Variability in milk and milk products due to •variation in fat and SNF content of cow and buffalo milk •variation due to difference in cattle feed, fodder, genetic composition, breed, etc. 3. Seasonality •lack of milk production in lean season due to lack of green fodder •lack of milk production due to lack of drinking water for the cattle in lean season •use of skimmed powder for production of milk and milk products in lean season •high production of milk and milk products in flush season
Product extrinsic attributes	1. Source and history of milk and other inputs such as fodder, vaccination, type of cattle or buffalo, storage, handling and transportation conditions, etc. 2. Processing technology and resources used in production process of dairy products
Process attributes	1. Mixing of raw milk collected from different sources before processing makes the traceability difficult. 2. Complex production scheduling due to sequential continuous and discrete production

special characteristics of products and processes, whose recording is essential for ensuring transparency (Trienekens et al. 2012). Table 5 elaborates special characteristics of a typical dairy supply chain. Attributes of dairy products and processes are grouped as intrinsic and extrinsic. This classification is in line with Trienekens et al. (2012) for the food products.

Different actors have their own perspectives regarding attributes of dairy products and processes, and their alignment in the dairy supply chain. Grunert et al. (2005) has defined four attributes of food products from the perspective of the end consumers, namely sensory attributes, health attributes, convenience attributes and process attributes. Attributers of dairy products along with their implications for dairy process management and associated impacts on consumers, government and dairy processors are elaborated in Table 6. Inherent complexities of the dairy supply chain present various forms of disturbances which impact operations along the underlying supply chain. Disturbances are characterized by elements such as frequency of occurrence, possibility of detection and impact on the dairy supply chain. Causes of disturbances are related to factors such as volume and quality, service, time, sustainability and cost. The underlying dairy supply chain is robust if it can withstand routine disturbances. The objective is the identification of various sources of vulnerability that explain various disturbances which affect the robustness and eventually increases the vulnerability of the underlying dairy supply chain.

Table 6 Dairy product attributes and their implications

Attributes	Demands from consumers and government	Implications
Intrinsic attributes		
Sensory attributes	–Taste, odour, colour and freshness	Various types of flavours and tastes are formulated to give sensory appeal to milk products.
Health attributes	–Infant and baby foods. –Less fat content milk products—such as toned milk and skimmed milk powders. –Milk-based health drinks and foods.	–Implementation of food standards and strict compliance and monitoring. –Hygienic and safe packaging with proper labelling –Milk and milk products are regarded as rich source of protein for infants, growing children, aged and vegetarian population.
Convenience attributes	–Ready-to-consume milk-based products such as yogurt, flavoured milk, cheese balls and slices, etc. –Traditional Indian dairy products and sweets such as shrikhand, lassi, chanch, kulfi, rasogolla, gulabjamun, etc.	–Implementation of food standards and strict compliance and monitoring. –Increase in sales of ready-to-use milk products Factory production of traditional Indian dairy products
Extrinsic attributes		
Process technology	–Energy conservation and use of energy-saving technology in chilling, storage, transportation and processing of milk. –Use of permitted pesticides and additives in cattle feed production. –Use of permitted levels of additives in processing of dairy products. –Animal welfare and care. Input materials (fertilizers, pesticides, etc.) –Disease prevention in milch animals –Biotechnology.	–Green production. –Implementation of food standards and scientific-based permitted levels of pesticides and additives in milk and milk products. –Legislations on animal welfare. –Cattle vaccination programmes and cattle disease-free zones
Impact on environment, people and society	–Environment, global warming, damage to ozone layer by increasing methane levels from cattle. Deforestation, soil erosion due to grazing. –Packaging material. –Labour rights, working conditions and safety. –Entrepreneurships and working conditions. –Dairy cooperative legislations.	–Sustainable dairy supply chains. –Reverse logistics for collection and recycling of packaging materials used in tetra packs, poly packs, glasses and aluminium foils.

(continued)

Table 6 (continued)

Attributes	Demands from consumers and government	Implications
Intrinsic attributes		
	–Community Social and economic dimensions of dairy farming.	
Supply and demand	–Mixed farming system. –Cultural preferences for milk and milk products. –Dairy cooperatives. –Reduction in non-tariff trade barriers and subsidies for free world trade of dairy products. –Fair trade practices, Obscure contracts.	–Corporate social responsibility (CSR) –Dairy cooperative reforms –Post-WTO regime and free global trade –Elimination of unethical trade practices. –Increasing consumer demand for health benefit-rich products.

4.5 Performance for Dairy Supply Chains

Various key performance indicators (KPIs) belonging to four dimensions financial, customer, internal business processes and learning and growth have been derived though exploratory research and discussions with the domain experts. Table 7 depicts these KPIs. These KPIs have been validated through survey of a select group of respondents associated with dairy supply chain, namely dairy farmers, milk collectors, processors, distributors as well as retailers and the consumers. The respondents were asked to rate the identified KPIs on a scale of 1 (not important) to 5 (very important) to find out their ranking. The respondents were also asked to rate each BSC perspective by assigning a score ranging from 0 to 10. This approach is in line with Bigliardi and Eleonora (2010).

4.6 Achieving Sustainability

Sustainability challenges arise owing to causes such as ineffective utilization of resources, poor logistics, inadequate process management, poor methods of dairy farming, challenges arising due to properties of milk and poor product management. Figure 3 depicts a fishbone diagram that represents difficulties. High wastages, continuously increasing quality norms, difficulties in use of resources to achieve broad social needs and difficulty in balancing social–economic welfare contribute towards ineffective utilization of resources. Increased fuel consumption due to high distance travel, inadequate green logistics options and variable cost structure of transportation modes contribute towards poor logistics. Poor flexibility of processes, poor responsiveness in terms of lead time and delivery, poor traceability of

Table 7 KPIs for measurement of performance (based on priority) (N = 307)

Internal process perspective (Importance rank:1) *(To satisfy our shareholders and customers, what business processes must we excel at?)*
Implementation of HACCP and other quality control measures
Use of clean technologies
Product development cycle for new value-added products
Use of IT and ERP systems for real-time sharing of data
Capacity utilization of plant
Product variety and range
Factory production of indigenous dairy products
Quality check and traceability of incoming milk and raw materials
Backward integration of processing units in milk procurement and production
Customer perspective (Importance rank:2) *(To achieve our vision, how should we appear to the consumers?)*
Compliance to national and international food quality and codex standards
Quality of delivery goods
Distribution network
Post-transaction measures of customer service
Supply chain integration
Customer perceived value of product
Flexibility of service system to meet particular customer needs
Financial perspective (Importance rank:3) *(To succeed financially, how should we appear to the consumers?)*
Net price of products
Transportation and distribution cost
Total supply chain cost
Net sales
Terms of payment
Return on investment
Supplier cost-saving activities
Learning and growth perspective (Importance Rank:4) *(To achieve our vision, how will we sustain our ability to change and improve?)*
Strong research and development (R&D)
Innovations
Public–private partnership (PPP) model in infrastructure development and cattle extension services
Supply chain robustness
Sustainability of operations

(continued)

Table 7 (continued)

Supply chain collaboration
Training and human resource development (HRD)
Transparency in supply chain
Corporate social responsibility (CSR)

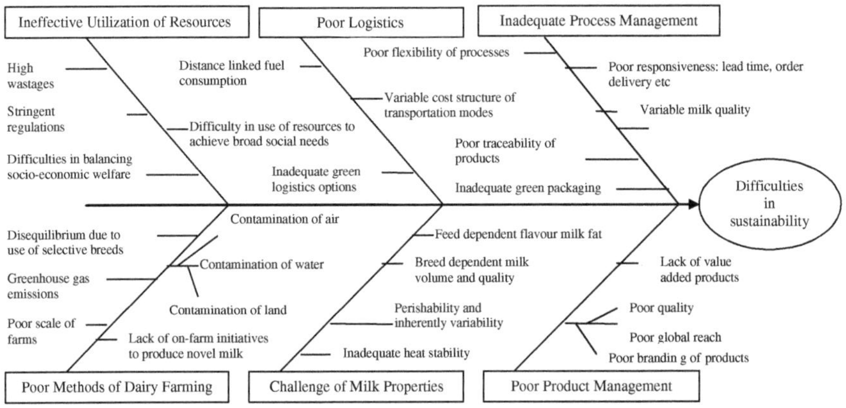

Fig. 3 Cause and effect diagram depicting various difficulties in achieving sustainability

products and inadequate green packing contribute towards inadequate process management. Greenhouse gas emissions, poor scale of farms, lack of on-farm initiatives to produce novel milk, disequilibrium arising due to selective use of animal breeds and contamination of air, water and land contribute towards poor methods of existing dairy farming. Feed-dependent flavour and milk fat, breed-dependent volume and quality of milk, perishable and variable nature of milk and inadequate heat stability contribute towards challenges of milk properties. Lack of value-added milk products, poor quality of products, poor global reach and poor branding of products contribute towards poor product management.

5 Conclusions

5.1 Implication for Theory

This paper explores the dairy supply chains from the perspective of supply chain management, and in doing so, it integrates operations and marketing processes for enhanced level of value creation. This approach is in line with Karmakar (1996) and Malhotra and Sharma (2002). To the best of author’s knowledge, this would be first of a kind study exploring sustainability for dairy operations in an Indian context.

5.2 Implication for Practise

Based on ensuing findings, a comprehensive framework has been proposed for sustainable growth of the Indian dairy sector. The proposed framework is anchored on public–private partnership (PPP) for inclusive and sustainable growth. Figure 4 depicts the proposed framework. The constituents of the framework are global marketing and brand development of indigenous dairy products, research and development for improvement in products, processes and development of clean technologies, infrastructure development to support operations, utilization of resources, nurturing of entrepreneurship, quality control and policy support. A similar approach has been adopted by Manzini and Accorsi (2013) for assessment of logistics of food products. Long-term sustainability of the dairy industry will be based on economic models which generates higher pay-offs through increased access to foreign markets coupled with continuous value addition through technological know-how-driven innovations across the supply chain (Augustin et al. 2013). Performance of sustainable supply chain requires inclusion of non-efficiency-based indicators (Shepherd and Gunter 2006) as inclusion of only efficiency-based indicators can produce misleading picture (Chen and Paulraj 2004). Bourlakis et al. (2014) has derived efficiency, flexibility, responsiveness and product quality as indicators of sustainable dairy supply chains. Indian dairy producers should focus primarily on milk-deficit markets in Southeast Asia, Middle East and Africa for export of milk and milk products; however, there is competition from European Union, Australia and New Zealand. Traditional Indian dairy products and sweets can do what Italian and Chinese food products have done for their country. There is a need for global marketing of the Indian dairy products and sweets. Commercialization of the Indian ethnic dairy products requires development of efficient mechanized systems and low-cost packaging solutions to increase their shelf life and make them price competitive. The public and private sectors have to put more investments in research for development of new dairy products for domestic and global market. Research is required to develop efficient processing capabilities with the aim of reducing environmental impact. Balanced and nutritious cattle feed is essential for increasing productivity of milch animals. There is need to increase bulk cooling and storage capacity of milk at collection centres to prevent wastage and preservation of milk quality for longer duration. To enhance utilization of resources, governments (both at central and state level) and private firms need to develop time-bound programs for the improvement in productivity of local breed of the Indian cows by expanding coverage of artificial insemination and systematic genetic improvement programs by supply of semen of evaluated bulls to various milk sheds. There is a need for the formation of cooperatives of small dairy processors (CSDP) on the lines of existing dairy cooperative societies (DCS) by unorganized small dairy units and halwaiis. This approach is similar to collective action and public–private partnership employed by the cooperatives of small grape growers in Maharashtra state of India and their private marketing partner Mahagrapes for accessing the global export market meeting the stringent food

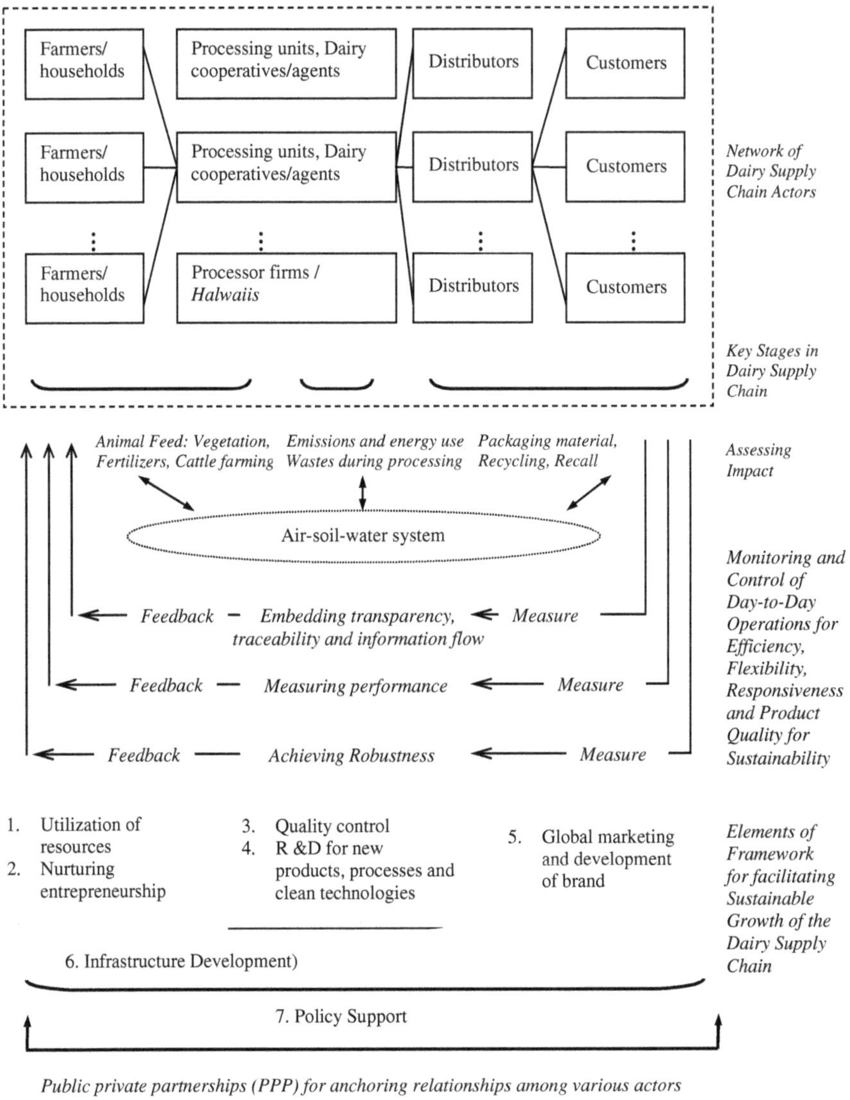

Fig. 4 Framework for sustainable growth of the Indian dairy supply chain

standard requirements (Narrod et al. 2009). To arrest adulteration and enhance quality control of milk and milk products, government has recently introduced FSS (2011). The aim of the new FSS is to make Indian Hazard Analysis and Critical Control Points (HACCP) compliant for milk and milk products. Taxes on dairy products in India are among the highest in the world. In addition there is multiplicity of taxes on dairy products in different states ranging from Octroi, sales tax, collection tax, custom and exercise duties on equipment and plant machinery etc., there is a need to reduce these taxes to promote dairy sector.

5.3 Implication for Social Sustainability

Livestock and livelihoods are closely related in India and the ownership of livestock is much more egalitarian. Successful growth of dairy sector in India in last four decades has made dairy farming an important source of income and employment generation in the rural areas in India. Similar developments have been happening in other developing economies of Asia. Poverty is a major cause of food insecurity, during global food crises livestock production has played an important role in economy, both at the rural and the national level. Milk being a near complete food, is a good source of nutrition to the rural poor in the world. Also it is main source of protein to the majority of vegetarian population of India.

5.4 Scope for Future Study

The study may be taken as a foundation for future studies on modelling and simulation of various issues pertaining in the dairy supply chain. Design of a facility layout for collection and movement of milk products at the processor may be an interesting area of work. Network planning for procurement of milk could be another theme of study. Development of a distributed information system for linking of various actors of the dairy supply chain may also be investigated.

References

Abad E, Palacio F, Nuin M, Gonzalez de Zarate A, Jurros A, Gomez JM, Marco S (2009) RFID smart tag for traceability and cold chain monitoring of foods: demonstration in an intercontinental fresh fish logistic chain. J Food Eng 93(4):394–399

Ahumada O, Villalobos JR (2009) Application of planning models in the agri-food supply chain: a review. Eur J Oper Res 195(1):1–20

Alasalvar C, Grogor J, Zhang D, Quantick P, Shahidi P (2001) Comparison of volatiles, phenolics, sugars, antioxidant vitamins, and sensory quality of different colored carrot varieties. J Agri Food Chem 49:1410–1416

Angell LC, Klassen RD (1999) Integrating environmental issues into the mainstream: an agenda for research in operations management. J Oper Manage 17(5):575–598

Augustin MA, Udabage P, Juliano P, Clarke PT (2013) Towards a more sustainable dairy industry: Integration across the farm-factory interface and the dairy factory of the future. Int Dairy J 31:2–11

Aung MM, Chang YS (2014) Traceability in a food supply chain: safety and quality perspectives. Food Control 39:172–184

Barratt M, Choi TY, Li M (2011) Qualitative case studies in operation management: trends, research outcomes, and future research implications. J Oper Manage 29:329–342

Bechini A, Cimino M, Marcelloni F, Tomasi A (2008) Patterns and technologies for enabling supply chain traceability through collaborative e-business. Inf Softw Technol 50(4):342–359

Bhamra R, Dani S, Bunard K (2011) Resilience: the concept, a literature review and future directions. Int J Prod Econ 49(18):5375–5393
Bigliardi B, Bottani E (2010) Performance measurement in the food supply chain: a balanced score card approach. Facilities 28(5/6):249–260
Bourlakis M, Maglaras G, Gallear D, Fotopoulos C (2014) Examining sustainability performance in the supply chain: the case of the Greek dairy sector. Ind Mark Manage 43:56–66
Chand S, Saraiya A, Sridhar V (2010) Public private partnership in Indian dairy industry. Available at: www.technopak.com. (Accessed 17 June 2014)
Chen IJ, Paulraj A (2004) Towards a theory of supply chain management: the constructs and measurements. J Oper Manage 22(2):119–150
Chung-Chi H, Cheng-Han W (2008) Capacity allocation ordering, and pricing decisions in a supply chain with demand and supply chain uncertainties. Eur J Oper Res 184:667–684
Coley DA, Howard M, Winter M (2009) Local food, food miles and carbon emissions: A comparison of farm shop and mass distribution approaches. Food Policy 34(2):150–155
Dabbene F, Gay P, Tortia C (2014) Traceability issues in food supply chain management: A review. Biosyst Eng 120:65–80
Dalziell EP, McManus ST (2004) Resilience, vulnerability, and adaptive capacity: implications for system performance. University of Canterbury, Christchurch
Deimel M, Frentrup L, Theuvsen L (2008) Transparency in food supply chains: empirical results from German pig and dairy production. J Chain Network Sci 8(1):21–32
Dharni K, Sharma S (2008) Food processing in India: opportunities and constraints. The ICFAI Univ J Agri Econ 5(3):30–38
Dong M (2006) Development of supply chain network robustness index. Int J Serv Oper Info 1 (1/2):54–66
Dube L, Pare G (2003) Rigor in information systems positivist case research: current practices, trends and recommendations. MIS Q 27(4):597–635
Economic Survey (2013) Economic Survey 2012-13. Available at: www.indiabudget.nic.in. Accessed 10 March 2013
FSS (2011) Food safety and standards (Licensing and Registration of Food Businesses), regulations 2011. Food Safety and Standards Authority of India, New Delhi
Gallpoin GC (2006) Linkage between, vulnerability, resilience and adaptive capacity. Global Eng Change 16(3):293–303
Georgiadis P, Vlachos D, Iakovu E (2005) A system dynamics modelling framework for the strategic supply chain management of food chains. J Food Eng 70:351–364
Gerbens-Leenes PW, Moll HC, Uiterkamp AJMS (2003) Design and development of a measuring method for environmental sustainability in food production systems. Ecol Econ 42(2):231–248
GoI (2005) Strategy and action plan for food processing industries in India, Ministry of Food Processing Industries, Government of India, New Delhi
Grunert K, Fruensgaard JL, Risom JK, Sonne A (2005) Market orientation of value chains; a conceptual framework based on four case studies from the food industry, Eur J Mark 139 (5/6):429–455
Handfield RB, Walton SV, Seegers LK, Melnyk SA (1997) Green' value chain practices in the furniture industry. J Oper Manage 15(4):293–315
Hobbs J, Young LM (2000) Closer vertical co-ordination in agri-food supply chains: a conceptual framework and some preliminary evidence. Supply Chain Manag 5(3):131–139
Hofstedel CJ, Schepers H, Spaans-Dijkstra L, Trienekens J, Beulens A (2005) Hide or confide: the dilemma of transparency. Reed Bus Inf BV's, Gravenhage
IBEF (2010) Food processing, Ernst & Young. Available at: www.ibef.org. Accessed 12 April 2012
IBEF (2012) Food processing, Aranca. Available at: www.ibef.org. Accessed 25 August 2012

Kaplan RS, Norton DP (1992) The balanced scorecard-measures that derive performance. Harvard Bus Rev 70(1):71–79
Karmakar AS (1996) Integrative research in marketing and operations. J Mark Res 33(2):125–133
Ketikidis PH, Koh SCL, Dimitriadis N, Gunasekaran A, Kehajova M (2008) The use of information systems for logistics and supply chain management in South East Europe: current status and future direction. Omega 36:592–599
Ketokivi M, Choi T (2014) Renaissance of case research as a scientific method. J Oper Manage 32:232–240
Lai KH, Ngai EW, Cheng TC (2002) Measures for evaluating supply chain performance in transport logistics. Transport Res Part E l 38(6):439–456
Llbery B, Maye D (2005) Food supply chains and sustainability: evidence from specialist food producers in the Scottish/English borders. Land Use Policy 22(4):331–344
Malhotra MK, Sharma S (2002) Spanning the continuum between marketing and operations. J Oper Manage 20(3):209–219
Maloni MJ, Brown ME (2006) Corporate social responsibility in the supply chain: an application in the food industry. J Bus Ethics 68(1):35–52
Manzini R, Accorsi R (2013) The new conceptual framework for food supply chain assessment. J Food Eng 115:251–263
Matopoulos A, Vlachopoulou V, Manthou V, Manos B (2007) A conceptual framework for supply chain collaboration. Supply Chain Manage Int J 12(3):177–186
Meredith JR (1998) Building operations management theory through case and field research. J Oper Manage 16(4):439–452
Minegishi S, Thiel D (2000) System dynamics modelling and simulation of particular food supply chain. Simul Pract Theor 8(5):321–339
Ministry of Food Processing Industries (MoFPI), Government of India (2009) Annual Report (2008–09). Available at: www.mofpi.nic.in. Accessed 15 October 2014
Morgan TG (2007) Guide to good dairy farming practice. A joint publication of the International Dairy Federation and the FAO of the UN
Narrod C, Roy D, Okello J, Avendano B, Rich K, Throat A (2009) Public-private partnerships and collective action in high value fruit and vegetable supply chains. Food Policy 34:8–15
Ouyang Y, Daganzo C (2008) Robust tests for the bullwhip effect in supply chains with stochastic dynamics. Eur J Oper Res 185:340–353
Pagell M (2004) Understanding the factors that enable and inhibit the integration of operations, purchasing and logistics. J Oper Manage 22(5):459–487
Patil AP, Gawande SH, Nande MP, Gobade MR (2009) Constraints faced by the dairy farmers in Nagpur district while adopting animal management practices. Vet World 2(3):111–112
Reardon T, Codron J, Harris LB (2001) Global changes in agrifood grades and standards: agribusiness strategic responses to developing countries. Int Food Agribusiness Manag Rev 2 (3/4):421–435
Ruteri JM, Xu Q (2009) Supply chain management and challenges facing the food industry sector in Tanzania. Int J Bus Manage 4(12):70–80
Sarc A, Absi N, Dauzere-Peres S (2010) A literature review on the impact of RFID technologies on supply chain management. Int J Prod Econ 128:77–95
Sarkis J (2001) Manufacturing's role in corporate environmental sustainability. Int J Oper Prod Manage 21(5/6):666–686
Shepherd C, Gunter H (2006) Measuring supply chain performance: current research and future directions. Int J Prod Perform Manage 55(3/4):242–258
Trienekens JH, Wognum PM, Beulens AJM, van der Vorst JGAJ (2012) Transparency in complex dynamic food supply chains. Adv Eng Inform 26:55–65
Van der Vorst JGAJ, Dijk SJ, van Beulens AJM (2007) Leagile supply chain design in food industry; an inflexible poultry supply chain with high demand uncertainty. Int J Logistics Manage 12(2):73–85

Venugopal P (2008) Retail environment and economic development. Effect Executive, IUP, Hyderabad

Vlajic JV, van der Vorst JGAJ, Haijema R (2012) A framework for designing robust food supply chains. International Journal of Production Economics 137(1):176–189

Wagner SM, Bode C (2006) An empirical investigation into supply chain vulnerability. J Purchasing Supply Manage 12:301–312

Wagner BA, Macbeth DK, Boddy D (2002) Improving supply chain relations: an empirical case study. Supply Chain Manage Int J 7(4):253–264

Wee HM, Yang PC (2004) The optimal and heuristic solutions of a distribution network. Eur J Oper Res 158:626–632

Yin RK (1994) Case study research: design and methods, 2nd edn. Sage Publ, Newbury Park, CA

Zhang J, Liu L, Mu W, Monga LM, Zhang X (2009) Development of temperature managed traceability system of frozen and chilled food during storage and transportation. J Agri Environ 7(3/4):28–31

Zhou H, Benton WJ (2007) Supply chain practice and information sharing. J Oper Manage 25 (6):93–100

Waste Reduction in Meat Processing Industry: The Application of MFCA (ISO 14051)

Watcharin Chaiwan, Chawis Boonmee and Chompoonoot Kasemset

Abstract This chapter presents application of material flow cost accounting (MFCA) analysis technique in meatball production. MFCA was used to analyze 6 processes that focused on the cost of positive and negative product occurred along the target production line. The results of MFCA analysis for existing process showed that the main negative product cost was material cost (MC). Thus, the improvement solution was proposed to reduce the cost of negative product of MC. The main material waste occurred at the first process of raw material preparation when melted ice occurred during the temperature reduction during raw material preparation step. To reduce this wastewater from melted ice, the proposed solution was to add a refrigerator instead of using ice for temperature reduction. By this method, the results from MFCA analysis showed that negative material quantity at raw material preparation step was reduced from 47.77 to 3.62 % that gave effects on negative MC reduction from 4.98 to 4.20 %. Consequently, this solution can help in saving production cost as 67,406.67 Baht per year when considering only this product.

Keywords Material flow cost accounting (MFCA) · Meatball production · Case study

1 Introduction

Material flow cost accounting (MFCA), a method of environmental management accounting, was developed in Germany. Along with study on the MFCA's approach and its effectiveness, MFCA has been introduced into many industries. MFCA is

---

Approximately 32.74 Baht = 1USD as of January, 2015.

W. Chaiwan · C. Boonmee · C. Kasemset (✉)
Department of Industrial Engineering, Chiang Mai University, Chiang Mai, Thailand
e-mail: chompoonoot.kasemset@cmu.ac.th

V. Kachitvichyanukul et al. (eds.), *Toward Sustainable Operations of Supply Chain and Logistics Systems*, EcoProduction,
DOI 10.1007/978-3-319-19006-8_12

being highly appraised and rapidly disseminated as a powerful method to simultaneously realize "reduced environmental impacts" and "improved business efficiency" by increasing transparency of material losses (Environmental Industries Office (Japan) 2011).

The company case study is one food industry located at Chiang Mai, Thailand. The selected product is one size of meatball product. The studied process of the meatball consists of six processes. In each process, a lot of waste such as food scraps and wastewater are released. Thus, MFCA is applied for classifying costs of both positive and negative products. At the process with the highest negative product cost, the improvement solution is provided in order to reduce the negative product cost that can be implied to the reduction of wastes for this process.

2 Literature Reviews

The original concept of MFCA is based on the work of the German "Institut für Management und Umwelt" (institute for management and environment) in the late 1990s which initialized few pilot projects in the German industry (Schmidt et al. 2014). However, MFCA's breakthrough had been reached in Japan. Due to the great success of first implementation in the year 2000, MFCA was strongly promoted thereafter and more than 300 Japanese companies adopted the method by now (Kokubu et al. 2004).

MFCA was one technique that can help a company enhance both its environmental and financial performance, through the analysis of material flows and energy usage (Chompu-inwai et al. 2013). MFCA focused on both the costs of products and the costs associated with material losses (Kokubu and Nakajima 2004). This technique was used to analyze positive and negative product cost and identify opportunities to reduce material usage and losses to improve the efficiency of materials and energy usage and to reduce adverse environmental impacts. The cost of positive product was the cost of quantity of materials made into company products, while the cost of negative product was the cost from overall materials wasted.

The cost of positive and negative product can be separated into 4 parts: material cost (MC), system cost (SC), energy cost (EC), and waste cost (WC). The negative costs of all parts were considered to be reduced at the improvement state.

Despite minor differences in terminology and the system boundaries used, the process of implementing MFCA has been consistently represented in extant literature. For ease of understanding, a flowchart displaying the main steps involved in this process is provided in Fig. 1.

MFCA was applied in many companies for waste reduction. The application of MFCA and design of experiment (DOE) concepts was proposed by Chompu-inwai et al. (2013) to reduce material consumption and minimize waste at wood product company. The results were the reduction of wood material losses in the cutting process that reflected to a proportion of total wood materials used from approximately 69–54 %. Kasemset et al. (2013) proposed the application of MFCA

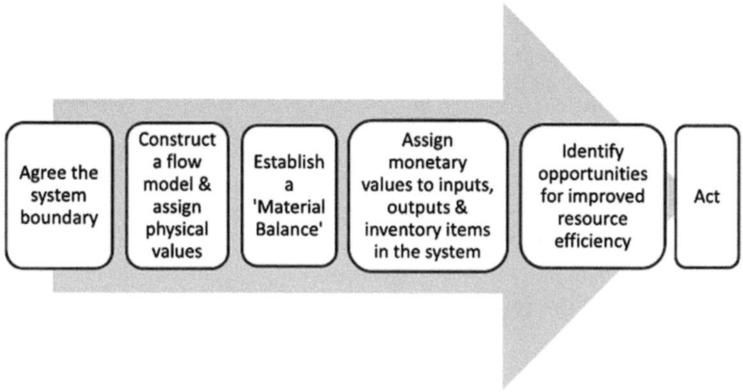

Fig. 1 The MFCA process (Christ et al. 2014)

analysis in process improvement. The production line of plastics packaging was analyzed, and the results showed that the highest negative product cost occurred at blow molding process. Then, the operations' flow at blow molding process was analyzed using motion study and ECRS concept in order to eliminate production defects. The improvement solution showed that the defects were reduced to 26.07 % from previous negative product cost.

In textile factory, there were also a lot of material losses generated from a cutting process. MFCA was proposed to one small textile factory in Thailand to identify the material losses in the company's production line. Improvement solutions were evaluated based on MFCA concept and proposed to the company for implementation (Kasemset et al. 2014).

The advantage of MFCA technique was to help in clearly identifying production waste in term of cost, and this can help management person to understand and realize on the effect of production waste to the company's cost.

3 Research Methodology

The research methodology was proposed based on 7 steps of MFCA implementation as follows.

1. Data Collection and Preparation
   During this step, target product/process was selected and the data related to this product/process were also collected including MC, EC, wage, SCs, and waste management cost.
2. Material Flow Process Chart Creation
   Material flow process chart was created using the data from the previous step to show input material and waste of each process along the target process.

3. MFCA Calculation for the existing process
   During this step, the costs of positive and negative products were identified using mass balance method and costs were also presented as 4 types: MC, SC, EC, and WC.
4. Problems Finding
   Pareto chart was used to analyze the locations of negative product cost. Then, problem analysis tools as cause and effect diagram and why-why analysis were used to identify possible causes of negative product cost.
5. Improvement Planning
   During this step, the improvement procedure was designed and planned for implementation.
6. Evaluating Improvement Plan by MFCA analysis
   The improvement solution was evaluated by re-analyzing MFCA. The results from MFCA calculation were finally compared with the existing situation.
7. Conclusion and Implementation
   The improvement solution was proposed to the company to plan for the implementation.

4 Case Study and Results

The case study company was one of the food industries located at Chiang Mai province, Thailand. The target product/process was set as one size of meatball products. The results of the study can be presented as follows.

A. ***Basic Process Data and Material Flow Process Chart***

The process of case study consisted of 6 processes: raw material preparation, emulsion, forming, cooking, temperature reduction, and packing. The Material flow process chart presented input and output of each process, including main material, sub-material, and auxiliary material, as shown in Fig. 2.

B. ***MFCA Calculation for Existing Process***

The concept of mass balance was used to allocate quantity of positive and negative material. Then, to allocate positive and negative product costs of each SC, EC, and WC were carried out using the percentage from mass balance (shown in Table 1). The percent allocation was also used to distribute positive and negative cost for each process and presented as MFCA diagram (showed in Fig. 3). For example, at raw material preparation process, there were input materials as meat, lard, and ice as total quantity as 15,280.00 g. The total material input at this process has become only 7980 g (52.23 %) that was positive product sent to the next process and wasted material 7300 g (47.77 %). Thus, mass balance concept was applied 52.23 and 47.77 % to separate the cost of positive and negative product for SC and EC at this process, respectively. For WC, this cost was only included in the cost of negative product.

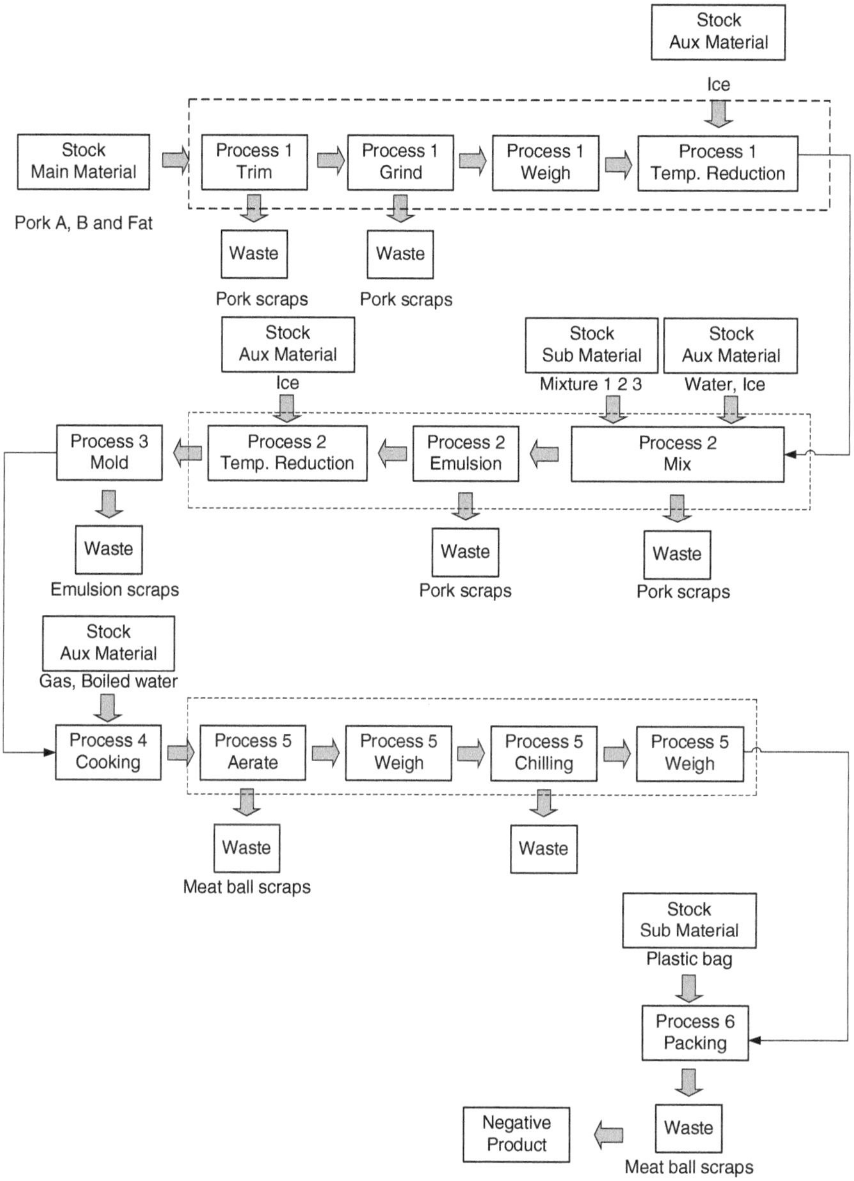

Fig. 2 Material flow process chart of study process

The conclusion of MFCA analysis for current production line of meatball was presented in Table 2. The total cost of meatball production was 1378.36 Baht consisted of MC as 1025.33 Baht, SC as 108.36 Baht, EC as 243.13 Baht, and WC as 1.55 Baht. The total cost can be as the cost of positive and negative product as

Table 1 Percentage used in cost allocation from mass balance and MC of each process

Process	Positive material cost	Negative material cost	Total
1. Raw material preparation	7980.00 g	7300.00 g	15,280.00 g
	(52.23 %)	(47.77 %)	(100.00 %)
	820.80 Baht	43.00 Baht	863.80 Baht
2. Mincing and mixing	11,150.00 g	400.00 g	11,550.00 g
	(96.54 %)	(3.46 %)	(100.00 %)
	882.58 Baht	36.76 Baht	919.34 Baht
3. Forming	11,103.00 g	47.00 g	11,150.00 g
	(99.58 %)	(0.42 %)	(100.00 %)
	878.86 Baht	3.72 Baht	882.58 Baht
4. Cooking	12,091.76 g	511.24 g	12,603.00 g
	(95.94 %)	(4.06 %)	(100.00 %)
	887.86 Baht	5.89 Baht	893.75Baht
5. Temperature reduction	11,354.43 g	474.51 g	11,828.94 g
	(96.00 %)	(4.00 %)	(100.00 %)
	852.34 Baht	35. 62 Baht	887.96 Baht
6. Packing	11,336.73 g	41.70 g	11,378.43 g
	(99.63 %)	(0.37 %)	(100.00 %)
	897.04 Baht	3.30 Baht	900.34 Baht

1212.78 and 165.58 Baht, respectively. Considering the cost of negative product, the largest portion was identified as MC as 77.48 % from the total cost of negative product as presented in Fig. 4.

C. ***Problem Finding***

As shown in Fig. 4, the highest value of negative product cost was identified as MC. The Pareto diagram was developed to show the source of material waste in meatball production line as in Fig. 5. From the chart, the first process, raw material preparation, was the place that generated the biggest portion of material waste as 33.52 % from the total material waste.

From the observation at raw material preparation process, there were two types of material wastes that were water from melted ice and meat scraps. For meat scraps, it can be reused as raw material again or sold to customers. Thus, the main problem was identified as wastewater from melted ice.

Cost item		1. Raw Mat. Prep.	2. Emulsion	3. Forming	4. Cooking	5. Temp. Reduction	6. Packing
Newly input total cost							
	Total	885.16	160.82	5.11	176.16	70.01	79.56
	Newly input MC	863.8	98.54	0	15	0	48
	Newly input SC	16.672	43.53	4.64	11.16	2.043	30.31
	Newly input EC	4.69	18.75	0.47	150.00	67.97	1.25
Total cost handed over from previous process							
previous process	**Total**	0.00	831.96	953.48	954.55	1,115.19	1,137.69
	MC from previous process	0.00	820.80	882.58	878.86	887.96	852.34
	SC from previous process	0.00	8.71	50.43	54.84	63.32	62.75
	EC from previous process	0.00	2.45	20.47	20.85	163.91	222.61
Process total of input cost							
	Total	885.16	992.78	958.59	1,130.71	1,185.20	1,217.25
	Input MC	863.8	919.34	882.58	893.86	887.96	900.34
	Input SC	16.672	52.24	55.07	66.00	65.36	93.06
	Input EC	4.69	21.20	20.94	170.85	231.88	223.86
Percent quanlity of negative product		47.77%	3.46%	0.42%	4.06%	4.00%	0.37%
Percent quanlity of positive product		52.23%	96.54%	99.58%	95.94%	96.00%	99.63%
Positive product cost total							
	Total	831.96	953.48	954.55	1,115.19	1,137.69	1,212.78
	Positive product MC	820.80	882.58	878.86	887.96	852.34	897.04
	Positive product SC	8.71	50.43	54.84	63.32	62.75	92.71
	Positive product EC	2.45	20.47	20.85	163.91	222.61	223.03
Negative product cost							
	Total	53.95	39.32	4.04	15.55	47.56	5.15
	Negative product MC	43.00	36.76	3.72	5.89	35.62	3.3
	Negative product SC	7.96	1.81	0.23	2.68	2.61	0.34
	Negative product EC	2.24	0.73	0.09	6.94	9.28	0.83
	Waste treatment cost	0.75	0.02	0	0.048	0.048	0.68
Sales of byproducts and recycled meterials	Selling prices	0	0	0	0	0	0

Fig. 3 MFCA analysis

Table 2 Cost allocation of meatball production line before improvement

	MC	SC	EC	WC	Total
Total cost	1025.33	108.36	243.13	1.55	1378.36
	74.39 %	7.86 %	17.64 %	0.11 %	100.00 %
Positive cost	897.04	92.71	223.03	–	1212.78
	73.97 %	7.64 %	18.39 %	–	100.00 %
Negative cost	128.29	15.64	20.10	1.55	165.58
	77.48 %	9.45 %	12.14 %	0.93 %	100.00 %

Cost unit Baht

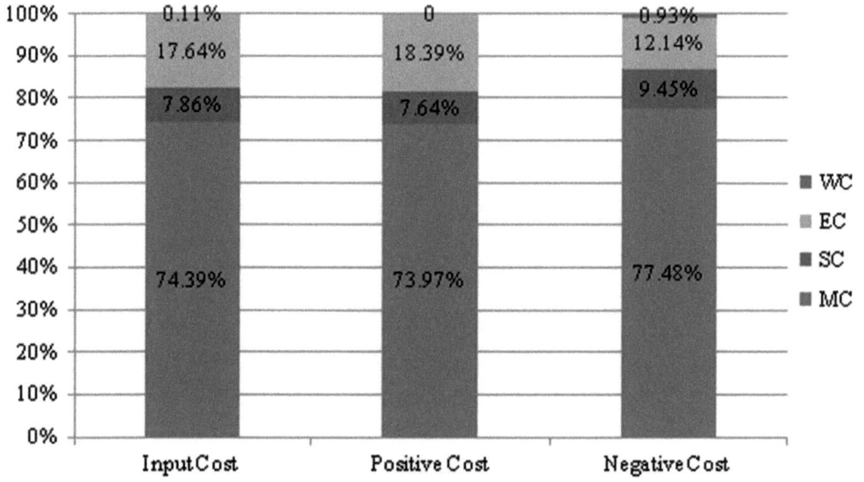

Fig. 4 Cost diagram of meatball production line

D. ***Improvement Planning***

Ice was used during preparation step for cooling meat. Wastewater from melted ice was too dirty to be reused in any activity, and it was drained into wastewater treatment system that increased the cost of waste treatment and directly played a part of the company's cost, as well. From data collection, 7000 g of ice was used per one production batch, and this entire amount has become to wastewater. To get rid of this material waste, a refrigerator was introduced to the preparation step. Using a refrigerator for cooling meat helped in reducing material loss from melted ice without the investment because the company already has its own refrigerator.

E. ***Evaluating Improvement Plan by MFCA Analysis***

The improvement solution, using refrigerator instead of ice in preparing step, was evaluated using MFCA again, and the results are listed in Table 3.

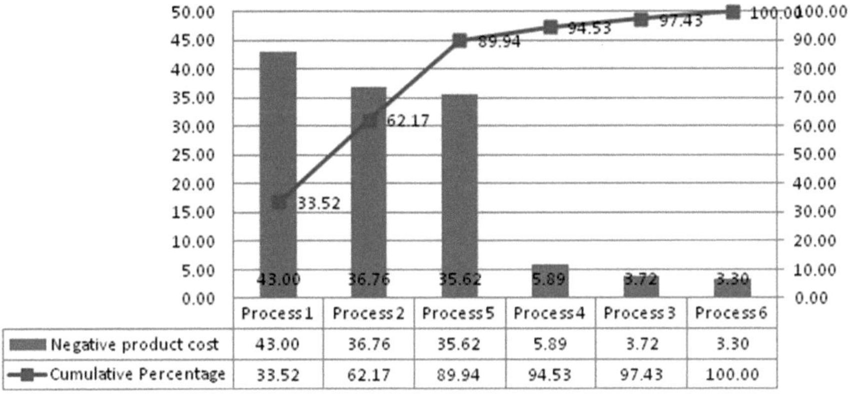

Fig. 5 Pareto diagram of meatball production

Table 3 Cost of meatball production after improvement

	MC	SC	EC	WC	Total
Input cost	1019.00	108.36	299.38	1.48	1428.21
	71.35 %	7.59 %	20.96 %	0.10 %	100.00 %
Positive cost	897.71	99.21	272.68	–	1269.60
	70.71 %	7.81 %	21.48 %	–	100.00 %
Negative cost	121.29	9.15	26.70	1.48	158.61
	76.47 %	5.77 %	16.83 %	0.93 %	100.00 %

Cost unit Baht

Comparing between Tables 2 and 3, the total cost was increased from 1378.36 to 1428.21 Baht. Although MC was reduced from 1025.33 to 1019.00 Baht, saving ice's cost, EC was increased from 243.13 to 299.38 Baht due to increasing of used electricity for the refrigerator.

From Fig. 6, after improvement, cost of input and negative products of MC and SC was decreased. For EC, it was increased due to adding the refrigerator to the production line. For positive product cost, SC and EC were increased, but MC was decreased due to the reduction in input cost of ice. The reduction of ice cost was a small part comparing with other raw MCs. When input cost was reduced in small value at the same time as the main waste of MC remaining approximately the same, the percentage of positive product cost of MC was relatively reduced.

Table 4 showed results comparison of material quantity and product cost at raw material preparation step. From this table, negative material quantity can be reduced at this process from 47.77 to 3.62 % that gave effects on negative MC reduction from 4.98 to 4.20 %.

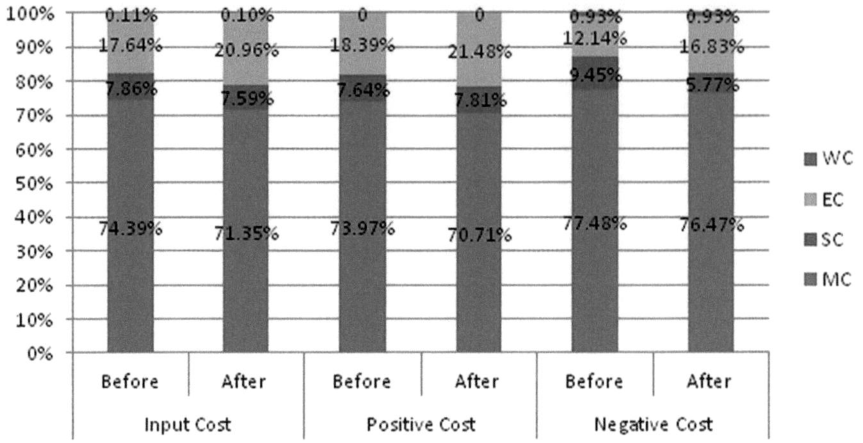

Fig. 6 Cost allocation comparison

Table 4 Cost of meatball production after improvement

MC at raw mat. prep.		Input	Positive	Negative
Material quantity (g)	Before	15,280	7980	7300
		100.00 %	52.23 %	47.77 %
	After	8280[a]	7980	300
		100.00 %	96.38 %	3.62 %
Product cost (Baht)	Before	863.80	820.80	43.00
		100.00 %	95.02 %	4.98 %
	After	856.80[a]	820.80	36.00
		100.00 %	95.80 %	4.20 %

[a]Cost of ice was 7 Baht for 7000 g
Cost unit Baht

5 Conclusion and Discussion

This study presented application of MFCA analysis technique in meatball production. MFCA was used to analyze 6 processes that focused on the cost of positive and negative products. The results showed that the largest cost of negative product was identified as MC as 77.48 % from the total negative product cost. Thus, the improvement solution was proposed to reduce the negative product cost of MC. Pareto chart was used to find the root cause, and the first process of raw material preparation step was identified. To reduced negative product cost of MC, waste material should be reduced at this process. The improvement solution was to use refrigerator for reducing temperature of raw material instead of using ice. After improvement, MFCA analysis showed the results as reducing in input and negative product costs of MC. This result was affected to reducing the amount of wastewater generated from this production line, as well.

Although adding one refrigerator to the preparation step was affected in increasing of used electricity, the cost of EC was increased from 4.69 to 60.94 Baht. This solution can be accepted because the refrigerator can be used for 20 batches of products per time, so the cost per batch is approximately 3.05 Baht comparing with the cost of used ice per batch that is 7.00 Baht. Thus, this solution can help in saving production cost as 67,406.67 Baht per year when considering only this product.

References

Chompu-inwai R, Jaimjit B, Premsuriyanunt P (2013) Gaining competitive advantage in an SME using integration of material flow cost accounting and design of experiments: the case of a wood products manufacturing company in northern Thailand. In: EMAN-EU 2013 conference on material flow cost accounting conference proceedings, pp 141–144

Christ KL, Burritt RL (2014) Material flow cost accounting: a review and agenda for future research. J Cleaner Prod. doi:10.1016/j.jclepro.2014.09.005 (in press)

Environmental Industries Office (2011) Environmental policy division, industrial science and technology policy and environment bureau, ministry of economy, trade and industry, Japan. Guide for material flow cost accounting. Retrieved from http://www.meti.go.jp

Kasemset C, Chernsupornchai J, Pala-ud W (2014) Application of MFCA in waste reduction: case study on a small textile factory in Thailand. J Cleaner Prod. doi:10.1016/j.jclepro.2014.09.071 (in press)

Kasemset C, Sasiopars S, Suwiphat S (2013, January) The application of MFCA analysis in process improvement: a case study of plastics packaging factory in Thailand. In: Proceedings of the institute of industrial engineers Asian conference 2013. Springer, Singapore, pp 353–361

Kokubu K, Nakajima M (2004) Material flow cost accounting in Japan: a new trend of environmental management accounting practices. In: Proceeding of 4th Asia pacific interdisciplinary research in accounting conference, Singapore, pp 1–16

Schmidt A, Götze U, Sygulla R (2014) Extending the scope of material flow cost accounting–methodical refinements and use case. J Cleaner Prod. doi:10.1016/j.jclepro.2014.10.039

A Suggestion of an Effective Reverse Logistics System for Discarded Tires in Japan

Kuninori Suzuki, Nobunori Aiura and Yutaka Karasawa

Abstract This chapter analyzes the current issues and proposes an effective reverse logistics system for discarded tires, which considers collection, transportation, and intermediate treatment enterprises, elements that are not involved in forward logistics. The possibility of constructing a reverse logistics network over a wide area is examined from a viewpoint based on previous studies and practices concerning the construction of a forward logistics network. The result revealed that these improvement measures for the reverse logistics can be effectively functioned.

Keywords Reverse logistics · Modal shift · Cooperative logistics · Discarded tires

1 Introduction

1.1 The Purpose of This Study

This study aims to analyze the collection and transport system for used tires in a series of flows in a reverse logistics network. Many discarded automobile tires are collected from gas stations, tire dealers, and the like. In this study, the possibility of constructing a reverse logistics network over a wide area and the integration or aggregation of logistical bases is examined. This simulation includes a collection system algorithm and saving method. This procedure consists of three factors. The first factor is the collection of discarded tires and their transportation to factories as thermal fuels. The second factor is the improvement of the actual reverse logistics

K. Suzuki (✉)
Nihon University, 1-2-1, Izumicho, Narashino City, Chiba Prefecture, Japan
e-mail: QWL02764@nifty.ne.jp

N. Aiura
Hokkai School of Commerce, Sapporo, Japan

Y. Karasawa
Kanagawa University, Yokohama, Japan

V. Kachitvichyanukul et al. (eds.), *Toward Sustainable Operations of Supply Chain and Logistics Systems*, EcoProduction,
DOI 10.1007/978-3-319-19006-8_13

system for discarded tires. The final factor is the design of the improved reverse logistical system. Here, the possibility of integrating the reverse logistical bases through a computer simulation with a collection system algorithm and saving method is discussed using local research.

2 Definition of Reverse Logistics

2.1 Reverse Logistics in Previous Studies

Schultz (2002, pp. 3–4), in his description of reverse logistics, indicated that "although the collection process was recently added to a reference model for supply chain operations, it will be more important for supply chain management (SCM) in the future." de Brito and Dekker (2003) described that "reverse logistics has spread across the world, and it will be involved in a hierarchy of various supply chain divisions." The range of reverse chain or reverse logistics is defined as a series of processes, including collection, recycling, and final disposal of products after leaving the site of waste generation. Fleischmann et al. (2003) pointed out that linking reverse logistics to a recycling system creates difficulty in defining the range of reverse logistics.

2.2 Definition of Reverse Logistics of Discarded Tires

In Japan, for the security, the reuse of automobile tires is currently not popular. However, after collection from gas stations and wreckers, many discarded tires are given an intermediate treatment for conversion to fuel chips, and the fuel chips are used in boilers in cement factories and paper and steel mills as a form of thermal recycling.

In 1994, Japan Automobile Tire Manufacturers Association (JATMA) unified the names of various used tires into "discarded tires." Since fuel chips are valuable materials, their transportation and distribution belong to forward logistics instead of the reverse logistics performed by the collection and transportation enterprises.

Figure 1 shows a basic logistical flow for discarded tires. When enterprises such as gas stations and car shops are sources of discarded tires, the discarded tires are treated as industrial wastes. When collecting discarded tires, a person in charge of discarded tires in the source enterprise is obliged to be present at the collection point and the collection is performed using a manifesto, or an industrial waste management slip. Therefore, the time required for collecting discarded tires in reverse logistics tends to be longer than that required for distribution in forward logistics, in which tires are only delivered. Enterprises that collect and transport discarded tires must establish a collection plan that considers these matters.

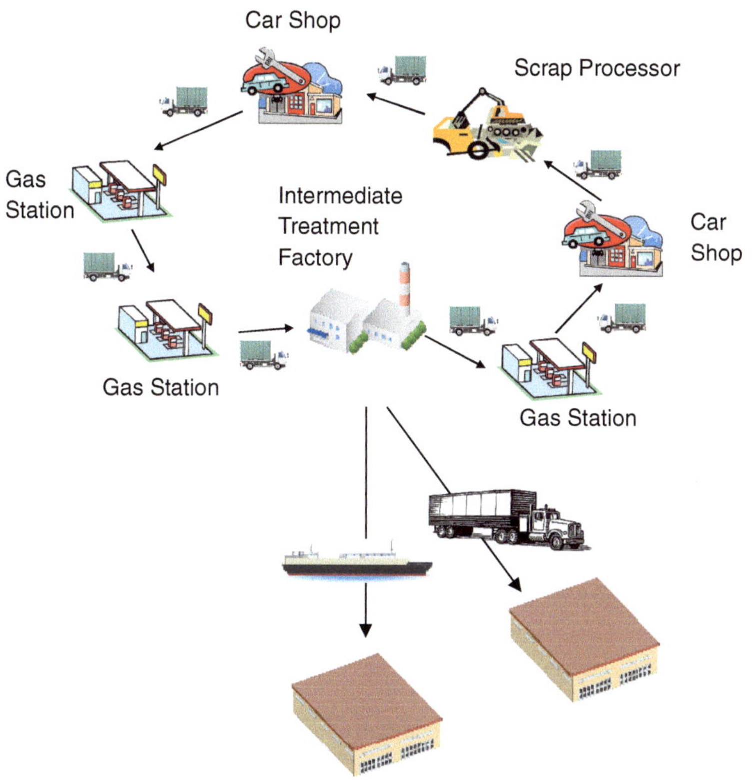

Fig. 1 Logistical flow of discarded tires

In the case where the same enterprise performs collection, transportation, and intermediate treatment of discarded tires, as well as distribution of fuel chips derived from discarded tires to paper and steel mills as thermal recycling, the enterprise must be engaged in the area of reverse logistics as well as of forward logistics, including storage and distribution of the recycled products.

3 Outlines of Analysis and Simulation

Reproducing the present status of a Japanese company A's collection system for discarded tires and performing a scenario analysis of the future possibility of the collection system allow investigation of the present situation. It also allows exploration of further possibilities for the collection system for discarded tires and the measures needed to realize an efficient reverse logistics network.

Company A, which recycles discarded tires at Kanagawa Prefecture in Japan, has licenses for collection, transportation, and intermediate treatment of discarded tires, and for transportation in forward logistics. The company recycles discarded tires into fuel chips, which it then sells to major paper and steel mills. Discarded tires are collected from tire dealers, auto wreckers, gas stations, and auto accessory stores in the Kanto district and then converted into fuel chips in the recycling factories.

The fuel chips are then directly transported to paper mills using 10-ton trucks or to steel mills using ships from the Port of Kawasaki City in Kanagawa Prefecture. The Kanto district includes parts of Tokyo Metropolis, Chiba Prefecture, and Saitama Prefecture, centering on Kanagawa Prefecture.

A 4-ton truck collects 1000 discarded tires from tire dealers, gas stations, car shops, and auto accessory stores every day. Since these tires cannot be collected all at once, they are collected two or three times. The weight of a tire is approximately 8 kg, and the tires are collected using boxes. Although a maximum of 200 tires can be put into each box, in many cases only about 150 tires are put into the box because the tires are not stacked neatly.

The business scheme of the company can be divided into two systems such as a collection system for discarded tires and the transportation and distribution system for fuel chips after collection and intermediate treatment of discarded tires for thermal recycling.

As Fig. 2 shows, the simulation model used in this study is composed of three echelons: the collection complex, the transshipment and storage complex, and the intermediate treatment factory.

The following three desirable situations are investigated from the viewpoints of the environment, load, and cost of the entire collection system, using a simulation model:

(a) Desirable situations of complexes (nodes) belonging to each echelon.
(b) Desirable situation of movement (links) between complexes belonging to the same echelon.
(c) Desirable situation of movement (links) between complexes belonging to different echelons.

Collection complexes belonging to echelons located at the upper part of Fig. 2, such as gas stations, car shops, and auto accessory stores, are used as subjects, and the study investigates a collection device at each complex in the same echelon, a collection route between collection complexes, and routes to different echelons, such as transshipment and storage complexes and intermediate treatment factories.

Intermediate treatment factories belonging to echelons located at the center part of Fig. 2 are used as subjects, as locations of intermediate treatment factories including their elimination and consolidation, for shifting to other intermediate treatment factories in the same echelon, for new establishment of a transshipment and storage complex, and for introduction of a modal shift between a transshipment and storage complex and an intermediate treatment factory.

When the entire area of Chiba Prefecture is assumed to be the future collection area, establishing a transshipment and storage complex (temporary storage complex) in Chiba Prefecture and performing batch transportation between the

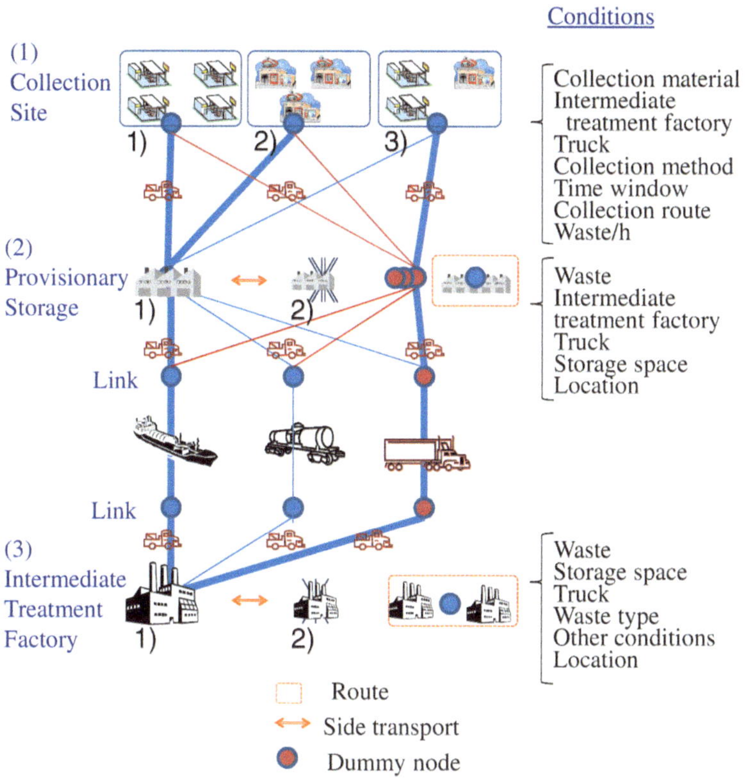

Fig. 2 Setting of echelons for discarded tires collection system

temporary storage complex and an intermediate treatment factory in Kanagawa Prefecture may be more efficient than consolidating collection complexes into one or two in Kanagawa Prefecture and performing real-time collection.

Figure 2 also shows conditions for these complexes. For example, the collection frequency, collection amount, type of vehicle used for collection, existence of transshipment and storage complex, identification of intermediate treatment factory, time window, and route group arrangement are described for collection complexes.

4 Route Search Model and Algorithm for Computer Simulation

The saving method by this simulation is used as an axis for collection course decisions for the total network in Fig. 3.

The saving method is a practical technique to pick the most suitable course in the transportation plan approximately from various transport paths. It does not deliver

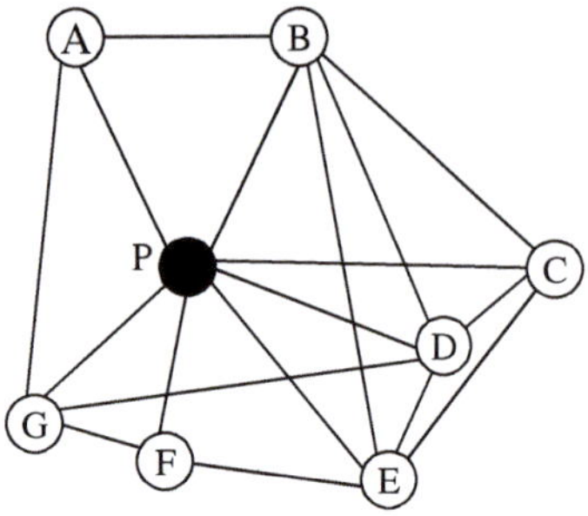

Fig. 3 The image of the total network

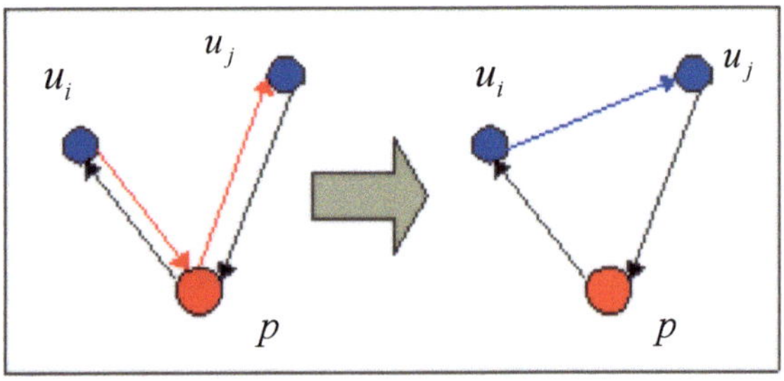

Fig. 4 Calculation for the saving method

the saving technique to two places of delivery separately, and it is the principle to plan improvement of the efficiency by delivering it as one route.

For a network to show in Fig. 1, two routes can be merged into one route, with the calculation for the differences between the distances or expenses in each delivery as in Fig. 4. It should be done for all the combination of the routes for an optimized delivery network, while a distance saving is generated.

The total distance can be shortened by a difference of the length for two red arrows and the length of the blue arrow in Figs. 5 and 6; savings s_{ij} can be calculated by $S_{ij} = d(u_i p) + d(p u_j) - d(u_i u_j)$.

The algorithm of the saving method is as follows:

1. The route between stores and intermediate treatment facilities is fixed as an initial route.
2. After calculating all the routes between the pair of stores, arrange the store pairs in the largest order of the saving level.

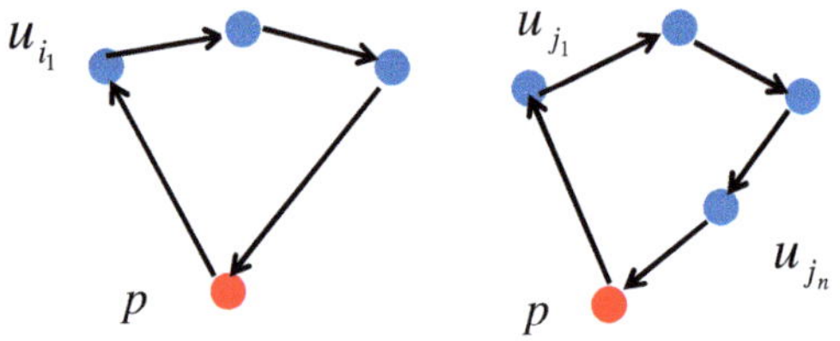

Fig. 5 Generation with the saving method

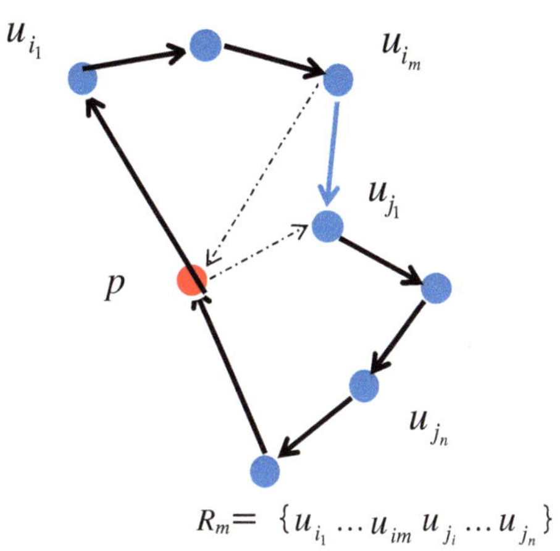

Fig. 6 Integration with the saving method

3. Choose the store pairs from the big candidate to unify sequentially of the saving level. If the saving level is not plus or a store pair to choose disappears, this process is finished.
4. If the store *i* and store *j* are located next to the intermediate treatment facility on each collection course, go to (5), and if not return to (3).
5. If the sums of demand quantity on the collection course including the store *i* and the demand quantity on the collection course including the store *j* are less than the loading capacity of the collection vehicle, go to (6). If not, return to (3).
6. Integrate the collection course as the store *j* is included in with the collection course where store *i* is included in and go back to (3).

5 Data on Collection Points

At present, company A has 150 collection points for discarded tires in the Kanto district as in Fig. 7 by Google Maps, centering on Tokyo Metropolis and Kanagawa Prefecture. In the future, the company will introduce a cooperative collection system and extend the collection area for discarded tires, concentrating on Chiba Prefecture. Assuming an increase in the number of collection points to 650 as in Fig. 8 by Google Maps, data on potential collection points are prepared in addition to data on existing collection points. A simulation is performed based on consolidation of complexes and information on collection points in this collection area, to be incorporated into the cooperative collection system, and the effect of the increase in the number of collecting points on the operational efficiency is examined.

Fig. 7 Actual collection site

Fig. 8 Potential collection site

6 Case Study

6.1 Numerical Results

To grasp the actual situation, using 150 site data, the program was run. In the mileage result, 9–11 of 4-ton trucks and 5 of 10-ton trucks are used.

In Table 1, the numerical results of truck types, driving mileage, driving time, working time (at collection sites), the number of discarded tires, and the number of trucks are shown. "DATE" means the day of the week, i.e., DATE: 0 means Monday, and DATE: 5 means Saturday. Concerning cost results by the week, the personnel expenses, the variable expenses, the fixed expenses, and the total cost are also calculated.

The scale merit and cost merit, however, cannot be found with the increase of the number of collection sites. The mileage distance increases compared to the number of the trucks and collection lots of discarded tires. The number of collection points that a truck could visit was restricted by the time required for the travel and the number of discarded tires. In the case where a certain number of collections were fixed, the number of vehicles was not affected by the number of collection points.

6.2 Cost-Reduction Measures

As the numerical results in Table 2, the cost merit of the increase in the number of collection points cannot be seen. That is to say, other cost-reduction measures should be considered. It is possible to reduce the working time (at collection sites).

Table 1 Mileage result of the actual network

Date	Truck type (ton)	Driving milage (km)	Driving time (h)	Working time (at collection sites)(h)	The number of discarded tires	The number of trucks
DATE: 0	4	1499	50	45	1999	11
DATE: 1	4	1212	40	44	1624	10
DATE: 2	4	1203	40	45	2048	9
DATE: 3	4	1499	50	45	1999	11
DATE: 4	4	1212	40	44	1624	10
DATE: 5	4	1203	40	45	2048	9
DATE: 0	10	1057	35	8	1638	5
DATE: 1	10	824	27	10	1220	5
DATE: 2	10	802	27	8	1298	5
DATE: 3	10	1057	36	8	1638	5
DATE: 4	10	824	27	10	1220	5
DATE: 5	10	802	27	8	1298	5
	total	13,193	440	321	19,654	90

Table 2 Cost result of the potential network

DATE	Truck type (ton)	Variable expense (driving) (yen)	Variable expense (at collection site) (yen)	Personnel expense (yen)	Fixed expense (yen)	Total (yen)
DATE: 0	4	127,270	26,182	401,400	168,000	722,852
DATE: 1	4	145,803	25,389	435,800	186,000	792,992
DATE: 2	4	152,394	23,535	441,200	189,000	806,129
DATE: 3	4	127,270	26,182	401,400	168,000	722,852
DATE: 4	4	145,803	25,389	435,800	186,000	792,992
DATE: 5	4	152,394	23,535	441,200	189,000	806,129
DATE: 0	10	12,913	1325	41,250	25,000	80,488
DATE: 1	10	25,127	1757	69,000	40,000	135,884
DATE: 2	10	28,914	1586	74,500	45,000	150,000
DATE: 3	10	12,913	1325	41,250	25,000	80,488
DATE: 4	10	25,127	1757	69,000	40,000	135,884
DATE: 5	10	28,914	1586	74,500	45,000	150,000
	Total	984,842	159,548	2,926,300	1,306,000	5,376,690

The collection of discarded tires requires lots of time at the sites. However, the effective measures, such as collection infrastructure, can reduce its time and costs. The possible measures are as follows:

- Introduction of RFID system for the inspection of discarded tires at the collection sites.
- Introduction of Electronic Manifesto and Advanced Shipment/Collection Notice System for efficient procedures of collecting discarded tires at the collection sites.
- Introduction of removable containers for reducing the loading time and waiting time at the collection sites.

Based on these cost-reduction measures, the scenarios for the simulations are made. The numerical results are given in Table 3. Compared to the actual network, the costs for each scenario can be reduced. "Full complement" includes RFID system, electronic manifesto and ASN, and removable containers.

Table 3 Cost result of the measures

Cost-reduction measure	Total cost (yen)
Actual state	1,306,250
RFID system	1,221,332
Electronic manifesto and ASN	1,128,018
RFID system, electronic manifesto, and ASN	1,087,960
Removable containers	1,021,432
Full complement	928,032

7 Conclusion

This study used actual data for the collection system of discarded tires to perform a simulation that clarified the actual state of the collection system and that pointed out problems that reduced the efficiency of reverse logistics. When a model for the consolidation of distribution complexes, which has been an effective tool to improve the distribution efficiency in forward logistics, was directly applied to reverse logistics, the expected results were not realized, because the time required for tasks at the complexes is longer in reverse logistics than in forward logistics. A cooperative reverse logistics network will require the establishment of a new scheme while utilizing accumulated knowledge in forward logistics.

In the future, the transportation and distribution system for fuel chips should be analyzed for cost reduction, while considering a reduction in environmental load. A simulation for the transportation and distribution system of fuel chips should include a number of factors including collection complexes in overseas regions such as China and South Korea in addition to Japan, intermediate treatment complexes, and consumption areas for the recycled goods.

References

de Brito MP, Dekker R (2003) A framework for reverse logistics. In: Reverse logistics. Springer, Berlin, pp 3–27

Fleischmann et al (2003) Reverse logistics network design. In: Reverse logistics. Springer, Berlin, pp 68–69

Schultz G (2002) The recurring side of supply chain management. APICS Perform Advant 12 (10):35–39

A Study on the Location Selection of Industrial Wastes Treatment Facilities: A Case of Intermediate Treatment Facilities in Chiba, Japan

Sarinya Sala-ngam, Kuninori Suzuki, Jun Toyotani, Keizou Wakabayashi and Akihiro Watanabe

Abstract This chapter attempts to raise the awareness of the recycling-based society through the creation of efficient strategy for reverse logistics by searching for the optimal location for collecting centers of the intermediate treatment facilities. Generally, the efficiency of collecting center can be improved by relocating the facilities and by changing the collection and transport routes. The optimal location of collecting centers for treatment facilities are found by the density method. The case study searches for the optimal location of the collecting centers for the intermediate waste treatment facilities in Chiba prefecture. The results show that the efficiency of waste collection can be improved with leading collection routs from the collecting center to all other treatment facilities.

Keywords Intermediate treatment · Waste treatment · Collection and transport · Computer simulation

1 Introduction

Recently, the environmental problems have become increasingly amplified, due to a consequence of fast growing economic activities of mass production and consumption. Customers strongly expect business operators or companies to reduce the environmental impact of their products and processes. Hence, the recycling-based society is attracting wide interest of many companies. Also, product recovery is becoming increasingly important problem in industrialized countries. Accordingly, reverse logistics comprising business about industrial waste treatment is an

S. Sala-ngam (✉) · K. Suzuki · J. Toyotani · K. Wakabayashi · A. Watanabe
Industrial Engineering and Management, Graduate School of Industrial Technology, Nihon University, Chiba 275-8575, Japan
e-mail: cisa14001@g.nihon-u.ac.jp

V. Kachitvichyanukul et al. (eds.), *Toward Sustainable Operations of Supply Chain and Logistics Systems*, EcoProduction,
DOI 10.1007/978-3-319-19006-8_14

important role and meaning for the companies in an era. However, the proper treatment of industrial wastes and strategy of reverse logistics are insufficient and discussed. Thus, the business of industrial waste treatment requires more efficient reverse logistic network. According to the quantitative models for reverse logistics by Fleischmann (2001), reverse logistics is defined as "The process of planning, implementing, and controlling the efficient, effective inbound flow and storage of secondary goods and related information opposite to the traditional supply chain direction for the purpose of recovering value or proper disposal" (Fleischmann 2001). A reverse logistics system comprises a series of activities that begin with collection of industrial wastes or used products, and treatment of the collected wastes and end with proper recovering of disposal. A detail of these activities is collection, disassembly, shredding, testing, sorting, storage, and transportation. In a case of transportation that can be divided into two main groups, the first group is reusable parts for recycling or remanufacturing, and this should be called as "redistribution." The second group is non-reusable parts for proper disposal.

Many intermediate treatment facilities where industrial wastes are collected and treated according to wastes can be represented as reuse option or should be disposed, can be found in Japan. However, after treating industrial wastes, the treatment facilities transport reusable parts to same recycling places and non-reusable parts to same disposal places each by themselves. From a result of this, the number of transport routes is excessive and the efficiency of transportation is not good as expected. Accordingly, the waste treatment facilities in Japan need to solve this problem and think of a new strategy for improving efficiency of transportation and collection. This problem can be solved simply with minimizing the number of transport routes.

Therefore, to improve the efficient collection and transportation of treatment facilities in Japan, we have determined to focus on the intermediate waste treatment facilities in Chiba, in order to minimize the number of routes by searching the optimal collecting center for treatment facilities. Recent research has occurred in the field of optimization. Barros et al. (1998) proposed a mixed integer linear programming (MILP) model based on considering the optimization of warehouse location problem for the sand and examine its optimization using heuristic procedures (Barros et al. 1998) (Roghanian and Pazhoheshfar 2014). Krikke et al. (1999) presented an MILP model based on a multilevel uncapacitated warehouse location model. The model was used to determine the location of the recovery facilities and also the transport links connecting various locations (Krikke et al. 1999) (Roghanian and Pazhoheshfar 2014). However, we have found The Destiny Function Method researched by Okabe and Suzuki (2008), which is very effective and easy in achieving the purpose of this study.

The purposes of this chapter are to raise the awareness of the recycling-based society through the construction and creation of efficient strategy for reverse logistics. The optimal relocation for the collecting center of intermediate treatment facilities in Chiba is selected by using the density method by Okabe and Suzuki (2008).

1.1 The Role of Intermediate Treatment Facilities

Industrial waste treatment processes and responsibilities differ from general waste treatment. In case of industrial waste, waste producers have to dispose their waste at their own responsibility, which is regulated by law. In this case, emitters have to use their own facilities for storage, collecting, and treating waste and then transport the treated waste to recycling place or disposal over a certain period of time with their own vehicles. In many cases, emitters are facing more difficulties for their own industrial wastes. Therefore, the emitters of industrial wastes can commit professional waste treatment operators who are committed from emitters to collect waste to treatment facilities and redistribute reusable parts or disposal. Incidentally, emitters of industrial waste do not need to get permission to treat waste; however, the permission is required from waste treatment operators to do process of treating industrial waste. However, emitters' responsibilities are still waste collection and redistribution or final disposal.

First, discharged industrial waste is collected and transported from factories for the purposes of recycling. The collected and transported industrial wastes are selected, graded, crushed, burned, melted, and dehydrated based on qualities of wastes in the immediate treatment facilities.

1.2 The Circumstance and Problems of Intermediate Treatment Facilities in Chiba Prefecture

Basically, intermediate waste treatment has been carrying out by each province. As most industrial waste producers are located outside of cities, waste collection and transportation requires longer time and distance across the province to reach the intermediate treatment facilities. As a result of this, emitters are seduced to risk illegal dumping. However, the risk of illegal dumping is decreasing due to tightened regulations and increased penalties.

From noted points above, the examination of how efficiency can be realized by collecting small and medium scale intermediate treatment facilities is focused and set to the purpose of this study. In Chiba prefecture, many of the small and medium intermediate treatment facilities can be found and the intermediate treatment is carried out after collecting industrial wastes from each local manufacturing industry or storage site of industrial waste. However, most of the intermediate facilities have a daily capacity of less than 10 tons. Consequently, the flow line of reverse logistics is longer than necessary when treated parts are redistributed for remanufacturing or disposal.

Therefore, in order to realize the efficiency of reverse logistics, the largest scale of existing intermediate treatment facilities in Chiba will be created to utilize as collecting center when collecting industrial waste from the small facilities first.

Next, the density function model by Okabe and Suzuki will be used in order to search the optimal location for the collecting center of waste treatment facilities in Chiba. Moreover, the collecting routes from intermediate treatment facilities to the collecting center will be found by using a development programming, and a result of routes will be shown by computer simulation. Increasingly, collecting distances, units of transportation (quantity and traffic of transportation), will be calculated.

2 Problem Statement

2.1 Scope of Study

In this study, Chiba prefecture in Japan is centered and many intermediate treatment facilities can be found (A public web site of Chiba, approach 2014). In order to achieve the purposes of this study, the position of thirty intermediate treatment facilities that treat only plastics waste and metals are investigated.

In Fig. 1, as a result of this investigation, the largest intermediate facility can be found in 1 place and following in the small scale of facilities can be found in 29 places.

Fig. 1 The positions of the intermediate treatment facilities in Chiba prefecture

Table 1 Use of data

Data	
Number of object intermediate treatment facilities	30 places
Chiba area	About 5156 km^2
Truck	10 tons of truck
Collecting form	Round ways (cyclic formula)

2.2 *Data for the Case Study*

In order to search the collecting center for treatment facilities in Chiba, the necessary data are shown in Table 1.

In addition, the position coordinates of treatment facilities will be required to search location of the collecting center by using the density function model. Therefore, the geocoding Web site is used to obtain the position coordinates of the intermediate treatment facilities in Chiba by changing the addresses to latitude and longitude. Table 2 shows a result of this. Moreover, a daily capacity of treatment is also investigated, and the result of this investigation is shown in Table 2. However, 60 % of waste treatment capacity will be calculated to search a location that would be the collecting center.

3 Methodology

The methods that used to achieve the purposes of this study are described as follows.

3.1 *Formulation of the Density Function Researched by Okabe and Suzuki*

The density function model by Okabe and Suzuki (Okabe and Suzuki 2008) is consulted to search the optimal location for the collecting center of treatment facilities. The density function model by Okabe and Suzuki showed that the location center (hub) can be found simply by calculating average value of position coordinates of its surrounding spokes using the density function.

Indices

j is an index of intermediate treatment facilities ($j = 1, 2, 3, \ldots, N$)
a is an index of specific area ($a = 1, 2, 3, \ldots, A$)

Table 2 Latitude and longitude of the intermediate treatment facilities and capacity of waste treatment (ton/day)

No.	Facilities	Latitude	Longitude	Treating capacity (60 %)
1	A	35.937476	139.903159	165.31
2	B	35.932859	139.904654	169.10
3	C	35.873652	140.481955	199.20
4	D	35.826767	140.066712	314.12
5	E	35.817412	140.067057	169.92
6	F	35.779161	140.642758	150.00
7	G	35.772611	140.618517	216.00
8	H	35.749535	140.719026	249.60
9	I	35.734437	140.128079	239.64
10	J	35.715329	140.823249	240.00
11	K	35.712641	140.686651	259.20
12	L	35.697265	140.486745	240.00
13	M	35.688540	140.044723	154.63
14	N	35.688282	139.951437	384.00
15	O	35.678300	139.932379	192.48
16	P	35.676779	140.246000	360.60
17	Q	35.674406	139.962661	499.48
18	R	35.665471	139.925918	402.53
19	S	35.641764	140.559058	203.04
20	T	35.629931	140.545204	184.20
21	U	35.587871	140.289829	198.24
22	V	35.524814	140.094149	150.60
23	W	35.510680	140.063087	216.00
24	X	35.472955	140.283402	297.60
25	Y	35.458070	139.999475	696.00
26	Z	35.374908	139.899930	515.76
27	AA	35.369428	139.898804	182.40
28	AB	35.329306	140.116953	169.62
29	AC	35.121033	139.919692	208.62
30	AD	34.984843	139.847719	576.00
Total	1068.626526	4206.108982	8203.89	

Parameters and Variables

N is the number of intermediate treatment facilities
A is the number of specific area
x is the position coordinate of the collecting center (latitude)
y is the position coordinate of the collecting center (longitude)
$\bar{x}_j$ is the position coordinate of j treatment facility (latitude)
$\bar{y}_j$ is the position coordinate of j treatment facility (longitude)

p_a is the number of treatment facilities in area a
d_j is the distance from the collecting center to j treatment facility
t_a is the total distance from the collecting center to other treatment facilities in specific area a
T is the total distance from the collecting center to all other treatment facilities
m is the total number of times collected from treatment facilities which are in specific area to the collecting center

The location problem of collecting center for treatment facilities can be formulated as follows:

$$N = \iint f(x, y)\mathrm{d}x\mathrm{d}y \tag{1}$$

Incidentally, a certain specific point $(\bar{x}_j, \bar{y}_j)$ and the number of treatment facilities in site area a are given by the following formula:

$$p_a = \iint f(\bar{x}_j, \bar{y}_j)\mathrm{d}x\mathrm{d}y \tag{2}$$

According to formula 2, when the point on a map of the collecting location is set to (x, y), the distance from the collecting center to other treatment facilities is given be the next formula.

$$d_j = \sqrt{(x - \bar{x}_j)^2 + (y - \bar{y}_j)^2} \tag{3}$$

Moreover, if all the distances from the collecting center to other treatment facilities are added, the total distance T is obtained. And the distance from the collecting center of one time of delivery to another treatment facilities is shown by t_j.

Therefore, T shows the sum total distance of all the circumferential delivery records.

$$t_a = d_j p_a = \sqrt{(x - \bar{x}_j)^2 + (y - \bar{y}_j)^2}\,\frac{1}{N} \tag{4}$$

$$T = \sum_{j=1}^{N}\sum_{a=1}^{A} d_j p_a = \sum_{J=1}^{N} \sqrt{(x - \bar{x}_j)^2 + (y - \bar{y}_j)^2}\,\frac{1}{N} \tag{5}$$

In each loading collection record, to add the number of times of collection as one time per matter to all the operation records, the value of density function for one collection record is 1/*N*. Moreover, when collecting wastes from the same place, as times in the unit of month or year, the number of treatment facilities in site area can be found as follows.

$$p_a = \iint f(\bar{x}_j, \bar{y}_j)\mathrm{d}x\mathrm{d}y \cong \frac{1}{N} \times m \tag{6}$$

$$T = \sum_{j=1}^{N} \sum_{a=1}^{A} d_j p_a = \sum_{j=1}^{N} \left(\sqrt{(x - \bar{x}_j)^2 + (y - \bar{y}_j)^2} \right) \left(\frac{m}{N} \right) \tag{7}$$

Next, in order to find the location of the collecting center, what is needed is the shortest total distance from the collecting center to other treatment facilities. Namely, if the total distance can be defined as formula 7, it becomes possible to calculate the shortest value by optimization formula 7 which turn out to be 0; from this, formulas 8 and 9 are obtained as follow.

$$\frac{\partial T}{\partial x} = 0 \tag{8}$$

$$\frac{\partial T}{\partial y} = 0 \tag{9}$$

And for this value to be the minimum value, the following condition needs to be fulfilled.

$$\frac{\partial^2 T}{\partial x^2} > 0 \tag{10}$$

$$H = \frac{\partial^2 T}{\partial x^2} \frac{\partial^2 T}{\partial y^2} - \frac{\partial^2 T}{\partial x \partial y} \frac{\partial^2 T}{\partial y \partial x} > 0 \tag{11}$$

Therefore, if formulas 10 and 11 are solved, and the answers are checked to be positive, the following formula will be obtained.

$$\frac{\partial T}{\partial x} = \sum_{j=1}^{N} (2x - 2\bar{x}_j) = 0 \tag{12}$$

$$\frac{\partial T}{\partial y} = \sum_{j=1}^{N} (2y - 2\bar{y}_j) = 0 \tag{13}$$

And the following formula is obtained by solving formulas 12 and 13 in more detail.

$$x = \frac{\sum_{j=1}^{N} \bar{x}_j}{N} \tag{14}$$

$$y = \frac{\sum_{j=1}^{N} \bar{y}_j}{N} \tag{15}$$

Therefore, the position coordinate can be solved by formulas 14 and 15. To find the optimal collecting location for intermediate treatment facilities, formulas 14 and 15 will be used.

3.2 Route-First, Cluster-Second Method

In order to find collecting routes from the collecting center to all other treatment facilities, route-first cluster-second method is consulted in this study.

From consulting route-first cluster-second method (Kubo 2007), first, Google Maps is used to find the distances from the collecting center to all intermediate facilities and the distances between the treatment facilities with other treatment facilities. Secondly, a list of distances ranging from shortest to longest is created. Next the scope is divided like a form of sectorial partitioning scheme by considering a list of distances ranging from shortest to longest which is mentioned above. And finally, the collecting routes can be found using a development program, which wrote by consulted route-first cluster-second method.

3.3 Quantity of Transportation

Quantity of transportation is a derived unit calculated as transported quantity divided by unit distance. Thus, quantity of transportation can be calculated by the following formula;

$$\mathrm{TQ} = \sum D \times \sum q \tag{16}$$

Parameters

TQ is the transportation quantity
D is the collecting distance
q is the transported quantity (which is shown in Table 2).

4 Results

4.1 Result-1

In this section, the result of utilizing the largest scale of existing intermediate treatment facility as the collecting center is shown. As a result of investigation

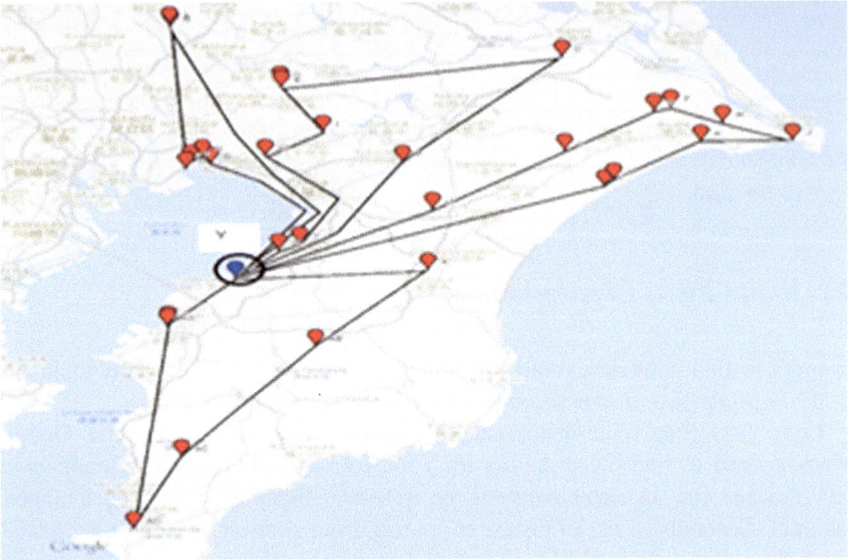

Fig. 2 The location of Y treatment facility (*blue point*) and simulation of collecting routes from Y treatment facility to other facilities

which is shown in Table 1, Y treatment facility is able to treat industrial wastes at the highest capacity of 696 ton a day. Therefore, Y treatment facility is considered to be the largest facility and considered to utilize as collecting center. Increasingly, collecting routes from the collecting center to all other intermediate treatment facilities are found by using a development program.

Figure 2 shows the results of the collecting routes from the Y treatment facility to all other treatment facilities in Chiba.

From the above result, the total distance of the collecting routes which are shown in Fig. 2 is 617 km., and the transportation quantity is about 506.2×10^4 t km.

4.2 Results-2

The result of using the density function researched by Okabe and Suzuki in order to search the optimal location for the collecting center is shown in this section.

In order to search the optimal location of the collecting center for the intermediate treatment facilities in Chiba, the density function researched by Okabe and Suzuki is consulted. From the density function model, formulas 14 and 15 are used to find the position of the collecting center by calculating the average value of position coordinates of treatment facilities (which is shown in Table 2) which is given as follows:

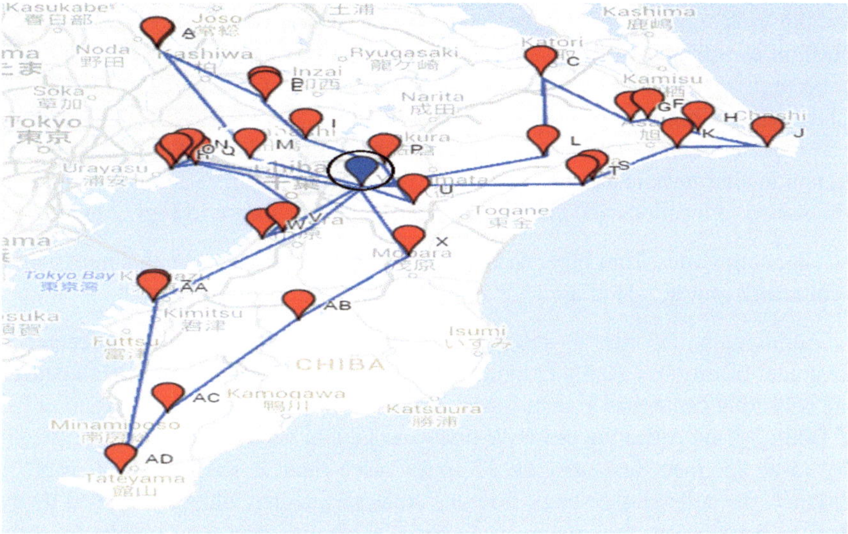

Fig. 3 The position of collecting center (*blue point*) and simulation of collecting routes from the collecting center to all other treatment facilities

$$x = \frac{1033.16846}{29} = 35.6264985$$

$$y = \frac{4066.109507}{29} = 140.2106727$$

From the result above, the position coordinate of collecting center can be found at 35.6264985, 140.2106727. Therefore, the collecting center location would be at 35.6264985, 140.2106727. This position can be searched by using the geocoding Web site in Fig. 3. And the collecting route from this location to all treatment facilities can be found by development program, and the result is shown in the following figure.

Resulting from the simulation above, the total distance of the collecting routes is about 493.1 km, and quantity of transportation is about 404.5×10^4 t km.

5 Conclusions

From this case study, the purpose of the study can be achieved by creating the largest scale of existing treatment facility in Chiba prefecture to utilize as the collecting center first. Moreover, from the density function, the optimal relocation for the collecting center can be found by calculating the average value of the position coordinates of treatment facilities. The optimal position coordinate of the collecting is 35.6264985, 140.2106727. This place is Yato-cho, Wakaba-ku, and Chiba-shi in Chiba prefecture.

As can be seen from the result, the examination revealed that efficiency of collection in intermediate treatment facilities can be improved from current situation by simulation of locations and collecting routes which are shown in Figs. 2 and 3.

Collecting wastes from other facilities to the collecting center with simulation of collecting routes, which are shown in Figs. 2 and 3.

Additionally, in order to utilize Y treatment facility as the collecting center, Y treatment facility has a daily capacity of waste treatment at only 696 tons; however, the collecting center has to treat waste more than 8000 tons per day. Thus, to utilize Y facility as the collecting center, Y treatment facility needs to increase the capacity of waste treatment and area by 12 times more than at present. As a result of distance, the collecting distance from the collecting center, which was found by the density function to other treatment facilities is shorter than the distance from the existing location of Y facility by about 20 %.

Therefore, 1266 Yato-Cho, Wakaba, Chiba Prefecture (the blue point in Fig. 3) is considered to be the best relocation for the collecting center of intermediate treatment facilities in Chiba.

To continue this study, the location selection of the collecting center for intermediate treatment facilities in 2 places and 3 places are be considered as future work. Moreover, a result of collecting distances, quantity of transportation, and value of carbon dioxide emission will be used for considering the number of the collecting center that should be proper for utilizing as the collecting center of treatment facilities in Chiba.

References

A public web site of Chiba prefecture. http://www.pref.chiba.lg.jp/haishi/shorigyou. Approach Sept 2014

Barros AI, Dekker R, Scholten V (1998) A two-level network for recycling sand: a case study. Eur J Oper Res 110(2):199–214

Cook W (2012) In pursuit of the traveling salesman: mathematics at the limits of computation. Princeton University Press, Princeton

Fleischmann M (2001) Quantitative models for reverse logistics. Lecture notes in economics and mathematical systems, vol 501, p 6

Japanese Geocoding web site. https://www.geocoding.jp. Approach Sept 2014

Krikke HR, van Harten A, Schuur PC (1999) Business case Oce: reverse logistic network re-design for copiers. OR-Spektrum 21(3):381–409
Kubo M (2007) Applied mathematics of logistics. Kyoristu 167–176
Okabe A, Suzuki A (2008) Mathematics for optimal relocation. Asakura bookstore 52–79
Roghanian E, Pazhoheshfar P (2014) An optimization model for reverse logistics network under stochastic environment by using genetic algorithm. J Manufact Syst 33(3):348–356

Part III
Manufacturing and Production Logistics

Performance Measurements Related to Lean Manufacturing that Affect Net Profit of SMEs in the Manufacturing Sector of Thailand

Panutporn Ruangchoengchum

Abstract This chapter aimed to assess performance measurement related to lean manufacturing and net profit of small and medium enterprises (SMEs) in Thailand, by interviewing 100 SME entrepreneurs, particularly in the manufacturing sector, and analyzing data by applying factor analysis and multinomial logistic regression. The results found that performance measurement for SMEs in the manufacturing sector in Thailand agreed with most variables at a high level, except for product usage and service at a medium level. When considering each entrepreneur individually, they had medium net profit at 11–15 %. When comparing performance measurements related to lean manufacturing, it was found that factors affected a high net profit of 15 % and they should continually work on it in order to ensure product quality, process quality, waste, defects, suppliers, work-in-process, output, lead time, delivery lead time, and inventory.

Keywords SMEs in the manufacturing sector · Net profit · Performance measurement · Lean manufacturing

1 Introduction

In Thailand at present, it is accepted that small and medium enterprises (SMEs) have a role important in the country's economic expansion. It can be seen from the expansion rate of the SME gross domestic product (GDP) is a key economic indicator for a macroeconomic overview. It is likely to be higher than the expansion rate of the country (2.9 %) in 2014 [The Office of Small and Medium Enterprise Promotion (OSMEP) 2014]. Meanwhile, the percentage of GDP of SMEs increased from 37.0 % of GDP of SMEs in 2013 to 37.4 % of GDP of SMEs in 2014

P. Ruangchoengchum (✉)
College of Graduate Study in Management, Khon Kaen University, Khon Kaen 40002, Thailand
e-mail: rpanut@kku.ac.th

V. Kachitvichyanukul et al. (eds.), *Toward Sustainable Operations of Supply Chain and Logistics Systems*, EcoProduction,
DOI 10.1007/978-3-319-19006-8_15

(OSMEP 2014). Indeed, SMEs, particularly those in the manufacturing sector, are the engine of growth in the economic system of Thailand and directly benefit the economic growth of the country. It can be seen when comparing the GDP's ratio of SMEs in the manufacturing sector against the service sector and the retail sector in 2014 that SMEs in the manufacturing sector had the significantly highest growth rates in GDP at 32.9 % of GDP compared with the service sector and the retail sector at 32.5 and 13.0 %, respectively (OSMEP 2014).

Despite these successes and the status of major SMEs in the manufacturing sector, they have failed to achieve the SME entrepreneurship confidence index. The Office of SMEs Promotion (2014) reports that the SME entrepreneurship has low confidence index in terms of the costs such as production cost, labor cost, and logistics and transportation cost, and the competitive marketplace is also related to the low confidence index. Low confidence index results in this sense affect business profits and the survival of business operations.

Government sectors are trying to promote and support SME operations, particularly in the manufacturing sector, in order to find a strategy on how to reduce costs and how to gain the highest profit margins, and also to create the potential for competitive advantage. For example, the Bank of Small- and Medium-sized Enterprises development of Thailand (SME Bank) organizes activities to strengthen SMEs by allowing easier access to capital. In addition, the Office of SMEs Promotion (2013) has conducted training programs to develop the potential of SMEs, and the federation of Thai Industries has provided a framework for cooperation between the public and private sectors for the development of SMEs in Thailand and to develop competitiveness capacity in both domestic and international markets (International Institute for Trade and Development 2013). But SME entrepreneurship might still have weak points in getting accurate data for analysis and business development planning for the right decision in operations. Performance measurement is then a management tool that is a vital part of helping entrepreneurs know the results of business operations in the past and predict business, as well as to know the impact on the profitability of the business and in which way SME entrepreneurs should improve or enhance performance to be suitable for increasing net profits and reducing costs, particularly unnecessary production costs or waste solutions, which can possibly be eliminated and lead to business profits. If there is no potential for SME business improvement, it would not be able to compete. Therefore, SMEs should have proper performance measurement to find weaknesses and to improve and strengthen their businesses.

For this reason, the researcher was interested in performance measurements related to lean manufacturing that affect net profit of SMEs in the manufacturing sector. Thus, it is the intention of this research project to find out to what extent this occurs. The following are significant research objectives: (1) to measure SMEs' performance in the manufacturing sector in Thailand; (2) to study SMEs' net profit in the manufacturing sector in Thailand, and (3) to study performance measurements related to lean manufacturing that affect net profit of SMEs in the manufacturing sector in Thailand. This would appear to be a promising method to alleviate industrial problems. The results of this research would be useful for SME entrepreneurs, particularly in the manufacturing sector, and other Thai government agencies, to be used as guidelines to reduce costs, to develop competitiveness in world markets, and to enhance the competitive advantage as preparation for continued economic expansion. To achieve research outcomes, we provide an in-depth review of the literature underpinning the research objectives toward the development of an evaluative framework with which to undertake the research project.

2 Reviewed Literature

This section reviews the literature concerning the parameters of research under four main headings: SMEs, Net Profit of SMEs, Performance Measurement, and Lean Manufacturing. These attributes build upon the literature of the research project's posited objectives and elucidate the key issues underpinning the research methodology.

2.1 Beyond the Established Theoretical Perspectives: SMEs

The coordination and project management of the Office of SMEs Promotion (2014) gave the definition of SMEs that cover business in the manufacturing sector, the wholesale sector, the retail sector, and the service sector, with the following meanings.

- The manufacturing sector is characterized by production industrial enterprises of all types. By definition, the internationalization of production is to transform the object into a new type of mechanical or chemical product, regardless of whether the work is done by machine or by hand. The manufacturing operations include the simple processing of agricultural products with industrial processes, and community enterprises, as well as household manufacturing.
- The wholesale sector means services for trading, and selling new and used goods to retailers, industrial, commercial, institutional, and professional users, and includes sales to the wholesalers themselves.
- The retail sector means the sale without transformation of both new and used goods to the general public for personal consumption or use by a particular household. This trade also includes being a broker or dealer, gas stations, and consumer cooperatives.
- The service sector means education, health, entertainment, transportation, construction and real estate, hotels and hostels, restaurants, selling food and beverage sales of restaurants, rental, entertainment and recreation, personal service, household services, services for business, all kinds of repair, and travel- and tourism-related businesses.

In sum, this research studied SMEs in the manufacturing sector. The sample groups were covered by the definitions above. The issue researched and analyzed is the net profit of SMEs.

2.2 Beyond the Established Theoretical Perspectives: Net Profit of SMEs

Net profit of SMEs will be measured by the survival level of companies with net profits from net income of approximately from less than 5 to 15 % and above as the researcher has defined, as the OSMEP (2013) set equations to predict the survival of SMEs in Thailand into 5 levels as follows:

- Level 5 strongly survives (net gain of more than 15 %);
- Level 4 survives as normal or firm (net gain of about 11–15 %);
- Level 3 not good enough, there are certain profits, but no loss (net profit of around 8–10 %);
- Level 2 quite poor, barely surviving, but circulation of funds is possible (net gain of around 5–7 %); and
- Level 1 awful, cannot survive, loss and no circulation of funds (net gain of less than 5 %).

Collectively, these attributes of the survival of SMEs suggest an approach to research, which is an issue that will be analyzed further in performance measurement.

2.3 Beyond the Established Theoretical Perspectives: Performance Measurement

Performance measurement is an indicator of results, in which the researcher studied documents and related research on performance measurements that affected net profit on various variables as follows.

Fred (2011) has classified performance measurement quantitatively using financial ratios as a basis for assessing corporate strategies, including return on investment (ROI), return on equity (ROE), profit margin, market share, debt to equity, earnings per share (EPS), sales growth, asset growth, and qualitative performance measurements such as in human resources, marketing, finance/accounting, accounting research and development (R&D) or management information systems.

Hudson et al. (2001) divided performance measurement into 6 areas: quality: reliability of products delivered or waste from the production service; time: the reduction in waiting time and resource utilization; flexibility: productivity and introduction of new products; finance: sales, cash flow, and inventory; customer satisfaction: delivery reliability, service, and market share; and human resources: quality of work, labor efficiency, and staff learning.

Inman (2014) mentioned that the performance measurement can be divided into two types: performance on achievement and productivity on output or outcomes, such as competitiveness and financial performance, and measurement operations to focus on the factors of achievement including quality and flexibility, resource utilization, and innovation. The basic performance measurements of the business process consist of currency (from profits), productive relationships: output/input or productivity to cater to the customer, innovation and adaptation to change, and human resources, with standard indicators in performance measurement which were productivity measures, quality measures, inventory measures, and lead time measures.

Training Resources and Data Exchange (TRADE), US Department of Energy (2005) noted that the performance measurement is generally divided into six categories; however, some organizations may develop metrics appropriate for type of business, depending on the organization's mission and including effectiveness of the product, efficiency of the production process, quality of product or service, timeliness, productivity, and safety at work of employees.

However, the issue will be researched and analyzed further in performance measurement related to lean manufacturing.

2.4 Beyond the Established Theoretical Perspectives: Lean Manufacturing of SMEs

The ways of performance measurement related to lean manufacturing will be significant by being without waste in the process of production. Ohno (1998) studied waste elimination such as overproduction, time delay or failing to meet the delivery lead time by suppliers or transporting, inappropriate processing, unnecessary inventory, excess motion and defects or return of the product, which can help companies significantly cut costs, particularly waste reduction. There are unnecessary production costs, which can be possibly eliminated, known as seven wastes. There are overproduction, time delay, transporting, inappropriate processing, unnecessary inventory, excess motion, and defects (Ohno 1998).

Lean manufacturing of SMEs will be defined by reducing costs and wasteful practices, and increasing quality (Ohno 1998; Melton 2005). These may help to increase net profit of SMEs. Likewise, Wilson et al. (2008) studied lean manufacturing or lean production of SMEs in New Zealand, which refers to processing and production management among various resources in order to decrease waste, inefficient activities and work-in-process while offering quality products on time and at least cost. In addition, Nasser Mohd et al. (2009) studied a review on lean manufacturing practices in SMEs, Malaysia. Those viewed as being more important are to be resources such as finance, manpower, and machines of SMEs. The benefits of SMEs to apply lean manufacturing will be the net profit of SMEs, which is the major significant measure of a company's profitability (OSMEP 2013).

In summary, the researchers have applied the concepts above, by focusing on the concepts of Ohno (1998), Hudson et al. (2001), Wilson et al. (2008), and Nasser Mohd et al. (2009) to specify the criteria for performance measurement related to lean manufacturing, and in all twenty-four variables under all the five performance measurement perspectives (quality, time, finance, customer satisfaction, and human resources) as follows:

- **Quality**: 5 variables include product quality, process quality, waste, defects, and suppliers;
- **Time**: 4 variables include work-in-process, output, lead time, and delivery lead time;
- **Finance**: 7 variables include inventory, orders/receipts, profit, turnover, costs and expenditure, cash flow, and sales/value added;
- **Customer Satisfaction**: 5 variables include user problems, product usage, customer service, returns, and complaints; and
- **Human Resources**: 3 variables include safety, staff turnover, and personnel development.

3 Conceptual Framework and Proposed Research

Based on the proposed theoretical background and the literature studied, we derive the conceptual framework as in Fig. 1, in which the prime variable of study is the performance measurement related to lean manufacturing. We studied twenty-four variables (as mentioned above) under all the five performance measurement perspectives (quality, time, finance, and customer satisfaction), and the fixed variable was SMEs' net profit in the manufacturing sector, Thailand.

These variables have been further explained in the framed hypotheses. Hence, we posit the following:

- **Hypothesis 1**: SMEs in the manufacturing sectors have different net profit, and
- **Hypothesis 2**: SMEs in the manufacturing sector's profit depend on performance measurements related to lean manufacturing.

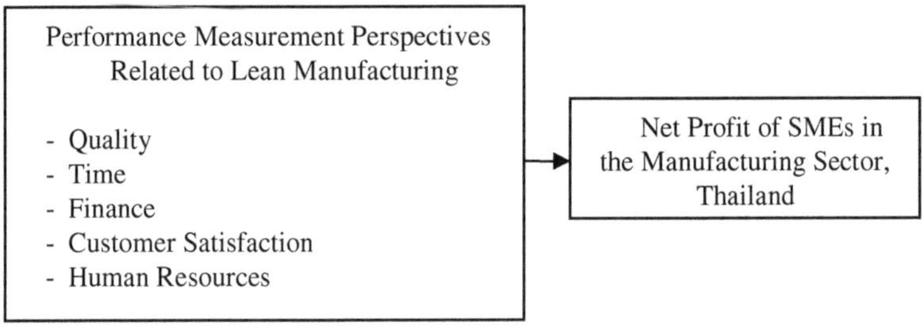

Fig. 1 Research conceptual framework

4 Research Methodology

This section provides an overview of the research methodology employed to resolve the research objectives concerning the framed hypotheses as mentioned above.

4.1 To Measure SMEs' Performance in the Manufacturing Sector in Thailand

4.1.1 Sampling and Data Collection

The samples used in this study were SMEs in the manufacturing sector, which are located in the northeastern region of Thailand, which is a major hub for rural industries such as the agricultural processing in the northeast region (Glassman and Sneddon 2003). Thus, the majority of entrepreneurs in the northeastern region are engaged in agriculture. It can be seen from more than 80 % of industrial firms in the northeast of Thailand being focused on rice milling and the remainder relying on the processing of agricultural produce and other primary manufacturing industries (Parnwell and Khamanarong 1996). We selected the area of Khon Kaen Province, which is the center of the northeastern region and has the densest SMEs, contributing 7.56 % to the total number of SMEs of Thailand (OSMEP 2014). Khon Kaen Province is then attractive to investors; both domestic and foreign investors are pouring in (Suwannaporn 2013).

In this study, the researchers focus on SMEs in the manufacturing sector, the sampled group size was 100 SMEs, and data were collected by quota sampling. With the above sampling groups, a questionnaire was used as a tool for data collection to create a questionnaire checklist by studying twenty-four variables under all the five performance measurement perspectives (quality, time, finance, and customer satisfaction) for performance measurement of SMEs in the manufacturing sector in the province.

4.1.2 Data Analysis and Statistics Used in the Research

The research data were analyzed using SPSS for Windows Version 17.0. Descriptive statistics used were arithmetic mean (AM) and standard deviation (SD) to interpret the comment level by applying an average score in each level on a statistical basis which is to find the range, and to use a formula for calculating the width of the interval to determine the average level of opinion as follows:

- Average of 4.21–5.00 is the highest opinion level;
- Average of 3.41–4.20 means there are a lot of comments;

- Average of 2.61–3.40 means there is moderate comment;
- Average of 1.81–2.60 means there is less comment; and
- Average of 1.00–1.80 means the least number of opinions.

4.2 To Study SMEs' Net Profit in the Manufacturing Sector in Thailand

4.2.1 Sampling and Data Collection

The samples were selected according to 4.1.1. A questionnaire was used as a tool for data collection and for studying net profits, which were divided into 5 levels: Level 5 strongly survives with a net gain of more than 15 %, Level 4 survives as normal or firm with a net gain of about 11–15 %, Level 3 not good enough as there are certain profits but no loss (net profit of around 8–10 %), Level 2 is quite poor, barely surviving, but circulation of funds is possible, with a net gain of around 5–7 %, and Level 1 awful, cannot survive, makes a loss and has no circulation of funds with a net gain of approximately less than 5 %.

4.2.2 Data Analysis and Statistical Methods Used in the Research

The research measured the level of survival of the SMEs from the net profit that was less than 5 % to over 15 % net profit by using descriptive statistics for percentage of acquisition characteristics and tested hypotheses of different SMEs with net profit different or not different with statistical significance.

4.3 To Study Performance Measurement Related to Lean Manufacturing that Affects Net Profit of SMEs in the Manufacturing Sector in Thailand

4.3.1 Statistic and Data Analysis

In a study of performance measurement related to lean manufacturing that affects net profit of SMEs in the manufacturing sector in Thailand, the researchers used factor analysis to find the factors. The study extracted factors using the principle component method and forecasted identify factors that affect net profit by using multinomial logistic regression analysis and grouping patterns of business operations of SMEs with statistical analysis.

4.3.2 Assessment of Research Tools

The researchers assessed the accuracy and reliability of the questionnaire by examining the content reliability from experts with updates and corrections and the test reliability by finding the discrimination power by using item-total correlation, which measures the performance of SMEs in the discrimination (r) range from 0.329 to 0.731. To test the tools' reliability, Cronbach's alpha coefficient was used, and the performance measures of SMEs had alpha coefficients ranging from 0.738 to 0.850.

5 Results

This section analyzes the outcomes of the research project with reference to the framed hypotheses, which guide the research methodology and structure the analysis of the research outcomes.

Three significant features of performance measurement emerge as relevant issues in an analysis of the participants' responses. These are the performance measurements of SMEs, net profit of SMEs, and performance measurements related to lean manufacturing that affect net profit of SMEs in the manufacturing sector in Thailand.

5.1 Performance Measurement of SMEs in the Manufacturing Sector in Thailand

Overall, it was found that the majority of SMEs in the manufacturing sector in Thailand had operated for a period of 5 years and 6–10 years in similar proportions of 30 % and 30.75 %, respectively, mainly employing only 25 staff or less (91.75 %) and having fixed assets of no more than 30 million baht (80.75 %), while most, 69.25 %, had registered commercially and had individual proprietorship.

All SMEs have been classified into manufacturing sector. We found that SME entrepreneurs overall had performance measurement at a high level in all variables, except for the product usage and service which was at the medium level as shown in Table 1.

5.2 SMEs' Net Profit in Manufacturing Sector in Thailand

From Table 2, SMEs in the manufacturing sectors had different net profits, which were statistically significant at the 0.05 level, and profitability as normally survive (net profit is about 11–15 %) was mainly in the manufacturing sector (37.24 %).

Table 1 Performance measurements for SMEs in manufacturing sector

Performance measurement	Manufacturing sector		
	$\overline{X}$	S.D.	Opinion level
Product quality	4.40	0.60	High
Process quality	4.30	0.70	High
Waste	4.30	0.60	High
Defects	4.30	0.60	High
Suppliers	4.40	0.60	High
Work-in-process	4.20	0.60	High
Output	4.20	0.60	High
Lead time	4.20	0.50	High
Delivery lead time	4.20	0.60	High
Inventory	3.90	0.50	High
Orders/Receipts	3.80	0.60	High
Turnover	3.80	0.50	High
Costs and expenditure	3.80	0.60	High
Cash flow	3.80	0.60	High
Profit	4.00	0.50	High
Sales/Value added	4.10	0.50	High
User problems	3.80	0.50	High
Product usage	3.20	0.70	Medium
Customer service	3.30	0.70	Medium
Returns	4.20	0.60	High
Complaints	3.80	0.50	High
Safety	3.80	0.60	High
Staff turnover	3.80	0.60	High
Personnel development	3.90	0.60	High

Table 2 SMEs' net profit in the manufacturing sector

Net profit	Manufacturing sector *n* (%)
Strongly survive (net gain of more than 15%)	16
Survive as normal or firm (net gain of about 11–15 %)	54
Not good enough, there is uncertain profit or loss (net profit of around 8–10 %)	27
Barely survive, but circulation of funds is possible (net gain of around 5–7 %)	2
Cannot survive (net gain of approximately less than 5 %)	1(100.00)

The entrepreneurs with a profit at a terrible loss, with no circulation of funds (net profit of less than 5 %), had only one entrepreneur in the manufacturing sector. These results support Hypothesis 1.

5.3 Performance Measurements Related to Lean Manufacturing that Affect Net Profit of SMEs in the Manufacturing Sector in Thailand

5.3.1 Factor Analysis

From SMEs performance measurement related to lean manufacturing in the province, it was found that there were several independent variables. The researchers wanted to know whether any independent variables affected the net profit of these SMEs. The analysis showed some independent variables were multicollinear; therefore, a statistical technique called factor analysis was used to group the independent variables together, which then led to the creation of new variables' equation in the following order.

When factor analysis has been done to determine the factors that affect the survival of SMEs in the province, the factors were extracted using principle component to see how many of the 23 variables could be factors. It was considered by eigenvalue that exceeds 1.0; the eigenvalue is indicative of the ability of the emerging factors to explain the variability of the original variables. Besides, in this research, we also applied the Varimax rotation method and the KMO statistics, which are used to measure the suitability of the information available, and KMO > 0.6 would be considered suitable data to use for factor analysis techniques. The results showed that the KMO = 0.8123, which was over 0.6, so the information was appropriate to use technical analysis. The results showed there were five factors that had eigenvalue over 1.0, so the analysis grouped the factors into five factors as in Table 3.

5.3.2 Multinomial Logistic Regression Analysis

The dependent variables were divided into 4 groups; the group with net profit less than 5 % was added to the 5–7 % group because there was just one case. Multinomial logistic regression analysis techniques were applied with the added group, earning less than 5 % to the 5–7 %, as a baseline category as in Table 4.

The probability that the entrepreneur would have the opportunity to gain a level3 as it is not good enough but has no loss (net profits of around 8–10 %) was considered. When compared to the opportunity profitable level 2 or level 1, the results showed that there were no factors affecting the net profit of the SMEs in the province as in Table 4.

From Table 5, the results of the entrepreneurs grouping showed the following. Group 1: Most entrepreneurs had normal net profit level which was sufficient or a net gain of about 7.5 %, and this group had very different comment level from medium to high level. Group 2: Entrepreneurs with enough net profit, had a net gain of about 8–15 %, and had high-level comments. Group 3: Entrepreneurs with stable profit level, had a net profit of over 15 %, and had high to highest level opinion.

Table 3 Factor analysis of five factors

Order and factors of variance performance measurement	% of variance	Performance measurement
F1. lean manufacturing	23.73 %	–Product quality
		–Process quality
		–Waste
		–Defects
		–Suppliers
		–Work-in-process
		–Output
		–Lead time
		–Delivery lead time
		–Inventory
F2. cash performance	12.99 %	–Orders/Receipts
		–Turnover
		–Cash flow
F3. customer service and human resource management	11.93 %	–Returns
		–Complaints
		–Safety
		–Staff turnover
		–Personnel development
F4. value-added resources	11.80 %	–Costs and expenditure
		–Profit
		–User problems
		–Sales/Value added
F5. product information service/consulting services	10.39 %	–Product usage
		–Service
Total	70.84 %	Total of 70.84 % that the 5 factors could explain variability of the original 23 variables

Likewise, with a significantly high net profit (over 15 %), entrepreneurs should have the highest opinion concerning lean manufacturing. These results support Hypothesis 2.

6 Discussion and Conclusion

The outcomes of this research project contribute significantly to the knowledge base of the discussion, and it can be concluded as follows.

Most SME entrepreneurs in the manufacturing sector had high level of opinion on performance measurement in product quality, process quality, waste, defects, suppliers, work-in-process, output, lead time, delivery lead time, inventory, orders/receipts, turnover, costs and expenditure, cash flow, profit, sales/value

Table 4 Factors affecting SMEs' net profit

Factors affecting SMEs' net profit	Coefficient	*P*-value	Marginal effect
Net profit level 3			
Constant	7.083	0.356	–
Factor 1	2.860	0.377	0.124
Factor 2	−4.753	0.117	0.001
Factor 3	7.521	0.349	−0.088
Factor 4	0.158	0.964	−0.367
Factor 5	−1.222	0.406	0.042
Net profit level 4			
Constant	2.615	0.726	–
Factor 1	2.052	0.521	−0.190
Factor 2	−4.795	0.110	−0.023
Factor 3	1.803	0.245	−0.031
Factor 4	2.274	0.507	0.409
Factor 5	−1.401	0.333	−0.010
Net profit level 5			
Constant	−1.179	0.883	–
Factor 1	2.885	0.379	0.080
Factor 2	−4.787	0.117	−0.025
Factor 3	2.665	0.134	0.129
Factor 4	1.353	0.700	−0.033
Factor 5	−1.635	0.277	−0.040
Pseudo R^2 = 0.0892			

added, user problems, returns, complaints, safety, staff turnover, and personnel development as business operation was an essential process in which business must be able to manage an organization operation system, personnel, and teamwork in order to drive the business process to reach the goal. This is relevant to the work of Hudson et al. (2001), which was about SME performance measurement systems. To be in line with the business strategy, this research studied the performance measures in five perspectives, which cover all aspects of the organization's operations including the results of operations (dimensions of quality, dimension of time and flexibility, financial results, how to make the company recognized externally (customer dimension), and work environment (human resources dimension) which was consistent with the concepts of Parmenter (2007) who noted that from assessment of the performance of SMEs, most of them used KPI measures and focused on the operational aspects of the organization. The things most important to the success and impact of the organization now and in the future were consistent with the concepts of Wiwakanond et al. (2004) who mentioned that there is a critical need to manage the organization's processes such as competitively low cost, quality or customer satisfaction, capacity or service of process, and fast and

Table 5 Performance measurement in each factor classified by group of entrepreneurs in the manufacturing sector

Performance measurement for each factor	$\overline{X}$	S. D.	Level opinion
Group 1			
Factor 1 lean manufacturing	3.81	0.45	High
Factor 2 cash performance	3.29	0.48	Medium
Factor 3 customer service and human resource management	3.30	0.43	Medium
Factor 4 value-added resources	3.61	0.43	High
Factor 5 product information service	3.13	0.62	Medium
Group 2			
Factor 1 lean manufacturing	4.19	0.31	High
Factor 2 cash performance	3.81	0.35	High
Factor 3 customer service and human resource management	3.97	0.28	High
Factor 4 value-added resources	4.09	0.29	High
Factor 5 product information service	3.83	0.44	High
Group 3			
Factor 1 lean manufacturing	4.56	0.33	Highest
Factor 2 cash performance	4.19	0.41	High
Factor 3 customer service and human resource management	4.26	0.38	High
Factor 4 value-added resources	4.50	0.35	High
Factor 5 product information service	4.34	0.41	High

accurate delivery. These are consistent with the company's operating philosophy, UM limited by Thailand Productivity Institute (2004) which noted that a company's operating philosophy should be "To maintain integrity, quality, punctuality," which focused on producing quality products according to customer requirements and being on time with delivery.

In order to accomplish the work with this philosophy, the business is committed to the development of production processes and product quality to meet customer requirements. These were applied in operation planning and production design, including a focus on suppliers, with having meetings and planning together all the time in order to work effectively. In addition to satisfying the customer, catering to the delivery, and processing, the company also focused on employees in the organization as it considered them as the most precious resource in the operation. The executives had to create a good atmosphere in the workplace, encourage learning and collaboration, create employee satisfaction, emphasize teamwork, and collect data to measure and analyze the performance as well.

Although entrepreneurs in the manufacturing sector had opinions on performance measurement on each variable at a high level, they had medium-level opinions in product usage and service such as providing free workshops on product use, and how to impress customers in order to retain customers in the manufacturing sector. This shows that most of entrepreneurs in the manufacturing sector may lack skills of product information service.

6.1 Performance Measurement Affected SMEs' Net Profits as Follows

Most of SMEs in the manufacturing sector had net profit: level 4 as normal survival or net profit of approximately 11–15 %. Factors that affected net profits to survive in level 4 included the following:

- Customer service and personnel development factors consisted of product usage including providing products introduction services, such as training in how to use the product for free, service to impress and retain customer, taking complaints or satisfaction from customers to make a better organization performance, promoting safety and staff welfare apart from that required by the law such as health insurance, creating good atmosphere with clean and safe workplace, staff turnover having efficient recruitment process to recruit good people who work hard and stay for a long time, and having low or no staff turnover each year. And for personnel development, factors were to encourage staff to develop their knowledge and skills by allowing staff to study further or providing staff training to enhance skills and knowledge, both in and out of the workplace, and having performance assessment of the parties involved;
- Value-added resource factors were costs and expenditure, cost reduction of production of goods/services such as raw materials purchased for production, production costs, overhead supplies in the production process, labor cost in services, cost of service goods, sales and administrative expenses, ability to increase profits as per goals, sales/value-added ability to increase sales volume and having service center to provide counseling whether clients have problem with the product/service; and
- Cash Performance factors were order/receipts, facilitating orders/services such as customers can order products/services by telephone or Internet, also for facilitating the payment for goods/services to the customers via the internet, ATM or credit card, turnover ability to manage inventory more effectively, such as the ability to remove products inactive for more than six months or to manage product storage or to reduce the inventory need and cash flow, to use resources cost effectively and to reduce costs and expenses of the company.

Factors that affected net profits at a strong survival level 5 (high net profit of 15 % or more) included lean manufacturing factors, which were product quality, quality test of raw materials used in process prior to the production of goods, and raw materials used in accordance with the standard set. Process quality is the quality check at every step, which leads to high-quality production of goods/services. Waste, minimal or no waste focuses on the use of resources for maximum benefit. Defects focus on delivering quality products and product insurance if it is damaged as a result of the delivery process. Suppliers with vendor selection and quality production factor and Work-in-process with a focus on reducing the amount of time (lead time) used in manufacturing with warranty products/services (outputs). If the customer is not satisfied with the product/customer service, the customer can choose to return the product or get new products to replace it or a refund within the prescribed period. In addition, human resource management factor focuses on personnel development and involves developing staffs' knowledge and customer service skills. Moreover, if human resource management, from the recruitment process, had an effective selection, and evaluation process, personnel welfare and safety apart from the law, service quality guarantee, as well as protect to return of goods, and facilitation for ordering service, then these factors would lead to business success. The entrepreneur should focus on performance measurement in customer satisfaction as it is the basic retention strategy to keep repeat and return customers (loyalty customers) with the organization as long as possible and to attract new customers to the business.

6.2 Factors of Lean Manufacturing

The entrepreneurs focused on the factors of lean manufacturing (product quality, process quality, waste, defects, suppliers, work-in-process, output, lead time, delivery lead time that should be near the operation plant to save transportation costs and have feedstock at the plant throughout the year and inventory management) as the greater the focus on performance measurement, the higher possibility of net profit. The entrepreneurs realized the importance from the selection of raw materials, production quality, and quality control process at every step to shorten the process time, minimize or remove waste, and have a damage warranty. These factors were particularly important, resulting in customers being highly satisfied with the products and services, leading to an increasing sales volume and net profits for the business. These were consistent with the concepts of Inman (2014), who mentioned business performance measurement focusing on the determinants of the results such as quality, which is basic for customers to focus on and links to financial results, in accordance with the concept of Olsen (2007) who stated that lean operations and waste were important factors to develop competitive pricing positions and net profit.

In addition, if the SMEs entrepreneurs in manufacturing focused on these factors it would result in a work process that moved the business forward such as the entrepreneur that provided consulting services involving product information service to clients when a problem with the product occurred, having information on service support for product guarantee with unsatisfactory services, or even taking complaints or assessing customer satisfaction to improve organization efficiency.

In conclusion, the analyses and discussions refer to the conceptual framework, hypothesis, and concepts derived from the literature review to elucidate the significance of entrepreneurs' responses and their relationship to the research outcomes. It is concluded emphatically that the SMEs' entrepreneurs in manufacturing should focus on performance measurement related to lean manufacturing factors (product quality, process quality, waste, defects, suppliers, work-in-process, output, lead time, delivery lead time, and inventory). These factors affected a high net profit of 15 % or more.

However, other possible directions for future research that might build on this research project include the following:

- The impact or the other factors affecting SMEs' net profit should be investigated, and the data from this study can be applied for operations management to create effective performance.
- Problems and obstacles that affect performance measurement of SMEs' business should be studied to find guidelines for business development, as there are changes of environment both inside and outside the company all the time.

Acknowledgments This study was supported by the Esaan Center for Business and Economic Research, Faculty of Management Science Research, Khon Kaen University, Thailand.

References

Fred RD (2011) Strategic management concepts and cases. Person Education Inc, New Jersey

Glassman J, Sneddon C (2003) Chiang Mai and Khon Kaen as growth poles: regional industrial development in Thailand and its implications for urban sustainability (Electronic version). Ann Am Acad Polit Soc Sci 590:93–115

Hudson M, Smart A, Bourne M (2001) Theory and practice in sme performance measurement systems (Electronic version). Int J Oper Prod Manag 21(8):1096–1114

Inman, R.A.(2014). *Performance Measurement*. Retrieved November 20, 2009, from http://www.referenceforbusiness.com/management/Or-Pr/Performance-Measurement.html#ixzz32VTndS61

International Institute for Trade and Development (Public Organization). (2013). *ASEAN SMEs*. Retrieved March 20, 2014, from http://www.tpa.or.th/tpanews/upload/mag_content/71/ContentFile1422.pdf

Melton, T. (2005). *The Benefits of Lean Manufacturing : What lean thinking has to Offer the Process Industries*. Chemical Engineering Research and Design

Nasser Mohd, A., Baba Md. Deros, R. and Nizam Abdul Rahman, M. (2009). *A Review on Lean Manufacturing Practices in SMEs*. Retrieved December 20, 2014, from http://www.academia.edu/942964/A_REVIEW_ON_LEAN_MANUFACTURING_PRACTICES_IN_SMALL_AND_MEDIUM_ENTERPRISES

Ohno T (1998) Toyota Production System: Beyond Large-Scale Production. English translation by Productivity Press, USA

Olsen,E. (2007). *Strategic Sourcing and Lean Manufacturing with Value Stream Roadmapping*. Retrieved March 20, 2014, from https://www.ventureoutsource.com/contract-manufacturing/benchmarksbestpractices/lean-outsourcing/lean-manufacturing-and-outsourcing

Parmenter D (2007) Key Performance Indicators. John Wiley & Sons Inc, Hoboken, New Jersey

Parnwell M, Khamanarong S (1996) Rural Industrialisation in Thailand: Village Industries as a Potential Basis for Rural Development in the North-East. In: Parnwell M (ed) Uneven Development in Thailand. Avebury, Aldershot, UK

Suwannaporn, C. (2013). *Evaluation of Thai SMEs for AEC*, Tourism Authority of Thailand and ASEAN Economics Community. Retrieved August 12, 2013, from http://th.aectourismthai.com/tourismhub/167

Thailand Productivity Institute (2004) SMEs Good Practices of Medium and Small Scale. Eastern Printing (Public Company Limited, Bangkok

The Office of Small and Medium Enterprises Promotion (2013) The studies of SMEs economics indicators (SMEs economics indicators). Retrieved 12 Dec 2013, from http://www.sme.go.th/Lists/EditorInput/DispF.aspx?List=15dca7fb-bf2e-464e-97e5-440321040570&ID=2001

The Office of Small and Medium Enterprises Promotion (2014) The situation report of small and medium enterprises. Retrieved 15 Jan 2015, from www.sme.go.th

Training Resources and Data Exchange (TRADE), U.S. Department of Energy (2005) How to measure performance: a handbook of techniques and tools. Oak Ridge Associated Universities

Wilson M, Heyl J, Smallman C (2008) Lean manufacturing initiatives in New Zealand. Retrieved 5 Jan 2013, from http://researcharchive.lincoln.ac.nz/handle/10182/6289

Wiwakanond K, Anakamanee K, Kitcharapoom K (2004) Process management. Innographics, Bangkok

The Survey on the Challenges of Organization of Automotive Component Remanufacturing in Small-sized Companies in Poland

Paulina Golinska-Dawson, Monika Kosacka and Anna Nowak

Abstract Remanufacturing of automotive components is a sector in which a big number of small- and medium-sized companies (SMEs) operate. The challenges of remanufacturing process are described in the literature, but the empirical studies in this domain are still limited. This chapter compares the theoretical findings with the empirical results. The overview of the literature analysis on remanufacturing process challenges is presented. The results of the pilot surveys in Polish small-sized companies that are involved in automotive part remanufacturing are presented along with the characteristics of the respondents, the main problems which appear by remanufacturing of automotive parts.

Keywords Remanufacturing process · Cores · Remanufacturing operations

1 Introduction

Remanufacturing in the automotive sector is a common business practice. The body of the literature provides an extended discussion which can be named as remanufacturing. The most common characteristics that appear in the previous research (Sundin 2004; Lund 1996; Parkinson and Thompson 2003; Hammond et al. 1998) are as follows:

- It is an industrial process,
- It transforms used or broken-down products,
- It focused on restoring to useful life, and
- It aims to achieve as good as new product's functionality.

P. Golinska-Dawson (✉) · M. Kosacka · A. Nowak
Faculty of Engineering Management, Poznan University of Technology, Strzelecka 11, Poznan, Poland
e-mail: paulina.golinska@put.poznan.pl

V. Kachitvichyanukul et al. (eds.), *Toward Sustainable Operations of Supply Chain and Logistics Systems*, EcoProduction,
DOI 10.1007/978-3-319-19006-8_16

Remanufacturing companies especially small- and medium-sized face problems to achieve adequate economy of scale of their operations and an operational excellence (Golinska and Nowak 2015). This paper presents the overview of the literature analysis on remanufacturing process challenges. The results of the pilot surveys in Polish SMEs that are involved in automotive component remanufacturing are presented. The paper consists of five sections. In Sect. 2, the theoretical background is presented. In Sect. 3, the authors discuss the research methodology. The survey results are presented in Sect. 4. Final conclusions are stated in Sect. 5.

2 Theoretical Background—Challenges for Remanufacturing of Automotive Parts

The remanufacturing process is more complex and more variable than traditional manufacturing. Remanufacturing process is a sequence of activities required to bring the worn-out product to "like a new" conditions. The authors present a remanufacturing as a flowchart in Fig. 1.

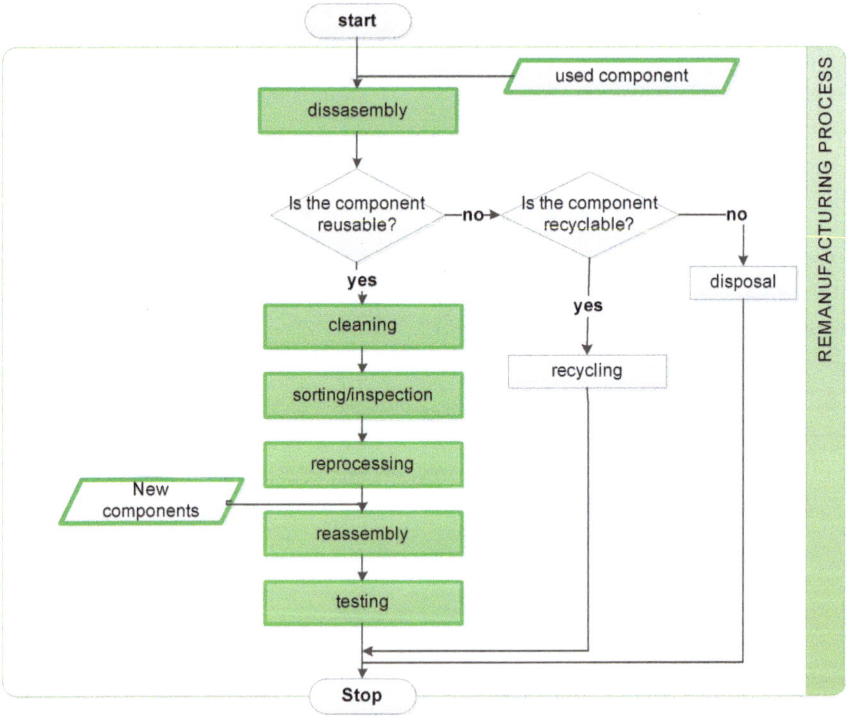

Fig. 1 Remanufacturing process

Remanufacturing process consists of six main activities including the following (Golinska and Kosacka 2014):

(a) Disassembling—products are disassembled to the level of a part. Reusable parts are passed to next operations. Other elements may be recycled or they are disposed.
(b) Cleaning—removing all contamination, including degreasing, derusting, and removal coatings of the surface as a paint.
(c) Inspection and sorting—sorting items into groups with assessing the parts' reusability and possibilities of reconditioning.
(d) Reprocessing—it includes milling, turning, grinding, material deposition, heat treatment, welding, powder coating, chroming, and painting.
(e) Reassembling—it is a process of assembling with used components (sometimes with new elements).
(f) Testing—every remanufactured part is tested to preserve efficiency at the level of 100 %. Remanufactured parts obtain a warranty.

The challenges appear at the process entry when used products (known as cores) are collected from the consumers and then delivered to the remanufacturing facility. Cores are the equivalent of the raw materials in the traditional manufacturing. A core has multiple modules that are materially recycled, reused, refurbished, or disposed (Jayaraman 2006). According to Souza (2008), the input to the remanufacturing process (cores) is uncertain in quality, quantity, and timing of arrival; furthermore, some cores are unsuitable for remanufacturing and might be scraped or recycled. Also Guide (2000) addressed the same problem, and he stated that there is the uncertain timing and quantity of returns; furthermore, there is also the uncertainty in materials recovered from return items. The cores (return products) availability depends on phase of products life cycle. Usually it is difficult to provide on regular basis big volume deliveries of particular model. The problems regarding the cores' supplies which were identified in the literature (by, e.g., Ostlin and Ekholm 2007; Hammond et al. 1998, or Rubio and Corominas 2008) are as follows:

- insufficient availability of the good quality cores,
- high product variability,
- variation of the rate of materials recovered, and
- materials' matching problems.

An efficient core management is the backbone of all remanufacturing processes (Subramoniam et al. 2010). Furthermore Lind et al. (2011) stated that it is one of the main challenges in the remanufacturing to achieve a steady flow of cores. The variability of products to be remanufactured is very high. The remanufacturers have to deal with small batches of wide range of product variants and generations, which complicates tool changing, disassembly, and assembly processes (Seitz and Peattie 2004). Most of the products caused problems in remanufacturing system because they are not designed for disassembly (Sundin and Bras 2005). The products are not usually built from easy to disassembly modules. Furthermore, OEMs use very often permanent techniques for joining elements such as welding or gluing. It results in

Challenges for remanufacturing in SMEs		
Inputs - quality of cores - availability of cores - variable delivery times - small batches - multimodels	**Process execution** - time consuming dissassembly - small batches - variable reprocessing lead time - time consuming reprocessing - frequent set-ups - time consuming cleaning operations - variable customers' requirments - materials matching restrictions	**Outputs** - matching demand&supply - not enough demand - difficulties in distribution of remanufactured products

Fig. 2 Challenges for remanufacturing in small- and medium-sized companies

lowering of the material recovery rate because some cores might be destroyed during disassembly. Another problem is also the fact that very often there is a lack of full information on technical parameters of the original product.

It is very difficult (almost impossible) to estimate lead times (Guide 2000). The lead times of remanufacturing operations are not standardized. Reprocessing may include different operations for the same type of returns depending on their conditions. Some operations/tasks are known with certainty but appearance of others might be probabilistic. The routings and operations lead times are highly variable.

The challenge is also very small batch size, which might require frequent setups. Customer requirements might differ for reprocessing of particular components. Returns with different quality levels have different lead times, costs, and capacity usage.

From the perspective of the remanufacturing process output, the challenge is to balance returns with demands (Guide 2000). Very often, customers are also not aware about the availability of remanufactured products so there are not enough orders to achieve the economy of scale. The summary of the challenges in the remanufacturing process in SMEs is presented in Fig. 2.

3 Research Methodology

The scope of remanufacturing process which is taken into consideration in this survey is presented in Fig. 3. The focus is placed on the "inputs" (cores' supplies) and "process execution" perspective.

A questionnaire was used to carry out the survey. To create an online survey form was used Google Docs application. The advantage of this online tool is that results can be presented as statistical charts, cumulative summary of responses, with the possibility of sharing the results obtained so far from all respondents. The questionnaire consisted of 20 questions (open, closed, and semi-open). The semi-

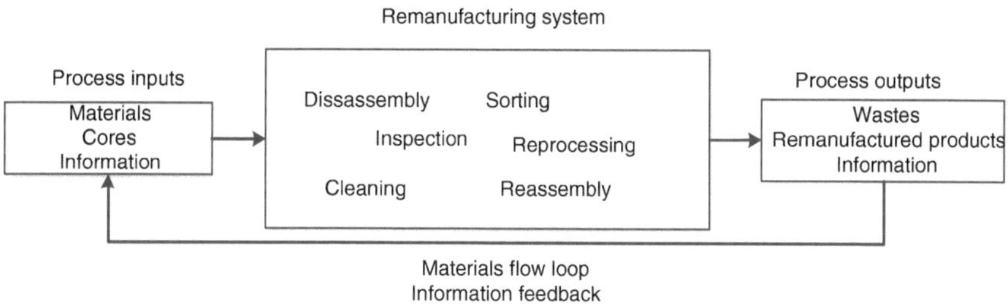

Fig. 3 System under study

open and open questions were structured depending on the topic as single-choice, multiple-choice, and Likert scale questions. Survey aims to verify the theoretical findings with the facts observed in the companies. Respondents belong to the group of companies specializing in the remanufacturing of automotive components. Findings are presented in the form of aggregated statistical charts. The respondents gave the answers anonymously.

The first part of the questionnaire focuses on characteristics of respondents, including the location, the size of the company, and major groups of products. The next part covers the cores supplies organization, including:

- the sources of cores supplies,
- the relations between remanufacturer and suppliers,
- quality issues of supplies, and
- problems in organization of supplies.

In the next section, respondents were asked about the characteristics of their remanufacturing process including the following:

- standard lot size,
- the average duration of the remanufacturing cycle,
- workload of various stages of remanufacturing process,
- the level of recovery of remanufacturing,
- difficulties of cleaning products with dangerous substances,
- problems with missing spare parts, and
- inventory management issues.

The companies for the pilot study were selected based on Internet search with key words as follows: remanufacturing, automotive parts, and regeneration. Among companies specializing in remanufacturing of automotive parts randomly were selected group of 70 companies. The original method of research was supposed to be the electronic survey, which was sent on indicated by the respondent (during a brief phone call) e-mail inbox. Because the response rate was not satisfactory (approx. 4.5 %), the companies were contacted again. The companies indicated that they would like to rather answer the questions from the online questionnaire during the phone conversations (they indicated that they are too busy to do it online, or that they are not familiar with online surveys). In the second round, the research team was entering the answers on behalf of the respondents to online form during phone conversation. In the second round, we have received the responses from 40 companies. The results of the study are presented in the next section.

4 Survey Results

In the survey were included responses from 40 enterprises specializing in remanufacturing of automotive components. In Table 1 are presented the general data about companies.

Respondents are mainly companies with over 10 years of experience (28 companies) in remanufacturing of automotive parts. The sizes of companies were as follows: 89 % are micro-enterprises employing up to 9 persons, rest of the companies were declaring that their employment is below 49 people.

The analyzed companies were specialized in remanufacturing of turbochargers (16 companies), steering gear (9 companies), and cylinder heads (7 companies). Among these can be distinguished companies that specialize in one type of products' group, or those whose operations enhance additional products. The answers obtained from companies confirmed the theoretical statement that in remanufacturing there is a very high variety of products models (products proliferation). The respondents explained that despite remanufacturing of single products' group (e.g., turbochargers) they still have to deal with hundreds of different variants of products, which belonged to different products' generation.

In order to additionally verify the problem of products proliferation, the companies were asked about the standard size of lot in their remanufacturing process. As presented in Fig. 4, majority of respondents (87.5 %) remanufactured mainly one-piece orders. Only 5 % of respondents are able to achieve the lot sizes of 6–20 pieces. Such situation influences organization of the cores because companies have to deal with high variety and complexity. This remark is consistent with the results of previous research.

The next problem that is addressed in the survey is quality of cores (see Fig. 5). The interviewers suggested several typical difficulties that were described in the literature (multiple-choice question):

Table 1 Experience of companies in remanufacturing and the employment level

	Less than 5 years (%)		5–10 years (%)		More than 10 years (%)	
Small enterprise (up to 49 employees)	0	0	1	12.5	3	11
Micro-enterprise (up to 9 employees)	4	100	7	87.5	25	89

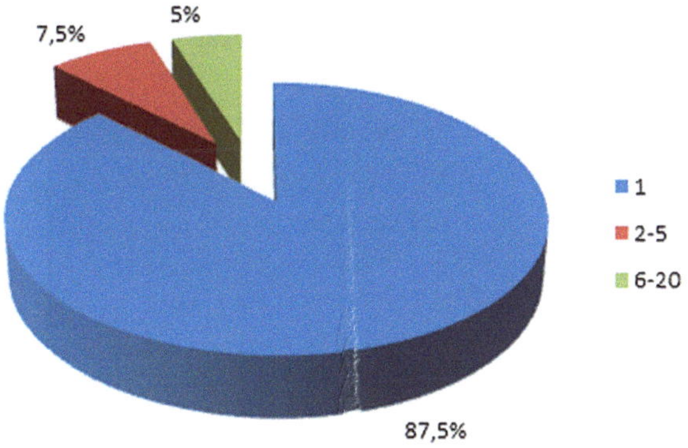

Fig. 4 Typical lot size in analyzed remanufacturing companies

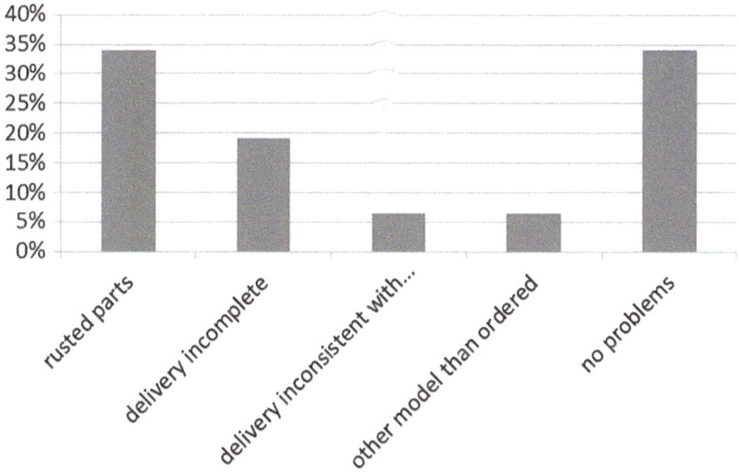

Fig. 5 Quality problems in cores' delivery

- rusted parts,
- incomplete delivery,
- delivery inconsistent with the contract/order,
- other model than ordered, and
- no problems with quality.

Companies have confirmed that frequently appearing problem is corrosion of parts and the incomplete deliveries. A significant proportion of respondents declared that there are "no problems." In addition to rusted parts, it is the most common answer among respondents. This finding might result in the fact that when providing remanufacturing services and dealing with one piece lot size companies are used to remanufacture any core disregarding its quality.

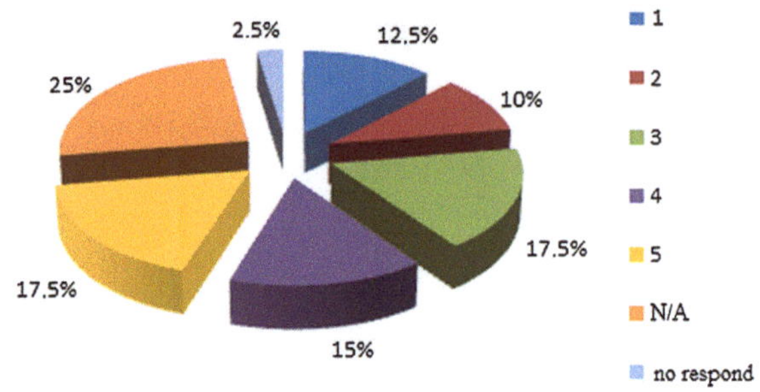

Fig. 6 Problems with matching parts in cores

The last factor that strongly affects remanufacturing process is the lack of spare parts used for reassembly stage. The problem is related to the time of launching the product on the market. Among the available models of cores are both old and new models. They have a limited number of spare parts, as partially or completely withdrawn them from the market already (old products) or have not yet been introduced to the market (new products).

We have examined the importance of this issue. The respondents have assessed this problem on Likert scale, where 1 means low importance of this problem and 5 means very high importance of this problem. The results are presented in Fig. 6. Majority of companies confirm that the matching parts problem is important regarding cores' management (answers on Likert scale from 3 to 5). Very few of the respondents (2.5 %) denied to answer this question. A big group of companies (25 %) stated that this problem is not applicable to them. That finding might result from the fact that remanufacturing companies, which provide remanufacturing services, try to fulfill any order they received and try to make in-house missing components.

The respondents provide the empirical data in domains as follows:

- identification of cores' supply,
- problems by cores' supply,
- products proliferation,
- quality of cores, and
- material matching problems.

The respondents were also asked about the typical batch size. The results are summarized in Fig. 6. The most typical is one-piece batch size (87.5 %). The batches bigger than 5 pieces are only 5 % of total production. This result confirms the theoretical statements from Sect. 2 (Fig. 7).

In order to further verify the finding from the literature review, the lead times of particular remanufacturing operations were investigated. The respondents were asked how long does it take to perform particular operations. The results are presented in Fig. 8. The most time-consuming operation in analyzed companies was reprocessing of components. The companies confirmed also statement that cleaning is time consuming.

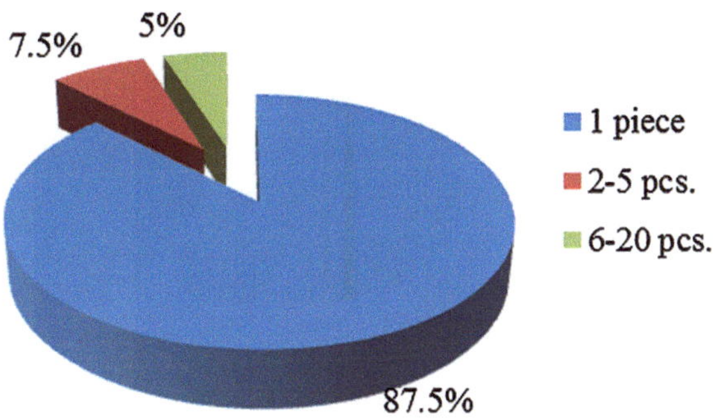

Fig. 7 Typical size of the batch for remanufacturing

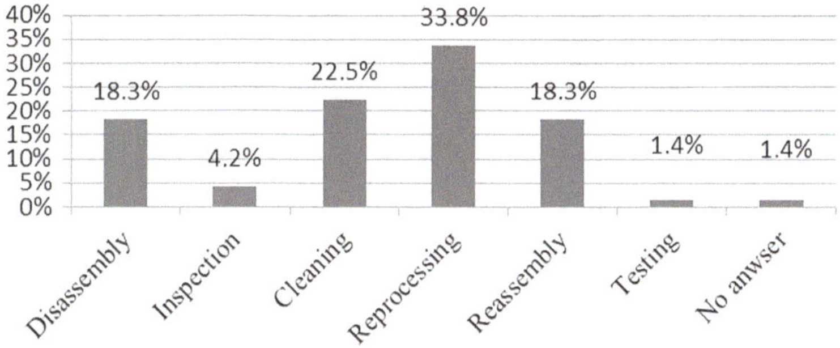

Fig. 8 The most time-consuming operations of remanufacturing

The respondents were also asked about the possibility of automation of the remanufacturing operations. According to the respondents, it is very difficult to introduce the atomization of the remanufacturing process. The answers to the question "Is it possible to automatize the following operations" were as follows:

- disassembly 100 % "no" answers,
- cleaning 92 % "no" answers,
- reprocessing 95 % "no" answers, and
- reassembly 100 % "no" answers.

We have examined the complexity of cleaning operations. The respondents have assessed this problem on the Likert scale, where 1 means low importance of this problem and 5 means very high importance of this problem. The results are presented in Fig. 9. Majority of companies in the survey (over 85 %) stated that cleaning operations are not very difficult (no problem or answers 1–2).

More problems the companies faced with the material matching restriction for reassembly. Half of the companies have assessed this problem on the Likert scale as an important one (assessment 3–5). The results are presented in Fig. 10.

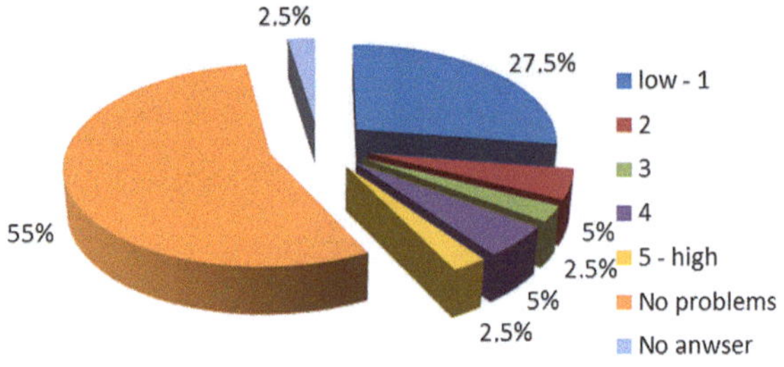

Fig. 9 The difficulty of cleaning operations

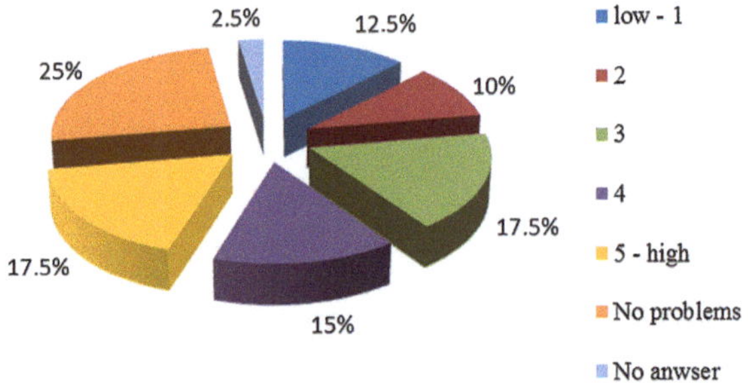

Fig. 10 Materials matching restrictions for reassembly

5 Conclusions

Authors in this paper aim to verify the theoretical findings in the empirically pilot study, which was conducted in small-sized companies involved in remanufacturing of automotive components.

Authors in the survey try to receive in depth inside in remanufacturing process organization regarding the problems described in the literature, such as

- insufficient availability of the good quality cores,
- high product variability,
- variation of the material recovery rate because of materials' matching problems, and
- time consumption of the operations.

Table 2 presents comparison table between problems which were found in the literature with those from the survey.

Most of the respondents remanufacture one piece lot sizes. The variety of products is very high. About 50 % of companies confirm that material matching restrictions are an important problem (assessment 3–5 in Fig. 10). Regarding quality problems, companies mainly have to deal with rusted parts and incomplete deliveries. Significant group of companies declare that quality of core problem is not relevant to their case. During phone survey, they explain that such statement is based on the fact that they do not own the cores. The further research step will include the extension of the respondents group and continuation of the pilot study with focus on medium-sized companies.

Table 2 Comparison of the literature findings and survey results

Domain	Problem	Status in survey
Inputs	Quality of cores	Not sufficient data—most of companies provide remanufacturing services
	Availability of cores	Not sufficient data—most of companies provide remanufacturing services
	Variable delivery times	Not sufficient data—most of companies provide remanufacturing services
	Small batches	Confirmed by respondents
	Multimodels	Confirmed by respondents
Process execution	Time-consuming disassembly	Confirmed by respondents
	Small batches	Confirmed by respondents
	Variable reprocessing lead time	Confirmed by respondents
	Time-consuming reprocessing	Confirmed by respondents
	Frequent setups	Confirmed by respondents
	Difficulty of cleaning operations	Not confirmed by respondents
	Time-consuming cleaning operations	Confirmed by respondents
	Variable customers' requirements	Confirmed by respondents
	Materials matching restrictions	Confirmed by respondents

Acknowledgments The authors would like to thank the Narodowe Centrum Badan i Rozwoju NCBiR (National Centre for Research and Development) for financing the research. This paper refers to the research financed by the NCBiR in the framework of the German-Polish cooperation for sustainable development, project "Sustainability in remanufacturing operations (SIRO)," grant no WPN/2/2012.

References

Golinska P, Kosacka M (2014) Environmental friendly practices in the automotive industry. In: Environmental issues in automotive industry. Springer, Berlin, pp 3–22

Golinska-Dawson P, Nowak A (2015) The survey on cores supplies in the sme in automotive remanufacturing sector. LogForum 11(1):51–61

Guide VDR Jr (2000) Production planning and control for remanufacturing. J Oper Manage 18:467–483

Hammond R, Amezquita T, Bras B (1998) Issues in the automotive parts remanufacturing industry: a discussion of results from surveys performed among remanufacturers. Eng Des Autom 4(1998):27–46

Jayaraman V (2006) Production planning for closed-loop supply chains with product recovery and reuse: an analytical approach. Int J Prod Res 44(5):981–998

Lind S, Olsson D, Sundin E (2011) Exploring inter-organizational relationships within the remanufacturing of automotive components. In: Proceedings of 1st international conference on remanufacturing, Glasgow, UK, 26–29 July 2011

Lund R (1996) The Remanufacturing industry-hidden giant. Research Report, Manufacturing Engineering Department, Boston University.

Ostlin J, Ekholm H (2007) Lean production principles in remanufacturing a case study at a toner cartridge remanufacturer. In: Proceedings of the 2007 IEEE international symposium on electronics and the environment, IEEE

Parkinson HJ, Thompson GA (2003) Analysis and taxonomy of remanufacturing industry practice. Part E J Process Mech Eng 217: 243–256

Seitz M, Peattie K (2004) Meeting the closed-loop challenge: the case of remanufacturing. Calif Manag Rev 46(2):74–89

Sundin E, Bras B (2005) Making functional sales environmentally and economically beneficial through product remanufacturing. J Clean Prod 13(9):913–925

Sundin E (2004) Product and process design for successful remanufacturing. Linköping Studies in Science and Technology Dissertation No. 906, Production Systems, Department of Mechanical Engineering Linköpings Universitet, Sweden

Subramoniam R, Huisingh D, Chinnam RB (2010) Aftermarket remanufacturing strategic planning decision-making framework: theory & practice. J Clean Prod 18(16):1575–1586

Souza GC (2008) Closed-loop supply chains with remanufacturing. Tutor Oper Res 130–153 (INFORMS, Hanover)

Rubio S, Corominas A (2008) Optimal manufacturing-remanufacturing policies in a lean production environment. Comput Ind Eng 55(1):234–242

Makespan Minimization for Scheduling Unrelated Parallel Machine with Sequence-Dependent Setup Time

Karn Moonsri and Kanchana Sethanan

Abstract This chapter presents an investigation of a scheduling problem of unrelated parallel machines with sequence-dependent setup time from hard disk drive industries. It specifically focuses on the testing process and setup changes of the testing program according to the product type required. When the demand of customers is greater, the setup of the testing program increases, causing a reduction in production efficiency. The objective is to sequence and allocate the jobs for the unrelated parallel machines to minimize the makespan. The problem is formulated as a mixed integer linear programming (MILP) model, and a heuristics algorithm is developed to find near-optimal solution. The results from the heuristics are compared with the optimal solutions to evaluate the effectiveness of the heuristics algorithm.

Keywords Scheduling problem · Makespan · Sequence-dependent setup time

1 Introduction

The scheduling problem is a crucial problem in all industrial production systems. Scheduling of production serves essential roles which can be regarded as the indicator of overall production efficiency. With the intense competition of industries, suitable methods are needed in order to maximize the efficiency of the production management system to be able to compete in the industrial market. The common scheduling problem of production is exceedingly complicated, especially

K. Moonsri (✉) · K. Sethanan
Research Unit on System Modeling for Industry, Industrial Engineering Department, Faculty of Engineering, Khon Kaen University, Khon Kaen 40002, Thailand
e-mail: hotphai@gmail.com

K. Sethanan
e-mail: ksethanan@gmail.com

V. Kachitvichyanukul et al. (eds.), *Toward Sustainable Operations of Supply Chain and Logistics Systems*, EcoProduction,
DOI 10.1007/978-3-319-19006-8_17

for unrelated parallel machine scheduling with numerous purposes such as to minimize completion time and the delay of submission and to maximize the output or minimize total cost.

According to the unrelated parallel machine scheduling, there is 'n' modified as job (job = 1, 2, 3, …, n) which is produced on single stage (stage: $s = 1$), while the workstation consists of machines with amount 'm' ($m = 1, 2, 3$). Each job can be processed by only one machine; similarly, each job can process only one step and there is no relation between the processing times of a job on the different machines. In addition, the scheduling problem of production on parallel machines can be divided according to the type of problem, such as uniform and identical machines cases. Additionally, problem types are described for a scheduling problem as $\alpha|\beta|\gamma$: α represents the machine environment; β represents constraints, processing, and characteristics; and γ is the objective function. Thus, Pinedo (2012) considered the problem of the unrelated parallel machine scheduling problem with sequence-dependent setup times which can be noted as $R|S_{ijm}|C\text{max}$. The proposed problem is categorized as an NP-hard problem.

The present study focuses on the hard disk drive assembly process of the head gimbal assembly (HGA) which is the process of assembling the head gimbal with the suspension in order to be assembled with other components. When the reader is assembled on the assembly line, the magnetic system is tested by the tester. In addition, due to the diversity of product features, of testing program, and of the tester, it causes the setting problem for the product test. The problem occurring in the current testing process is the inappropriate setting of the testing program each day. Consequently, the tester needs to be converted continuously, causing capacity loss due to the unit change of tester in order to enable the tester to work fully with the testing programs. Furthermore, the conversion of the tester settings takes a long period to convert the program, which is the sequence-dependent setup time (SDST), and the duration of machine setting is based on the job. That can be viewed as a cause of damage due to the conversion of the tester setting on both the loss of production capability and time-consuming which might affect production; production cannot meet the requirements of customers.

Hence, this study aimed to explore the methods to schedule the production process of the reader's magnetic test with its problems of unrelated parallel machine scheduling and its limitations on SDST. The objective was to minimize completion time. The solution was given in the four sections: (1) developing a mathematical model by using the mixed integer linear programming (MILP) model for optimal solution; (2) developing an heuristics algorithm to find near-optimal solutions for massive and complicated problems, as well as for shortening the duration for finding the answers; and (3) comparing the answer of heuristics algorithm development and optimal solution. Finally, (4) concludes the paper with the findings of the research.

2 Literature Review

The review of literature focused on research on parallel machine problems which developed a mathematical model and employed various methods of heuristics for problem-solving; also, the studies on scheduling parallel machines with the limitation on SDST were also reviewed. Ramezanian and Saidi-Mehrabad (2012) studied the multi-product unrelated parallel machine scheduling problem with rework processes. The research developed the mathematical model on mixed integer nonlinear programming (MINLP) and employed various methods of heuristics for problem-solving and making a comparison to find the best solution for production scheduling and to minimize makespan, including random, shortest processing time (SPT), longest processing time (LPT), modified shortest processing time (MSPT) and modified longest processing time (MLPT). Ruiz-Torres et al. (2013) carried out research on parallel machine scheduling to minimize the makespan with sequence-dependent deteriorating effects. The mathematical model developed for mixed integer linear programming (MILP) was tested to find the best answer, by comparing list schedule algorithms with the simulated annealing (SA) method in order to find the most efficient scheduling. Principally, the development of heuristics is accompanied with an optimal solution in order to compare the efficiency of finding the solution by the heuristics. Edis et al. (2013) explained the components of parallel machine production scheduling in order to check and analyze related problems on PMS to which resources were added and to estimate the result for further study by analyzing the strengths and weaknesses of studies in the literature with the focus on the machine environment, the increase in resources, objective function, complication, solution, and the importance of that literature report.

Tahar et al. (2006) have developed a linear programming approach for identical parallel machine scheduling with job splitting and SDSTs, by developing a mathematical model compared with the efficiency of the heuristics' initial solution. The objective was to find the minimum makespan. Vallada and Ruiz (2011) used a genetic algorithm (GA) solution for the unrelated parallel machine scheduling problem with sequence-dependent setup times to provide a mixed integer programming (MIP) mathematical model, and the MIP model will be used later in the computational experiments. Logendran et al. (2007) found a way to minimize the weighted tardiness of jobs in unrelated parallel machining scheduling with sequence-dependent setups and dynamic machine availability for jobs released during the planning horizon. They presented a statistical model developed for comprehensively addressing search algorithm tabu search and comparison efficiency. Lee and Pinedo (1997) studied a number of jobs to be processed on a number of identical machines in parallel and developed a heuristic to minimize the sum of the weighted tardiness through SDSTs. They used the SA method and found SA can be adapted to be suitable for the total weighted tardiness problem without setup times. Kuo et al. (2011) studied an unrelated parallel machine scheduling problem with setup time and learning effects simultaneously. They considered the total absolute deviation of job completion times and the total load on all machines as scheduling measures and studied an unrelated parallel machine problem simultaneously with SDSTs for the presented problem to remain polynomial solvable. Behnamian et al. (2009) proposed a hybrid metaheuristic for the minimization of makespan in scheduling problems with parallel machines and SDSTs. They developed a hybrid of population-based evolutionary searching ability of ant colony optimization (ACO), a SA for solution evolution, and a variable neighborhood search (VNS) finding heuristic and comparison for near-optimal solution. Ying and Cheng (2010) studied the dynamic parallel machine scheduling problem with SDSTs by a developed greedy heuristic and the corresponding lower bound (LB). Cappadonna et al. (2013) have developed a MILP model for optimally solving the problem in the unrelated parallel machine scheduling problem with limited human resources and developed a GA for finding minimization completion time.

Moreover, the scheduling problem can be applied to other fields. Kaplan and Rabadi (2012) focused on the principle of parallel machine production scheduling on identical machines. The principle was applied to the refueling of in-flight aircraft without interruption. The development of the mathematical model was compared with the efficiency of a heuristic solution and SA methods in order to maximize the efficiency of schedules for in-flight aircraft refueling. The aircraft containing fuel acts as machine, while the acquired aircraft seems as a job to meet the purpose which is to minimize total weighted tardiness.

According to the comparison of the findings on heuristics, it was found that the use of heuristics is suitable to solve a huge problem as well as to reduce the acceptable length of time lesser than the mathematical model.

3 Problem Statement

This study focused on the problems of the testing process with N products as shown in Fig. 1. Each product was separated for testing with a PG program on M machines. For each machine, the length of time for conversion was dependent on SDST; each job could be tested on multiple machines with differences of duration as well as of working, and there was a limitation of time for each machine.

4 Mixed Integer Linear Programming (MILP) Model

A MILP model is formulated for the addressed unrelated parallel machine scheduling problem with sequence-dependent setup times in manufacturing ($R|S_{ijm}|$ Cmax) which is based on an adapted standard form of the unrelated parallel machine scheduling problem. We first need notation which is defined as follows.

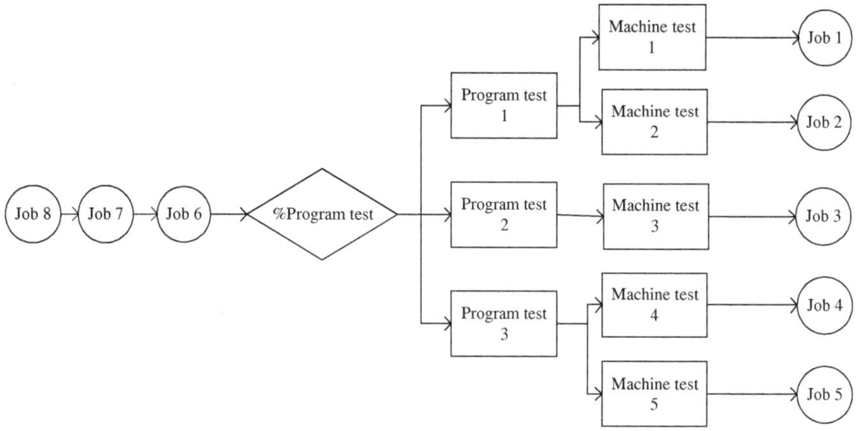

Fig. 1 Characteristic of the studied problem

4.1 Notation Definition

Indexes

i, j Product index $i, j = 1, 2, 3, \ldots, N$

g, f Program test index $g, f = 1, 2, 3, \ldots,$ Pg

m Machine index $m = 1, 2, 3, \ldots,$ MC

Parameters

N The number of products

MC The number of machines

Pg The number of program tests

DP_{ig} The quantity demanded for product i program test g

PT_{igm} The processing time for product i program test g on a machine $MTA_{m,}$ the processing time limit on machine m

S_{ijgfm} Program test of machine-based sequence-dependent setup time on machine when processing job j, after having processed job i

M A large enough number

Decision variables

$C_{\max}$ Makespan

C_{igm} Completion time for product i program test g on machine m

PS_{igm} Quantity testing product i program test g on machine m

Binary variables

$x_{igm} = 1$ if product i program test g assigned to machine m and 0 otherwise

$y_{igfm} = 1$ if product i program test g assigned to machine m before product i program test f to machine m and 0 otherwise

4.2 Mathematical formulation

The objective function is

$$\text{Min}\, C_{\max} \tag{1}$$

and the constraints are

$$\sum_{m=1}^{MC} x_{igm} = 1 \quad \forall i, g \tag{2}$$

$$\sum_{j=1}^{N} \sum_{g=1}^{\mathrm{Pg}} y_{ijgfm} = 1 \quad \forall i, m;\; i \neq j \tag{3}$$

$$\left(x_{igm} * M\right) - \mathrm{PS}_{igm} \geq 0 \quad \forall i, g, m \tag{4}$$

$$\sum_{m=1}^{MC} \mathrm{PS}_{igm} = \mathrm{DP}_{ig}, \quad \forall i, g \tag{5}$$

$$\begin{aligned} C_{jgm} + M\left(1 - y_{igfm}\right) &\geq C_{igm} + S_{ijgfm} \\ &+ \mathrm{PS}_{igm} * \mathrm{PT}_{igm} \quad \forall i, g, m;\; i \neq j \end{aligned} \tag{6}$$

$$x_{igm} - u_{0jm} - \sum_{i=1}^{N} y_{ijgfm} = 0 \quad \forall m, g \tag{7}$$

$$x_{igm} - w_{0im} - \sum_{j=1}^{N} y_{ijgfm} = 0 \quad \forall m, g \tag{8}$$

$$\sum_{j=1}^{N} u_{0jm} = 1 \quad \forall m, g \tag{9}$$

$$\sum_{i=1}^{N} w_{0im} = 1 \quad \forall m, g \tag{10}$$

$$C_{igm} \leq \mathrm{MTA}_m \quad \forall i, g, m \tag{11}$$

$$C_{\max} \geq C_{igm} \quad \forall i, g, m \tag{12}$$

$$C_{\max}, C_{igm}, \mathrm{PS}_{igm} \geq 0 \quad \forall i, g, m \tag{13}$$

$$x_{igm}, y_{igfm} \in \{0, 1\} \tag{14}$$

The objective function (1) is to minimize the maximum completion time or makespan. Constraints (2) ensure that every job and each program test is assigned to exactly one machine. Constraints (3) ensure that each program test can be processed on only one machine. Constraints (4) are for $x_{igm} = 1$ when job i program test g on machine m. Constraints (5) ensure that the quantity testing product equals the quantity demanded for the product. Constraints (6) are to control the completion times of the jobs at the machines. Basically, if a job j is assigned to machine m after job i, its completion time C_{ifm} must be greater than the completion time of job i, C_{igm}, plus the setup time (changed program test) between i and j and the processing time of i. If $y_{igfm} = 0$, then the big constant M renders the constraint redundant. Constraints (7) ensure that, except for the first product, a product scheduled on a machine must be immediately preceded by exactly one different product. Constraints (8) ensure that, except for the last product, a product scheduled on a machine must be immediately followed by exactly one product. Constraints (9)–(10) a machine can have exactly one first and one last product. Constraints (11) ensure that the completion time of each job must not exceed the processing time limit of machines. Constraints (12) ensure that the makespan is equal to or greater than the completion time of the products. Constraints (13)–(14) are the basic restrictions on the decision variables.

To solve the problem using a mathematical model is only suitable for small problems. Using such a model for a massive problem would be difficult or time-consuming to find the optimal solution because the massive problem, such as the problem on a production sequence, is too complicated. The essential limitation is the technological limitations on finding the solution. The cause of the inefficient finding of an optimal solution may lie in the limitation of variables.

5 Heuristics Algorithm

Heuristics are the process of finding solutions without rules as well as testing the decision from trial and error which is the guide for finding a solution. Finding solutions through heuristics is different from other methods, because heuristics do not take into account any possible solutions; instead, it chooses the answer which is suitable for the method. This kind of process benefits finding answers from a massive and complicated source of data. However, heuristics still have their own disadvantages; the solution from such a method is only a near-optimal solution.

Thus, the present study aimed to develop the algorithm named unrelated parallel machine with sequence-dependent setup times (UPSDST) in order to resolve the massive problem and gain a near-optimal solution. The details of UPSDST and variables and symbols used to create UPSDST are defined below.

i, j	Product index
g, f	Program test index
m	Machine index
N	Set number of products; $N = \{DP(i, g)\}$
DM	Set of jobs that max demand; DM = MAX $\{DP(i, g)\}$
MC	Set number of machines
$ST(i, g, f, m)$	Set setup time of program test on machines m
$PT(i, g, m)$	The processing time for product i program test g on machine m
$PS(i, g, m)$	Quantity testing product i program test g on machine m
$C(i, g, m)$	Completion time for product i program test g on machines
Cmax	Makespan MAX $\{C(i, g, m)\}$

The UPSDST algorithm can be divided into 8 stages. Details are given below

Step 1. Consider overall job needs by choosing the most needs on set N job indicated/specified as DM = MAX $\{DP(i, g)\}$

Step 2. Take into account the overall machines of set MC and produce the set of machines compatible with DM indicated/specified as MC(dm)

Step 3. Select the machine in MC(m) for tasks/jobs of set DM by choosing m; $m \in$ MC(m)

3.1. Firstly select the machine with the minimum test time which is m'; $m' = \text{MIN}\{PT(i, g, m)\}$; $\forall\, m$ and choose the setup time with minimum time S'; $S' = \text{MIN}\{ST(i, g, f, m)\}$, then make a check for $m' \cap S'$ on each sequence and turn it into $m'' = \{PT(i, g, m)\}$.

3.2. Focus on the equality of the minimum test time. If it is equal, then following No. 4.1.3; if not, skip to step 4

3.3. Choose minimum test time machine as well as fastest test time on readiness of working: m''; $m'' = \text{MIN}\{PT(i, g, m)\}$; $\forall\, m$, then follow step 4

Step 4. Consider the quantities of DM production for m machine from step 3 as follows:

4.1. Check for remaining time of machine: $MTA(m) > DP(i, g)*PT(i, g, m) + ST(i, g, f, m)$; if it is over, the amount of production should be based on $PS(i, g, m)$; $PS(i, g, m) = DP(i, g)$, then skip to step 4; if not, follow no. (4.2)

4.2. Produce according to the maximum amount specified in m machine which is PS(i, g, m); PS(i, g, m) = MTA(m)/PT(i, g, m), then follow no. (4.3)
4.3. Update the data of set DM = DM\{PS}, MC(dm) = MC(dm)\{m}, then back to step 3

Step 5. Update the data of set N; N\{DP(i, g) ∈ DM}, MC; MC\{m}, then follow step 6
Step 6. Check the correctness of set $N = Ø$; if it is correct, forward to step 7; if not, back to step 1
Step 7. Calculate the completion time of each job; $C(i, g, m)$ = PS(i, g, m) * PT(i, g, m) + ST(i, g, f, m), then follow step 8
Step 8. Calculate the completion time of makespan; Cmax = MAX{$C(i, g, m)$}

6 Comparison study

In order to make the comparison between the finding from the algorithm and the answer from the mathematical model, the sample problem was created with 6 different sizes, each size with 5 different subsizes of problem. The optimal solution was concluded through CPLEX/MPL Modeling System for Windows Student Version on 2.40 GHz PC (4 GB RAM), while a heuristic was developed through the software named MATLAB Version 7.14.0.739 in order to find the solution. The comparison of algorithm and optimal solution and the result of such comparison are shown in Tables 1, 2, 3, 4, 5, 6, and 7.

Table 1 The comparison of algorithm and optimal solution (number of job, number of program test, number of machine)

Product size (4, 2, 3)	Optimal	UPSDST	Error	% error
1	643	684	41	6.38
2	547	547	0	0.00
3	702	720	18	2.56
4	683	692	9	1.32
5	651	686	35	5.38
			Average	3.13

Table 2 The comparison of algorithm and optimal solution (number of job, number of program test, number of machine)

Product size (4, 3, 8)	Optimal	UPSDST	Error	% error
1	521	574	53	10.17
2	568	594	26	4.58
3	549	584	35	6.38
4	574	596	22	3.83
5	484	497	13	2.69
			Average	5.53

Table 3 The comparison of algorithm and optimal solution (number of job, number of program test, number of machine)

Product size (3, 2, 10)	Optimal	UPSDST	Error	% error
1	280	297	17	6.07
2	326	356	30	9.20
3	317	363	46	14.51
4	395	406	11	2.78
5	350	369	19	5.43
			Average	7.60

Table 4 The comparison of algorithm and optimal solution (number of job, number of program test, number of machine)

Product size (2, 2, 10)	Optimal	UPSDST	Error	% error
1	142	163	21	14.79
2	164	173	9	5.49
3	159	166	7	4.40
4	157	185	28	17.83
5	174	192	18	10.34
			Average	10.57

Table 5 The comparison of algorithm and optimal solution (number of job, number of program test, number of machine)

Product size (5, 4, 6)	Optimal	UPSDST	Error	% error
1	546	595	49	8.97
2	521	532	11	2.11
3	593	603	10	1.69
4	502	516	14	2.79
5	549	563	14	2.55
			Average	3.62

Table 6 The comparison of algorithm and optimal solution (number of job, number of program test, number of machine)

Product size (5, 4, 5)	Optimal	UPSDST	Error	% error
1	463	486	23	4.97
2	484	491	7	1.45
3	492	512	20	4.07
4	587	619	32	5.45
5	593	612	19	3.20
			Average	3.83

Table 7 Result of comparison

Maximum % error	17.83
Minimum % error	0.00
Average % error	5.71

The comparison of completion time between algorithm and the 30 most suitable solutions was performed through the use of the mathematical problem of the production of 2–5 products tested by 2–4 testing program(s) and 3–10 machines. The results indicate that the completion time of UPSDST calculated as a percentage is 5.71 %. This reveals that UPSDST has met the criteria of working performance.

7 Conclusions and future research

According to the study of unrelated parallel machine scheduling under the limitation of sequence-dependent setup times, the completion time is minimized. The devised mathematical model can offer an optimal solution. Nevertheless, due to the difficulty of using a mathematical model with a massive problem or the limitation of the software, it was unable to find the answer. Therefore, the algorithm UPSDST needs developing so as to mesh with a massive problem. Based on the comparison of efficiency, it can be seen that the algorithm UPSDST can provide the closest and most near-optimal solution as well as the percent error of 5.71 %. This algorithm can be simply applied for production scheduling without complication. It can shorten the length of production planning as well as decrease the cost of production.

Further research is to focus on the production factors on machine restriction and setup time which are caused by changes of product, but not by changes of program. However, the program test used in this study will be used in further studies. The research will develop the form of problem considering the duration of long-term planning in order to simulate the problem and statistical test. The study will draw a comparison between metaheuristics and the UPSDST algorithm in order to find near-optimal and suitable answers.

Acknowledgments The authors would like to thank the anonymous reviewers for their helpful comments. In addition, this research is supported by the Research Unit on System Modeling for Industry, Khon Kaen University, Thailand.

References

Behnamian J, Zandieh M, Ghomi SF (2009) Parallel-machine scheduling problems with sequence-dependent setup times using an ACO, SA and VNS hybrid algorithm. Expert Syst Appl 36(6):9637–9644

Cappadonna FA, Costa A, Fichera S (2013) Makespan minimization of unrelated parallel machines with limited human resources. Procedia CIRP 12:450–455

Edis EB, Oguz C, Ozkarahan I (2013) Parallel machine scheduling with additional resources: notation, classification, models and solution methods. Eur J Oper Res 230(3):449–463

Kaplan S, Rabadi G (2012) Exact and heuristic algorithms for the aerial refueling parallel machine scheduling problem with due date-to-deadline window and ready times. Comput Ind Eng 62 (1):276–285

Kuo WH, Hsu CJ, Yang DL (2011) Some unrelated parallel machine scheduling problems with past-sequence-dependent setup time and learning effects. Comput Ind Eng 61(1):179–183

Lee YH, Pinedo M (1997) Scheduling jobs on parallel machines with sequence-dependent setup times. Eur J Oper Res 100(3):464–474

Logendran R, McDonell B, Smucker B (2007) Scheduling unrelated parallel machines with sequence-dependent setups. Comput Oper Res 34(11):3420–3438

Pinedo ML (2012). Scheduling: theory, algorithms, and systems. Springer, Berlin

Ramezanian R, Saidi-Mehrabad M (2012) Multi-product unrelated parallel machines scheduling problem with rework processes. Scientia Iranica 19(6):1887–1893

Ruiz-Torres AJ, Paletta G, Pérez E (2013) Parallel machine scheduling to minimize the makespan with sequence dependent deteriorating effects. Comput Oper Res 40(8):2051–2061

Tahar DN, Yalaoui F, Chu C, Amodeo L (2006) A linear programming approach for identical parallel machine scheduling with job splitting and sequence-dependent setup times. Int J Prod Econ 99(1):63–73

Vallada E, Ruiz R (2011) A genetic algorithm for the unrelated parallel machine scheduling problem with sequence dependent setup times. Eur J Oper Res 211(3):612–622

Ying KC, Cheng HM (2010) Dynamic parallel machine scheduling with sequence-dependent setup times using an iterated greedy heuristic. Expert Syst Appl 37(4):2848–2852

Single Machine Scheduling for Minimizing Earliness/Tardiness Penalties with Sequence-Dependent Setup Times

Changseong Ko, Sooyong Kim, Byungnam Kim and Shiegheun Koh

Abstract This chapter deals with a single machine system with job release times and sequence-dependent machine setup times. The objective of the problem is to minimize weighted sum of earliness and tardiness penalties. A mixed integer programming formulation for the problem is first presented, and using this formulation, one can easily find optimal solutions for small problems by using a commercial optimization software. Since the problem is NP-hard and the size of a real problem is large, a number of heuristic solution procedures are proposed including genetic algorithm to solve the practical big-sized problems in a reasonable computational time. To assess the performance of the algorithms proposed, a computational experiment was conducted and the results demonstrate that the heuristic algorithms show different performances as the problem characteristics are changed and a heuristic shows much better performance than genetic algorithm for the case when the number of jobs is relatively large.

Keywords Scheduling · Earliness/tardiness penalty · Sequence-dependent setup

1 Introduction

After the mid-1950s, there were thousands of studies dealing with the scheduling problem for manufacturing facilities (Allahverdi et al. 2008), and especially, since 1970s, the scheduling problems considering earliness/tardiness penalties have been studied because the just-in-time (JIT) concept was spreading over the world (Baker and Scudder 1990). In this paper, we study a single machine scheduling problem in

S. Kim · B. Kim · S. Koh (✉)
Pukyong National University, Yongsoro 45, 608-737 Namgu, Busan, South Korea
e-mail: sgkoh@pknu.ac.kr

C. Ko
Kyungsung University, Suyoungro 30945, 608-736 Namgu, Busan, South Korea

V. Kachitvichyanukul et al. (eds.), *Toward Sustainable Operations of Supply Chain and Logistics Systems*, EcoProduction,
DOI 10.1007/978-3-319-19006-8_18

which the weighted sum of earliness and tardiness penalties is minimized (abbreviated to SET; Single machine problem with Earliness and Tardiness penalties).

There were many SET literatures after Sidney (1977), and they are well summarized by Baker and Scudder (1990), in which it was emphasized that though most of the earliness/tardiness studies avoid the issue of the idle time inserted between jobs either by assuming a common due date for all jobs or by restricting the solution to be a nondelay schedule, the essence of the earliness/tardiness problem lies in its nonregular performance measure, and so, imposing the arbitrary restriction that there be no idle time declines the importance of this objective. Kanet and Sridharan (2000) focused on the inserted idle time (abbreviated to IIT) in scheduling problems and reviewed the earliness/tardiness literature with IIT.

To solve the SET with IIT, according to Baker and Scudder (1990), a procedure consisting of two steps which are sequencing jobs and assigning idle times between jobs is needed. To determine the job sequence, Fry et al. (1990) proposed an adjacent pairwise exchange heuristic in which nine precedence relationship rules are used to reduce the number of candidates for interchanging jobs, while a straightforward linear programming formulation is proposed to find the job starting times. Yano and Kim (1991) and Kim and Yano (1994) found the optimal job sequence using a branch-and-bound algorithm and performed optimal idle time assignment between jobs using the dynamic programming. Davis and Kanet (1993) studied the case that the penalties are general convex functions of earliness and tardiness and proposed a pseudopolynomial algorithm. Lee and Choi (1995) proposed an algorithm finding the optimal job starting time to minimize the earliness and tardiness costs when the job sequence is given and, using this algorithm, developed a genetic algorithm to find the optimal job sequence. Szwarc and Mukhopadhyay (1995) showed the solution is composed of several clusters of uninterrupted jobs separated by idle time and provided an efficient two-stage procedure for identifying clusters and then determining their starting times. For the problems with due windows, Wan and Yen (2002) proposed a Tabu search-based procedure using an algorithm similar to Lee and Choi (1995) to find the job starting time.

Setup is very common work to prepare the machine in the industry, and it is an important issue in the scheduling theory. Allahverdi et al. (2008) introduced a number of studies that insisted on the importance of sequence-dependent setup times. To reflect this importance, since the mid-1960s, many scheduling literatures considering setup times independently of the processing times have been published. For the SET with sequence-dependent setup times, however, we could not find many. Coleman (1992) firstly formulated an SET with sequence-dependent setup times. Zhu and Heady (2000) developed a mixed integer programming model for the nonuniform parallel machine system with earliness/tardiness penalties and sequence-dependent setup times and, using that model, found an optimal solution for 3-machine and 9-job problem in a reasonable computation time.

To raise reality of the problem, meanwhile, recent studies dealing with the scheduling problems tend to consider the release time of each job. For the SET with the release time, Keyser and Sarper (1991) presented a target starting time heuristic to minimize the sum of earliness, tardiness, and waiting time costs. The heuristic permits machine idle times between jobs and it is improved by an adjacent pairwise interchange algorithm. Sridharan and Zhou (1996) proposed a heuristic algorithm for the SET with job release times that identified soon-to-arrive jobs and kept the machine deliberately idle for them. Later, Mazzini and Armentano (2001) proposed the moving heuristic procedure for the same problem. This heuristic consists of two steps, the first step for finding feasible solutions and the second step for applying a local search algorithm. Recently, Tsai (2007) proposed and compared a number of heuristics including genetic algorithm for the SET with job release times. For the most general problem dealing with earliness/tardiness penalties, Balakrishnan et al. (1999) studied a parallel machine earliness/tardiness problem with sequence-dependent setup times as well as job release times. They presented a mixed integer formulation that has fewer zero-one variables than other formulations and gave computational experiences to solve small-sized problems.

In this paper, we study an SET with sequence-dependent setup times and job release times. First of all, we present a mixed integer programming model for the problem. But it is well known that the SET is NP-hard (Baker and Scudder 1990; Garey et al. 1988), and then, a big-sized problem cannot be solved in a reasonable computation time. Therefore, we concentrate on developing heuristics, by which not optimal but good solution for realistic big-sized problem can be acquired in a reasonable computing time.

2 Model

The starting point of the study is representing the problem as a mathematical model. In this section, therefore, we formulate the problem as a mixed integer linear programming model. First, notation for the formulation is presented as follows:

Parameters

i and j	indices representing jobs ($i, j = 1, 2, \ldots, n$)
r_i	zero if the job i arrived already, the projected arrival time of the job, otherwise
p_i	processing time of job i
d_i	due date of job i
α_i	cost per unit earliness of job i
β_i	cost per unit tardiness of job i
s_{0i}	setup time for job i if it is the first job
s_{ij}	setup time for job j when the job i is processed right before it

Decision variables

x_i	starting time of job i (or to be exact, the setup starting time of job i)
f_i	completion time of job i
ε_i	earliness of job i
τ_i	tardiness of job i
z_{0i}	one, if job i is the first job, zero, otherwise
z_{ij}	one, if job j is processed right after job i, zero, otherwise

Using the above parameters and variables and referring to the models proposed by Zhu and Heady (2000) and Joo and Kim (2012), we can represent our problem to an integer linear programming (ILP) form as follows:

$$\text{Minimize} \sum_{i=1}^{n} (\alpha_i \varepsilon_i + \beta_i \tau_i) \tag{1}$$

subject to

$$x_i \geq r_i \quad \text{for all } i \tag{2}$$

$$x_i + \sum_{\substack{j=0 \\ j \neq i}}^{n} s_{ji} z_{ji} + p_i = f_i \quad \text{for all } i \tag{3}$$

$$x_j + M(1 - z_{ij}) \geq f_i \quad \text{for all } i, j \tag{4}$$

$$f_i - \tau_i + \varepsilon_i = d_i \quad \text{for all } i \tag{5}$$

$$\sum_{j=1}^{n} z_{0j} = 1 \tag{6}$$

$$\sum_{\substack{j=0 \\ j \neq i}}^{n} z_{ji} = 1 \quad \text{for all } i \tag{7}$$

$$\sum_{\substack{j=1 \\ j \neq i}}^{n} z_{ij} \leq 1 \quad \text{for all } i \tag{8}$$

z_{0i} 0/1 integer for all i
z_{ij} 0/1 integer for all i, j

Equation (1) means that this problem is to minimize weighted sum of earliness and tardiness penalties. Equation (2) assures that a job can start its processing (or to be exact, the setup) only after it becomes available. Equation (3) shows the relationship between the starting time, setup time, processing time, and the finishing time of a job. Equation (4) ensures a job can start its processing only after the preceding job is finished. Equation (5) gives the earliness and the tardiness values for each job. Equation (6) says that only one job can be assigned as the first job. According to Eq. (7), a job is either the first job or a successor of another job.

Similarly, Eq. (8) says that a job should be a predecessor of another job. But this equation includes inequality because of the case when the job i is the last job.

To validate the formulation, we provide a test problem whose data are as follows:

$$\begin{aligned} &\mathrm{n} = 9,\ r_i = (0, 0, 0, 0, 5, 8, 12, 17, 22),\ p_i = (5, 3, 4, 3, 6, 2, 2, 3, 5),\\ &d_i = (7, 12, 35, 18, 22, 29, 25, 38, 45),\ \alpha_i = (3, 5, 4, 6, 5, 9, 5, 2, 1),\\ &\beta_i = (3, 6, 7, 8, 7, 9, 4, 4, 3),\ s_{0i} = (2, 1, 1, 3, 2, 2, 1, 1, 2), \end{aligned}$$

and

$$s_{ij} = \begin{bmatrix} 0 & 1 & 3 & 2 & 1 & 2 & 2 & 1 & 1 \\ 2 & 0 & 1 & 2 & 1 & 3 & 2 & 2 & 1 \\ 1 & 2 & 0 & 1 & 1 & 2 & 2 & 1 & 3 \\ 2 & 1 & 2 & 0 & 2 & 1 & 1 & 2 & 2 \\ 1 & 2 & 2 & 1 & 0 & 2 & 3 & 2 & 1 \\ 3 & 1 & 1 & 2 & 1 & 0 & 1 & 2 & 1 \\ 2 & 1 & 2 & 1 & 2 & 1 & 0 & 1 & 1 \\ 2 & 2 & 1 & 3 & 2 & 1 & 2 & 0 & 2 \\ 1 & 1 & 2 & 1 & 2 & 3 & 1 & 1 & 0 \end{bmatrix}.$$

Using LINGO system with these data and 15 s of running time in 2.6 GHz PC, we could easily find an optimal solution, in which the optimal job sequence is (1, 2, 5, 4, 7, 6, 3, 8, 9) and the optimal schedule for each job is (0, 2, 7), (7, 8, 11), (11, 12, 18), (18, 19, 22), (22, 23, 25), (26, 27, 29), (29, 30, 34), (34, 35, 38), and (38, 40, 45), where the three values in each parenthesis are (setup starting time, main process starting time, and finishing time) for the corresponding job. The result is summarized in Table 1. The character 'F' of 'early/tardy' column in the table means that the corresponding job finishes just on its due date, in other word, the job is neither early nor tardy, while E(k) and T(k) mean the job is early and tardy by k days. For example,

Table 1 Optimal schedule of the example problem in tabular form

Order	Job	Setup start	Main start	Finish	Due	Early/tardy	Cost
1	1	0	2	7	7	F	0
2	2	7	8	11	12	E(1)	5
3	5	11	12	18	22	E(4)	20
4	4	18	19	22	18	T(4)	32
5	7	22	23	25	25	F	0
6	6	26	27	29	29	F	0
7	3	29	30	34	35	E(1)	4
8	8	34	35	38	38	F	0
9	9	38	40	45	45	F	0
Total cost							61

one can see in the table that the job 5 is scheduled on the third position, its setup begins at 11, the main process begins at 12 (because the preceding job is 2, the setup time is s_{25}, or 1), and it finishes at 18. Since the due date of job 5 is 22, the job finishes 4 days earlier than due date, and then, we can see E(4). Now, the cost for each job can be calculated. The cost for job 5 is 20 because the cost per unit earliness of job 5 is 5. Finally, the total earliness/tardiness penalty is 61.

Although we could find an optimal solution for this small example problem in rather short time, any IP solver such as CPLEX as well as LINGO cannot provide an optimal solution for a practical-sized problem in a reasonable CPU time because the problem is known as NP-hard. Hence, we focus on heuristic algorithms including genetic algorithm to obtain near-optimal solutions of the problem in a reasonable CPU time.

3 Idle Time Between Jobs

As stated earlier, Baker and Scudder (1990) noticed two-step procedures, that are a procedure for job sequencing and another procedure for assigning idle times between jobs, are needed to solve an SET. In this section, we focus firstly on the procedure for assigning IIT. Though a number of earlier studies proposed the procedures, we propose another algorithm that assigns idle times between jobs when the job sequence is given. The algorithm serves the same result as Lee and Choi (1995) and Tsai (2007), but it is very intuitive and easier to understand than them. Detailed steps of the algorithm are as follows (in the algorithm, 'Ear,' 'Fit,' and 'Tar' mean the early-finished job, the on-time-finished job, and the tardy job, respectively):

Algorithm *INSERT_IDLE*

Step 0. According to given job sequence, make an earliest starting schedule, in which starting time of the first job is max{0, release date of the job} and starting time of the ith job is max{release date of the job, finishing time of the $(i - 1)$th job} for $i = 2, 3, \ldots, n$.

Step 1. If the last job in the schedule is 'Ear,' move it to the right until its finishing time equals to its due date. And set $i = n - 1$.

Step 2. If $i = 0$, stop the algorithm.

Step 3. If the ith job is 'Tar' or 'Fit,' set $i = i - 1$ and go to Step 2.

Step 4. If the finishing time of the ith job equals to the starting time of the $(i + 1)$th job, go to Step 6.

Step 5. Move the ith job to the right until one of two conditions given below is met. If the first condition is satisfied earlier than the second, set $i = i - 1$ and go to Step 2. Otherwise, go to Step 6.

1. finishing time of the ith job = due date of the ith job
2. finishing time of the ith job = starting time of the $(i + 1)$th job

Step 6. Let C be the set of jobs that consists of the ith job and its right-connected jobs. Here, the 'right-connected job' means that the job is later than the ith job and there is no idle time between the job and the ith job. Calculate EP and TP values for the set C, which mean the sum of earliness penalties for 'Ear' jobs in the set and the sum of tardiness penalties for 'Tar' and 'Fit' jobs in the set, respectively. If TP ≥ EP, set $i = i - 1$ and go to Step 2.

Step 7. Move all the jobs in the set C to the right until one of the following two conditions is met. Go to Step 6.

1. finishing time of the last job in C = starting time of the earliest job later than the jobs in C. (If there is no job later than the jobs in C, this condition should be ignored.)
2. finishing time of an 'Ear' job in C = due date of the job

4 Heuristic Procedures

Once we have a job sequence, the optimal schedule including IIT can be obtained by the algorithm INSERT_IDLE proposed in the previous section. On the other hand, the sequencing problem with the sequence-dependent setup times is well known as an NP-hard problem, and then, an optimal sequence for the problem cannot be found in a reasonable computation time. In this section, to find a good solution for the problem, we propose a number of heuristic procedures, each of which includes the INSERT_IDLE algorithm for assigning idle times between jobs. The first one is using the well-known heuristic rule EDD (earliest due date), and the procedure is as follows:

Algorithm *EDD*

Step 1. Assign jobs in increasing order of the due date, d_i.

Step 2. Assign idle times between jobs using the INSERT_IDLE algorithm.

Though EDD is very easy to understand and to get a solution, one can think its performance may not be very good because it is one-path algorithm; in other words, there is no improvement once an assignment was made. From this idea, we propose another heuristic algorithm, EDD-PI algorithm, which is an improvement of EDD. In this algorithm, we use a pairwise interchange (PI) method for adjacent jobs to improve the quality of a solution from EDD. The starting point of EDD-PI algorithm is a resulting schedule of EDD algorithm. The procedure is as follows:

Algorithm *EDD-PI*

Step 0. Start with a schedule that is a result of EDD algorithm.

Step 1. Let $S = \{1, 2, \ldots, n - 1\}$ and cur_obj be the objective value for the current schedule.

Step 2. Calculate δ_i for $i \in S$ as follows:

Case (1) If the job [i] is 'Ear,' and the job [i + 1] is 'Tar' where [i] means the ith job in the current job sequence, $\delta_i = \alpha_{[i]} + \beta_{[i+1]}$.
Case (2) If the job [i] is 'Ear' and the job [i + 1] is 'Ear' or 'Fit,' $\delta_i = \alpha_{[i]} - \alpha_{[i+1]}$.
Case (3) If the job [i] is 'Tar' or 'Fit' and the job [i + 1] is 'Tar,' $\delta_i = \beta_{[i+1]} - \beta_{[i]}$.
Case (4) Otherwise, $\delta_i = 0$.

Step 3. If $\delta_i \leq 0$ for all $i \in S$, stop the algorithm.
Step 4. Exchange the sequence of [i^*] and [i^* + 1], where i^* is the index of maximum δ_i. Apply the INSERT_IDLE algorithm and calculate the objective function value. If the value is smaller than cur_obj, go to step 5; otherwise, go to step 6.
Step 5. Let the current schedule be the schedule after the exchange and cur_obj be the objective value of the new schedule. Let $S = \{1, 2, \ldots, n - 1\} - \{i^*\}$. Go to Step 2.
Step 6. Maintain the current sequence before exchange. Let $S = S - \{i^*\}$ and go to Step 2.

Besides the EDD heuristic, another famous heuristic for the scheduling problem considering due dates is 'Min-Slack' heuristic, in which the job with smallest slack time is processed first. The slack time for a job is defined as due date of the job subtracted by processing time of the job. Based on the 'Min-Slack' heuristic, we propose another algorithm, MS (Min-Slack), as follows:

Algorithm *MS*

Step 0. Set $S = \{1, 2, \ldots, n\}$.
Step 1. Calculate slack time for each job $i \in S$ as follows:

1. $ST_i = d_i - (s_{0i} + p_i)$, if $r_i \leq 0$,
2. $ST_i = d_i - (r_i + s_{0i} + p_i)$, otherwise.

Step 2. Let i^* be the index of job having the smallest slack time. Assign the job i^* into the first job to be processed and set $S = S - \{i^*\}$. Set $CT = s_{0i^*} + p_{i^*}$, if $r_{i^*} \leq 0$, or $CT = r_{i^*} + s_{0i^*} + p_{i^*}$, otherwise.
Step 3. If $S = \varnothing$, apply the INSERT_IDLE algorithm to the current job sequence, and stop the algorithm.
Step 4. Calculate slack time for job $j \in S$ as follows:

1. $ST_j = d_j - (CT + s_{i^*j} + p_j)$, if $r_j \leq CT$,
2. $ST_j = d_j - (r_j + s_{i^*j} + p_j)$, otherwise.

Step 5. Let j^* be the index of job having the smallest slack time. Assign the job j^* into the next job to be processed and set $S = S - \{j^*\}$. Set $CT = CT + s_{i^*j^*} + p_{j^*}$, if $r_{j^*} \leq CT$, or $CT = r_{j^*} + s_{i^*j^*} + p_{j^*}$, otherwise. Set $i^* = j^*$ and go to Step 3.

In the earlier proposed two heuristics (EDD and MS), a job is attached after the last job of current partial sequence when it is selected by the corresponding rule (due date, slack time, or objective value). They are simple to make solutions but the performances of the solutions are not likely to be very good. So we propose another heuristic, mEDD algorithm, in which a job is selected by the earliest due-date rule and it is not attached after the last job of the current sequence but inserted into a proper position between jobs in the current sequence. Though the calculation quantity of the algorithm increases, we can anticipate better performance from the algorithm. The detailed procedure is as follows:

Algorithm *mEDD*

Step 0. Let S be the set of all jobs and Ω be the current partial sequence, and then $S = \{1, 2, \ldots, n\}$ and $\Omega = \varnothing$.
Step 1. Sort jobs in S in increasing order of the due date, d_i.
Step 2. Let k be the first job in S. Assign the job k into the first job to be processed and set $S = S - \{k\}$ and $\Omega = \{k\}$.
Step 3. If $S = \varnothing$, apply the INSERT_IDLE algorithm to the current job sequence, and stop the algorithm.
Step 4. Let k be the first job in S and m be the number of jobs in Ω. Using the algorithm ParCost, calculate C_i, which is the cost of a new partial sequence that is made by inserting job k into the position before ith job in Ω, i = 1, 2, …, m + 1. The case of $i = m + 1$ means the position after the last job in Ω.
Step 5. Let i^* be the index of the smallest C_i value. Insert job k into the position before (i^*)th job in Ω. If $i^* = m + 1$, insert job k after the last job in Ω. Set $S = S - \{k\}$ and go to Step 3.

In step 4 of mEDD heuristic, consider an algorithm to calculate the objective function value for the given partial sequence. Its name is ParCost algorithm and the procedure is as follows:

Algorithm *ParCost*

Step 0. Let a be the number of jobs in a partial sequence. Set $i = 1$ and Cost = 0.
Step 1. Calculate the finishing time of job $[i]$, which is the ith job in the partial sequence, as follows:

1. $f_{[i]} = s_{0[i]} + p_{[i]}$, if $r_{[i]} \leq 0$,
2. $f_{[i]} = r_{[i]} + s_{0[i]} + p_{[i]}$, otherwise.

Step 2. Calculate cost value as follows:

1. $Cost = Cost + \alpha_{[i]}(d_{[i]} - f_{[i]})$, if $f_{[i]} \leq d_{[i]}$,
2. $Cost = Cost + \beta_{[i]}(f_{[i]} - d_{[i]})$, otherwise.

Step 3. Set $CT = f_{[i]}$ and $i = i + 1$.
Step 4. If $i = a + 1$, stop.
Step 5. Calculate the finishing time of job i as follows:

1. $f_{[i]} = CT + s_{[i-1][i]} + p_{[i]}$, if $r_{[i]} \leq CT$,
2. $f_{[i]} = r_{[i]} + s_{[i-1][i]} + p_{[i]}$, otherwise.

Step 6. Go to Step 2.

Finally, like the EDD heuristic, the other heuristics (MS and mEDD) can be improved by the adjacent pairwise exchange method that is used in the EDD-PI algorithm. We call these improved versions of the two heuristics as MS-PI and mEDD-PI. We will not explain about these new heuristics because they are exactly the same as the EDD-PI algorithm except step 0.

5 Genetic Algorithms

The proper representation of a solution plays a key role in the development of a genetic algorithm. We use the random keys representation. Since the representation method eliminates the infeasibility of the offspring chromosomes as well as randomly generated chromosomes by representing solutions in a soft manner, this representation is applicable to a wide variety of sequencing optimization problems. In our representation scheme, we use n random numbers from (0, 1) interval, by which a job sequence is made, and using this sequence and the INSERT_IDLE algorithm, a schedule can finally be generated. The decoding (i.e., converting a chromosome to corresponding solution) procedure is as follows:

Algorithm *DECODE*

Step 1. To generate the job sequence, sort the genes in a chromosome in ascending order. The job corresponding to the smaller gene value is assigned earlier.
Step 2. Assign idle times between jobs using the INSERT_IDLE algorithm.

We use the easiest and the most popular method that is referred to as roulette wheel reproduction where each current chromosome in the population has a roulette wheel slot sized in proportion to its fitness.

The crossover operator takes two chromosomes and swaps a part of their genetic information to produce new chromosomes. The easiest and the most classical method to crossover is to choose a random cut-point and generate the offspring by combining the segment of one parent to the left of the cut-point with the segment of the other parent to the right of the cut-point. When some representation methods, such as the permutation representation, are used in the sequencing problem, this one-cut-exchange crossover can hardly be applied since the resulting chromosomes from the crossover may be illegal. But, since our random keys representation has no legality problem, we can use the one-cut-exchange crossover.

Mutation produces spontaneous random changes in various chromosomes. This genetic operation serves the crucial role of replacing the genes lost from population during the selection process so that they can be tried in a new context, or providing the genes that were not present in the initial population. In our algorithm, we also use the simplest method, in which a gene is selected by a given probability and replaced with another random number from (0, 1) interval.

Adopting the elitist strategy, two best chromosomes are excluded from the crossover procedure, and they are directly copied to the next generation. And moreover, the best chromosome is excluded from the mutation procedure. The detailed procedure to generate the next generation is as follows:

Algorithm *EVOLVE*

Step 1. Copy two best chromosomes to the next generation.

Step 2. Select two chromosomes by the roulette wheel method and generate a random number from (0, 1) interval. If this random number is less than P_C, generate two new chromosomes from the selected two chromosomes by the one-cut-exchange crossover, or copy the two chromosomes to next generation directly, otherwise. Repeat this step until a new population is constructed. Here, P_C is a predetermined crossover probability that can be obtained from some experiments.

Step 3. For every gene in the chromosomes generated at step 2, generate a random number from (0, 1) interval. If this random number is less than P_M, replace the gene with another random number from (0, 1) interval. Here, P_M is a predetermined mutation probability that can be obtained from some experiments.

6 Computational Experiments

To evaluate the performances of the heuristics including genetic algorithm proposed earlier, we conduct a computational experiment in which the results of the heuristics are compared with the result of genetic algorithm using a number of randomly generated problems. Since the complexity of the problem depends on the number of jobs (n), we control this parameter to five levels (20, 40, 60, 80, and 100). For each value of n, we randomly generate 10 problems. To generate a problem for an n value, one has to generate r_i, p_i, d_i, α_i, β_i, s_{0i}, and s_{ij} for $i, j = 1, 2, \ldots, n$. For the value of r_i, we assume that half the jobs are ready to begin (i.e., $r_i = 0$) and the others are $U[0, 10]$, where $U[a, b]$ means a randomly generated integer from interval [a, b]. It is assumed that the processing time (p_i) is $U[2, 7]$ and the values of α_i and β_i are $U[0, 10]$. We also assume that s_{0i} and s_{ij} are $U[0, 5]$. And finally, the due date (d_i) is assumed to be $r_i + U[0, p_i + s_{0i}]$.

Each problem generated is solved by six heuristic algorithms (EDD, EDD-PI, MS, MS-PI, mEDD, and mEDD-PI) and genetic algorithm. The result of the computational experiments is shown in Table 2. The number in the table is the fraction value whose numerator is the objective value from the corresponding algorithm and the denominator is the objective value from the genetic algorithm. As one can see, the table consists of 40 solid line cells according to the five levels of

Table 2 Result of computational experiments

n		EDD	EDD-PI	MS	MS-PI	mEDD	mEDD-PI
20	Mean	5.67	2.74	6.59	2.57	1.44	1.38
	Min	2.99	1.27	3.57	1.90	1.03	1.03
	Max	14.32	4.30	15.92	3.75	2.04	1.71
40	Mean	7.51	4.56	9.45	4.20	1.18	1.08
	Min	4.56	2.78	5.29	2.54	0.97	0.90
	Max	12.26	6.14	16.80	6.48	1.44	1.34
60	Mean	5.92	3.33	7.01	3.38	0.78	0.74
	Min	3.99	2.02	5.35	1.95	0.63	0.63
	Max	13.04	8.48	14.58	7.94	1.04	1.04
80	Mean	5.06	3.05	6.06	3.15	0.61	0.58
	Min	3.09	2.07	3.90	2.20	0.42	0.42
	Max	6.83	4.32	7.11	4.50	0.85	0.79
100	Mean	4.19	2.59	5.34	2.68	0.47	0.44
	Min	3.45	2.09	4.43	1.88	0.37	0.35
	Max	5.03	3.25	5.95	3.77	0.68	0.60

n and the eight varieties of heuristics. Each cell contains three values, which are the average, the minimum, and the maximum fraction values of the 10 test problems that are randomly generated.

After observing the table carefully, we found some characteristics of the heuristic algorithms as follows:

1. The results of pairwise exchange algorithms (i.e., EDD-PI, MS-PI, and mEDD-PI) are better than those of their corresponding original heuristics (i.e., EDD, MS and mEDD, respectively).
2. Regardless of the number of jobs (n), the MS heuristic shows the worst performance.
3. The mEDD shows much better performance than the other three heuristics (i.e., EDD and MS).
4. When the number of jobs (n) is large, the mEDD heuristic is better than even the genetic algorithm. In particular, one can find that, when n is 100, the average objective values of the mEDD and the mEDD-PI heuristics are just 47 and 44 % of the genetic algorithm, respectively. This is very surprising result because the genetic algorithm could find near-optimal solution in many scheduling problems.
5. The difference between the mEDD and the mEDD-PI heuristics is smaller than the other three pairs of heuristics. This is because the performance of the mEDD is so good that the effect of the pairwise exchange is relatively small.

In summary, when the number of jobs (n) is small, the genetic algorithm shows the best performance, but the performances of the mEDD and the mEDD-PI heuristics are not much worse than genetic algorithm. When the value of n is large, on the other hand, the mEDD and the mEDD–PI are much better than the other heuristics including the genetic algorithm. And moreover, the difference between the mEDD and the mEDD-PI heuristics is very little, and the mEDD heuristic can be used to solve the problem when n is large because its procedure is much simpler than the mEDD-PI heuristic.

Now we consider the performances of the proposed algorithms in terms of the computation time. We measured the computation time when the algorithms were coded in C++ and run on a PC of 2.60 GHz G620 processor with 4-GB memory. When the number of jobs (n) is 20, 40, 60, 80, and 100, the CPU times for the genetic algorithm were 0.02, 0.12, 0.28, 0.57, and 0.89 s, respectively, while the times are less than 0.01 s for all other heuristics (i.e., EDD, EDD–PI, MS, MS-PI, mEDD, and mEDD-PI) and for all values of n (i.e., 20, 40, 60, 80 and 100). Note that the mEDD and the mEDD-PI heuristics are better than the genetic algorithm from the aspect of computation time as well as objective value.

7 Conclusions

In this research, we developed scheduling algorithms for a single machine system with job release times and sequence-dependent machine setup times. This research was motivated by the practical scheduling work in small-sized companies that are specialized in the plastic injection molding. To minimize weighted sum of earliness and tardiness penalties, we first presented a mixed integer linear programming (MILP) model. Since the optimization model is an NP-hard problem, we developed eight heuristic procedures and designed a genetic algorithm to solve the practical big-sized problems in a reasonable computational time. Through a series of computational experiments, we finally showed that one of the proposed scheduling algorithms could provide solutions of very good quality in a short time.

Several directions for future research are apparent from this study. First, the model considering a parallel machine system can be examined. Second, some constraints such that one operator can handle the machine group can be added to the parallel machine model.

Acknowledgment This research was supported by Basic Science Research Program through the National Research Foundation of Korea (NRF) funded by the Ministry of Education, Science and Technology (NRF-2012R1A1A2039063).

References

Allahverdi A, Ng CT, Cheng TE, Kovalyov MY (2008) A survey of scheduling problems with setup times or costs. Eur J Oper Res 187(3):985–1032

Baker KR, Scudder GD (1990) Sequencing with earliness and tardiness penalties: a review. Oper Res 38(1):22–36

Balakrishnan N, Kanet JJ, Sridharan V (1999) Early/tardy scheduling with sequence dependent setups on uniform parallel machines. Comput Oper Res 26(2):127–141

Coleman B (1992) Technical note: a simple model for optimization the single machine early/tardy problem with problem with sequence-dependent setups. Prod Oper Manage 1(2):225–228

Davis JS, Kanet JJ (1993) Single-machine scheduling with early and tardy completion costs. Naval Res Logistics (NRL) 40(1):85–101

Fry TD, Armstrong RD, Rosen LD (1990) Single machine scheduling to minimize mean absolute lateness: a heuristic solution. Comput Oper Res 17(1):105–112

Garey MR, Tarjan RE, Wilfong GT (1988) One-processor scheduling with symmetric earliness and tardiness penalties. Math Oper Res 13(2):330–348

Joo CM, Kim BS (2012) Non-identical parallel machine scheduling with sequence and machine dependent setup times using meta-heuristic algorithms. Ind Eng Manage Syst 11(1):114–122

Kanet JJ, Sridharan V (2000) Scheduling with inserted idle time: problem taxonomy and literature review. Oper Res 48(1):99–110

Keyser TK, Sarper H (1991) A heuristic solution of the E/T problem with waiting costs and non-zero release times. Comput Ind Eng 21(1):297–301

Kim YD, Yano CA (1994) Minimizing mean tardiness and earliness in single-machine scheduling problems with unequal due dates. Naval Res Logistics (NRL) 41(7):913–933

Lee CY, Choi JY (1995) A genetic algorithm for job sequencing problems with distinct due dates and general early-tardy penalty weights. Comput Oper Res 22(8):857–869

Mazzini R, Armentano VA (2001) A heuristic for single machine scheduling with early and tardy costs. Eur J Oper Res 128(1):129–146

Sidney JB (1977) Optimal single-machine scheduling with earliness and tardiness penalties. Oper Res 25(1):62–69

Sridharan V, Zhou Z (1996) A decision theory based scheduling procedure for single-machine weighted earliness and tardiness problems. Eur J Oper Res 94(2):292–301

Szwarc W, Mukhopadhyay SK (1995) Optimal timing schedules in earliness-tardiness single machine sequencing. Naval Res Logistics (NRL) 42(7):1109–1114

Tsai TI (2007) A genetic algorithm for solving the single machine earliness/tardiness problem with distinct due dates and ready times. Int J Adv Manuf Technol 31(9–10):994–1000

Wan G, Yen BPC (2002) Tabu search for single machine scheduling with distinct due windows and weighted earliness/tardiness penalties. Eur J Oper Res 142(2):271–281

Yano CA, Kim YD (1991) Algorithms for a class of single-machine weighted tardiness and earliness problems. Eur J Oper Res 52(2):167–178

Zhu Z, Heady RB (2000) Minimizing the sum of earliness/tardiness in multi-machine scheduling: a mixed integer programming approach. Comput Ind Eng 38(2):297–305

Part IV
Optimization Problems in Warehousing, Distribution and Transportation

Designing of Relief Network for Disaster Response Operation

Reinny Patrisina, Nikorn Sirivongpaisal and Sakesun Suthommanon

Abstract This chapter deals with the design of relief network to improve efficiency and effectiveness of relief operation. About 80 % of disaster response operation involves logistics activities. The problem is modeled as a location-allocation problem in distributing relief supplies to disaster victims. The location of local distribution centers (LDCs) and the amount of relief supplies that will be delivered through the relief network are determined by minimizing logistics cost. Equity among demand points are considered by giving a penalty cost for unfair distribution. A numerical example is conducted to illustrate how the model operates. The results show the effect of unfairness cost on performance of relief system, particularly on logistics cost and demand satisfied.

Keywords Humanitarian logistics · Facility location · Relief network · Disaster response operation

1 Introduction

None can predict exactly when, where, what, and how big a disaster would happen. Based on EM-DAT data, there were 6669 disasters, 1,149,920 deaths, and 2167 million affected people, and causing about US$ 1572 billion of disaster costs between 2003 and 2012 (IFRC 2013). Three deathless disaster of decade were the 2004 Indian Tsunami (226,408 deaths), the 2010 earthquake in Haiti (222,570 deaths), and Cyclone Nargis in Myanmar in 2008 (138,375 deaths).

According to United Nations International Strategy for Disaster Reduction (UNISDR), disaster is a serious disruption of the functioning of community or a

R. Patrisina (✉) · N. Sirivongpaisal · S. Suthommanon
Business Logistics and Supply Chain Management Research Team,
Department of Industrial Engineering, Faculty of Engineering,
Prince of Songkla University, 90112 Hatyai, Thailand
e-mail: reinny.patrisina@gmail.com

V. Kachitvichyanukul et al. (eds.), *Toward Sustainable Operations of Supply Chain and Logistics Systems*, EcoProduction,
DOI 10.1007/978-3-319-19006-8_19

society involving widespread human, material, economic or environmental losses and impacts, which exceeds the ability of the affected community or society to cope using its own resources (UNISDR 2009). The disaster does not only cause loss of lives but also influences social and economic fundamental of the affected areas at the same time. Disaster management (DM) could be applied to reduce disaster risk. The DM is a continuous process which consists of four phases: mitigation, preparedness, response, and recovery. Based on the time, the activity could be classified into pre-disaster activities including mitigation and preparation and post-disaster activities including response and recovery.

An effective logistics management has a significant effect to the success of disaster management since it provides effective and efficient relief operation that result in reducing loss live and severity at post-disaster (Moe and Pathranarakul 2006). Logistics regarding humanitarian is called humanitarian logistics. According to Thomas and Kopczak (2005), humanitarian logistics is defined as the process of planning, implementing, and controlling the efficient, cost-effective flow, and storage of goods and materials, as well as related information, from the point of origin to the point of consumption for the purpose of alleviating the suffering of vulnerable people. Even though logistics activities take much portion in disaster response operation, its role is not directly acknowledged by humanitarian organizations. Before the 2004 Asian Tsunami, logistics in relief operation is regarded as a necessary expense than a critical issue of their operation (Van Wassenhove 2006).

The disaster relief environment gives additional complexity to humanitarian logistics comparing to business logistics as follows (Balcik and Beamon 2008):

1. Demand uncertainties;
2. Communication and coordination difficulties due to communication line damage and a large number of actors involved in the relief operation;
3. Hard to achieve efficient and timely delivery; and
4. Resources scarcity (supply, people, and transportation capacity).

For supporting disaster relief operation, relief distribution network design plays a significant rule in order to obtain an efficient aid delivery (Balcik and Beamon 2008; Azimi et al. 2012). The complexity and uncertainty of relief environment become challenge to relief organization in establishing effective and efficient relief networks (Balcik and Beamon 2008). The other challenge in the relief operation is how to allocate the relief supplies to beneficiaries at the affected areas since there is limited number of aid supplies. This situation will trig unsatisfied demand, but the unsatisfied demand is prevalent in disaster relief environment. Regarding this situation, mechanism provided an efficient resource allocation for minimizing severity of the victims and achieving equity in distributing relief aid to all targeted demand is utmost importance. With respect to each people, everyone should have the same chance to be survived so that fairness issue is supposed to be concerned by any relief actors in distributing the relief supplies (Holguín-Veras et al. 2013). By considering fairness, the model ensures that required relief supplies will be delivered to all affected areas (Torre et al. 2012).

Our study focuses on selecting location of temporary warehouses (local distribution centers) at post-disaster for facilitating an efficient and effective response operation and determining amount of relief supplies that will be delivered through the relief network in order to minimize cost as well as maximize customer satisfaction under uncertainty environment. The proposed relief network encompasses three tiers: multi-collection centers, candidates of local distribution centers (LDCs), and demand points. It should be noted that the majority of papers address to historical information to select the distribution center locations. Our model is applied for designing relief planning based on the future disaster predicted by scientist and agreed by disaster stockholders.

The remainder of this paper is organized as follows. Section 2 presents a review of related literatures. Section 3 provides the problem description. Section 4 explains the model and assumptions. Numerical study is given in Sect. 5. The computational results are reported in Sect. 6, followed by conclusions and future works in Sect. 7.

2 Literature Reviews

There are some papers discussing about designing relief network. Jia et al. (2007) proposed a maximal covering model to select location for storing medical supplies for encountering large-scale emergencies. Different to Jia et al. (2007), focused solely on location problem, other literatures such as Balcik and Beamon (2008) and Duran et al. (2011) integrated location and inventory problems. Balcik and Beamon (2008) determined number and location of the fixed-capacity distribution centers and amount of relief supplies that will be stored in each location to maximize beneficiary satisfaction under budget limitation. Duran et al. (2011) developed a mixed integer programming inventory-location model to obtain relief network design that minimizes the average response time for a given investment. Actually, both models are close, but Duran et al. (2011) excluded post-disaster operation cost and consider the effect of multi-disasters to time replenishment. While Balcik and Beamon (2008) and Duran et al. (2011) proposed Location-inventory model, Ukkusuri and Yushimito (2008) developed a location-routing model. The model is aimed to determine locations for storing the relief prepositioning stocks by maximizing probability of demand points served by at least one facility and route of vehicles under unreliable relief network and budget restriction.

Other literatures are addressed to location and allocation problems such as Tzeng et al. (2007), Bozorgi-Amiri et al. (2013). They considered 3-tier of the relief network design: collection centers–distribution centers–demand points. Tzeng et al. (2007) formulated a multi-objective model to decide the location of distribution centers (transit points) functioning as a bridge between collection points and demand points in affected areas and amount of relief supplies that will be distributed through the relief network. They consider three objectives, namely total cost, travel time, and demand satisfaction; then, the model is solved using Fuzzy approach. Bozorgi-Amiri et al. (2013) modeled a multi-objective stochastic robust programming model. Relief distribution centers decisions regarding location, amount of relief prepositioning stocks, and demand allocation are made under demand, supply, and cost uncertainty by minimizing logistics cost as well as maximizing customer satisfaction. The uncertainty is applied through scenarios of disaster. They integrate pre- and post-disaster operation by considering procurement and relief transportation to relief distribution centers during pre- and post-disaster. Their model also presents a possibility of some facilities such as distribution centers as well as roads destroyed in the event. Actually, the model is not considered as a location model for after disaster situation, since the facilities are functioned for storing prepositioning stocks for encountering disaster. Though Bozorgi-Amiri et al. (2013) try to maximize demand satisfaction, their model could not ensure that no location gets over or under supplies than others so the uneven distribution is possible to occur. In disaster relief distribution, it is considered as utilitarian rather than egalitarian. Egalitarian policies provide maximizing equality among demand points in terms of amount of items delivered or speed, while utilitarian policies maximize the amount of demand satisfied regardless equality in distributing the relief items (Torre et al. 2012). Our study tries to balance between efficiency and equity in terms of quantity in relief logistics planning. Both Tzeng et al. (2007) and Bozorgi-Amiri et al. (2013) proposed that all demand points are served merely from transit points (either one or more transit points). Since the transit points could be supplied from one or more collection points, it would be more efficient if each demand point was assigned to one of the transit points. Consequently, it would be easy to control the distribution of the relief supplies since communication and coordination are hard to be performed during the disaster. In addition, sometimes the demand points are located close to collection points so that it would be also more efficient to deliver the aid supplies directly from the collection points to beneficiaries as compared to the use of transit points.

Mete and Zabinsky (2010) and Rawls and Turnquist (2010) introduced two-stage stochastic programming model that provide location, allocation, and routing decisions in relief logistics. Mete and Zabinsky (2010) formulated a location-inventory model under different possible disaster types and magnitude to determine warehouse locations for storing medical supplies and amount of supplies that will be stored at the warehouses for facing large-scale emergency at the first stage. Then at the second stage, they proposed a transportation model to determine amount of supplies that will be distributed to hospitals. Rawls and Turnquist (2010) decide location and amount of prepositioning stocks of the warehouses with possibility that the warehouses will damage in the disaster decisions at the first stage, while routing is provided in the second stage.

All previous literatures addressed to determine locations of permanent facilities for storing relief prepositioning stocks for encountering disaster established before the disaster. Some literatures related to relief network design focusing on temporary facility established during the disaster response operation are such Yi and Ozdamar (2007) and Lin et al. (2012). Yi and Ozdamar (2007) developed a location-routing model to select locations of temporary medical centers and to assign vehicle fleet for transporting both relief supplies and wounded people to the medical center.

Lin et al. (2012) considered using temporary depot for storing relief supplies from one collection point before transporting to beneficiaries during response operation. They developed a deterministic model for making location and allocation decisions during response operation by minimizing operational cost which is resulted in four types of penalty cost. First, it is related to the demand that have not been served by any depots. Second, represented backordered cost and unsatisfied demand after the operation are terminated. Third associates with the delivery cost to demand points. The last cost regards to unfairness of services among demand points. The fixed costs for establishing the depot are not considered in Lin et al. (2012), while it affects to relief operation cost. Frequently, there are more than one collection center used to collect the aid supplies from donors before transporting to beneficiaries at the affected areas. Accordingly, this study considers multi-collection centers and the demand points is allowed to receive aid supplied from either collection centers or local distribution centers that would be established during disaster response time.

3 Problem Description

Immediately after disaster occurs, a large amount of aid supplies is required by affected people in affected areas. Many humanitarian relief organizations, governments, communities, and privates give aid supplies in order to reduce severity of disaster victims. The aid supplies are often collected in some locations that are pointed as collection centers such as in location of disaster coordination center, airports, harbors, or other locations before it is distributed to the victims. The collection centers are usually located in the capital cities or big cities and possibly distant from affected areas. Consequently, it takes long travel times to deliver the supplies to the affected areas and travel back to the collection centers. Therefore, opening local distribution centers (LDCs) functioned as a bridge between the collection centers and the beneficiaries in the affected areas during disaster response operation would be helpful for improving efficiency of relief operation. The proposed design of relief network is shown in Fig. 1.

This study models the location-allocation problem in disaster relief operation for deciding location of LDCs and the amount of supplies that will be delivered

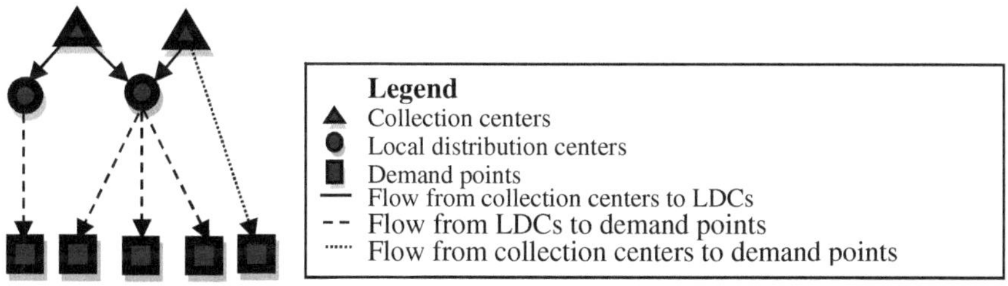

Fig. 1 Problem description

through the disaster relief network. This model has an objective to minimize logistics cost included facility opening cost, transportation cost, shortage cost, and unfairness cost. Unfairness cost is a penalty cost regarding uneven distribution among beneficiaries since equity is principal in the relief operation. By involving fairness, we expect no demand points get oversupplies than others.

There are some assumptions used in this model as follows:

1. There are multi-relief supplies that will be delivered through the relief network;
2. Each demand point is supplied by either one of collection centers or one of the LDCs;
3. There is no inventory cost in LDCs since the relief supplies should be delivered to the affected people quickly;
4. There is no inventory in demand point due to scarcity of relief supplies; and
5. The location of the candidate LDCs and the refugee shelters are located close to the high way.

4 Model Development

We developed a mixed integer programming (MIP) model that minimizes total cost by considering equality in distributing relief supplies to all targeted demand.

4.1 Notation and Definition

The following notations are used in our MIP model.

Sets

I Set of collection centers indexed by $i \in I$, $i = 1,2,\ldots,l$,
J Set of candidate LDCs indexed by $j \in J$, $j = 1,2,\ldots,m$,
K Set of refugee shelters (demand point) indexed by $k \in K$, $k = 1,2,\ldots,n$, and
P Set of relief items indexed by $p \in P$, $p = 1,2,\ldots,o$.

Parameters

f_j Fixed cost for opening the LDC at location j,
d_{pk} Amount of relief supplies p required by demand point k,
s_{pi} Amount of available supplies p at the collection centers i,
M A big positive number,
c_{pia} Transportation cost of the relief supplies p from the collection center i to destination a ($a \in J \cup K$),
c_{pjk} Transportation cost of the relief supplies p from the LDC j to the demand point k,
h_p Shortage cost for the relief supplies p, and
r_p Penalty cost for the relief supplies p relates to unfair distribution.

Decision variables

X_{pia} Amount of item type p is delivered from source i to destination a ($a \in J \cup K$),

X_{pbk} Amount of the relief supplies p is delivered from point b ($b \in I \cup J$) to demand point k,

Z_j Binary variable that equals 1 if LDC at location j is selected, and 0 otherwise,

L_{bk} Binary variable that equals 1 if demand point k is supplied by point b ($b \in I \cup J$), and 0 otherwise,

U_{pk} level of satisfaction to the relief supplies p in the demand point k, and

E_p the maximum difference of satisfaction level for the relief supplies p due to the unfair distribution between two demand points.

4.2 Objective Function and Constraints

Minimize

$$\sum_{j \in J} f_j Z_j + \sum_{p \in P} \left(\sum_{i \in I} \sum_{a \in J \cup K} X_{pia} c_{pia} + \sum_{j \in J} \sum_{k \in K} X_{pjk} c_{pjk} \right) + \sum_{p \in P} h_p \left[\sum_{k \in K} \left(d_{pk} - \sum_{b \in I \cup J} X_{pbk} \right) \right] + \sum_{p \in P} r_p E_p \quad (1)$$

Subject to

$$\sum_{i \in I} X_{pij} = \sum_{k \in K} X_{pjk} \quad \forall p \in P, \forall j \in J \quad (2)$$

$$\sum_{a \in J \cup K} X_{pia} \le s_{pi} \quad \forall p \in P, \forall i \in I \quad (3)$$

$$X_{pbk} \le L_{bk} d_{pk} \quad \forall p \in P, \forall b \in I \cup J, \forall k \in K \quad (4)$$

$$\sum_{p \in P} \sum_{k \in K} X_{pjk} \le M Z_j \quad \forall j \in J \quad (5)$$

$$\sum_{k \in K} L_{jk} \le M Z_j \quad \forall j \in J \quad (6)$$

$$\sum_{b \in I \cup J} L_{bk} = 1 \quad \forall k \in K \quad (7)$$

$$U_{pk} = \frac{d_{pk} - \sum_{b \in I \cup J} X_{pbk}}{d_{pk}} \quad \forall p \in P, \forall k \in K \tag{8}$$

$$E_p \geq \left| U_{pv} - U_{pw} \right| \quad \forall p \in P, \forall (\mathrm{v}, w) \in K, \ \mathrm{v} \neq \mathrm{w} \tag{9}$$

$$X_{pia} \geq 0 \quad \forall p \in P, \forall i \in I, \forall a \in J \cup K \tag{10}$$

$$X_{pjk} \geq 0 \quad \forall p \in P, \forall j \in J, \forall k \in K \tag{11}$$

$$Z_j \in \{0, 1\} \quad \forall j \in J \tag{12}$$

$$L_{bk} \in \{0, 1\} \quad \forall b \in I \cup J, \forall k \in K \tag{13}$$

The objective of the proposed model, Eq. (1), is to minimize relief logistics costs including cost for opening LDCs (the first part), cost for transporting the supplies through the relief distribution network (the second part), penalty cost for shortage demand (the third part), and penalty cost for unfair distribution among demand points. Equation (2) is the flow conservation to ensure that total inflow to LDC j is the same as total outflow from the LDC j. Equation (3) represents that total item transported from each collection centers should be less than or equal to amount of the available supplies at the collection centers i. Equation (4) describes that amount of the supplies transported to demand point k is less than or equal to number of supplies required at the demand point. Equations (5) and (6) ensure that no supplies could be delivered from LDC j to any demand points and no demand point could be assigned to the LDC j at least the LDC j opens. Equation (7) explains that the demand point is assigned to only one source point either one of collection centers or one of the LDCs. Equations (8) and (9) are constraint for measuring level of unsatisfaction of demand point and unfairness distribution among demand points. Equations (10) and (11) are nonnegative constraints, while the last two equations (Eqs. 12 and 13) are non-integer constraints.

5 Case Study

To illustrate the proposed model, we present the Mentawai megathrust scenario developed in contingency plan for encountering tsunami triggered the earthquake in West Sumatra province, Indonesia. Sieh (2006) has predicted that the giant earthquake caused of tsunami will attack West Sumatera within next 50 years, shown in Fig. 2. To prepare to this event, Sumatera Barat local agency for disaster management (BPBD Sumatera Barat) together with other government institutions, military, experts, and non-governmental organizations has developed a contingency plan for this future disaster. The plan is based on the possible worst-case scenario of the disaster. The 8.8 scale of earthquake located at 150 km distance of south west of

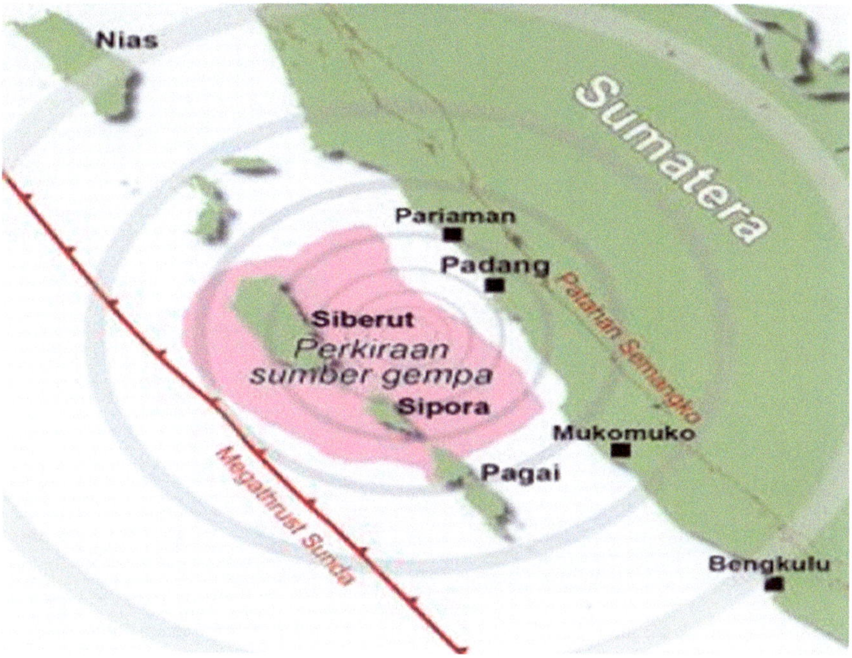

Fig. 2 Possible source of the future earthquake (*Source*) Earth Observatory of Singapore 2015

Padang city, West Sumatera, and at 30 km depth is predicted affecting to almost one million people lived in west coastal cities of West Sumatera, namely Pesisir Selatan, Padang, Padang Pariaman, Pariaman, Agam, Pasaman Barat, and Mentawai (BPBD Sumatera Barat 2012). Our study will consider all the affected cities in West Sumatera province excluded Mentawai because this city has different characteristics to the other affected cities, illustrated by Fig. 3. While the other cities could be reached by truck from the capital city of West Sumatera (Padang), but to Mentawai, we have to use boat or helicopter.

A total of 34 locations of demand points are determined based on the subdistrict or sector of affected areas. Amount of demand is determined based on the percentage of people estimated save in the disaster and as described in West Sumatera contingency plan for encountering tsunami (BPBD Sumatera Barat 2012). The plan also encompassed estimation of safe people moved to another area or no longer stay at the shelters in all period of emergency response (assumed 30 days). In case the disaster victims leave the shelters, they will not be considered as beneficiaries anymore. Since the time that they would move are not known, amount of demands during response period involves uncertainty. In case the victims leave the shelters at the first period of response time, amount of demand during the response period would be less than when the victims leave the shelters at the middle or at the end of

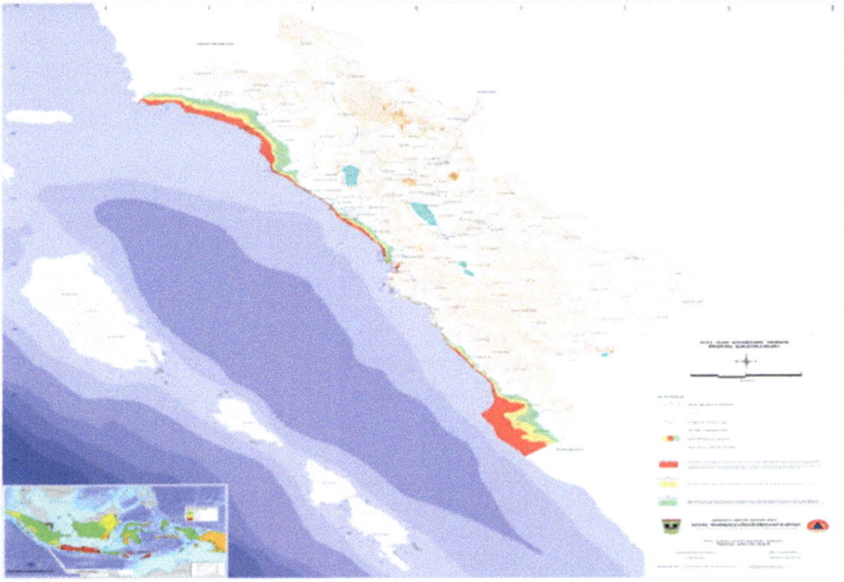

Fig. 3 Tsunami hazard map of Mentawai megathrust (*Source*) BPBD Sumatera Barat 2012

the period. Then to determine the number of affected people required of aid supplies during the response period is performed in two steps. First, the number of affected people stayed in the shelter during the response period was determined using a dynamic model of Hadiguna and Wibowo (2013). Second, using this number, distribution of required demand during the response period is determined. Number of affected people and distribution of required demand during the response period are shown in Table 1.

Each affected people is provided rice, noodle, and preserved food per day as much as 400 gram, 95 gram, and 200 gram, respectively (Kusumastuti et al. 2013). Then, Pusdalops Sumatera Barat, Social Welfare of West Sumatera province, and Social Welfare of Bukittinggi are considered as collection points of the relief supplies from donor before distributing to the affected areas.

The number of people that need relief supplies in the predicted disaster (551,155 people) is almost the same as the number of affected people that require for aid supplies in the 2009 West Sumatera earthquake (573,985 people) as presented in BPK Sumatera Barat (2010) regarding the report of the 2009 West Sumatera earthquake. Therefore, we assumed that amount of available supplies that will deliver to the affected areas during the response period is as much as amount of supplies that was received by BPBD Sumatera Barat during the 2009 West Sumatera earthquake response period (32 days), provided in BNPB daily report in

Table 1 Number of affected people and demand

Affected area	Number of demand points	Number of affected people (people)	Safe people (%)	Move (%)	Demand
Pesisir Selatan	10	212,870	65	1	Normal (136,234; 1,685)
Padang	9	487,193	60	1	Normal (286,004; 3760)
Padang Pariaman[a]	6	24,861	60	1	Normal (15,853; 210)
Pariaman[a]	3	56,743	70	1	Normal (40,093; 418)
Agam[a]	3	29,059	60	1	Normal (21,710; 246)
Pasaman Barat	3	95,775	60	1	Normal (57,291; 638)
Total	34	906,501			

[a]Number of affected people is adjusted to the recent population data, published by BPS of the city
Source BPBD Sumatera Barat 2012; BPBD Padang 2014; BPBD Pasaman Barat 2014; BPBD Pesisir Selatan 2014; BPBD Padang Pariaman 2012; BPBD Pariaman 2012; BPS Agam 2014

Table 2 Number of available supplies (kg)

Collection centers	Rice	Noodle	Preserved food
Pusdalops	1,253,124	222,621	196,476
Social welfare province	9,065,053	15,450	32,527
Bukittinggi	3,506,505	112,483	225,116

October, 2009 (BNPB Indonesia 2009), and BPK report of the 2009 West Sumatera earthquake (BPK Sumatera Barat 2010). The amount of supplies that will be delivered from the collection centers is presented in Table 2, while the relief distribution network is shown in Fig. 3.

A total of 17 locations of demand points are also proposed as candidate of LDCs that would be selected. Fixed cost for opening the LDCs is US$500 (1US$ = 10,000 IDR) referred to Kusumastuti et al. (2013). The transportation costs from collection centers to any destination (either LDCs or refugee shelters) and from LDCs to the shelters per kg per km are assumed US$0.0034 and US$0.0042. The distances between locations are measured using Google map. Shortage costs (in US$) for rice, noodle, and preserved food are 6, 5, and 8, respectively, while penalty costs for unfair distribution are set a million time of shortage costs. Locations of collection centers, candidate LDCs, and demand points are shown in Fig. 4.

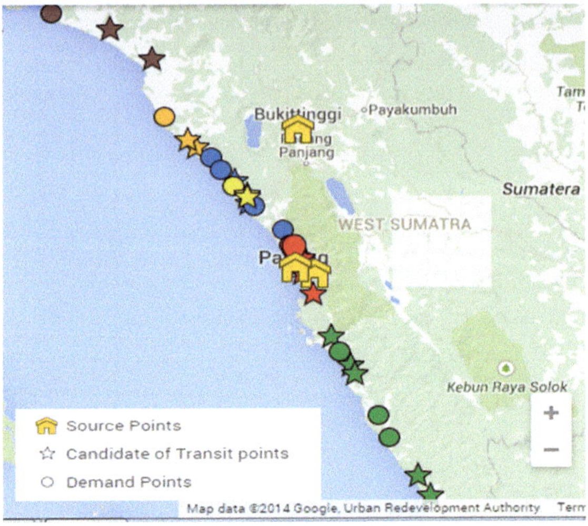

Fig. 4 Collection centers, candidate of LDCs, and demand points

6 Results and Analysis

The purposed relief network problem consists of 3 collection centers, 17 candidate LDCs, and 34 demand points. The model was tested 10 times (replications) where each replication has a different number of demands, while the other parameters are set at the same number. Two scenarios were applied. First, penalty cost for unfairness distribution is set to certain value for considering equality in distributing relief supplies to all targeted demand. Second, the model is applied by regardless equity among demand points; then after the results are obtained, the unfairness is calculated to see the effects of the uneven distribution to the total costs. The global optimal solution was found using the branch and bound algorithm provided in Lingo 14.0, presented in Table 3.

From Table 3, we can see that the total cost for scenario 1 (considering fairness) is an average of 0.52 % higher than that of scenario 2^{a} (regardless fairness and penalty cost). Although equity issue makes the relief operation less efficient, it could increase satisfaction of disaster victims. In case the value of unfairness is not equal to zero, it indicates that at least one demand point get over supplies than others. If the value is one, then at least one demand point gets fully satisfied (no shortage) and at least one demand point is totally unsatisfied (receive no supplies).

For scenarios 1 and 2^{b} are obtained US$ 30,905,397 and US$ 49,746,778 of the average total logistics cost and US$ 288,393 and US$ 288,004 of standard deviation, respectively. Considering fair distribution from the first time (scenario 1) could save almost 40 % of disaster response operation budget comparing to give the penalty to unfair distribution later (scenario 2^{b}).

Table 3 Results

Replication	Total cost ($US)			Unfairness for each items (rice, noodle, and preserved food)	
	Scenario 1	Scenario 2[a]	Scenario 2[b]	Scenario 1	Scenario 2
1	30,738,750	30,580,530	49,580,530	(0, 0, 0)	(1, 1, 1)
2	31,274,670	31,115,630	50,115,630	(0, 0, 0)	(1, 1, 1)
3	30,898,160	30,739,410	49,739,410	(0, 0, 0)	(1, 1, 1)
4	30,987,920	30,829,100	49,829,100	(0, 0, 0)	(1, 1, 1)
5	31,196,640	31,037,660	50,037,660	(0, 0, 0)	(1, 1, 1)
6	30,475,750	30,317,730	49,317,730	(0, 0, 0)	(1, 1, 1)
7	31,243,040	31,084,020	50,084,020	(0, 0, 0)	(1, 1, 1)
8	30,914,250	30,755,470	49,755,470	(0, 0, 0)	(1, 1, 1)
9	30,467,540	30,309,530	49,309,530	(0, 0, 0)	(1, 1, 1)
10	30,857,250	30,698,700	49,698,700	(0, 0, 0)	(1, 1, 1)

[a]Before calculating penalty cost for unfair distribution
[b]After calculating penalty cost for unfair distribution

7 Conclusion and Future Research

This study proposed a location-allocation model in disaster relief operation. The 3-tier of relief network location is considered (multi-collection centers–LDCs–demand points). Locations of LDCs are selected among a number of candidate LDCs, and amount of relief supplies that would be delivered through the relief network is determined by minimizing logistics cost. The logistics cost compasses cost for opening the LDC, transportation cost, shortage cost, and penalty cost for unfair distribution. Involving equality in distributing the relief supplies since the first time drives the demand satisfied and the minimum logistics cost.

Acknowledgments This research was funded by the Ministry of Research and Higher Education of Republic of Indonesia and Prince of Songkla University of Thailand.

References

Balcik B, Beamon BM (2008) Facility location in humanitarian relief. Int J Logistics Res Appl 11 (2):101–121

BNPB Indonesia (2009) Daily Report of PUSDALOPS BNPB on October 1–November 1, 2009 (in Indonesia language

Bozorgi-Amiri A, Jabalameli MS, Mirzapour Al-e-Hashem SMJ (2013) A multi-objective robust stochastic programming model for disaster relief logistics under uncertainty. OR Spectrum. doi:10.1007/s00291-011-0268-x

BPBD Padang (2014) Tsunami evacuation path document of Padang city (in Indonesia lan-guage)

BPBD Padang Pariaman (2012) Contingency plan for encountering earthquake and tsunami in Padang Pariaman regency (in Indonesia language)
BPBD Pariaman (2012) Contingency plan for encountering earthquake and tsunami in Pa-riaman city (in Indonesia language)
BPBD Pasaman Barat (2014) Tsunami evacuation path document of Pasaman Barat regency (in Indonesia language
BPBD Pesisir Selatan (2014) Tsunami evacuation path document of Pesisir Selatan regency (in Indonesia language)
BPBD Sumatera Barat (2012) Contingency plan for encountering tsunami in West Sumatera province (in Indonesia language)
BPK Sumatera Barat (2010) Investigation report of disaster management activities for earthquake during the response period conducted by West Sumatera government, financial year of 2009 (in Indonesia language)
BPS Agam (2014) Number of population by district and sex, year of 2012
Duran S, Gutierrez MA, Keskinocak P (2011) Pre-positioning of emergency items for CARE international. Interfaces 41(3):223–237
Earth Observatory of Singapore. West Sumatra tectonics and tsunami hazard. Earth Obser-vatory of Singapore: Nanyang Technology University of Singapore. Retrieved on March 15, 2015 from http://www.earthobservatory.sg/outreach/natural-hazard-outreach/west-sumatra-tectonics-and-tsunami-hazard
Hadiguna RA, Wibowo A (2013) Simulation of aid logistics system for earthquake-tsunami disaster: case study in Padang city (in Indonesia Language). Jurusan Teknik Industri Universitas Muhamadiyah Malang 13(2):116–125
Holguín-Veras J, Pérez N, Jaller M, VanWassenhove LN, Aros-Vera F (2013) On the ap-propriate objective function for post-disaster humanitarian logistics models. J Oper Manag. doi:10.1016/j.jom.2013.06.002
IFRC (2013) World disaster report 2013: Focus on technology and the future of humanitarian action. Retrieved on September 20, 2014 from http://www.ifrc.org/PageFiles/134658/WDR%202013%20complete.pdf
Jia H, Ordóñez F, Dessouky MM (2007) Solution approaches for facility location of medical supplies for large-scale emergencies. Comput Ind Eng 52(2):257–276
Kusumastuti RD, Wibowo SS, Insanita R (2013) Modelling facility locations for relief lo-gistics in Indonesia, in: Zeimpekis V, Ichoua S, Minis I (Eds.) (2013) Humanitarian and relief logistics. Oper Res/Comp Sci Interfaces Series, Springer New York 183–205
Lin Y-H, Batta R, Rogerson PA, Blatt A, Flanigan M (2012) Location of temporary depots to facilitate relief operations after an earthquake. Socio-Econ Planning Sci 46(2):112–123
Mete HO, Zabinsky ZB (2010) Stochastic optimization of medical supply location and distribution in disaster management. Int J Prod Econ 126(1):76–84
Moe TL, Pathranarakul P (2006) An integrated approach to natural disaster management: Public project management and its critical success factors. Disaster Prev Manag. doi:10.1108/09653560610669882
Naji-Azimi Z, Renaud J, Ruiz A, Salari M (2012) A covering tour approach to the location of satellite distribution centers to supply humanitarian aid. Eur J Oper Res 222(222):596–605
Rawls CG, Turnquist MA (2010) Pre-positioning of emergency supplies for disaster re-sponse. Transp Res Part B 44(4):521–534
Sieh K (2006) Sumatran megathrust earthquakes: from science to saving lives. Phil Trans R Soc A 364(1845):1947–1963
Thomas AS, Kopczak LR (2005) From logistics to supply chain management: the path forward in the humanitarian sector. Fritz Institute
Torre LE, Dolinskaya IS, Smilowitz KR (2012) Disaster relief routing: integrating research and practice. Socio-econ Planning Sci 46(1):88–97
Tzeng GH, Cheng HJ, Huang TD (2007) Multi-objective optimal planning for designing relief delivery systems. Transp Res Part E Logistics Transp Rev 43(6):673–686

Ukkusuri SV, Yushimito WF (2008) Location routing approach for the humanitarian prepositioning problem. Transp Res Rec J Transp Res Board 2089(1):18–25

UNISDR (2009) Terminology on disaster risk reduction. Retrieved on November 15, 2014 from http://www.unisdr.org/eng/terminology/terminology-2009-eng-html

Van Wassenhove LN (2006) Blackett memorial lecture humanitarian aid logistics: supply chain management in high gear. J Oper Res Soc 57(5):475–489

Yi W, Özdamar L (2007) A dynamic logistics coordination model for evacuation and support in disaster response activities. Eur J Oper Res 179(3):1177–1193

Integrated Relief Supply Distribution and Evacuation: A Stochastic Approach

Wapee Manopiniwes and Takashi Irohara

Abstract This chapter presents a stochastic optimization model for disaster management planning. In particular, the focus is on the integrated decisions about the distribution of relief supplies and evacuation operations. The proposed decision-making approach recommends the best relief distribution centers to use as storage locations and determines their optimal inventory levels. The model also incorporates the priorities for the evacuation of particular communities, as well as specific disaster scenarios with estimates of the transportation needs and demand for aid. A case study is presented to determine the distribution of aid for a flood emergency in Thailand that uses a flood hazard map.

Keywords Facility location · Stochastic · Humanitarian logistics · Disaster relief

1 Introduction

Humanitarian logistics is central to disaster relief. In both rapid- and slow-onset disasters or emergencies, it is necessary to procure and move the required material (water, food, shelter, clothing, medicines, and other supplies) from storage locations to appropriate destinations in the affected area and to do this in the most efficient and effective way possible. It can thus be argued that the logistic capabilities required by aid agencies and others who deal with large scale, sudden-onset disasters, are similar to those required by commercial organizations. In a previous study (Manopiniwes et al. 2014), we described a situation in which decisions were based on a deterministic model in which a flood or other catastrophe could occur. However, humanitarian logistics can be defined as the ability to respond rapidly to unexpected changes in demand or supply conditions. Therefore, in this study, we introduce a stochastic approach for planning the storage and distribution of

W. Manopiniwes (✉) · T. Irohara
Sophia University, Chiyoda-ku, Tokyo 102-8554, Japan
e-mail: wapee@sophia.jp

V. Kachitvichyanukul et al. (eds.), *Toward Sustainable Operations of Supply Chain and Logistics Systems*, EcoProduction,
DOI 10.1007/978-3-319-19006-8_20

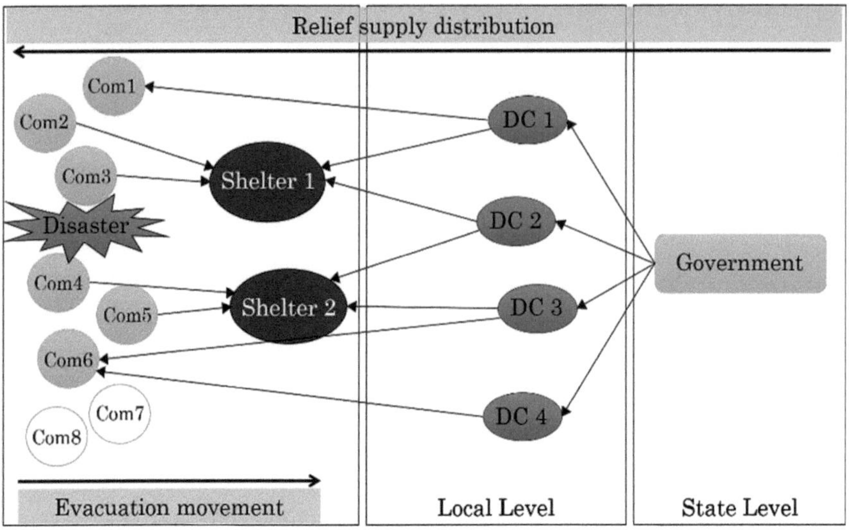

Fig. 1 Supply chain management for disaster relief

emergency aid packages in a flood disaster. We also consider how to make integrated decisions about the evacuation of various communities.

Our stochastic model provides a decision-making model for the preparedness stage of a disaster management system, and it considers the outcomes of the various possible responses. While past researches had considered supply distribution and evacuation process separately, we connect both planning contexts as the integrated decision as summarized in Fig. 1 corresponding not only to the demand at evacuation shelters but also to people who remain in the affected area. Our method is able to recommend which of the candidate distribution centers (DCs) will be the best storage locations and to determine appropriate inventory levels. The model incorporates the priorities of evacuating particular communities, as well as specific disaster scenarios, with transportation and demand estimates.

The remainder of this chapter is organized as follows. In Sect. 2, we provide an overview of the existing literature. Section 3 presents our stochastic approach to disaster preparedness, and Sect. 4 presents a case study of a flood disaster in Thailand. In Sect. 5, we evaluate and analyze the results of our approach, and finally, in Sect. 6, we summarize our contributions and discuss directions for further research.

2 Literature Survey

According to the steep increase in the number of articles on relief supply chain management in recent years, a broad overview of the issues is addressed ranging from a principle or conceptual framework to the analytical methodology and case study applied. Among a number of distributions in this field, the Operations Research (OR) encompasses the most objective challenges. Therefore, articles surveying disaster supply chain papers include an OR methodology survey by Altay and Green (2006) and a review of relief supply chain optimization by Manopiniwes and Irohara (2014). Most articles dealing with disaster supply chain are focusing either on the distribution of relief supplies or on the evacuation movement.

2.1 Stock Prepositioning and Distribution Management

There are many papers dealing with stock prepositioning and distribution management. Barbarosoglu and Arda (2004) offered a multi-modal network flow to plan the transportation of vital first-aid commodities to disaster-affected areas during emergency response. Jia et al. (2007) proposed models and solution approaches for determining the facility locations of medical supplies in response to large-scale emergency. The study is to investigate which facilities should be opened and the order in which facilities are opened can have a significant effect on the quality of coverage. Balcik and Beamon (2008) and Rawls and Turnquist (2010) created an optimization model whose solution provides a prepositioning strategy for facility locations and inventory decision under uncertainty. The mathematical models determine the number and locations of the distribution center in relief network and the amount of relief supplies to be stocked at each distribution center. Irohara et al. (2013) proposed a tri-level programming model for this integrated category. The top level addressed facility location and inventory decisions; the second level represents damage caused by the disaster, while the third level determines response and recovery decisions. A recent paper by Abounacer et al. (2014) presents a multi-objective location-transportation problem for disaster response aims at determining the number, the position, and the mission of required humanitarian aid distribution centers within the disaster region. Salman and Yücel (2015) propose an emergency facility location to maximize the expected demand coverage within a specified distance over all possible network realizations.

2.2 Sheltering and Evacuation Operations

We have also found several papers considering on evacuation operations. Kongsomsaksakul et al. (2005) studied the optimal shelter locations for the flood evacuation planning. They formulated a bi-level programming model for determining shelter locations with the upper level and traveler behavior with the lower level. Another bi-level optimization model was proposed by Liu et al. (2006) for optimal evacuation planning: The high level tries to maximize the throughput in the target evacuation clearance time, whereas the low level is to minimize total clearance time. Song et al. (2009) presented an optimization modeling technique to develop an evacuation plan for transit-dependent residents during the hurricane evacuation. Most recent work with respect to natural disasters has focused either on evacuation strategies or the distribution of relief supplies. Recently, article by Kelle et al. (2014) proposed the integrated evacuation and resource allocation which the decision criterion is to find the appropriate compromise between the worst-case and average cost consideration. While their paper is based on the assumption that all affected people are evacuated to some facilities, our current study formulates the integrated decision of evacuation and relief supplies distribution corresponding not only to the demand at evacuation shelters but also to the people who remain in the affected area.

2.3 Stochastic Programming

Stochastic programming is one of the most widely used tools for planning in the preparedness stage (Altay and Green 2006) due to its ability to handle uncertainty by probabilistic scenarios representing disasters and their outcomes (Barbarosoglu and Arda 2004; Balcik and Beamon 2008; Rawls and Turnquist 2010; Salman and Yücel 2015; Song et al. 2009; Kelle et al. 2014). The flood emergency logistics preparation problem with uncertainty is formulated by Chang et al. (2007) as two stochastic programming models that allow for the determination of a rescue resource distribution system for urban flood disasters by applying the data processing and network analysis functions of the geographic information system. Flooding potential maps can estimate the possible locations of rescue demand points and the required amount of rescue equipment. Mete and Zabinsky (2010) developed a stochastic model for the location and delivery of medical supplies in Seattle area with regard to the earthquake disaster. The randomness in the location and amount of demand, and in the available transportation routes and transportation times leads to different types of scenarios where the transportation routes and times are directly related to the location and amount of demand. Lately, Verma and Gaukler (2014) evaluated the performance of two models between deterministic and stochastic modeling approaches using large-scale emergencies caused by an earthquake in California. They found that the locations suggested by the stochastic model significantly reduce the expected cost of providing supplies. They also demonstrate that the cost advantage of the stochastic model over the deterministic model is especially large when only a few facilities can be placed.

Our study makes the following contributions: provides the new approach modeling for the integrated decision on the evacuation planning and relief supplies distribution for floods emergency response. With the stochastic formulation based on the flood hazard map, the value of decision performance is particularly great in a large-scale realistic with budget-constrained situations.

3 Proposed Integrated Stochastic Model

In this section, we formulate the model for disaster preparedness decision making, with consideration of the expected outcomes of the response stage. The decisions that are made before a disaster hits are usually made with consideration of all possible scenarios, while decisions that are made afterward must depend on the actual scenario. To allow for this uncertainty, we calculate the expected costs for each of several scenarios. Stochastic programming allows for several different decisions to be integrated.

Pre-disaster decisions:

- The selection of the locations of the local relief DCs, which are to be used for receiving the relief supplies from the state level and dispatching them to the demand points
- The inventory of relief supplies to be stored at each DC

Scenario-specific post-disaster decisions:

- The distribution assignment for each DC, such that relief supplies are delivered to the demand destinations, which include shelters as well as communities that have not been evacuated
- The evacuation operations for moving each affected community to the assigned shelters.

Let I be the set of communities in the affected area, J be the set of candidate DC locations, L be the set of possible shelters, and the scenarios be denoted $\omega \in \Omega$ in the formulation of the model.

Pre-disaster decision variables:

m_j amount of relief supplies to be stored at DC j

x_j

$$: \begin{cases} = 1, & \text{if DC } j \text{ is opened,} \\ = 0, & \text{otherwise;} \end{cases}$$

Post-disaster decision variables:

$q_{ij}^{C}(\omega)$ quantity of relief supplies satisfied by DC j that provides service to community i for scenario ω

$q_{lj}^{S}(\omega)$ quantity of relief supplies satisfied by DC j that provides service to shelter l for scenario ω

$e_{il}(\omega)$ total evacuees moved from community i to shelter l for scenario ω.

Pre-disaster parameters:

c_j capacity of DC j;

a unit volume of relief supplies;

f_j opening cost of DC j; and

g_j^D unit cost of shipping relief supplies to DC j from state level.

Post-disaster parameters:

$d_i(\omega)$ demand of relief supplies at community i for scenario ω;

$g_{ij}^C(\omega)$ unit cost of shipping relief supplies to community i from DC j for scenario ω;

$g_{lj}^S(\omega)$ unit cost of shipping relief supplies to shelter l from DC j for scenario ω;

$g_{il}^E(\omega)$ transportation cost to evacuate one person from community i to shelter l for scenario ω; and

prob(ω) probability of occurence of disaster scenario ω.

The mixed-integer stochastic programming formulation is given as:

$$\text{minimize} \sum_{j\in J} f_j x_j + \sum_{j\in J} m_j g_j^D + E_\Omega[Q(x_j, m_j, \omega)] \tag{1}$$

subjected to

$$E_\Omega[Q(x_j, m_j, \omega)] = \sum_{\omega\in\Omega} \text{prob}(\omega) Q(x_j, m_j, \omega) \tag{2}$$

$$\begin{aligned} Q(x_j, m_j, \omega) = {} & \sum_{i\in I}\sum_{l\in L} g_{il}^E(\omega) e_{il}(\omega) + \sum_{i\in I}\sum_{j\in J} g_{ij}^C(\omega) q_{ij}^C(\omega) \\ & + \sum_{l\in L}\sum_{j\in J} g_{lj}^S(\omega) q_{lj}^S(\omega) \quad \forall \omega \in \Omega \end{aligned} \tag{3}$$

$$\text{a}m_j \le c_j x_j \quad \forall j \in J \tag{4}$$

$$\sum_{i\in I} q_{ij}^C(\omega) + \sum_{l\in L} q_{lj}^S(\omega) \le m_j \quad \forall j \in J, \forall \omega \in \Omega \tag{5}$$

$$\sum_{j\in J} q_{ij}^C(\omega) = d_j(\omega) - \sum_{l\in L} e_{il}(\omega) \quad \forall i \in I, \forall \omega \in \Omega \tag{6}$$

$$\sum_{j\in J} q_{lj}^S(\omega) = \sum_{i\in I} e_{il}(\omega) \quad \forall l \in L, \forall \omega \in \Omega \tag{7}$$

$$q_{ij}^C(\omega), q_{lj}^S(\omega), e_{il}(\omega) \ge 0 \quad \forall i \in I, \forall j \in J, \forall l \in L, \forall \omega \in \Omega \tag{8}$$

$$x_j \in \{0, 1\} \quad \forall j \in J \tag{9}$$

Optimization (1) is to minimize, as the objective function, the cost of opening each DC, the cost to ship relief supplies from state level to each DC, and the expected cost of the response stage for each disaster scenario. The expected cost of the response stage is expressed by Eq. (2), where, for each scenario, the cost of the response stage includes the cost of transporting evacuees and the cost of shipping relief supplies from the DCs to the demand destinations (communities and shelters), as shown in constraint (3). Constraint (4) ensures that the amount of relief supplies maintained at any DC does not exceed the capacity of that DC, whereas constraint (5) states that the combined supply stock at the DCs is not less than the total demand of both the communities and the shelters. Constraint (6) ensures that the amount of relief supplies shipped from a DC to a community be equal to the demand that remains after any residents are evacuated. In addition, constraint (7) ensures that all evacuees at shelters receive relief supplies from a DC. The non-negativity constraint (8) indicates the number of shipments of relief supplies, and constraint (9) is a binary variable for each DC location.

However, we note that the cost of opening the shelters is not considered in the model formulation, because these are usually preexisting building, such as stadiums, and they are only temporarily used as evacuation shelters. Therefore, in this model, we assume that there is no opening or operating cost for the shelters.

4 Case Study of a Flood in Thailand

This section presents a case study regarding to the preparation and response to flood disaster in Chiang Mai Province in northern Thailand. Infrequent large floods usually occur in northern Thailand late in the May–October rainy season which is dominated by masses of moist air moving northeast from the Indian Ocean associated with tropical depressions moving westward from the South China Sea (Wood and Ziegler 2008).

Chiang Mai developed an early flood warning system which is able to detect by the real-time situation of Ping River as the gauging station P.1. When there is a heavy rainfall, it is able to evaluate the flood impact in the downtown by the level of Ping River at station P.1. In this study, we generate the possible scenarios for the need of relief supplies following to Chiang Mai flood hazard map studied by Civil Engineering Chiang Mai University Natural Disasters Research Unit: CENDRU (2015) as shown in Fig. 2. The flood hazard map is the analysis of different flooding risk area based on the historical data of station P.1 which can be constructed as 7 impact levels.

Using our stochastic approach, we used the flood hazard map for Chiang Mai to generate disaster scenarios that closely match real floods in this area. We considered seven different scenarios, and their probabilities of occurrence are calculated from historical data for the past three decades: 0.35, 0.20, 0.18, 0.12, 0.08, 0.05, and 0.02 for scenario 1–7, respectively. Each of these scenarios creates a different demand for relief, depending on how much of the area is affected.

Relief supplies are as identical as aid packages. We consider six possible relief DCs, two possible shelters, and sixteen communities with estimated demand amount at communities from the number of population record. Figure 3 marks the

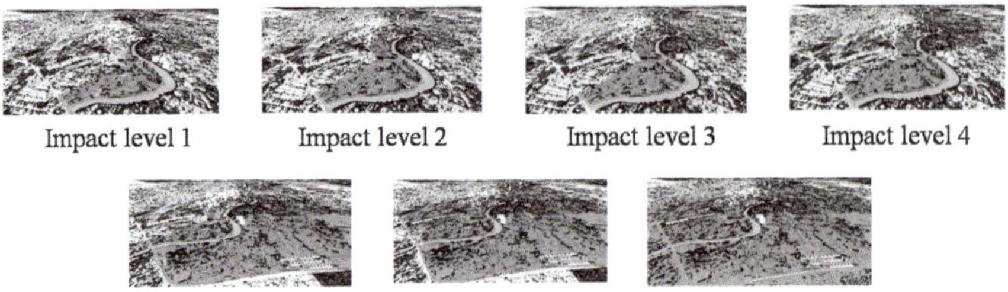

Fig. 2 Seven levels of Chiang Mai flood hazard map (CENDRU 2015)

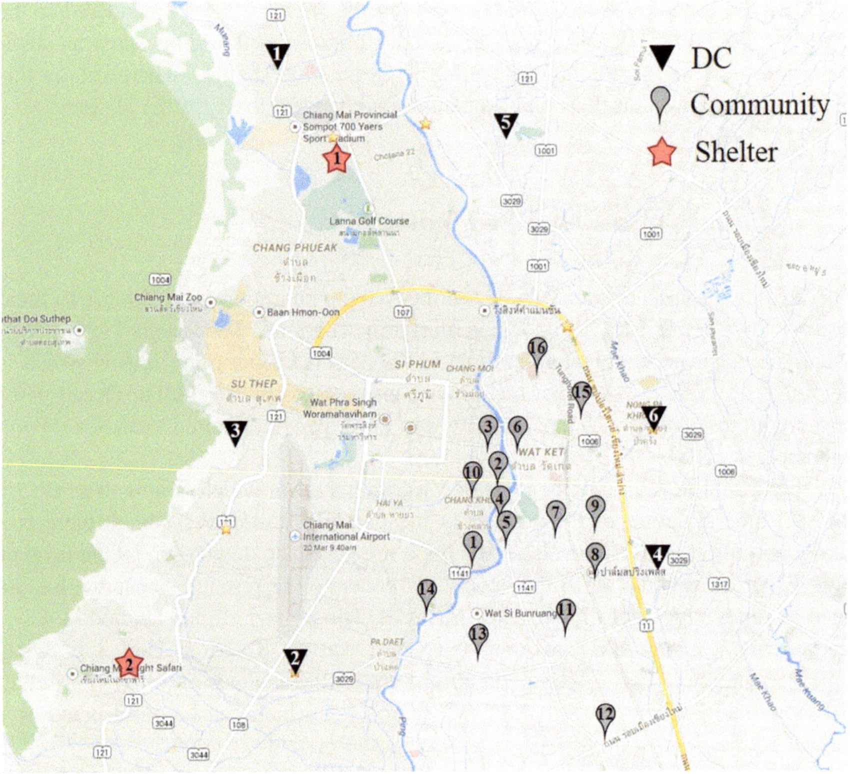

Fig. 3 Locations of affected communities, possible shelters, and relief DCs

locations of DCs, shelters, and communities given in this study. The transportation times for each scenario are determined by considering the effect of flood impact level. For example, the same route between DC 4 and community 1 does not take equal transportation time for distributing the relief supplies in each scenario. Transportation time is getting longer in the more severe impact level.

5 Computational Results

We solve the stochastic model using Gurobi Optimizer Ver. 6.0.0 mathematical programming solution software. All experiments were run on a personal computer with an Intel (R) Core (TM) i7-3770 CPU (3.40 GHz) and 16.0 GB of RAM. All the runs solved in less than 1 min.

The optimal solution to the stochastic model selected three relief DCs (2, 3, and 4) for storing emergency aid packages in preparation for a possible flood. The primary reasons that these locations were selected were their capacities and their distances from the communities and shelters. The opening cost and capacity of each DC were the same. For a successful storage plan, the total expected demand, which is based on the various possible scenarios, requires at least three relief DCs. On the left side of the Ping River, DCs 2 and 3 were chosen because they are closest to the demand destinations, not only the communities likely to be affected but also the evacuation shelters. DC 4 was chosen over DCs 5 and 6 to serve the communities on the right side of the Ping River.

We also calculated the results with the addition of one more constraint (10), which

$$\sum_{j \in J} x_j = k \quad k \in \{1, 2, 3, 4, 5, 6\} \tag{10}$$

limits the total number for DCs selected; this was done in order to determine their relative importance. As stated earlier, the expected demand requires at least three DCs for storing the relief. When fewer than three DCs are allocated, it was necessary to increase their capacity; otherwise, there was no feasible solution. We increased the cost of opening a DC so that it would reflect the larger capacity, and this is consistent with realistic situations. Figure 4 shows the selected DCs; the total number of DCs ranged from one to six, as in constraint (10). We found that the selection of three DCs minimized the total cost. We note that DC 2 was the best potential location, and it was selected each time; DCs 4 and 3, respectively, were the second and third most frequently chosen.

Because the purpose of the model is to make integrated decisions about the distribution of supplies and about evacuations, the cost of shipping supplies and the evacuation costs are some of the most critical factors in this approach. Figure 5 depicts the experimental results when varying the cost per unit of evacuation and the shipping cost per unit of supplies. The solutions that are shown satisfy the objective function, and the combined cost of evacuation, shipping to communities, and shipping to shelters. The cost of opening DCs and the cost of shipping supplies from the state level to the DCs are not included in the solutions shown in the figure, because the DCs all had the same opening cost, and the cost of shipping from the state level will be almost constant, since the distances to the DCs are similar.

It is clear that the left side of the figure represents a lower cost per unit of evacuation, while the right side represents a lower cost for shipping a unit of

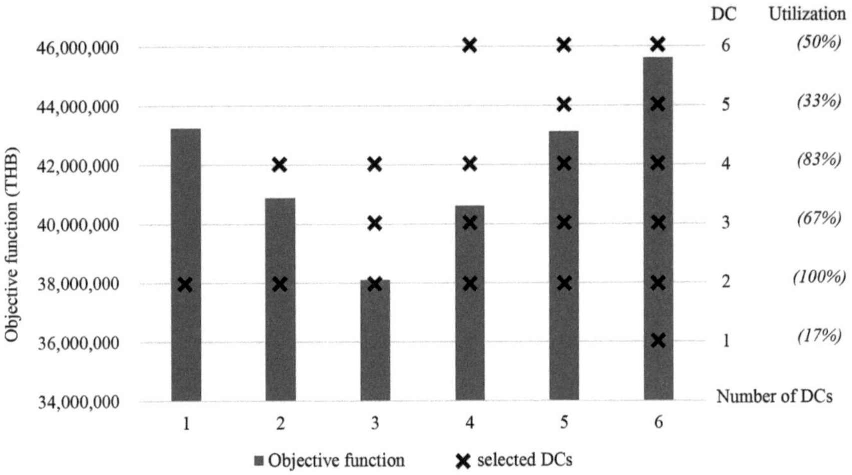

Fig. 4 Selection of relief DCs when the total number was specified

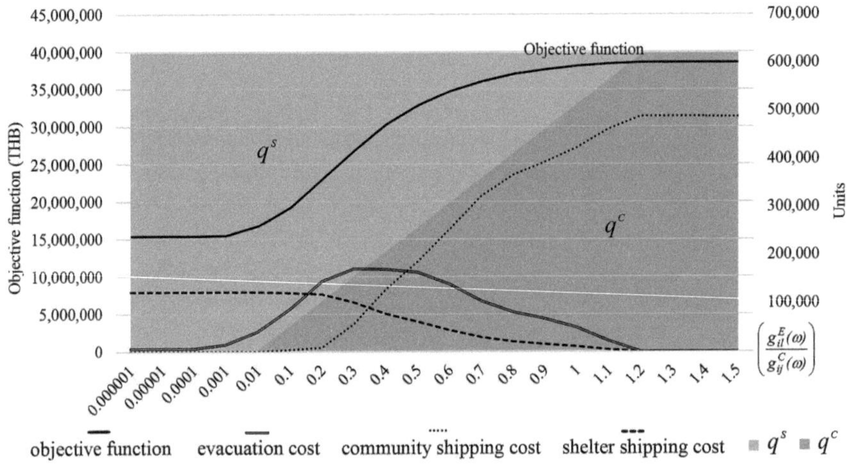

Fig. 5 Experimental results when varying the unit cost of evacuation and the cost for shipping a unit of relief supplies

supplies. Therefore, we can see that there is a steady increase in the amount of supplies shipped to shelters and evacuees when the unit cost of evacuation is significantly decreased. When the unit cost of evacuation is less, the model decides to evacuate more people from their communities to shelters, and thus more supplies have to be shipped to the shelters. The amount of supplies shipped to communities, on the other hand, increases when there is a higher cost per unit of evacuation; this continues until the unit cost of evacuation is 20 % higher than the unit cost of shipping supplies, and at that point, the model decides that everyone should remain in their community. The costs of shipping to the communities and shelters are directly related to the amount of supplies that are shipped to each.

We note that the cast of evacuation is important. Everyone is evacuated to a shelter, until the unit cost of evacuation is a thousand times smaller than the unit cost of shipping supplies; this is because the unit cost of evacuation is very small. Evacuations peak when the unit cost of evacuation is one-third of the unit cost of supplies, and they continuously decrease until they reach zero, when the unit cost of evacuations is 20 % higher than the unit cost of shipping supplies. Finally, the solutions provide a better value for the objective function when the evacuation rate is higher. Higher evacuation rates indicate a higher demand at the shelters, which reduces the expense of shipping supplies, because the distance from the DCs to the shelters is less than the distance to the communities.

6 Conclusion and Future Research

The present study proposed a decision-making model for the disaster management planning, which incorporate making integrated decisions about the distribution of relief supplies and about evacuation operations. Our stochastic approach determines the best storage locations from among a set of possible relief DCs, and it also determines their inventory levels. The model is also able to incorporate the priorities for evacuating particular communities, as well as specific disaster scenarios with estimates of the transportation needs and demands for relief.

There have been many studies of disaster supply chains, focusing on the distribution of relief supplies or on the evacuation of residents. To the best of our knowledge, no previous disaster management study has integrated these two aspects. Our stochastic model uses information about the expected magnitude of disasters to provide the optimal policy for the location and inventory of relief DCs and shelters.

Our future research will focus on developing a more sophisticated method for routing vehicles and a better scheduling system for the response stage of disaster management. With the results that we have obtained in this study, we can create an aggregated transportation plan with a detailed plan for routing vehicles, determining their loads of emergency aid packages, and ensuring the demand is met at each of the shelters and communities.

References

Abounacer R, Rekik M, Renaud J (2014) An exact solution approach for multi-objective location —transportation problem for disaster response. Comput Oper Res 41:83–93

Altay N, Green WG (2006) OR/MS research in disaster operations management. Eur J Oper Res 175(1):475–493

Balcik B, Beamon BM (2008) Facility location in humanitarian relief. Int J Logistics Res Appl 11 (2):101–121

Barbarosoglu G, Arda Y (2004) A two-stage stochastic programming framework for transportation planning in disaster response. J Oper Res Soc 55(1):43–53

Chang MS, Tseng YL, Chen JW (2007) A scenario planning approach for the flood emergency logistics preparation problem under uncertainty. Transp Res Part E 43:737–754

Civil Engineering Chiang Mai University Natural Disasters Research Unit (2015) Chiang Mai City Flood Preparedness System. http://cendru.eng.cmu.ac.th/cmflood/map.html. Approach 24 Jan 2015

Irohara T, Kuo YH, Leung JMY (2013) From preparedness to recovery: a tri-level programming model for disaster relief planning. In: Pacino D et al (eds) Computational logistics. Springer, Heidelberg, pp 213–228

Jia H, Ordonez F, Dessouky MM (2007) Solution approaches for facility location of medical supplies for large-scale emergencies. Comput Ind Eng 52(2):257–276

Kelle P, Schneider H, Yi H (2014) Decision alternatives between expected cost minimization and worst case scenario in emergency supply—second revision. Int J Prod Econ 157:250–260

Kongsomsaksakul S, Yang C, Chen A (2005) Shelter location-allocation model for flood evacuation planning. J Eastern Asia Soc Transp Stud 6:4237–4252

Liu Y, Lai X, Chang G (2006) Two-level integrated optimization system for planning of emergency evacuation. J Transp Eng 132:800–807

Manopiniwes W, Irohara T (2014) A review of relief supply chain optimization. Ind Eng Manage Syst 13(1):1–14

Manopiniwes W, Nagasawa K, Irohara T (2014) Humanitarian relief logistics with time restriction: thai flooding case study. Ind Eng Manage Syst 13(4):398–407

Mete HO, Zabinsky ZB (2010) Stochastic optimization of medical supply location and distribution in disaster management. Int J Prod Econ 126(1):76–84

Rawls CG, Turnquist MA (2010) Pre-positioning of emergency supplies for disaster response. Transp Res Part B Methodol 44(4):521–534

Salman FS, Yücel E (2015) Emergency facility location under random network damage: insights from the Istanbul. Comput Oper Res

Song R, He S, Zhang L (2009) Optimum transit operations during the emergency evacuations. J Transp Syst Eng Inf Technol 9(6):154–160

Verma A, Gaukler GM (2014) Pre-positioning disaster response facilities at safe locations: an evaluation of deterministic and stochastic modeling approaches. Comput Oper Res

Wood SH, Ziegler AD (2008) Floodplain sediment from a 100-year-recurrence flood in 2005 of the Ping River in northern Thailand. Hydrol Earth Syst Sci 12:959–973

Combining Simulation with Optimization to Evaluate Freight Logistics Performances for Developing a Corridor

Sirasak Tepjit and Thanyawan Chanpanit

Abstract The chapter presents an evaluation of freight logistics performances and recommends policy on logistics facility and related transport infrastructure investment along the corridor that links between Laem Chabang Port in Thailand and Port of Vung Ang in Vietnam via Laos PDR. A dry port and its transportation links to the seaport have been proposed for investment. There are five nominal routes within the network where an origin is located over the dry port hinterland and Guangzhou Port in China PDR is a destination. The combination of simulation model and goal programming has been applied to evaluate the corridor in terms of freight logistics operational performance. A discrete event model visualizes the desirable system. The weighted goal programming is used to find an optimal route. Finally, the policy prioritizes that investment must be made for the dry port and the connection to Laem Chabang Port by direct rail link. In order to encourage the dry port to be part of the network, the desirable logistics system requires overall transit cost should be around US$1100. Quality of services for on time performance should be maintained around 320 h of total transit with 21 h of traveling time variance and 54 h of operation time variance.

Keywords Corridor development · Goal programming · Discrete event simulation · Freight logistics performance

S. Tepjit (✉) · T. Chanpanit
King Mongkut's University of Technology North Bangkok, Rayong Campus, Bangkok, Thailand
e-mail: sirasakt@kmutnb.ac.th

T. Chanpanit
e-mail: thanyawanc@kmutnb.ac.th

V. Kachitvichyanukul et al. (eds.), *Toward Sustainable Operations of Supply Chain and Logistics Systems*, EcoProduction,
DOI 10.1007/978-3-319-19006-8_21

1 Introduction

Greater Mekong Subregion (GMS) was one of subregional economic cooperation that designed to enhance economic relation among five ASEAN country members including Thailand, Vietnam, Laos PDR, Myanmar and Cambodia, and two regions of the People's Republic of China (Ishida 2008). GMS countries had been already undertaking a number of major infrastructure investments, and more had been also planned. Somehow, many countries in the subregion had been inadequate transport infrastructure and high logistics service costs. It was a vital constraint for economic corridor development and integration. Development of economic corridors was a multisectoral development program of GMS strategy (ADB 2010).

Corridor development was an approach that stimulates regional integration in term of spatial, transportation infrastructure and logistics facilitation development (Reynaud 2003; Srivastava 2011; Rodrigue 2012). It was a holistic strategy that improved and enhanced investment in the subregion. However, transport infrastructure and logistics facilities such as rail links were the massive investment and made several impacts in long term. It was also hard to convince policy makers to decide on investment without strong evidences.

The project aimed to determine and evaluate the infrastructure development in case of a corridor that linked between ports in Thailand and Vietnam via Laos PDR. Freight logistics processed would be modeled as a conceptual model based on the existing situation adapted with the proposed infrastructures. In particular, when public policy analysis required a systematic sketch of the impacts of certain changes in the system, the development of simulation model should aid this decision making.

2 Literature Review

Corridors were a fundamental structure-shaping economic development (Srivastava 2011). The function of a corridor was to promote both internal and external trade by providing more efficient transport and logistics services (Rodrigue et al. 2013). The concept had become a substantive regional integration method for the European Union in a part of transport policy (Reynaud 2003). In order to stimulate the regional integration and spatial economic growth, the concept of development corridor had been applied to several regions; for example, Greater Mekong Corridors in GMS (ADB 2010), Delhi-Mumbai dedicated freight Corridor in India (Pangotra and Shukla 2012), Central Corridor in Sub-Sahara Africa (Arnold 2005), and CANAMEX trade corridor in North America (ERA 2001).

Basically, there were four components of a transport corridor that consisted of nodes, a link, flow, and a hub (Rodrigue et al. 2013). A node was a position of access to the transport network. A link was physical infrastructure that connected to nodes. There were at least two nodes and a link within a corridor. A link could contain one or more than one mode of transport and intermodal connection as well. Flow meant traffic circulate between nodes. A hub could be defined as a major node that covered several logistics activities such as distribution centers, warehouses, and business centers.

Logistics system was the process of planning, facilitating, implementing, integrating, and controlling the efficient and effective flow and storage of freight, and movement of people and information within and between logistics systems to enhance traders' competitiveness in order to increase national and/or regional competitive advantage (Rodrigue 2012). Freight logistics process was characterized by delivery of goods between point to point within a supply chain, generally defined as nodes. Performances of the services could be measured by total transit time and cost of services through a corridor (Banomyong 2008). These might be broken down into time and cost for specific links and nodes. In addition to JICA (2009) project, it converted transit time into cost of transport and assessed a corridor in terms of economic cost.

In order to evaluate logistics system, Yang et al. (2011) used a goal programming to analyze intermodal freight from China to Indian Ocean. There was the competitiveness of 36 alternative routes involved in the network. The goal programming was able to handle the conflicting objective functions such as minimizing transportation cost, transit time, and transit time variability while ensuring flow continuity for the entire route. Kengpol et al. (2012) developed a decision support system which enabled model evaluation to optimize multimodal transport route within GMS countries . The criteria were included to achieve the optimal transportation route within budget, time, and risk. Bookhinder and Fox (1998) applied a combination model to solve the two objective problem of minimizing time and cost of shipping for routes between Canada and Mexico. The competitive routes for a particular origin–destination were determined by non-dominated tradeoffs between cost and time.

3 Methodology

3.1 Research Framework

A methodology was developed to describe the desirable situation and then determined an optimal route. A combination of a simulation model and a goal programming to evaluate freight logistics performance had been used as a methodological tool. A discrete event model was built to simulate operation time and cost incurred in the logistics network. Then, a goal programming was constructed to determine an optimal route. The method could compromise the goals by minimizing unwanted deviational variables. Since the optimal solution could be reached, it probably had been not the desirable system. This would feedback to policy makers to adapt their objectives and reformulate their policies. Then, they would be repeatedly tested until obtaining the desirable system. Finally, policy recommendations on transport infrastructure and logistics facilities could be made.

3.2 Data Collection

The project illustrated a big picture of the logistics network which includes operations on sea ports, dry port, railway, road way, and logistics services. It could be impossible to collect operational data into detailed of each process. Basically, data include distances between each node, approximate traveling time and cost, and approximate operation time at each node. To complete the requirement, data were collected from both primary and secondary sources.

Firstly, some of the data were collected from secondary sources, such as websites like http://ports.com, www.adb.org/data/statistics, and www.thaibsaa.com, and publications from research organizations like ERIA, ADB, IDE, and JETHRI (Banomyong 2008; Kumagai et al. 2009; ERIA team 2009; Shibata, 2006). Secondly, some data were collected from primary sources like participating government agencies and private companies, such as customs agencies, Laem Chabang Port terminal operators, and freight forwarders.

4 Discrete Event Modeling

4.1 Assumptions

The conceptual idea of the corridor development plan could draw the network as shown in Fig. 1. The origination of the routes represented any plants where located in the catchment area of dry port hinterland in Thailand and the destination was Guangzhou Port in China. The following potential five nominal routes were demonstrated shipping cargoes which sent from a production plant to Guangzhou Port:

- **Route No. 1**: Goods were transported from a plant to Laem Chabang Port by truck and directly shipped to Guangzhou Port.
- **Route No. 2**: Goods were transported from a plant to Laem Chabang Port by truck, transited at Singapore Port and then shipped to Guangzhou Port.
- **Route No. 3**: Goods were transported from a plant to dry port by truck and modal shifted to Laem Chabang Port by train and directly shipped to Guangzhou Port.
- **Route No. 4**: Goods were transported from a plant to dry port by truck and modal shifted to Laem Chabang Port by train, transited via Singapore Port and then shipped to Guangzhou Port.
- **Route No. 5**: Goods were transported to Port of Vung Ang through Nakornphanom–Thakheg border and Napoa–Chalo border and directly shipped to Guangzhou Port.

Because dry port did not exist, assumed logistics operations at the dry port perform as same as the performance of Laem Chabang Port. The distance between the dry port and Laem Chabang Port was 400 km, approximately. An average rail

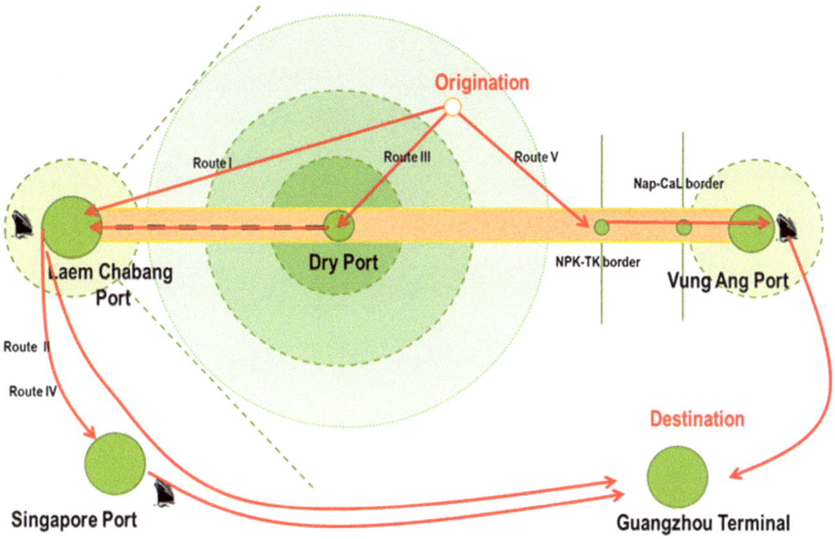

Fig. 1 The network assumption for a simulation model development

freight speed was 80 km per hours. Thus, an average traveling time between the dry port and Laem Chabang Port was approximately 6 h by rail. Meanwhile, Port of Vung Ang performs as bulk terminals. A container vessel might be not called at the port. In order for freight container to be shipped from the port, it was assumed that there were container shipping line services to Guangzhou Port, at least two vessels a week.

4.2 Simulation Model Description

The model was developed in Micro Saint Sharp software. It was a visual interactive modeling system. The structure of the model consisted of a task network of freight logistics operation along the corridor as shown in Fig. 2. A task was the basic building block of a model and contains timing information, execution constraints, effects of the task on the system, and routing information (MAAD 2009). Basically, timing information in each task represents traveling and operation time where nodes and links perform. It could be assumed as a normal distribution. An input data consisted an average time and standard deviation.

The routing information was showed in path tap. Paths describe the conditions under which particular paths were taken. There were three different decision types: Tactical, probabilistic, or multiple were used in Micro Saint Sharp software. The decision type will determine the path or paths that an entity should take. Decision

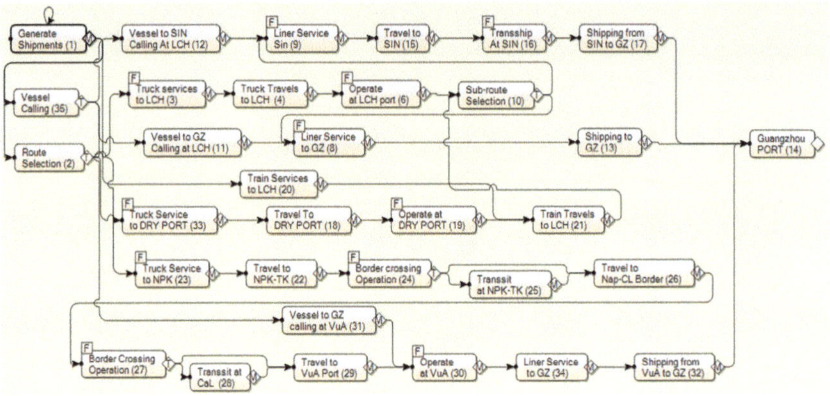

Fig. 2 The simulation network

nodes displayed with a diamond shape in the task network diagram and contained a letter inside of it (T, P, or M) indicating the type of decision.

The network could determine total transit time of each nominal route. For example, every single day shipments were generated from a production plant at task 1. Then, they were assigned to one of the nominal routes by calculating shortest time between a shipment arrival to a port and a vessel departure at task 2. Supposed Route no. 1 was selected, the shipments would move to task 3, truck serviced to LCH, and waiting time for a truck available would be recorded. Task 4, truck traveled to LCH, performed traveling time that shipments were transported from a plant to Laem Chabang Port. At this point, total transit time was an accumulation of waiting time for truck available and traveling time. The next task, shipments would be operated included handling, for instance, customs clearance, cargo inspection and loading to a vessel. Once operations at the port finished, operation time would be collected. At the last task, shipments were directly shipped from Laem Chabang Port to Guangzhou Port. The processes had been repeated until simulation time had been stopped. Outputs were the processing time and its variability of every single route.

5 Goal Programming Model

To combine the discrete simulation model to the goal programming, the simulated outputs including traveling and operation time at each node were arranged and put in the structure of a goal programming. The problem was applied by weighted goal programming. In order to obtain an optimal solution, Excel Solver on spreadsheet was applied.

5.1 Assumptions

Given the diversity of cost and time issues relating to various services of nominal routes, the objective of the model was the optimal route to transport shipments from a plant where located in the dry port hinterland in Thailand to Guangzhou Terminal in China. The feasible route was selected by minimizing unwanted positive deviations which related freight logistics performance measurements as following: (i) total transit time; (ii) total transit cost; (iii) traveling time variance; and (iv) operation time variance. The relative priorities of measurements were determined by weight of deviations.

5.2 Model Description

- **Decision Variable**

x_{ijk} was a binary decision variable for the goal program. It represented a link between two nodes. Whether shipments are shipped from node i to node j by transportation mode k, x_{ijk} was 1, otherwise it equaled to 0. These were algebraically defined as:

$$x_{ijk} = \begin{cases} 1, & \text{node } i \text{ and node } j \text{ can link to each other by mode } k \\ 0, & \text{otherwise} \end{cases}$$

- **Goal**

The following objective function was applied by percentage normalization. It was formulated to minimize total weight unwanted deviations of freight logistics performance measurements. They were total transit cost, total transit time, traveling time variance, and operation time variance, respectively. Each deviation was turned into percentage away from target value of the performance measurements as shown in Eq. (1).

$$\text{Min } z = \sum_{i=1}^{n} \frac{W_i(p_i + n_i)}{Q_i} \tag{1}$$

- **Constraints**

Equations (2) and (3) represented to total transit cost and transit time of every single route with deviation n_1 and p_1 equaled to the target value of total transit cost (Q_1). It included traveling cost (TC_{ijk}) and operation cost (OpC_i) at node i where shipments were transported from node i to node j by transportation modal k. The constraint maintains traveling time variance (TV_{ijk}) and total operation time variance (OpTV_{ijk}) at each link of entire its route with unwanted deviation equaled to target value as displayed in Eqs. (4) and (5), respectively.

To ensure every shipment was sent from an origination to a destination were expressed in Eqs. (6) and (7). Meanwhile, every single could be sent from node i to node j by node k and had to be not taken back in the same link were restricted by Eqs. (8) and (9). Finally, Eq. (10) controlled total weight of deviations equal to 1.

$$\sum_{i\in I}\sum_{j\in J}\sum_{k\in K} \mathrm{TC}_{ijk}x_{ijk} + \sum_{i\in I}\sum_{j\in J}\sum_{k\in K} \mathrm{OpC}_{i}x_{ijk} + n_1 - p_1 = Q_1 \tag{2}$$

$$\sum_{i\in I}\sum_{j\in J}\sum_{k\in K} T_{ijk}x_{ijk} + \sum_{i\in I}\sum_{j\in J}\sum_{k\in K} \mathrm{OpT}_{i}x_{ijk} + n_2 - p_2 = Q_2 \tag{3}$$

$$\sum_{i=I}\sum_{j=J}\sum_{k=K} \mathrm{TV}_{ijk}x_{ijk} + n_3 - p_3 = Q_3 \tag{4}$$

$$\sum_{i\in I}\sum_{j\in J}\sum_{k\in K} \mathrm{OpTV}_{ijk}x_{i} + n_4 - p_4 = Q_4 \tag{5}$$

$$\sum_{j\in J}\sum_{k\in K} x_{1jk} = 1 \tag{6}$$

$$\sum_{i\in LD}\sum_{k\in K} x_{iDk} = 1 \tag{7}$$

$$\sum_{i\in I}\sum_{j\in J}\sum_{k\in K} x_{ijk} = 1, \quad i \neq 1 \tag{8}$$

$$\sum_{i\in I}\sum_{k\in K} x_{ihk} = \sum_{i\in I}\sum_{k\in K} x_{hjk}, \quad h \neq i \neq j \tag{9}$$

$$\sum_{q=1}^{n} w_q = 1 \tag{10}$$

The variable x_{ijk} was defined as binary. All deviations n_q *and* p_q were defined as non-negative variables. All other variables were also non-negative.

6 Results and Experimental Analysis

6.1 Numerical Results

- **Simulation results**

The total transit time which simulated by the discrete event model was drawn in Fig. 3. It showed the traveling and operation time at nodes and links via every single route. The simulation results gave us Route no. 5 took the shortest total

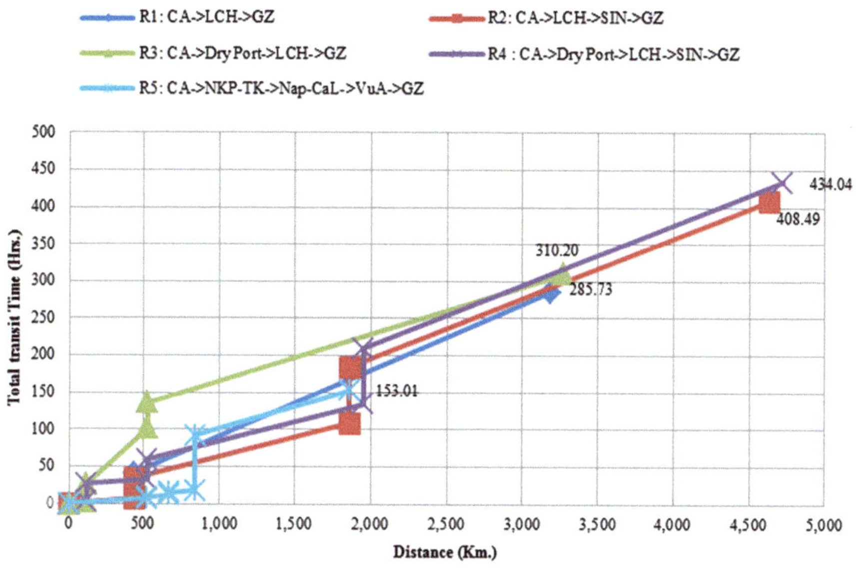

Fig. 3 Transit time versus distance line graph

transit time by 153.01 h. The longest transit time was Route no. 4. It was via the dry port and to transit at Singapore Port. Total transit time was 434.04 h. The most of proportion time consuming was to wait for logistics services. These were rail freight service at the dry port, shipping line service at Laem Chabang Port, and transshipment at Singapore Port. It contributed to 29.18 % of total transit time.

To analyze the corridor performance, processing time variance was one of the most importance measurements. It represented reliability of services. In this case, processing time variability had been divided into traveling time variance and operation time variance. Route no. 5 gave us both the minimum traveling time and operation time variances by 9.41 and 29.03 h, respectively. Contrast to Route no. 4 performed the greatest value of both traveling time and operation time by 35.36 and 109.37 h, respectively. The results could be seen in Table 1.

- **Results from the collected data**

Transit cost was collected by both secondary and primary data. To make the information consistency, freight price and related operation cost was arranged in cost-distance line graph. The results could be found that Route no. 5 gave us the greatest cost with 1860 km of the shortest path. Total transit cost was $US2660. Route no. 1 gave the smallest total transit cost. Nevertheless, this route gave the smallest cost but Route no. 2 was more often used owing to Singapore Port was a strategic position participated as a transshipment hub for southeast Asia (Tsuneish 2007). The port represented an active feeder shipping spot in Asia with a network service ranging from short to long routes. Whereas the dry port and rail link involved in the network, there was only $US90 of total cost higher than the routes

Table 1 Result summary

Route	Total cost ($US)	Total time (Hrs.)	Traveling time variance (Hrs.)	Operation time variance (Hrs.)
*R*1	1080	312.84	19.67	31.94
*R*2	1160	435.60	33.73	86.68
*R*3	1170	328.72	21.30	54.64
*R*4	1250	434.04	35.36	109.37
*R*5	2260	154.06	9.41	29.03
Min *Ri*	1080	154.06	9.41	29.03
Max *Ri*	2260	451.48	35.36	109.37

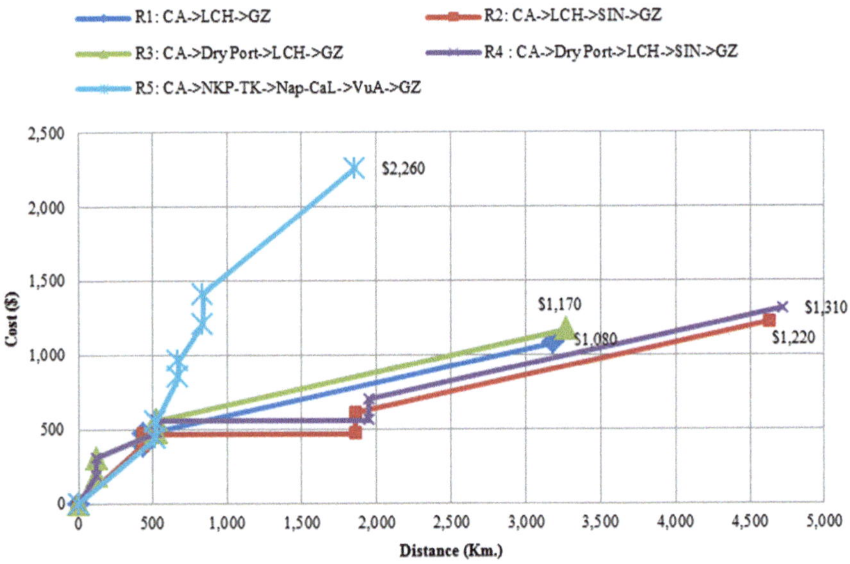

Fig. 4 Transit cost versus distance line graph

without it. It seemed to be indifferent among the network significantly. Total transit cost via Route no. 3 was $US1170 and $US1310 via Route no. 4. This implied that a dry port was high probable to develop. The result was depicted in Fig. 4.

6.2 Experimental Analysis

An experiment analysis was conducted on target values and deviation weight. Two scenarios could be tested. Firstly, given all of target values were the smallest of four logistics performance measurements and all of deviation weights were equally. Secondly, the scenario was to examine a number of target values that could promote a dry port involved to the network.

According to the results from the simulation model, an optimal route could be determined by given target value (Q_l) was the smallest value of corridor measurements among five competitive routes. The initial weights were set at all equal. The targets used in the goal program were as follows: a minimum average total transit cost of $US1080, total transit time of 154.06 h, traveling time variance of 9.41 h, and operation time variance of 29.03 h.

The above conditions gave Route no. 5 as an optimal solution with 0.27 of total weight deviation. The deviation of total transit cost was $US1070. To make all of the goals satisfied, total transit cost could be $US2250. Total transit time should maintain at 154.06 h with 9.41 h of traveling time variance and 29.03 h of operation time variance.

If one considers on time performances as a priority, Route no. 5 was a feasible choice. Meanwhile, Route no. 1 took the most advantage on cost. The dry port probably not included to both solutions. Consequently, if a dry port could be not invested, the area along corridor might be not attractive for an export-oriented investment. Therefore, production plants had been concentrated over an area connected to Laem Chabang Port. In addition, if shipments would be sent via Route no. 5, it might not take advantage on transit time so far and its transit cost also increased. The development could not change the existing situation. Thereby, the objective of this scenario was to adjust the set of target value which supported the dry port as a part of an optimal network.

Assumed deviation weights were equal. The goal programming could figure Route no. 3 as an optimal network since total transit cost was $US1100. Total transit time was 320 h. It was 21 h of traveling time variance and 54 h of operational time variance. The set of target values was adequately strong due to the optimal route had not changed since deviation weights had become unequal.

7 Conclusions and Recommendations

7.1 Conclusions

The corridors had become the structure of the region. Briefly, transport corridors will physically link territories or a region, whereas economic corridors would integrate economic activities over territories or a region. The project had demonstrated a combination of simulation model and goal programming to evaluate the corridor performance. This methodology assisted policy makers to formulate an adequate policy on logistics facility and transport infrastructure investment.

To visualize the desirable system, a discrete event model was used as a methodological tool. The simulation gave transit time and its variability of each competitive route. These outputs were put in the goal programming as well as collected transit cost. The weighted goal programming was used to figure an optimal route by compromising target values of freight logistics performance.

The findings suggested that Route no. 5 is the most time advantage but it was also the most costly; because Singapore Port is the main regional port in southeast Asia, Route no. 2 was a general route in the current situation; if one consider in multigoals, given total transit cost was slightly importance than total time performance, Route No. 1 was more advantage.

7.2 Recommendations

The optimal routes probably not included a dry port into the network. Thereby, Route no. 5 might not perform efficiently, if the dry port could be not considered to invest. Also, the corridor could not be improved since the territory along the corridor has less potential to develop industrial plants. Hence, in order to promote economic activities over the region along the corridor, a dry port was a critical logistics facility. It had to directly connect to seaports with high-capacity transport means such as rail links.

Priority should be given, the dry port had to invest and connect to Laem Chabang Port directly by rail link. In order to encourage the dry port into the network, the desirable logistics system required overall transit cost should be controlled by 1100$US. Quality of services on time performance should be maintained by 320 h of total transit with 21 h of traveling time variance and 54 h of operation time variance.

References

Asian Development Bank (2010) Strategy and action plan for the greater mekong subregion east-west economic corridor. Asian Development Bank, Philippines

Arnold J, Olivier G, Arvis JF (2005) Best practices in corridor management. World Bank, Washington DC

Banomyong R (2008) Logistics development in the north-south economic corridor of the greater mekong subregion. J GMS Dev Stud 4:43–58

Bookbinder JH, Fox NS (1998) Intermodal routing of Canada-Mexico shipments under NAFTA. Transp Res Part E Logistics Transp Rev 34(4):289–303

ERIA team (2009) Mekong-India economic corridor development. In: Concept paper Economic Research Institute for ASEAN and East Asia

ERA (2001) TASK III: transportation strategies and economic impact analysis. (CANAMEX corridor Plan working paper). The CANAMEX Corridor coalition. Accessed from http://www.canamex.org/wp-content/uploads/2013/02/WorkingPaper3.pdf)

Ishida M (2008) GMS economic cooperation and its impact on CLMV development. In: Sotharith C (ed) Development strategy for CLMV in the age of economic integration, ERIA research project report 2007-4. IDE-JETRO, Chiba, pp 115–140

JICA (2009) The research on the cross-border transport infrastructure: phase 3, Final report
Kengpol A, Meethom W, Tuominen M (2012) The development of a decision support system in multimodal transportation routing within greater mekong sub-region countries. Int J Prod Econ 140(2):691–701
Kumagai S, Gokan T, Isono I, Hayakawa K, Keola S (2009) Geographical simulation analysis for logistic enhancement in East Asia. ERIA Research Project Report, pp 7–2
MAAD (2009) In: Micro saint sharp: user guide
Pangotra P, Shukla PR (2012) Infrastructure for low-carbon transport in India: a case study of the Delhi-Mumbai dedicated freight corridor. Indian Institute for Management, Ahmedabad
Rodrigue JP (2012) The benefits of logistics investments: opportunities for Latin America and the caribbean. Inter-American Development Bank, Washington DC
Rodrigue JP, Comtois C, Slack B (2013) The geography of transport systems. Routledge
Reynaud C (Director) (2003 Nov 27) The concept of corridors and networks in developing Pan-European Infrastructure. *Seminar on transport infrastructure development for a wide Europe*. Lecture conducted from, Paris
Shibata T (2006) Asian logistics network map. Japan External Trade Organization (JETRO)
Srivastava P (2011) Regional corridors development in regional cooperation. Asian Development Bank Economics Working Paper Series, p 258
Tsuneishi T, 恒石隆雄 (2007) Thailand's economic cooperation with neighboring countries and its effects on economic development within Thailand
Yang X, Low JM, Tang LC (2011) Analysis of intermodal freight from China to Indian Ocean: a goal programming approach. J Transp Geogr 19(4):515–527

Time Management of Domestic Express Transportation Services

Punnarat Chanasit

Abstract This chapter explores time management of domestic express transportation services in Thailand. The express transportation services as known as courier that have significant roles in transportation industries. There are three key performance measures for customer satisfaction: time, speed, and prefer condition of the shipments, and they are related to the operation logistics in delivery measurement, delivery in-full on-time (DIFOT). The competition level in express transportation services is extremely high, and the core competitive advantage of this sector is time management and cost reduction. Almost all local express transportation services in Thailand only have the lower cost advantage. It means that they only compete on low price without the actual commitment on delivery date/time, proper speed, and good package. The main discussion is on the time management of express delivery service to improve service level. Data were acquired from four selected international transportation services companies in Thailand for a comparative study.

Keywords Time management · Express transportation service · Courier · Operation logistics

1 Introduction

1.1 Statement of the Problem

Nowadays, the express transportation services have significant grown toward household to corporate scale. It starts from sending the letters to each other through awkward parcels to the industrial factories. There are many service providers in this

P. Chanasit (✉)
Logistics Management, School of Business Administration, Bangkok University, Pathumthani 12120, Thailand
e-mail: punnarat.ch@gmail.com

V. Kachitvichyanukul et al. (eds.), *Toward Sustainable Operations of Supply Chain and Logistics Systems*, EcoProduction,
DOI 10.1007/978-3-319-19006-8_22

sector including local and international service providers. The express transportation service has been developed from postal activities, even so customers' demand have to be rapidly responded more than ever. Ultimately, the express transportation services would start upon that period. Entrepreneurs in this sector have to compete together due to there will be new comer into this sector easier. Almost local service providers are competing with international services providers especially that serve only domestic express transportation services. At present, there are big four express well-known companies: TNT Express, UPS, FedEx, and DHL Express.

Express transportation is a part of operation logistics that there are "couriers" who pick up and deliver the shipments from door-to-door. It consists of many components of logistics activities. Express transportation service is on the top from all of logistics and supply chain. Logistics activities blend all of them and finally to be the express transportation. In other word, express transportation is composed by function of transportation and warehouse that are main function in logistics and supply chain activities. The integrated express transportation service is displayed in Fig. 1.

Express service providers have faced the main problem that is time management besides the cost issues. Time management is not only trouble in express transportation, but it occurs to many businesses also. Time is extremely important factor of logistics and supply chain especially in express transportation. The big four express companies can manage limited time and improve their service level more and more. Meanwhile, the local service provides cannot manage time to deliver shipments to destinations as the big four do. They are able to wait the shipments full-truck-load only. Another big competitor can pick up shipments and deliver to end-customers within next day service.

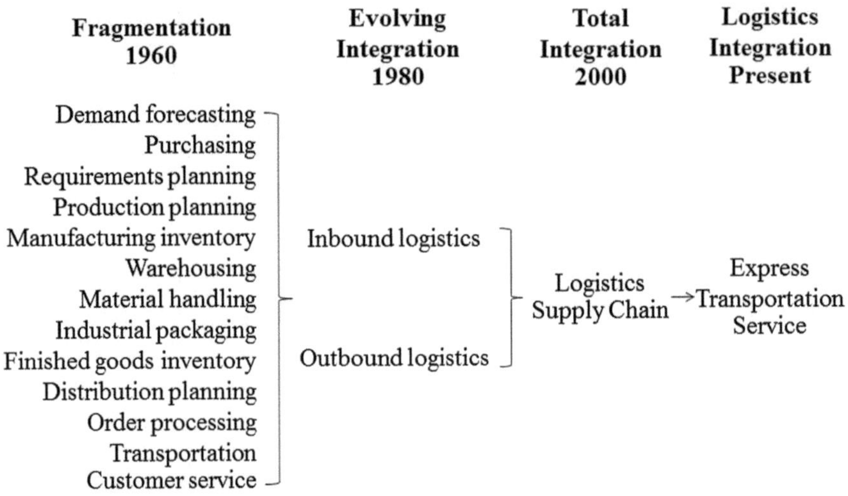

Fig. 1 Integrated express transportation service (a part from integrated logistics management *source* Center for Supply Chain Research, Penn State University.) (Coyle et al. 2009)

The researcher collected data that based on survey research method. The four selected samples focused on significant factors: their distribution center's (DCs) location, amount of couriers, types of the vehicle, how to plan day-to-day operation, and KPIs which were compared, and the data were analyzed in order to select the factor that affects the time management and can serve the service level as high as the international express companies do. Consequently, the local entrepreneurs of express transportation services can use and benchmark such significant factors to improve and manage their limited time.

1.2 Objective

- To study significant factors which affect time management in domestic express transportation services and
- To explore necessity data and success express companies' factors for improving service level and management limited time of local express companies in Thailand

1.3 Scope of Study

This research studied 4 international express companies that service the domestic express transportation. These express companies have transportation networks around Thailand. There are individual DCs and depots (drop points). Four samples are overseas company which they can operate the express transport in Thailand. This research compared main data that are mentioned in 1.1 and yields the outcome in statistics pattern.

1.4 Significance of the Research

- Exploring significant factors for managing delivery and pick up time in domestic express transportation services,
- The significant factors can be used or benchmarked to improve and manage efficiency time, and
- Investigating result of significant factors in statistic pattern in order to be outcome from KPIs target.

2 Literature Review

2.1 Overview of Time Management

Normally, time management in logistics and supply chain fields is concerned about just-in-time process. This research used the principle of just-in-time pioneered to apply into success the express companies' significance factors. Just-in-time principle is eliminating waste and improving product quality. The basic just-in-time principles are to make only what is needed, when needed, and in the amount needed. Therefore, the objective of just-in-time is to improve the process by eliminating waste (Nayab and Edwards 2010). Toyota's Fujio Cho defines waste as:

> Anything other than the minimum amount of equipment, materials, parts, and workers, which are absolutely essential to production.

The type of waste in just-in-time principle is following (in Fig. 2):

- Processing waste (over-processing),
- Waste due to idling time of machinery (waiting),
- Waste due to product defects (defects),
- Waste of motion or faulty working techniques (motion),
- Waste related to transportation (transportation),
- Waste from over-production (over-production), and
- Inventory waste (inventory).

In this research, 7 wastes have applied with the express transportation service in order to eliminate the waste time in door-to-door operations. Normally, express companies have time-definite services, and the delivery is guaranteed on a specific day or at a certain time of day (Coyle et al. 2011) depends on their service provided to the customers.

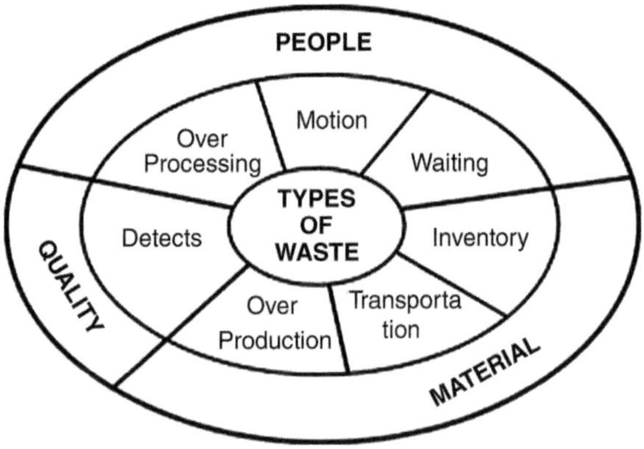

Fig. 2 Seven wastes in just-in-time principle (Adam n.d.)

Table 1 Comparison of modal capabilities

Mode	Strengths	Limitation	Primary role
Truck	- Accessible - Fast and versatile - Customer service	- Limited capacity - High cost	Move smaller shipments in local, regional, and national markets
Air	- Speed - Freight protection - Flexibility	- Accessibility - High cost - Low capacity	Move urgent shipments of domestics freight and smaller shipments of international freight

2.2 Transportation Principles

In the transportation principles, the mode of transportation are divided into 5 modes: truck, rail, air, water, and pipeline. Actually, the express transportation is intermodal transportation between truck and air mode. Both modes can compare the strengths, limitations, and primary role as in Table 1.

The express transportation selects good points of each mode especially the point of accessible, fast, speed, and flexible except cost issue. From these points, this research uses transportation mode selection principles: (1) accessibility, (2) capacity, (3) transit time, (4) reliability, and (5) safety (Coyle et al. 2011), for grouping significant factors to manage time in express transport.

3 Research Methodology

This research's objectives are to study significance factors which affect time management and explore necessity data and successful express companies' factors for improving service level and time management. At first, comparing four selected samples. The data were collected in year 2014 and primary data were collected by supervisory and operations level's interview and survey. Then, significance factors compare selected samples item by item. This research will explain why the selected samples have grown and improve service level in dimension of time better than local competitors. In the end of this chapter, the research summarize that the delivery performance arises from the significance factors.

Refer to type of 7 wastes in Sect. 2.1 and transportation modes selection principle in Sect. 2.2. In transportation section, they have delivery measurement that is delivery in-full on-time (DIFOT). The researcher had created the significance factors that link between 2 parts. It consists of (1) DC's location, (2) amount of couriers, (3) type of vehicle, (4) how to plan day-to-day operation, and (5) KPIs. Five factors are grouped with 7 wastes/mode selection and related to DIFO. Because of these factors should improve service level and make customer satisfy

Table 2 Grouping significant factors into 7 wastes and mode selection

No.	7 wastes	Mode selection	Significant factor
1	Over-processing	Reliability	Amount of couriers
2	Waiting	Transit time, reliability	DC's location and amount of courier
3	Defects	Capacity, reliability	Type of vehicle and how to plan
4	Motion	Safety	Amount of couriers, type of vehicle, and KPIs
5	Transportation	Accessibility, safety	Amount of couriers and type of vehicle
6	Over-production	Reliability	KPIs
7	Inventory	Reliability	KPIs

continuously if the companies do their jobs and follow the role of JIT (7 wastes) and transportation modes selection principle. These factors can be grouped into both principles as displayed in Table 2.

3.1 Significant Factors from International Express Companies

This topic will explain why this research selected those factors to study and make them as factors for managing time.

1. DC's location: This is first priority to locate hub, DCs, or depot. If the company locates DCs/hub on proper areas, the shipments can be delivered faster. Usually, DCs in express company has to set the warehouse type as a cross-docking. Due to fast responsive transportation system is necessary for this warehouse type. (Simchi-Levi et al. 2009)
2. Amount of couriers: The courier or driver is like ambassador of each express company, because they have to meet customers' whole shipments before another staff in the company. Even though, express company should have enough couriers for serving the incoming volume of shipments.
3. Type of vehicle: The shipment can be moved by the vehicle. Proper vehicle can deliver faster shipment especially in urban transportation system.
4. How to plan day-to-day operation: First 3 factors cannot be good for managing time unless there was no good plan for day-to-day operation. Mostly, supervisor in express company has to plan and set the routing before couriers are released for deliver and pick up of the shipments to/from customer.
5. KPIs: This is necessary factors that express company has to set up for its couriers. It means that if there is KPIs, then there will be reward/incentive closely. Entire big four express companies have their KPIs for stimulating courier deliver/pick up the shipment on-time.

Table 3 Express company profile (data in year 2014)

No.	Nationality	International express	Domestic express	Transportation mode in domestics	YTD volume (consignments)
1	Hong Kong	No	Yes	Truck (and air)	4,779,555
2	The Netherlands	Yes	Yes	Truck (and air)	1,117,428
3	USA	Yes	Yes	Truck	706,398
4	USA	Yes	Yes	Truck	439,985

3.2 International Express Company Profile

There are 4 samples. Their profiles have been compared in the Table 3.

3.3 Domestic Express Transportation Activities

Four samples have common express shipment activities and processes, but there is a few differences in some detail of each company. Common activities and processes are shown in the flowchart in Fig. 3.

3.4 Compare Significant Factors of 4 Samples

In this part, the selected significant factors from 4 samples are compared with each and explained the details as follows.

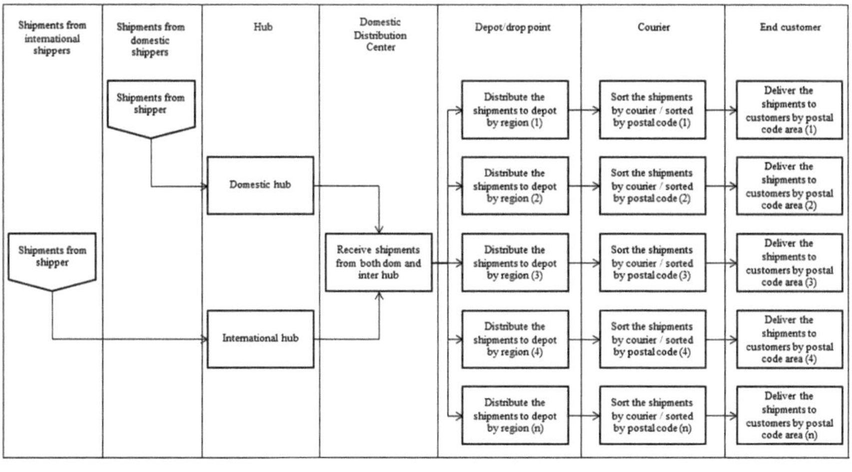

Fig. 3 Flow of express shipment activities and processes

Table 4 Distribution center's location (data in year 2014)

Sample	No. of hub	No. of DC, depot, drop point	Separated by region of Thailand				
			North	Northeast	East	Central	South
First	1	52	10	6	6	16	14
Second	2	16	3	3	1	5	4
Third	4	39	2	1	3	28	5
Fourth	2	2	0	0	0	2	0

1. DC's location: In Table 4, this factor is compared with the location of each sample. The first sample has 1 hub and 52 DCs. The second sample has 2 hubs, including international and domestic hub, and 16 depots, while there are 4 hubs and 39 drop points for the third, and the fourth has only 2 hubs and 2 drop points. Nevertheless, from the studying found that the third and fourth sampling have used the first sampling as subcontract in case of overcapacity or out of service area.
2. Amount of couriers: This factor is variable for the express transportation service and also including transportation industries. The turnover rate of couriers and motorbikers are high and continue every month. If express company can offer and provide good salary/compensate toward them, then the service level especially deliver/pickup time is better. Hence, the number of couriers of each sample is displayed in Table 5. This figure includes subcontract headcount.
3. Type of vehicle: Entire samples have same type of vehicle. They are 2-wheel vehicles and 4-wheel vehicles. The 2-wheel vehicle or motorcycles have been used for document or small parcels delivery services, while the 4-wheel vehicle or utility trucks (with container) have been used for the large parcel delivery services. However, the second sampling uses 6-wheel truck in urban area, because it is suitable for small manufactory in town (freight and awkward) as shown in Table 6.
4. How to plan day-to-day operation: Researcher interviewed the supervisory level of 4 samples about how to plan day-to-day operation and how to set the routing for their couriers. As common activities and processes have been mentioned in Fig. 3, the supervisor of 4 samples plans and prepares their manpower (courier), equipment (vehicle), and routing before releasing their subordinate to deliver/pick up the shipments in working hour. Additional data, supervisor, or manager of local transport company will plan the manpower and routing toward its driver when they get booking or request from customer. No more plan or over-prepare schedule for incoming or unexpected volume. It certainly differs

Table 5 Amount of couriers (data in year 2014)

Sample	No. of couriers or operation staff (on average)
First	1000–1500
Second	800–1200
Third	500–700
Fourth	300–500

Table 6 Type of vehicle

Sample	Vehicles		
	Motorbike	4-wheel pickup	6-wheel truck
First	Yes	Yes	No
Second	Yes	Yes	Yes
Third	Yes	Yes	No
Fourth	Yes	Yes	No

from express transport company. The details have been shown in Table 7. It shows depth detail of their operation plan.

5. KPIs: KPIs is the key success factor for domestic express transportation service. Every express company has to set up the KPIs for its couriers or operation staff. KPIs seem the result from all factors of time management. Each sample has the KPI that measure time such as delivery on-time (as known as DIFO, delivery in-full on-time) and cutoff pickup time. The list of KPIs of each sample has been

Table 7 How to plan day-to-day operation

Sample	Interview result: how to plan day-to-day operation
First	In the morning, they have to sort the shipment of each courier and separate them by postal code. After that, they focus on the estimate due date of each shipment, because if couriers can deliver those shipments to customer within estimate due date, they will get the incentive (money) from the KPIs target. During working days, the supervisor has to monitor delivery shipment per courier and catch up with concern parties to try to deliver to customers finally. In the evening before finish work, they will get alert mail from the hub that shows the detail about how many incoming volume by tomorrow. So, they can plan or prepare more manpower or labor in case of over-capacity volume into their station
Second	In the morning, the sorter consolidates the shipments from hub to depot. When the shipments are unloaded, couriers will sort the shipments into their route and separate by postal code. Supervisor briefs couriers about the problem and informs the daily performance and after that the couriers will check the shipments and scan the detail of each shipment into the company's system. At noon, the dispatcher of night shifts will check the incoming shipments which have the service on next day delivery
Third	Supervisor calls team and let couriers to sort their shipments into route separated by postal code. The courier will collect the undelivered shipment from the day before and combine the shipments together. Then, they focus on the document shipments and parcel shipments which have due date on that date first. On the way to customer, if customers call-in to request the courier to pick up their shipment, call center will generate data and send shipper details to courier directly via mobile equipment. However, if there are no pending shipments, the courier has to stand by for picking up other booking in area
Fourth	Couriers have been divided into 2 types, van and motorcycle. The shipments will be sorted by courier who is owner route. The document shipments or small parcel will be sorted to motorbiker, while the big parcel will be sorted to courier. Before this process, the shipments that shipper send to up-country will be sorted to subcontracts or agents. Normally, the courier delivers and picks up the shipments in Bangkok and Bangkok-greater area only. Almost shipments were shipped from overseas and send to destination in Thailand

Table 8 KPIs

Sample	KPIs	Result of delivery performance (%)
First	1. Deliver on-time	94
	2. Auto return shipments (in case of unable to delivery)	
	3. Return invoice or hard copies to shipper	
	4. Late data entry	
	5. COD (money) transferred	
	6. First delivery time	
	7. Truck accident	
	8. Missed pickup (booking) shipment	
	9. Measure actual dimension of the shipment	
Second	1. Deliver on-time	95
	2. Pickup on-time	
	3. Accident (truck and motorbike)	
	4. PAC scan (hardcopy, AWB, and invoice)	
	5. Update correct status code into the system	
	6. COD transferred	
Third	1. Delivery on-time	97
	2. Shipment loss or damage	
	3. COD transferred	
	4. Pick up and send the shipment to hub on-time	
	5. Accident	
Fourth	1. Delivery on-time	97
	2. Shipment loss or damage	
	3. COD and COP transferred	
	4. Accident	

compared in Table 8, and the result of KPIs in topic of time has figured in the end of each column.

The figure result of delivery performance shows the outcome (productivity) of time management from each sample. The first is 94%, the second is 95%, and the third and fourth is 97%, respectively. In statistics pattern of logistics performance index (LPI) in part of domestic LPI (shipments meeting quality criteria) in 2014, World Bank reported that Thailand gets 82.5% (The World Bank Group 2015). Thus, 4 samples have outstanding result, that is, more than domestic LPI. It means that these significance factors can improve service level and manage time accordingly.

4 Research Result

In this part, the researcher brought all 5 significance factors to interview 2 local domestic express companies in Thailand. The details are following:

4.1 Interview and Survey Result

The researcher had interviewed and surveyed 2 local domestic express companies. The selection criterions are following: (1) the owner must be Thai citizen, (2) to conduct a business as express transportation or mail service, (3) there are own vehicles, (4) there is (are) stations(s) or warehouse(s), and (5) service area is in urban. Therefore, SME express or transport companies were selected as the result details are in Table 9.

Table 9 Interview and survey result

Interviewed	1st	2nd
Factor 1	Agree (if there are many locations in network transportation, we or someone else can deliver more and faster than competitors in the market)	Agree (it related with transport and warehouse principle that if there are many locations, the transport cost will be reduced and will serve the service to customer faster and better than)
Factor 2	Agree (but depends on the volume of each company. If we hire drivers as much as big company do, for sure we loss profit and cannot serve job to our drivers enough)	Agree (manpower is an important resource in transport industry. Nowadays, turnover rate is higher, because they change jobs to be more salary and benefit especial in overseas companies)
Factor 3	Agree (almost express companies use motorbikes and 4-wheel truck for delivering and picking up the shipments in urban area. It means more drivers and more vehicles for delivering and picking up also)	Agree (normally, transport company in Thai uses 4-wheel or 6-wheel trucks and still wait almost full-truck load before deliver the shipments to consignees. SME scale concerns the cost issues more than another thing)
Factor 4	Agree (if there is a system or incentive program, we plan how to deliver the shipments to nearest customer to far customer respectively only)	Agree (leader will plan and dispatch the shipments to our drivers. Normally, drivers fixed in the same route. We do not change them unless they change job)
Factor 5	Agree (we set up KPIs for evaluating driver's performance and adjusting annual salary)	Agree (KPIs is an important tool in logistics industry. It will stimulate employee to do more, better, faster, and correct. We set KPIs, but there is a few to make drivers know their job and do it)

5 Conclusion and Suggestion

5.1 Conclusion

As mentioned in part 4, two Thai express transportation companies have absolutely conformed to 5 significant factors that it can manage their time and lead to improve service level. Moreover to use these factors in the service for SME and entrepreneur, cost majority is a concern because they cannot invest capital to locate warehouse, hire more drivers, and so on in starting phase. For cost limitation reason, the conclusion is considered by 5 significant factors without cost factor. They can benchmark these factors to improve their service level.

To sum up, there are 5 significant factors for time management of domestic express transportation services which are (1) DC's location, (2) amount of couriers, (3) type of vehicle, (4) how to plan day-to-day operation, and (5) KPIs. To make it more effective, all of these have to use together. It derived from customers' requirement. To apply these factors as a tool, it can manage time.

5.2 Suggestions

1. These significance factors should be used in normal situation. In case of uncontrollable situation, the cost issues will be concerned.
2. These factors are proper to manage time to transportation sector also.
3. Practically, there are many unexpected factors such as customs and intermodal. It means that there will be more hidden important factor in Thai transport industry.
4 The first and second factors are concerning cost issue directly, but in this research, the cost issue was not studied.

References

Adam P (n.d.) Daily management (part of Lean Leadership). Retrieved 1 Dec 2014, from http://sipocsolutions.com/daily-management.htm

Coyle J, Langley J Jr, Novack R, Gibson B (2009) Managing supply chains a logistics approach, 9th edn. South-Western, Cengage Learning, Canada, p 16

Coyle J, Novack R, Gibson B, Bardi E (2011) Transportation a supply chain perspective, vol 493, 7th edn. South-Western, Cengage Learning, Mason, OH, pp 344–347

Nayab N, Edwards G (2010) Principles of JIT. Retrieved 5 Dec 2014, from http://www.brighthubpm.com/methods-strategies/70470-elements-of-jit-and-lean-production/

Simchi-Levi D, Kaminsky P, Simchi-Levi E (2009) Designing and managing the supply chain: concepts, strategies and case studies, 3rd edn. McGraw-Hill/Irwin, Singapore, p 233

The World Bank Group (2015) Logistics performance index. Retrieved 13 Jan 2015, from http://lpi.worldbank.org/domestic/performance/2014/C/THA

Sugarcane Harvester Planning Based on the Vehicle Routing Problem with Time Window (VRPTW) Approach

Kallaya Kittilertpaisan and Supachai Pathumnakul

Abstract This chapter addresses the sugarcane mechanized harvesting problem. A mathematical model based on the vehicle routing problem (VRP) with time window (VRPTW) is formulated to obtain optimal harvesting plan for harvesters given specific time windows (TWs). The harvesters can operate only within the specific TW. The objective of the model is to determine the suitable sugarcane harvesting plan of a harvester. The model is applied to solve 18 generated sugarcane harvesting scenarios. Each scenario consists of 15 sugarcane fields and 20 harvesting periods. The solution comprises the harvesting sequences, traveling routes, harvesting periods, and harvest starting time. The model provides the optimal solutions which can be applied for sugarcane growers in Thailand and other similar regions.

Keywords Harvesting plan · Mathematical model · Sugarcane · Vehicle routing problem

1 Introduction

Sugarcane is an important crop in Thailand. It can be grown in almost all regions of the country. Its harvesting period is 10–12 months and can be harvested for 3–4 years (The Office of the Cane and Sugar Board 2011). One of the major problems in sugarcane industry is the shortage of harvesting labors. Recently, the use of harvesting machines (i.e., harvesters) has increased to replace the use of labor force. A harvesting machine can harvest 125 tons/harvester/day or it is able to substitute 100–125 workers. However, the sugarcane growers in Thailand cannot normally afford owning an expensive harvester, and most harvesters belong to a mill. In each crop year, the mill operator needs to set up a harvester utilization plan.

K. Kittilertpaisan · S. Pathumnakul (✉)
Supply Chain and Logistic Systems Research Unit, Department of Industrial Engineering, Faculty of Engineering, Khon Kaen University, Khon Kaen, Thailand
e-mail: supa_pat@kku.ac.th

V. Kachitvichyanukul et al. (eds.), *Toward Sustainable Operations of Supply Chain and Logistics Systems*, EcoProduction,
DOI 10.1007/978-3-319-19006-8_23

The plan includes the allocation of harvesters to the cane growing regions and the routing of each harvester throughout the harvesting season. A set of fields were managed to be harvested by one harvester. The objective of the plan is to effectively utilize the harvesters and to obtain a high sugarcane yield with the lowest harvester moving cost. The moving cost covers the cost of the harvester moving in field (i.e., harvesting cane in the field) and the cost of the harvester traveling from one field to the next.

However, to optimize the harvesting plan, sugarcane yield characteristic should be considered. That is, each field has its own specific period times suitable to be harvested. This demonstrates TW behavior. Furthermore, at the early period of the harvesting season begins, one sugarcane field may reach its best yield, while other fields may reach their peak yield periods in the middle or almost the end of the season (Khamjan et al. 2013). Therefore, a practical harvesting plan should identify when (which period), where (which field), and how (which route) to harvest for a harvester machine in order to minimize the harvester transportation cost and obtain high sugarcane yields.

Due to characteristics of the sugarcane harvesting problem, traveling from one field to the next can be considered as the VRP with time window (VRPTW) constraints, which has been extensively studied in the OR literature. The TW constraints represent the time span of a particular field that can be harvested. A field can only be harvested within an allowed TW, in which the yield may vary. In this chapter, the mathematical model based on VRPTW is developed to solve this specific problem. The objective of the model is to maximize total revenue of farmers whose share the same harvester. The revenue is estimated by the revenue of selling obtained cane minus the harvesting and harvester transportation costs. The revenue expresses in the function of quantity of cane harvested in a specific period.

2 Literature Review

The sugarcane harvesting operation has been explored by many researchers. Crop growth simulation and mathematical modeling were proposed by Piewthongngam et al. (2009) to forecast cane yield and to set up efficient cultivation plan to maximize overall sugar production. The extensive aspects in sugarcane value chain research were reviewed by Higgins et al. (2007). The authors informed that there have been many studies and researches in sugarcane harvesting planning problem and one of the potential tools normally employed in solving problem is the mathematical model. Higgins et al. (1998) described the use of an optimization model to maximize net revenue and sugar yield considering harvest date and crop age using a case study in Australia. This study was continued by conducting the sensitivity analysis of these responses to age distribution and crop class at harvest (Muchow et al. 1998). Jiao et al. (2005) presented the attempt to create better harvest scheduling for maximizing gain in CCS in a harvest season by proposing a second-order polynomial to fit CCS across sugarcane farm within a mill region and applying a linear programming (LP) model to solve the problem. Stray et al. (2012) created a decision support system for seasonal sugarcane harvest scheduling for growers in South Africa. The optimization models based on a time-dependent traveling salesman problem were formulated and solved by mean of a tabu search.

Salassi et al. (2002) determined the optimal harvest system of wholestalk harvester and combine harvester for sugarcane production in Louisiana, USA. The sugar yield prediction model and mixed-integer linear programming (MIP) model were used to maximize the net revenues considering farm costs to identify the most economic sugarcane harvest system. Higgin et al. (2004) built a framework for combining a complex sugarcane harvesting and transportation system for sugar production. One attempt of this study was to solve many existing inefficiencies resulting from excessive number of harvesting machines owned by growers and harvester contractors.

Research works relating to VRPTWs were also studied and reviewed. VRPTW is one of VRP that the routing or traveling can be scheduled and services can be conducted only in a specific interval time or in the TW. TW comprises the earliest service time (lower bound) and the latest service time (upper bound). When vehicles arrive before lower bound and after upper bound, mistakes can be occurred because the both situations result the waiting cost and opportunity loss cost. The basic VRPTW mathematical model is presented by many works such as works of Desrocher et al. (1988), Tan et al. (2001), Yeun et al. (2008), and El-Sherbeny (2010). This model is well known and used by many researchers combining with heuristic approaches to solve several problems. Ghoseiri and Ghannadpour (2010) presented a VRPTW model for formulating the problem of a multi-depot homogenous locomotive assignment and solved it by a hybrid genetic algorithm. Kritikos and loannou (2013) addressed the problem of heterogeneous fleet vehicle routing with overload and TWs. The model was formulated and solved it by applying a sequential insertion heuristic. Its computational results show that the proposed approach is effective in reducing vehicle cost with minimal violation of capacity.

3 Mathematical Model

In this section, the mathematical model is formulated to solve the addressed problem.

Sets and indexes

- i index of field i that has to be harvested ($i = 1, 2, \ldots, I$) where "$i = 1$" is a dummy starting field and "$i = I$" is a dummy ending field.
- j index of destination field j ($j = 1, 2, \ldots, I$) where "$j = 1$" is a dummy starting field and "$j = I$" is a dummy ending field.
- k index of harvester k ($k = 1, \ldots, K$).
- p index of harvesting period p ($p = 1, 2, \ldots, P$).

A set of arcs which are direct connections between the dummy starting field and other fields and among the fields.
I number of fields that have to be harvested including the dummy starting field and the dummy ending field.
K number of harvesters.
P number of harvesting periods.

Parameters

R_{ip} the revenue from harvesting field i at time period p (unit cost).
C_{ij} the transportation cost of a harvester traveling from field i to field j (unit cost).
H_i the cost of harvesting field i (unit cost).
S_i the time required to harvest field i (unit time).
t_{ij} the time required for a harvester to travel from field i to field j (unit time).
$[a_i, b_i]$ harvesting TW of field i.
$M^{[1]}$ the large positive number.
$M^{[2]}$ the constant number where $0.999 \leq M^{[2]} < 1$.

Variables

X_{ijkp} is 1 if the harvesting is operated at field i and leaves for field j by harvester k at time period p, and 0 otherwise.
Y_{ik} is the time period where the harvesting is operated at field i by harvester k.
$Y_{ik}^{[s]}$ is the starting time of harvester k in field i.

Mathematical Model

Maximizing

$$\sum_{i=1}^{I}\sum_{j=1}^{I}\sum_{k=1}^{K}\sum_{p=1}^{P} X_{ijkp}R_{ip} - \sum_{i=1}^{I}\sum_{j=1}^{I}\sum_{k=1}^{K}\sum_{p=1}^{P} X_{ijkp}C_{ij} - \sum_{i=1}^{I}\sum_{j=1}^{I}\sum_{k=1}^{K}\sum_{p=1}^{P} X_{ijkp}H_i \quad (1)$$

Subject to

$$\sum_{j=1}^{I}\sum_{k=1}^{K}\sum_{p=1}^{P} X_{ijkp} = 1, \quad \forall i \in I\backslash\{1, I\} \quad (2)$$

$$\sum_{j=1}^{I}\sum_{p=1}^{P} X_{1jkp} = 1, \quad p \geq a_j, p \leq b_j, \forall k \in K \quad (3)$$

$$\sum_{i=1}^{I}\sum_{p=1}^{P} X_{iIkp} = 1, \quad \forall k \in K \tag{4}$$

$$\sum_{i=1}^{I}\sum_{j=1}^{I}\sum_{p=1}^{P} X_{ijkp} = 0, \quad \text{for} \quad i = j, i = I, j = 1, \ \forall k \in K \tag{5}$$

$$\sum_{i=1}^{I}\sum_{p=1}^{P} X_{ifkp} - \sum_{j=1}^{I}\sum_{p=1}^{P} X_{fjkp} = 0, \quad \forall k \in K, \forall f \in I\backslash\{1, I\} \tag{6}$$

$$Y_{ik}^{[s]} + T_{ij} + S_i - M^{[1]}(1 - X_{ijkp}) \leq Y_{jk}^{[s]}, \quad \forall i \in I, \forall j \in I, \forall k \in K, \forall p \in P \tag{7}$$

$$Y_{ik}^{[s]} - M^{[1]}\left(1 - \sum_{j=1}^{I}\sum_{p=1}^{P} X_{ijkp}\right) \leq Y_{ik}, \quad \forall i \in I, \forall k \in K \tag{8}$$

$$Y_{ik} - M^{[1]}\left(1 - \sum_{j=1}^{I}\sum_{p=1}^{P} X_{ijkp}\right) \leq Y_{ik}^{[s]} + M^{[2]}, \quad \forall i \in I, \forall k \in K \tag{9}$$

$$\sum_{j=1}^{I}\sum_{p=1}^{P} pX_{ijkp} = Y_{ik}, \quad \forall i \in I\backslash\{I\}, \forall k \in K \tag{10}$$

$$\sum_{j=1}^{I}\sum_{k=1}^{K}\sum_{p=1}^{P} X_{ijkp} = 0, \quad \text{for} \quad p < a_i, p > b_i, \ \forall i \in I\backslash\{1\} \tag{11}$$

$$Y_{ik}^{[s]} \geq a_i, \quad \forall i \in I, \forall k \in K \tag{12}$$

$$Y_{ik}^{[s]} + S_i \leq b_i, \quad \forall i \in I, \forall k \in K \tag{13}$$

$$\sum_{j=1}^{I}\sum_{k=1}^{K}\sum_{p=1}^{P} X_{1jkp} \leq K \tag{14}$$

$$X_{ijkp} \in \{0, 1\}, \quad \forall i \in I, \forall j \in I, \forall k \in K, \forall p \in P \tag{15}$$

$$Y_{ik} \in \{\text{Integer}\}, \quad \forall i \in I, \forall k \in K \tag{16}$$

Objective function (1) is maximization of profit which is estimated by the difference between revenue and relating costs (harvesting cost and harvester traveling cost). Constraint (2) ensures that each field is visited and harvested exactly once. Constraints (3) and (4) ensure that each harvester must leave the dummy starting field and must arrive at the dummy ending field, respectively. Constraint (5) forbids looping from a field to itself and ensures that index i is not equal to "I + 1" ($i \neq I + 1$) and index j is not equal to "0" ($j \neq 0$). Constraint (6) enforces that after a harvester arrives at a field, it must leave for another destination. Constraints (7), (8), and (9) demonstrate the relationship between the harvest starting time of a field and its immediate successor. Constraint (10) estimates the traveling time of a harvester moving from field i to field j. Constraints (11), (12), and (13) ensure that the TWs are considered. Constraint (14) gives an upper bound on the number of harvesters. Finally, constraint (15) describes that the decision variable X_{ijkp} is binary and constraint (16) describes that the decision variable Y_{ik} is an integer.

4 Experiments

To identify the performance of the proposed mathematical model, it is employed to solve 18 sample problems following the actual situation of general sugarcane fields in Thailand. Each problem consists of 2 harvesters and 15 sugarcane fields. The sizes of fields are varied from 5 to 20 rai (rai is Thai unit area, 1 rai = 0.16 ha). The planning period is 20 periods. It is estimated that 1 harvesting period equals to 12 operating hours. The cost parameters used in the sample problems are also simulated following the practical situation. An example of sugarcane yield TW of 15 fields is as shown in Table 1. The fields number 1 and 17 are dummy fields represent the starting and ending fields, respectively. For example, field number 2, the harvester should harvest from periods 1 to 11. The peak yield of the field is at period 6.

Figure 1 illustrates the example of yield pattern (Field 2). It demonstrates the yields of eleven harvesting periods in its TW.

Table 1 The sugarcane yields of 15 fields in 20 harvesting periods

Field	Yield (ton/period/rai)																			
	1	2	3	4	5	6	7	8	9	10	11	12	13	14	15	16	17	18	19	20
1	0	0	0	0	0	0	0	0	0	0	0	0	0	0	0	0	0	0	0	0
2	14.1211	14.1232	14.1248	14.1260	14.1266	14.1268	14.1266	14.1258	14.1247	14.1230	14.1209	0	0	0	0	0	0	0	0	0
3	0	0	0	0	0	0	0	14.2065	14.2093	14.2114	14.2129	14.2138	14.2140	14.2136	14.2125	14.2108	14.2084	14.2054	0	0
4	0	0	0	0	0	0	0	0	0	14.8576	15.1216	15.3204	15.4541	15.5227	15.5261	15.4644	15.3375	15.1455	14.8884	14.5661
5	0	0	14.4565	14.4585	14.4599	14.4605	14.4608	14.4611	14.4608	14.4587	14.4568	14.4544	14.4514	0	0	0	0	0	0	0
6	0	0	0	0	0	0	13.7280	13.7299	13.7311	13.7313	13.7317	13.7319	13.7301	13.7278	13.7252	13.7220	13.7182	0	0	0
7	0	0	0	0	0	0	0	0	13.7046	13.7066	13.7080	13.7083	13.7087	13.7089	13.7079	13.7064	13.7043	13.7016	13.6984	0
8	0	0	0	0	0	14.1675	14.1686	14.1688	14.1691	14.1693	14.1695	14.1693	14.1687	14.1677	14.1662	14.1643	0	0	0	0
9	0	0	0	14.1312	14.1322	14.1330	14.1336	14.1339	14.1340	14.1339	14.1335	14.1330	14.1322	14.1311	0	0	0	0	0	0
10	12.5688	12.5726	12.5754	12.5772	12.578	12.5778	12.5766	12.5743	12.5711	12.5669	12.5616	0	0	0	0	0	0	0	0	0
11	0	0	0	0	0	0	0	0	0	14.2822	14.2831	14.2837	14.2840	14.2843	14.2849	14.2833	14.2825	14.2813	14.2799	14.2782
12	0	0	0	0	0	0	0	0	0	13.7445	13.7473	13.7494	13.7509	13.7518	13.7520	13.7516	13.7505	13.7488	13.7464	13.7434
13	0	0	0	0	0	0	0	0	14.2468	14.2474	14.2479	14.2482	14.2484	14.2486	14.2483	14.2480	14.2476	14.2470	14.2463	0
14	0	0	0	0	13.7897	13.7943	13.7977	13.7999	13.8005	13.8008	13.7995	13.7971	13.7935	13.7887	13.7827	0	0	0	0	0
15	0	0	0	13.6940	13.6971	13.6991	13.7002	13.7004	13.7009	13.6977	13.6949	13.6911	13.6864	13.6807	0	0	0	0	0	0
16	0	14.1225	14.1235	14.1243	14.1249	14.1253	14.1255	14.1254	14.1252	14.1248	14.1242	14.1234	0	0	0	0	0	0	0	0
17	0	0	0	0	0	0	0	0	0	0	0	0	0	0	0	0	0	0	0	0

Note Field 1 is the dummy starting field and Field 17 is the dummy ending field

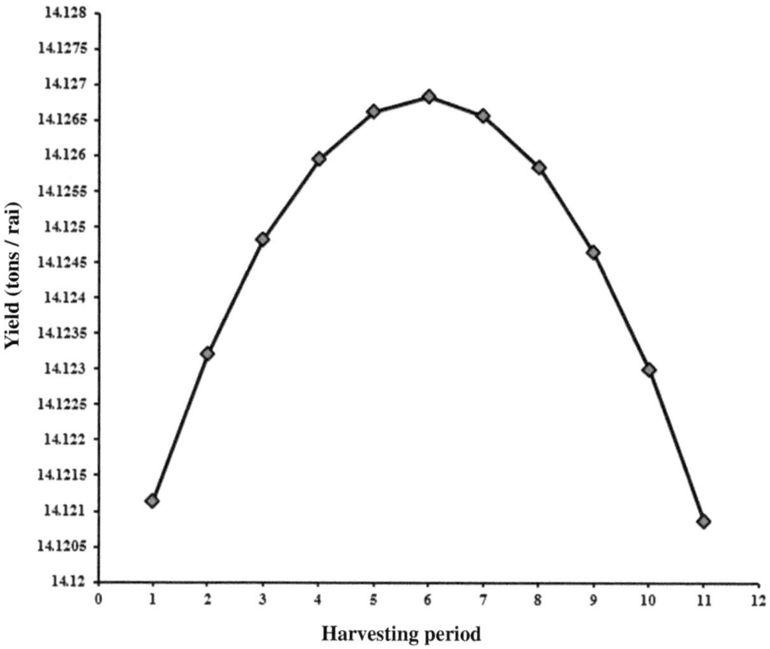

Fig. 1 The example of yield pattern of field No. 2

The problems are solved by using the mathematical software LINGO 13.0 on a PC with Intel (R) Core (TM) i7-4710 MQCPU@2.5 GHz and 8 GB of RAM.

5 Results

The results obtained from the mathematical model are shown in Table 2. Considering the average run times of the three problem groups (Problem group 1 average run time = 230 s, Problem group 2 average run time = 285 s, and Problem group 3 average run time = 358 s), the larger problem size, the longer computational time tends to be needed. Moreover, it can be observed that the complexity of field layout may also affect the computational run time. The more complicated layout would cause longer time to obtain its result. To illustrate the routing and timing solutions of the problem, the example solutions of problem number 10 is depicted in Table 3 and Fig. 2. Noted that in Fig. 2, the dummy starting field (field 1) and the dummy ending field (field 17) are assigned to share the same position and function similarly as a depot.

Table 2 The mathematical model results

Problem group	Problem No.	Maximum profit (unit cost)	Computational run time (s)	Remark
1 (5–10 rai area/ 120 rai total area)	1	2,117,501	242	Min run time = 201 s
	2	2,116,664	245	
	3	2,116,917	248	Max run time = 248 s
	4	2,116,785	224	
	5	2,117,208	201	Average run time = 230 s
	6	2,117,821	219	
2 (10–15 rai area/ 200 rai total area)	7	3,532,995	226	Min run time = 203 s
	8	3,533,673	222	
	9	3,532,406	460	Max run time = 460 s
	10	3,532,457	203	
	11	3,533,082	247	Average run time = 285 s
	12	3,533,145	353	
3 (15–20 rai area/ 120 rai total area)	13	4,772,141	195	Min run time = 195 s
	14	4,772,277	345	
	15	4,772,519	482	Max run time = 712 s
	16	4,771,454	712	
	17	4,771,801	206	Average run time = 358 s
	18	4,771,912	209	

Table 3 The routing and timing results of an example problem No. 10

K	Route (field-field)			Harvest period (Y_{ik})	Harvest starting time (YY_{ik}) (1)	Harvest time (S_i) (2)	Travel time (T_{ij}) (3)	Finish time (1) + (2) + (3)	Harvester waiting time (unit time)
1	1	–	10	–	–	0	0.0302	1.0302	4.9698
	10	–	16	6	6	0.60	0.0223	6.6223	0
	16	–	5	7	6.6223	0.44	0.0276	7.0899	0.9101
	5	–	8	8	8	0.52	0.0105	8.5305	2.4695
	8	–	6	11	11	0.60	0.0212	11.6212	0
	6	–	3	12	11.6212	0.48	0.0105	12.1117	0.2740
	3	–	7	13	12.3857	0.60	0.0153	13.0010	0.9990
	7	–	4	14	14	0.48	0.0096	14.4896	0.5104
	4	–	17	15	15	0.40	0.0251	15.4251	–
2	1	–	2	–	–	0	0.0075	1.0075	4.9925
	2	–	9	6	6	0.56	0.0177	6.5777	1.4233
	9	–	15	9	8.0010	0.52	0.0274	8.5484	0.4516
	15	–	14	9	9	0.60	0.0094	9.6094	0.3906
	14	–	13	10	10	0.52	0.0080	10.5280	2.9436
	13	–	12	14	13.4716	0.52	0.0094	14.0010	0.4138
	12	–	11	15	14.4148	0.56	0.0252	15	0
	11	–	17	15	15	0.60	0.0374	15.6374	–

Note Field 1 is the dummy starting field and Field 17 is the dummy ending field

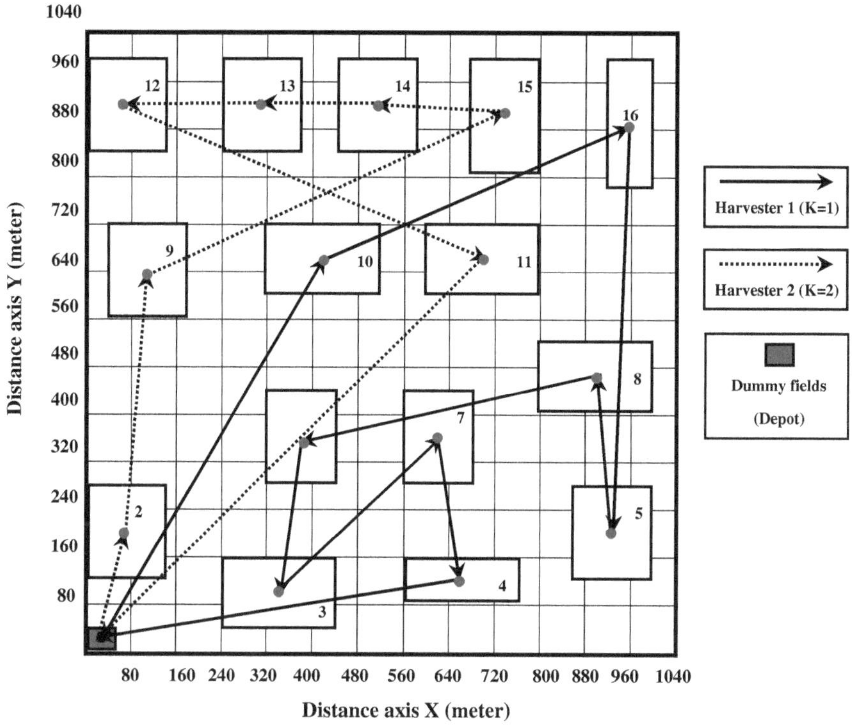

Fig. 2 The optimal harvesting sequence and traveling route of problem No. 10

6 Conclusion and Discussion

In this study, the mathematical model based on the VRPTW model is developed to solve the sugarcane harvester planning problem. The model provides practical solutions with the maximum profit for the sample 18 problems. The solutions provide the optimal harvesting sequences, harvesters' traveling routes, harvesting periods, and harvest starting time for a set of sugarcane fields using harvesters. Nevertheless, the larger problem size and the complexity of field layout may result the longer computational time. For large-size problems, the heuristic may be required to be developed to solve the problems. Moreover, harvester's waiting time observed in Table 3 may cause the cost of harvesting opportunity loss. In the future work, the opportunity loss cost may be considered in order to illustrate the more realistic of sugarcane harvesting problem.

Acknowledgment This research was funded by Khon Kaen University and Sakon Nakon Rajabhat University. The authors also acknowledge the invaluable guidance from all supportive assistances and useful information and data from all sources.

References

El-Sherbeny NA (2010) Vehicle routing with time windows: an overview of exact, heuristic and metaheuristic methods. J King Saud Univ Sci 22(3):123–131

Ghoseiri K, Ghannadpour SF (2010) A hybrid genetic algorithm for multi-depot homogenous locomotive assignment with time windows. Appl Soft Comput 10(1):53–65

Higgins AJ, Muchow RC, Rudd AV, Ford AW (1998) Optimising harvest date in sugar production: a case study for the Mossman mill region in Australia: I. Development of operations research model and solution. Field Crops Res 57(2):153–162

Higgins A, Antony G, Sandell G, Davies I, Prestwidge D, Andrew B (2004) A framework for integrating a complex harvesting and transport system for sugar production. Agric Syst 82 (2):99–115

Higgins A, Thorburn P, Archer A, Jakku E (2007) Opportunities for value chain research in sugar industries. Agric Syst 94(3):611–621

Jiao Z, Higgins AJ, Prestwidge DB (2005) An integrated statistical and optimisation approach to increasing sugar production within a mill region. Comput Electron Agric 48(2):170–181

Khamjan W, Khamjan S, Pathumnakul S (2013) Determination of the locations and capacities of sugar cane loading stations in Thailand. Comput Ind Eng 66(4):663–674

Kritikos MN, Ioannou G (2013) The heterogeneous fleet vehicle routing problem with overloads and time windows. Int J Prod Econ 144(1):68–75

Lenstra JK, Desroches M, Savelbergh MWP, Soumis F (1988) Vehicle routing with time windows: optimization and approximation. Elsevier Science Publishers, North-Holland, pp 65–83

Muchow RC, Higgins AJ, Rudd AV, Ford AW (1998) Optimising harvest date in sugar production: a case study for the Mossman mill region in Australia: II sensitivity to crop age and crop class distribution. Field Crops Res 57(3):243–251

Piewthongngam K, Pathumnakul S, Setthanan K (2009) Application of crop growth simulation and mathematical modeling to supply chain management in the Thai sugar industry. Agric Syst 102(1):58–66

Salassi ME, Breaux JB, Naquin CJ (2002) Modeling within-season sugarcane growth for optimal harvest system selection. Agric Syst 73(3):261–278

Stray BJ, Van Vuuren JH, Bezyidenhout CN (2012) An optimization-based season sugarcane harvest scheduling decision support system for commercial growers in South Africa. Comput Electron Agric 83:21–23

Tan KC, Lee LH, Zhu QL, Ou K (2001) Heuristic methods for vehicle routing problem with time windows. Artif Intell Eng 15(3):281–295

The Office of the Cane and Sugar Board (OCSB) (2010/2011) The ministry of industry of Thailand, planting season

Yeun LC, Ismail WR, Omar K, Zirour M (2008) Vehicle routing problem: models and solutions. J Qual Meas Anal 4(1):205–218

A Memetic Algorithm Approach for Solving the Truck and Trailer Routing Problem

Soon-Kyo Lee and Taesu Cheong

Abstract This chapter considers a truck and trailer routing problem (TTRP) where the fleet consists of trucks and trailers, and some customers can be served by a truck pulling a trailer, while the other may be accessible by a truck only. A memetic algorithm is proposed by dividing the TTRP into two VRPs in the construction phase and then uses the alternating edge crossover with several local search methods. It is tested on TTRP benchmark instances in the literature and corresponding numerical results are presented.

Keywords Truck and trailer routing problem · Metaheuristics · Memetic algorithms · Vehicle routing problems

1 Introduction

In this paper, we consider a variant of the vehicle routing problem (VRP), the truck and trailer routing problem (TTRP). VRP is a problem that is widely studied and researched in the field of combinatorial optimization and operations research. The goal of solving the VRP is to design a route which has the minimum total distance while not violating the capacity of vehicles. TTRP is similar to VRP except that it has more constraints including that some customers cannot be served by a complete vehicle (truck pulling a trailer), but can be served only by a truck.

TTRP has many applications in practice because such a constraint is common in the real world due to government regulations, limited maneuvering spaces, or road

S.-K. Lee · T. Cheong (✉)
School of Industrial Management Engineering, Korea University, 5-1 Anam-Dong, Seongbuk-Gu, Seoul 136-713, South Korea
e-mail: tcheong@korea.ac.kr

S.-K. Lee
e-mail: myeva81@korea.ac.kr

V. Kachitvichyanukul et al. (eds.), *Toward Sustainable Operations of Supply Chain and Logistics Systems*, EcoProduction,
DOI 10.1007/978-3-319-19006-8_24

conditions. Some real-world examples of TTRP were given by Gerdessen (1996) presenting an application in the distribution of dairy products by the Dutch dairy industry, and Semet and Taillard (1993) discussing the case in a food chain store in Switzerland. There is another recent application of the TTRP in a steel factory in China. According to Bo (2009), the trucks and trailers of a steel factory called Heng Yang Steel Co. in China have to transport raw materials and finished steel products to and from different sub-factories. However, due to limited maneuvering space, a truck has to unload the trailer for having access to some factories. Wide applications of TTRPs demand effective and efficient methods to solve the TTRP.

We now provide the problem definition of TTRP. The VRP was first introduced by Dantzig and Ramser (1959). It consists of a central depot, a set of customers with known demand, and a fleet of vehicles with known capacity. Figure 1 gives a graphical illustration of VRP. The objective of the problem is to service the customers with the fleet of vehicles so that three conditions are met:

1. Each customer is visited once.
2. Assuming that the cost of transporting is proportional to the distance traveled, the total distance traveled is minimized.
3. The capacity of each vehicle is not exceeded.

The TTRP extends the VRP in which the vehicle is categorized into two types—the trucks and trucks pulling trailers. A single truck is called a *Pure Truck*, and a truck with a trailer is called a *Complete Vehicle*. The customers are divided into two types as well: a customer who is accessible by both a pure truck and a complete vehicle is called a *Vehicle Customer*, and a customer who is only accessible by a pure truck is called a *Truck Customer*. Its graphical description is shown in Fig. 2.

TTRP is formally defined on a graph $G = (V, A)$ with $V = \{V_D, V_1 \ldots V_n\}$ and $A = \{(V_i, V_j): V_i, V_j \in V, i \neq j\}$. The vertex V_D represents the central depot and there are m trucks and n trailers at the depot ($n < m$). A complete vehicle can uncouple the

Fig. 1 Graphical illustration of vehicle routing problem (Lin et al. 2009)

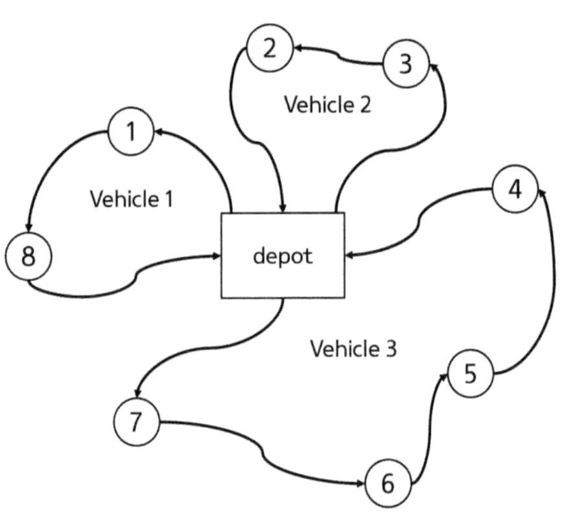

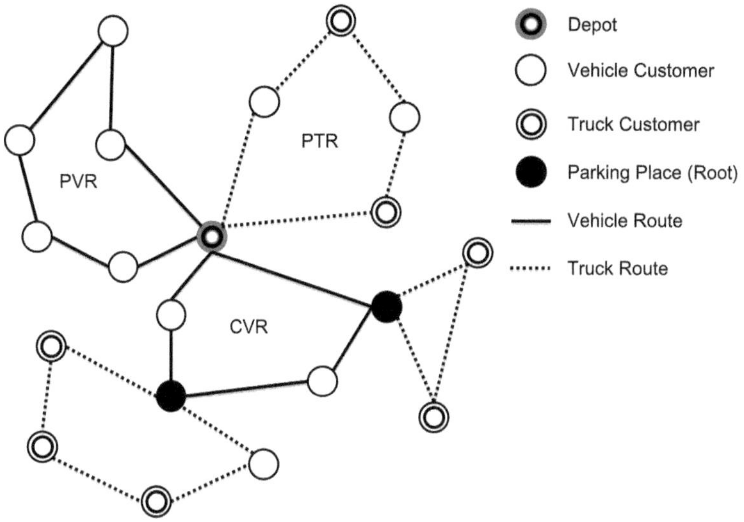

Fig. 2 Graphical illustration of TTRP (Lin et al. 2009)

trailer at any vehicle customer and starts to serve truck customers. This is a sub-tour in a main tour. A main tour is traveled by a complete vehicle and a sub-tour or pure truck tour is traveled by a pure truck only as shown in the following graph. We refer the readers to Villegas et al. (2013) for the mathematical formulation of TTRP.

The VRP is a computational complex problem or NP-hard problem. When the number of customers increases, solving the TTRP is computationally intractable. Being more complicated than the VRP, the TTRP is computationally more difficult to solve, making it at least NP-hard. Hence, it is only logical to make use of heuristic methods to tackle it as we cannot solve it exactly in reasonable amount of time if the number of customers is large.

The memetic algorithm is one of the metaheuristics that can be used to find a reasonably good solution in many NP-hard problems. It is gaining popularity in recent years because it makes use of a combination of evolutionary algorithm and local searches to explore the solution space more thoroughly and efficiently.

The remainder of the paper is organized as follows: Sect. 2 is a literature review of the methods employed by past authors to solve the TTRP. Section 3 is a description of the basic memetic algorithm framework that we adapt to tackle the truck and trailer routing problem. Experimental results of the proposed algorithm are presented in Sects. 4 and 5, we make some discussion regarding the current implementation of the proposed algorithm and future research directions.

2 Literature Review

For the last few decades, research on VRP has attracted a lot of attention along with traveling salesman problem which is similar to VRP in many ways. However, the research on TTRP is relatively scarce. Most people treat TTRP as a variant of VRP and had not given it a name until 2002 when Chao (2002) first named such a problem as TTRP and used a tabu search heuristic to solve 21 TTRPs adapted from the classical 7 VRPs of Christofides et al. (1979). Since then, this topic draws increasing notice and many more heuristics and metaheuristics were developed to solve this problem.

Chao (2002) developed a sophisticated tabu search method consisting of an initial construction heuristic and a tabu search improvement heuristic to search the neighborhood of an initial solution and hence refine them. The method contains standard VRP moves as well as a new TTRP-specific root-defining move where the roots of sub-tours are changed to new customers. Four years later, Scheuerer (2006) extended Chao's work to develop a better tabu search heuristic that outperforms the Chao's method. Scheuerer made use of two new construction heuristics called the T-Cluster and T-Sweep. The new construction heuristics proved to successfully find good initial solutions for TTRP. He was able to obtain better results in each of the 21 instances Chao (2002) adapted.

Lin et al. (2009) approached the TTRP using a specific indirect search approach. The search in the auxiliary search space is guided by a simulated annealing control and uses three simple string operators: swap, insertion, and changing service type. Its main characteristics are the combination of a two-level solution representation with the use of dummy depots, and the use of the random neighborhood structure which utilizes three different types of moves. Caramia and Guerriero (2010) broke the TTRP into two separate problems called customer-route assignment problem (CAP) and route definition problem (RDP). He first solved two mathematical formulations and then used a local search algorithm that acts on the solutions, so found either to possibly obtain a better solution itself or to let the mathematical models find a better solution in the next iteration. Yu et al. (2011) proposed a two-stage approach where an ant colony system constructs feasible routes in the first stage, followed by improving 2-opt, swap, and insertion moves in the second stage.

In recent years, Villegas et al. (2011) presented a hybrid metaheuristic which is the state-of-the-art method for the TTRP. This approach combines different concepts such as the greedy randomized adaptive search procedure (GRASP) using a route-first cluster-second construction heuristic and variable neighborhood search (VNS) in the improvement phase with root-refining move and path relinking with respect to a pool of elite solutions. Finally, Derigs et al. (2013) customized a flexible metaheuristic framework to solve various variants of the truck and trailer routing problem. This method combines local searches and large neighborhood searches under the standard metaheuristic control along with the attribute hill climber method. Computational experiments show that his approach is competitive to state-of-the-art approaches for the TTRP.

Table 1 A chronological list of TTRP-related research

Reference	Approach
Chao (2002)	Hybrid tabu search/threshold accepting
Scheuerer (2006)	Tabu search
Lin et al. (2009)	Simulated annealing
Caramia and Guerriero (2010)	Mathematical programming-based heuristics
Yu et al. (2011)	Ant colony problem
Villegas et al. (2011)	GRASP with evolutionary path rethinking
Villegas et al. (2013)	Set-partitioning formulation of TTRP
Derigs et al. (2013)	Hill climber method which is specific to implementation of tabu search

A chronological list of TTRP-related literature is shown in Table 1.

3 Memetic Algorithm for TTRP

Memetic algorithms (MA) are a class of stochastic global search heuristics in which evolutionary algorithm-based approaches are combined with problem-specific solvers where the latter is mostly implemented as local search heuristics. It is assumed that the evolutionary algorithm explores the wider solution space while the local search can zoom in on the basin of attraction of promising solutions. The key concern that is addressed by MA is the balance between the global search and the local search—i.e., the strategy that evolution might need to implement so as to gain benefit from a successful trade-off between global exploration and local exploitation.

A top-level design is illustrated in Fig. 3. Note that local search is applied to many steps including construction heuristics, crossover, mutation, and selection. This is to prevent any unexploited space missed in each of the global search heuristic steps.

3.1 Solution Representation

We use a data structure to represent a unique solution because there are many fields that should be included in a solution. These fields include the route (represented by an array of numbers) as well as the total distance traveled for that particular route. A data structure should be most suitable for this representation because it has to contain not only a vector but also a number. The data structure is organized as follows. It contains two fields:

Fig. 3 Framework of the memetic algorithm

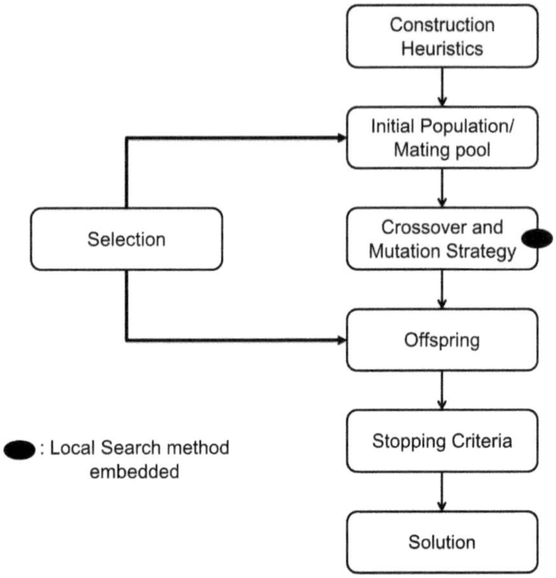

- Route: Each route starts from the depot (denoted by 0) and ends with the depot. When the capacity is full, it returns to the depot.
- Distance: Total distance traveled for this route will be the fitness of each solution.

Let us consider the following small TTRP of 15 customers as an illustrative example where the truck capacity is 100 and the trailer capacity is 50 (Table 2).

Table 3 is a unique solution to the TTRP in Table 2. It consists of a pure truck route (Route 1) and two complete vehicle routes (Route 2 and 3). It means that (i) a truck begins from a depot to serve 4-2-9 sequentially and then returns and (ii) a vehicle (truck pulling a trailer) serves 8-3-4 sequentially and then returns. Finally, a vehicle serves 5 first where the trailer will be parked, and the truck will then serve 11-10-13 after it returns to 5 to get the trailer, then serves 15-6-7-1-12, and eventually returns to the depot.

A more representative graphical solution is illustrated here in Fig. 4 as well.

Table 2 Illustrative TTRP of 15 customers

Customer	1	2	3	4	5	6	7	8	9	10	11	12	13	14	15
Type	VC	TC	VC	VC	VC	VC	TC	TC	TC	VC	TC	VC	VC	VC	TC
Demand	5	20	15	40	15	10	5	100	40	30	35	5	35	35	10

Table 3 A solution to the illustrative TTRP

0	4	2	9	0	8	3	14	0	5	11	10	13	5	15	6	7	1	12	0
Route 1				Route 2				Route 3											
Pure truck route				Complete vehicle route				Complete vehicle route											

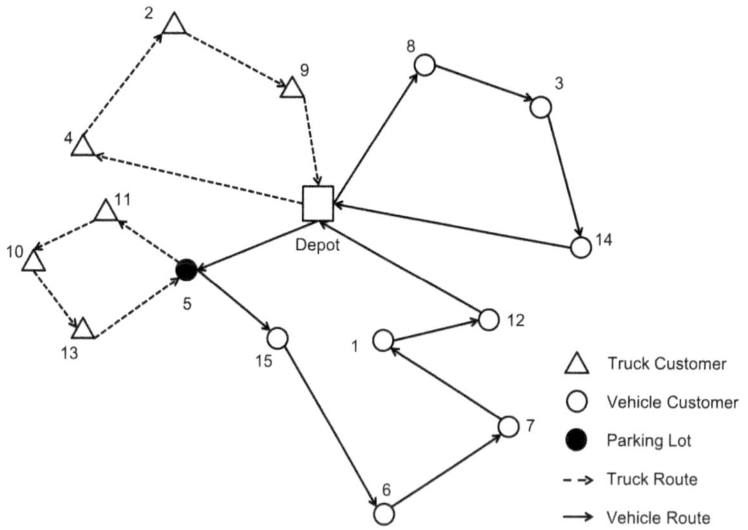

Fig. 4 Graphical representation of the solution in Table 2

3.2 Construction Heuristics

A construction heuristic is a heuristic that generates a starting pool of solutions so that the future evolution and mutation can start from this pool to converge to better solutions. A very good start mating pool cannot guarantee a very good final solution because the evolution and local search strategy play a more significant role. Hence, construction heuristics do not need to generate a very good solution to the TTRP, rather a reasonably good one will do. In this paper, we propose a construction method to deal with the problem. In this construction heuristics, a TTRP is divided into two VRPs: one consists of VC customers only and they are served by vehicles, and the other consists of TC customers only and they are served by trucks.

There is some randomness incorporated into the heuristics because we need to generate different solutions to form the mating pool. The first customer to serve in each of the two VRPs is selected randomly using a random number generator. Then, the nearest customer is added to the route using the cheapest insertion technique. The heuristic continues adding customer to a route until the capacity is not enough to meet the following one. In that scenario, the vehicle returns to depot and a new route is initiated.

3.3 Crossover Strategy

TTRP is an extension of a VRP, and hence, they share many things in common. An insight can be drawn from the crossover for VRP to extend to the TTRP. According to Puljic and Manger (2013), if a single crossover operator is used, the best performance is accomplished by AEX, which stands for alternating edge crossover where edges are copied to the offspring alternatively from each parent. This method conserves the edges of parents with some additional random choices in case of infeasibility. This idea can be also applicable to TTRP as its crossover strategy. For instance, let us consider the following two parents, *P*1 and *P*2:

*P*1	5	1	7	8	4	9	6	2	3
*P*2	3	6	2	5	1	9	8	4	7

The procedure starts by choosing the arc 5-1 from P1 as the first arc. So the child is initialized as

5	1	–	–	–	–	–	–	–

Next, an arc that goes out from 1 from *P*2, i.e., 1-9, is chosen. Thus, the child becomes

5	1	9	–	–	–	–	–	–

Then, the arc from *P*1 going out from 9 is added, and so on. After a few steps, the following partially formed child is obtained:

5	1	9	6	2	3	–	–	–

The arc going out from 3 should be chosen from *P*2, but such choice is infeasible since both 3 and 6 already become part of the partial tour in the child. To avoid this situation, one of the remaining unvisited vertices is picked up randomly, for instance 7, which results in:

5	1	9	6	2	3	7	–	–

From this point, the ordinary procedure can be resumed again by choosing the arc 7-8 from P1, and 8-4 from P2. The completed child looks as follows:

5	1	9	6	2	3	7	8	4

3.4 Local Search Strategy

In this section, we present several local search strategies, which are essential in the proposed memetic algorithm. They are illustrated with examples below.

1. Node Exchange
   The node exchange operator randomly selects two customers and exchanges their position, provided the capacity of the vehicle or truck is not violated. If the capacity is violated, new route is created.

0	4	0	8	1	15	13	9	0	2	3	7	14	5	6	0

   For instance, if the random number generates 2 and 6, then the 2nd customer 4 and 6th customer 15 are selected to exchange their position. The final solution becomes

0	15	0	8	1	4	13	9	0	2	3	7	14	5	6	0

2. Node Relocation
   The node relocation operator randomly selects a customer and inserts it to another route randomly.

0	4	0	8	1	15	13	9	0	2	3	7	14	5	6	0

   For example, if customer 4 is selected and a random position is chosen as 6, then customer 4 is inserted after the 6th position between customers 15 and 13, resulting in:

0	0	8	1	15	4	13	9	0	2	3	7	14	5	6	0

3. Root Redefining
   This local search method is specific to TTRP. This method explores shifting the root to other neighbor customers as the new root. For example, the black lines in Fig. 5 represent vehicle routes, while dotted lines represent truck routes. The lower half of the graph has both black lines and dotted lines, and hence, it is a

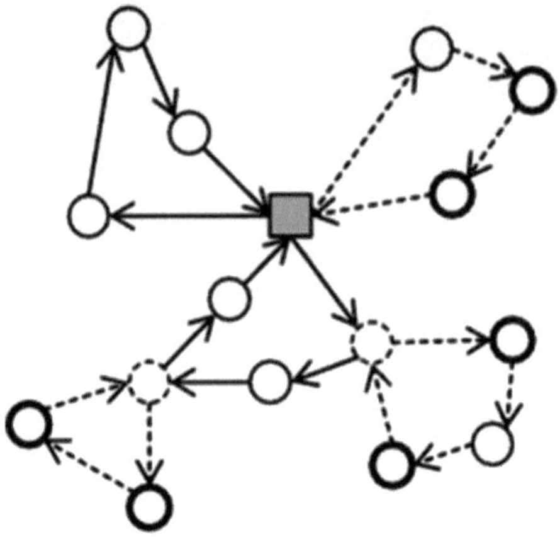

Fig. 5 Illustration of root-redefining method

mixed route including both vehicle routes and truck routes, where the truck routes are the sub-tours of the main vehicle routes. The root-redefining local search explores the possibility of shifting the base of the sub-tour to the nearest neighbor and checks whether it lowers the objective function value.

3.5 Overall Memetic Algorithm Procedure

We now present the overall memetic algorithm we propose. Before presenting the overall algorithm procedure, we mention parameters used to control the algorithm. Their values greatly affect the performance of the algorithm. They include the following:

- The number of iterations of the crossover algorithms, represented by M
- The number of iterations of the local search algorithms, represented by N
- The number of solutions in the pool of parents for crossover, represented by K
- The probability of using each of the local search methods, represented by Rand$L1$, Rand$L2$, Rand$L3$.

Figure 6 presents the overall algorithm procedures as follows:

4 Computational Results

In this section, we evaluate the proposed algorithm based on the memetic algorithm through well-known benchmark instances of the TTRP and make further discussions regarding the performance of the algorithm.

```
Step 1: Construct two routes for each of the VRP problems from a randomly chosen customers a
nd the order of other customers are selected using the Cheapest Insertion Heuristic. Combine
them into one construction solution.

Step 2: This step makes a pool of solutions as the mating pool and the mating pool contains K
solutions.

Step 3: Set n = 1, while n <= M
                {Set position = 1, while position <= K
                        {Offspring = Crossover (Parent1, Parent2);
                         Offspring is made feasible by Make_feasible function;
                         If Offspring is better than any of the Parents, replace parent.
                        }
                }

Step 4: Set n = 1, while n <= N
                {Set position = 1, while position <= K
                        {Generate r = random(0,1);
                          Case r < RandL1
                             Generate a new solution using the node exchange local search;
                             If the new solution is better than original one, replace it;
                         Case RandL1 < r < RandL2
                            Generate a new solution using the node reallocation local search;
                            If the new solution is better than original one, replace it;
                        Case r > RandL2
                           Generate a new solution using the root redefining local search;
                           If the new solution is better than original one, replace it;
                        }
                }

Step 5: Seek the minimum distance of all the solutions in pool. Output the best solution.
```

Fig. 6 Overall algorithm procedures for TTRP

4.1 Test Instances

The proposed memetic algorithm was tested on the 21 TTRP benchmark instances in Chao (2002), and these are available at http://web.ntust.edu.tw/~vincent/ttrp/. All the benchmark problems were derived from the 7 basic VRPs created by Christofides et al. (1979). In the benchmark instances, the number of customer ranges from 50 to 199 and the fraction of truck customers for each problem is 25, 50, and 75 %. To our knowledge, this is the only set of TTRPs available in the literature.

They are generated in the following manner. For each customer i in a CMT problem, the distance between i and its nearest neighbor customer is calculated and denoted by A_i. Each CMT problem was converted into three TTRPs. In the first problem, 25 % of the customers with the smallest A_i values are specified as truck customers. This percentage was increased to 50 and 75 % in the second and third problem, respectively. Table 4 shows the numbers and capacities of available trucks and trailers for all test problems and the ratio of total demand to total capacity reported by Chao (2002). The detailed descriptions of the 21 benchmark TTRPs are illustrated in Table 4.

Table 4 Details of the 21 benchmark TTRPs

Problem ID	Original problem	Number of customers		Trucks		Trailers		Ratio of demand to capacity
		VC	TC	Number	Capacity	Number	Capacity	
1	CMT1	38	12	5	100	3	100	0.971
2		25	25					
3		13	37					
4	CMT2	57	18	9	100	5	100	0.974
5		38	37					
6		19	56					
7	CMT3	75	25	8	100	4	100	0.911
8		50	50					
9		25	75					
10	CMT4	113	37	12	100	6	100	0.931
11		150	75					
12		100	112					
13	CMT5	150	49	17	150	9	100	0.923
14		100	99					
15		50	149					
16	CMT11	90	30	7	150	4	100	0.948
17		60	60					
IS		30	90					
19	CMT12	75	25	10	150	5	100	0.903
20		50	50					
21		25	75					

4.2 Numerical Experiments and Comparison

The proposed algorithm was implemented by Matlab 2012 and tested on a PC with Intel(R) Core(TM) i5-2450 M CPU@2.50 GHz. In the benchmark problems, distances between points were calculated using the Euclidean metric. Solutions in the following sections were obtained by running the algorithm for ten times. The best solution and the average solution were shown in Table 5 as well as the average running time. Best known solutions (BKS) for each instance were obtained from http://hdl.handle.net/1992/1141.

Table 5 compares the proposed memetic algorithm with previous methods from the literature. This table includes the best known solution of the 21 benchmark instances from Villegas et al. (2013), tabu search method by Scheurer (2006), simulated annealing method by Lin et al. (2009) as well as the memetic algorithm in this paper. This table reports the best solution and average solutions for 10 runs, the gap between this solution, and the best known solution as well as the running time to obtain each solution.

Table 5 Comparison of results with other methods in TTRP literature

Instance (Chao 2002)			Tabu search (Scheuerer 2006)				Simulated annealing (Lin et al. 2009)				Memetic algorithm			
ID	Customer	BKS[1]	Best	Avg	Gap[2] (%)	Time (min)	Best	Avg	Gap (%)	Time (min)	Best	Avg[4]	Gap (%)	Time (min)
1	50	564.48	566.8	567.98	0.59	9.51	566.82	568.86	0.74	9.51	578.44	581.23	2.97	3.09
2	50	611.53	615.66	619.35	1.28	9.60	612.75	617.48	0.97	9.60	622.45	625.24	2.24	1.61
3	50	618.04	620.78	629.59	1.87	11.24	618.04	620.5	0.40	11.24	631.06	633.85	2.56	3.30
4	75	798.53	801.6	809.13	1.33	18.49	808.84	817.71	2.40	18.49	863.49	866.28	8.48	3.28
5	75	839.62	839.62	858.98	2.31	15.16	839.62	858.95	2.30	15.16	849.51	852.30	1.51	10.61
6	75	930.64	936.01	949.89	2.07	18.62	934.11	942.6	1.29	18.62	968.06	970.85	4.32	3.94
7	100	830.48	830.48	832.91	0.29	33.60	830.48	838.5	0.97	33.60	899.84	902.63	8.69	18.18
8	100	870.94	878.87	881.26	1.00	25.66	875.76	882.7	1.35	25.66	947.33	950.12	9.09	20.16
9	100	912.02	942.31	955.95	4.82	30.47	912.64	921.97	1.09	30.47	997.61	1000.40	9.69	1925
10	150	1036.2	1039.23	1052.65	1.31	60.94	1053.9	1074.38	3.69	60.94	1142.41	1145.20	10.52	57.45
11	150	1091.91	1098.84	1107.47	1.29	56.17	1093.57	1108.88	1.55	56.17	1200.91	1203.70	10.24	54.68
12	150	1149.41	1175.23	1184.58	2.80	63.71	1155.44	1166.59	1.49	63.71	1301.61	1304.40	13.48	205.88
13	199	1284.71	1288.46	1296.33	0.71	165.41	1320.21	1340.98	4.38	165.41	1485.61	1488.40	15.85	190.18
14	199	1333.66	1371.42	1384.13	3.34	132.06	1351.54	1367.91	2.57	132.06	1573.51	1576.30	18.19	216.90
15	190	1416.51	1459.55	1488.71	479	154.10	1436.78	1454.91	2.71	154.10	1532.51	1535.30	8.39	216.25
16	120	1000.84	1002.49	1003	0.05	43.14	1004.47	1007.26	0.64	43.14	1127.41	1130.20	12.93	57.03
17	120	1026.17	1042.35	1042.79	1.62	33.73	1026.88	1035.23	0.88	33.73	1176.01	1178.80	14.87	55.47
I8	120	1098.15	1129.16	1141.94	3.99	31.78	1099.09	1110.13	1.09	31.78	1200.41	1203.20	9.57	53.66
19	100	812.69	813.5	813.98	0.08	28.84	814.07	823.01	1.27	28.84	895.07	897.80	1048	30.02
20	100	848.12	848.93	852.89	0.47	24.57	855.14	859.06	1.29	24.57	945.11	947.90	11.76	29.58
21	100	909.06	909.06	914.04	0.55	26.84	909.06	915.38	0.70	26.84	1049.21	1052.00	15.72	33.62

[1]Best known solution to date
[2]Gap between this solution and the best known solution
[3]Best solution of 10 runs
[4]Average value of 10 runs

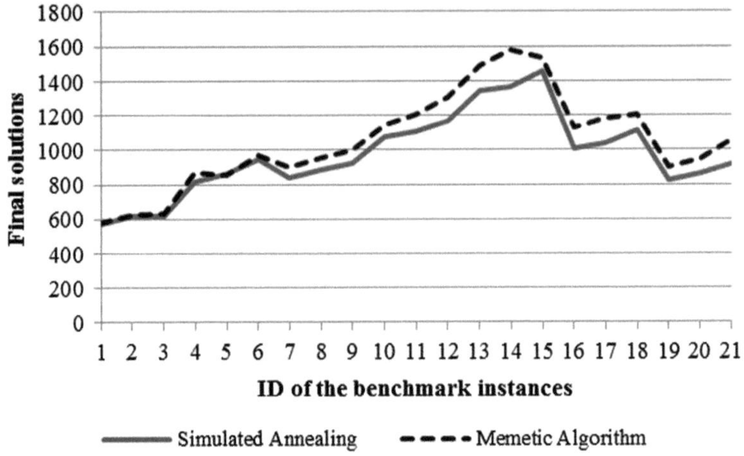

Fig. 7 Graphical comparison of solutions between Lin et al.'s (2009) and proposed algorithm

Figure 7 plots the solutions of the simulated annealing method in Lin et al. (2009) and the proposed algorithm for all the 21 benchmark TTRPs . We could see that for instances 1–3, the proposed algorithm could give very close solutions to those of Lin's ones. However, for the instances with larger number of customers such as instances 10–15, the gap between the proposed algorithm and Lin et al.'s (2009) one gets larger. This phenomenon may be due to several factors. A more detailed discussion will be given in the next section.

4.3 Analysis of Performance of Memetic Algorithm

The proposed memetic algorithm consists of the construction phase (breaking a TTRP into two VRPs) and the improvement phase (crossovers and local searches) and eventually gives a solution within the time constraints. We need to plot how the objective functions changes as time goes by to assess whether the algorithm converges to a local optimum. Hence, the trace of the objective function for instance 1 is plotted in Fig. 8, and we observe that the objective function value keeps decreasing. This graph is consistent with the logic of memetic algorithm where crossover and mutation allow the pool of solutions to become more fit. We could clearly see that it is easier to find a better solution initially and the difficulty increases as time goes by.

The proposed algorithm could not outperform the BKS in the literature. There are still some gaps to close between the BKS and the solutions we obtain. To understand the gaps better, the gaps with BKS and number of customers are plotted in Fig. 9, where the x-axis corresponds to the number of customers in each instance and the y-axis corresponds to the percentage gap with the BKS.

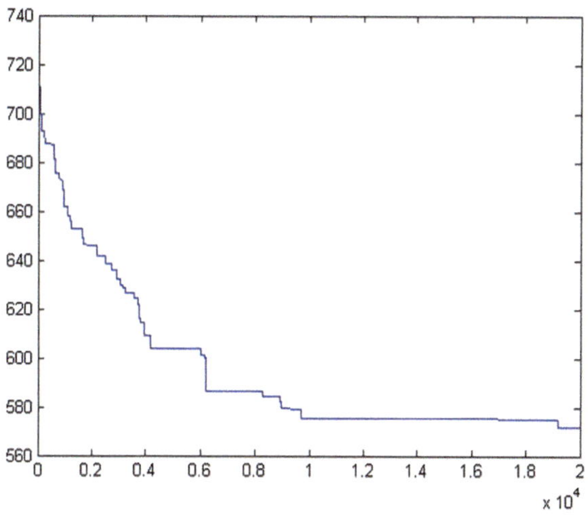

Fig. 8 Change in objective function as iterations goes for first instance (the x-axis is the number of iterations and the y-axis is the objective function)

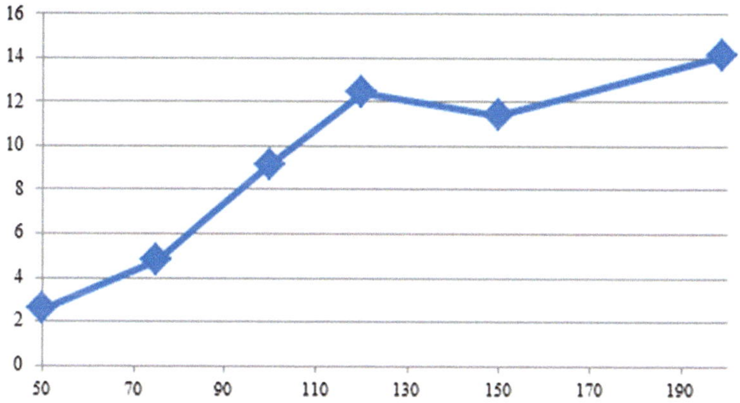

Fig. 9 Gap (%) with BKS versus no. of customers

The percentage gap varies between around 2 and 14 %. We could notice that the gap increases as the number of customers gets larger, indicating that the performance of the proposed algorithm deteriorates with larger instances, except for the instance with 150 customers. This shows that although the proposed algorithm gives a close solution to that of BKS of each instance, but there is still a room for improvement.

5 Concluding Remarks

In this paper, an algorithm based on the memetic algorithm scheme is proposed to solve the truck and trailer routing problem. The proposed algorithm firstly breaks the TTRP into two VRPs in the construction phase and then uses alternating edge crossover and several local search methods to allow parents to crossover and children to mutate which eventually give a constrained optimum solution.

It is tested using Chao's 21 benchmark instances and the solutions are compared with previous methods in the TTRP literature. Although the proposed algorithm could not outperform the best known solution of each instance, it can give a reasonably close solution to that of BKS. The reasons of the algorithm not giving very competitive solutions may be due to that the solution is trapped at a local optimum or the exploration space is not big enough. Possible ways to improve the proposed algorithm include refining the crossover and local search strategies to further expand the exploration space as well as tune the parameters of the algorithm to avoid trapping in a local optimum.

References

Bo L (2009) 大型钢铁企业厂内车辆循环甩挂运输模式研究. Master Degree Thesis in Logistics Engineering, Central South University, China

Caramia M, Guerriero F (2010) A heuristic approach for the truck and trailer routing problem. J Oper Res Soc 61:1168–1180

Chao I-M (2002) A tabu search method for the truck and trailer routing problem. Comput Oper Res 29:33–51

Christofides N, Mingozzi A, Toth P (1979) The vehicle routing problem. In: Christofides N, Mingozzi A, Toth P, Sandi C (eds) Combinatorial optimization. Wiley, Chichester, pp 315–338

Dantzig GB, Ramser JH (1959) The truck dispatching problem. Manage Sci 6(1):80–91

Derigs U, Pullmann M, Vogel U (2013) Truck and trailer routing—problems, heuristics and computational experience. Comput Oper Res 40:536–546

Gerdessen JC (1996) Vehicle routing problem with trailers. Eur J Oper Res 93(1):135–147

Lin S-W, Yu VF, Chou S-Y (2009) Solving the truck and trailer routing problem based on a simulated annealing heuristic. Comput Oper Res 36(5):1683–1692

Puljic K, Manger R (2013) Comparison of eight evolutionary crossover operators for the vehicle routing problem. Math Commun 18:359–375

Scheuerer S (2006) A tabu search heuristic for the truck and trailer routing problem. Comput Oper Res 33:894–909

Semet F, Taillard E (1993) Solving real-life vehicle routing problems efficiently using tabu search. Ann Oper Res 41:469–488

Villegas JG, Prins C, Prodhon C, Medaglia AL, Velasco N (2011) A GRASP with evolutionary path relinking for the truck and trailer routing problem. Comput Oper Res 38(9):1319–1334

Villegas JG, Prins C, Prodhon C, Medaglia AL, Velasco N (2013) A matheuristic for the truck and trailer routing problem. Eur J Oper Res 230:231–244

Yu VF, Lin T, Lu CC (2011) An ant colony system for solving the truck and trailer routing problem. J Chin Inst Transp 23(2):199–218

Modified DE Algorithms for Solving Multi-depot Vehicle Routing Problem with Multiple Pickup and Delivery Requests

Siwaporn Kunnapapdeelert and Voratas Kachitvichyanukul

Abstract This chapter proposes two DE algorithms for solving multi-depot vehicle routing problem (MDVRP) with multiple pickup and delivery requests (GVRP-MDMPDR). Two modified DE algorithms based on subgrouping of vectors and strategy switching concepts are developed and evaluated. In the first proposed algorithm, subgrouping of vectors is applied in the crossover process of the classical version of DE algorithm. The vectors are divided into two subgroups. The first subgroup applies exponential crossover process, while the other subgroup applies binomial crossover process. Experiences from two different crossover approaches are shared by allowing a target vector to randomly select vectors from both groups during the mutation process. The other algorithm is based on strategy switching concept applied in the crossover process. Two different crossover processes, exponential and binomial crossover processes, are used alternately. The results obtained from these two proposed algorithms are compared to those obtained by the classical DE algorithm. The results show that both proposed DE-based algorithms outperform the classic DE.

Keywords Differential evolution · Vehicle routing problem · Logistics · Strategy switching · Subgrouping of vectors

S. Kunnapapdeelert (✉)
Burapha University International College, 169 Longhard—Bangsean Rd, T.Seansook a.Muang, Chonburi 20131, Thailand
e-mail: siwapornk@buu.ac.th

V. Kachitvichyanukul
Asian Institute of Technology, Pathumthani, P.O. Box 4, Klong Luang, Pathumtani 12120, Thailand
e-mail: voratas@ait.asia

V. Kachitvichyanukul et al. (eds.), *Toward Sustainable Operations of Supply Chain and Logistics Systems*, EcoProduction,
DOI 10.1007/978-3-319-19006-8_25

1 Introduction

The classical vehicle routing problem (VRP) was first presented by Dantzig and Ramser (1959). The objective of the problem is to design the vehicle route to deliver the goods to the customers with the lowest cost. Since then, a large number of VRP variants are proposed and studied by many researchers by adding constraints that reflect closely to the real-world situations in transportation and logistics management. GVRP-MDMPDR is one of the problems that capture most constraints on logistic operation factors and reflect closest to the practical real-world situation, and it has become a focus of many recent research attempts.

VRP is considered an *NP*-hard problem that requires long computational time to solve by the exact algorithm if the problem size is large. As an alternative, metaheuristic approaches can provide good quality solution to *NP*-hard problem with a reasonable solution time and many metaheuristic approaches have been applied to deal with the variants of VRPs.

Differential evolution (DE) algorithm is one of the more effective metaheuristic algorithms for solving optimization problems. It was developed by Storn and Price (1997). DE uses a population of vectors to search for the optimal solution by perturbing vectors via the combination of several vectors randomly selected from the population. DE algorithm composes four main processes: (1) population initialization, (2) crossover, (3) mutation, and (4) selection processes. Based on its structures, DE algorithm is recognized as a method with easy concept, simple structure, fast convergence, and robust (Das and Suganthan 2011). DE algorithms have been successfully applied in many areas such as electrical power systems, control systems and robotics, bioinformatics, chemical engineering, and industrial engineering.

This paper presents the comparison of two enhanced DE-based algorithms. The first proposed algorithm is developed by applying subgrouping concept in the crossover process. Two commonly used crossover processes: exponential and binomial are used in the crossover process. The population of vectors is divided into 2 subgroups. Exponential crossover process is used in the first subgroup, while binomial crossover process is used in the other subgroup. The second proposed algorithm is developed by applying switching concept in the crossover process. In this case, the whole population of vectors starts with one crossover process. The same crossover process will continue to be used for as long as the solutions continue to improve. The crossover process will be switched to the other if the best solution does not improve for 25 consecutive iterations. Performances of the proposed algorithms are tested by using 30 benchmark test problem instances from Sombuntham and Kunnapapdeelert (2012). The computational results are then compared with the results obtained from the classical DE algorithm.

The remaining parts of this work are organized as follows. Section 2 reviews the literature of MDVRP and DE algorithms. In Sect. 3, the concept of DE for GVRP-MDMPDR is presented. The proposed DE algorithms with subgrouping and switching schemes for solving GVRP-MDMPDR are discussed in Sect. 4. Section 5 describes the solution representation and decoding procedure for solving GVRP-MDMPDR. The computational experiments for solving GVRP-MDMPDR of the proposed algorithms are discussed in Sect. 6. Lastly, summary and conclusions are discussed in Sect. 7.

2 Literature Review

VRP is an important decision in transportation and logistics management. VRP was firstly mentioned and studied by Dantzig and Ramser (1959) as the truck dispatching problem which is considered as the generalization of traveling salesman problem. Later, researchers have modified such problem by adding some conditions from the real-world applications into the VRP's model. Subsequently, VRP variants such as capacitated VRP (CVRP), heterogeneous fleet VRP (HVRP), VRP with time windows (VRPTW), VRP with simultaneous pickup and delivery (VRPSPD), and multi-depot VRP (MDVRP) are created. The focus of this paper is on the GVRP-MDMPDR since this variant of VRP presents the conditions closest to the practical real-world problems.

Single-depot VRPs are not appropriate for some real-world situation where the company has more than one depot. Consequently, MDVRP is proposed as extension to VRP to take into account the operations with more than one depot. In general, all customers are served by a corresponding vehicle from one of multiple depots that are assigned, and must return to the original depot. The MDVRP is more complicated than the single-depot VRP, but many of the solution methods are mostly based on those of single-depot VRP.

Renaud et al. (1996) proposed a tabu search algorithm based on three phases called FIND: Fast improvement, INtensification, and Diversification for solving MDVRP. The proposed algorithm was tested on a set of 23 benchmark problems. The results shown that the tabu search can be applied for solving MDVRP effectively. Ho et al. (2008) developed two hybrid genetic algorithms (HGAs: HGA1 and HGA2) to design the optimal route of MDVRP. The computational results showed that the HGA2 is superior to the HGA1 when applied to solve MDVRP in terms of solution quality. Later, GA was applied to design the optimal routes for MDVRP in Surekha and Sumathi (2011). Performance of GA was tested by Cordean's instances. The experimental results presented that GA can effectively solve MDVRP.

Performance of two different algorithms namely simulated annealing and tabu search for solving MDVRP with inventory transfer between depots in a three-echelon supply chain was studied by Fard and Setak (2011). Three-echelon supply chain composed of a single plant, multiple distribution centers and a set of retailers with deterministic demands. The results depicted that the tabu search approach outperformed the simulated annealing approach in terms of quality of solution.

Rahimi-Vahed et al. (2015) developed a new modular heuristic algorithm (MHA) to design the optimal routes of three VRPs, i.e., MDVRP, periodic VRP (PVRP), and multi-depot periodic VRP (MDPVRP). Performance of such algorithm was validated with the existing published instances. The computational results illustrated that the new proposed algorithm performs very well when applied to solve three problem classes.

DE has been successfully applied in many operational research problems, such as flow shop scheduling, job shop scheduling, and VRP. Some of the applications are briefly reviewed below.

Čičková and Števo (2010) presented DE to solve flow shop problem. Performance of the proposed algorithm was validated on available published instances. They concluded that DE is applicable to deal with flow shop problem. Later, Wisittipanich and Kachitvichyanukul (2011) proposed one-stage differential evolution algorithm (1ST-DE) to deal with job shop scheduling problem (JSP) and it was compared to one-stage particle swarm optimization (1ST-PSO). The results demonstrated that the 1ST-DE outperforms the 1ST-PSO in terms of both solution quality and computational time. Wisittipanich and Kachitvichyanukul (2012) further developed two new DE variants for improving searching efficiency of traditional DE for solving JSP. The first algorithm is called DE with subgrouping. The DE population was divided into subgroups. Each group of population performs its own searching strategy, and searching information from each group is shared across all groups to improve search ability. Another algorithm is called DE with switching strategy. The whole population is allowed to perform a mutation strategy as long as solution quality is improved. However, if the solution is not improved, the population switches its mutation strategy. The results confirmed that these two proposed algorithms are very efficient for solving JSP.

The applications of DE for solving VRPs are reviewed next. Cao and Lai (2007) presented an improved differential evolution (IDE) algorithm to design the optimal vehicle route in the VRP with simultaneous delivery and pickup service (VRP-SDP). The traditional DE was modified in two places. First, the mutation operation is modified by multiplying random number into each vector. Second, the crossover probability was time-varying in the crossover operation. The crossover probability changed from small to large with number of iteration. The results were compared to those obtained from GA and they presented that their proposed method yielded better solutions. Later, Rachman et al. (2009) applied DE for solving practical case of VRPs. It was used for designing optimal transportation route to deliver the pickup components from three suppliers to one car assembling plant. The results indicated that the proposed algorithm can significantly reduce the transportation cost by 25.21 %.

DE was applied to deal with the PDPTW by Kunnapapdeelert and Kachitvichyanukul (2013). The obtained results were compared to those obtained from PSO and the best known solutions. They indicated that DE is effective for solving the problems where the locations of the requests are clustered.

3 Generalized Vehicle Routing Problem with Multi-depot Multiple Pickup and Delivery Requests (GVRP-MDMPDR)

The detailed mathematical model of the problem is given in Sombuntham and Kachitvichyanukul (2010). The main objective is to design the vehicle route by minimizing total traveling distance. The assumptions considered in this study include the following:

1. The load of a vehicle does not exceed the vehicle capacity;
2. Every route starts and ends only at the given terminals;
3. The number of vehicles used must not exceed the number of available vehicles;
4. The total route time does not exceed the preset limit;
5. The routes of vehicle are constructed by attempting to minimize total routing cost, minimize number of vehicle, and maximize fulfilled demand under the following restrictions: Each request must be served exactly once by only one vehicle.

4 Modified DE Algorithms

Various researchers have tried to improve the performance of the DE algorithm by developing some different processes especially in mutation and crossover processes. Crossover process was modified into two different ways to enhance the searching ability of the DE algorithm in this work. The first improved algorithm is performed by using subgrouping concept in the crossover process to diversify the search. In the other proposed algorithm, switching concept is applied in the crossover process to enhance the searching ability of DE algorithm as explained below.

4.1 Subgrouping DE Algorithm

The classical version of DE algorithm consists of 4 main processes: (1) population initialization, (2) mutation process, (3) crossover process, and (4) selection process. Mutation process and crossover process are very important processes since these two processes govern the perturbation of DE vectors. The interest of this paper is on the crossover process. The objective of crossover process is to increase diversity of the perturbed vectors. New trial vector is constructed from the mixing of the elements from target vector produced in the previous generation and the elements from the mutant vector according to the chosen crossover process. There are two commonly used crossover processes in DE, namely exponential crossover and binomial crossover. In general, there is no single best crossover process for all problems. Consequently, the subgrouping of vectors' population is applied to include two crossover processes to improve the performance of DE algorithm. In this work, the population of vectors is divided into two subgroups. The first subgroup consists of 60 % of the vectors in the population. Exponential crossover is used in the first subgroup. The remaining vectors in the population belong to the second subgroup. Here binomial crossover process is applied. This ratio is determined empirically through experiments and it provides improved solution quality.

4.2 Switching DE Algorithm

The modification of the DE algorithm in this paper is occurring mainly in the crossover process. Two crossover processes used in this proposed algorithm are exponential crossover and binomial crossover. In subgrouping DE Algorithm, both crossover processes are used concurrently. However, in switching DE algorithm, the two crossover processes are used one at a time. For example, the exponential crossover process is used first. The crossover process will be switched to binomial crossover if the obtained solution does not improve within 25 consecutive iterations. In the same way, when the binomial crossover process is in use, whenever the computational results do not improve for 25 consecutive iterations, the crossover process will be switched back to the other. The reason for applying this concept in the algorithm is to produce different vectors and increase the chance to escape from stagnate solutions.

5 Solution Representation and Decoding Procedure

Previous works on GVRP-MDMPDR are reported by Sombuntham and Kachitvichyanukul (2012) and Sombuntham (2010). All these works are completed using the algorithms based on GLNPSO, a variant of PSO with multiple social learning terms. They proposed three solution representations and decoding procedures, namely SD1, SD2, and SD3.

According to the literatures, they found that on average, the computational results from SD1 provide the best solution quality but take much longer computational time than those of SD2 and SD3. Considering the solution quality of these three solution representation and decoding procedures, they found that solution quality obtained from SD1 and SD3 is comparable. Consequently, SD3 is used for decoding the data into the suitable format in this work.

The solution representation SD3 is described here. The solution vector consists of dimension $2r + 3m$, where r and m represent the number of requests and the number of vehicles, respectively. For each request, there is a corresponding pair of origin–destination which is used for generating pickup–delivery priority list.

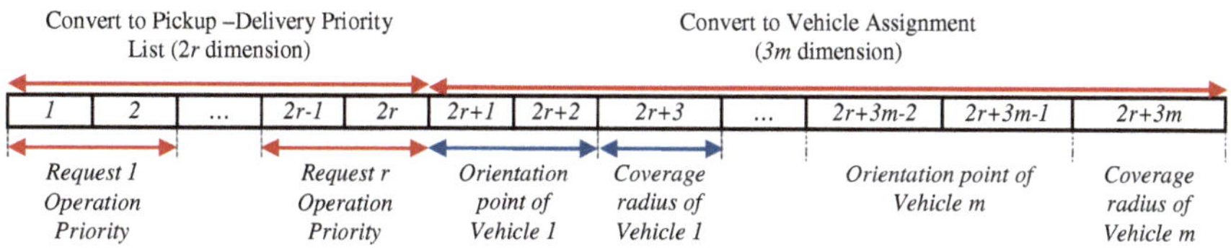

Fig. 1 Solution representation structure for SD3

The pickup–delivery priority list is generated by sorting values in $2r$ dimension from a vector in ascending order and then assigning the corresponding request number for pickup operation and delivery operation to the list, respectively.

For each vehicle, two positions are used to store the vehicle orientation point and one position to represent coverage radius of vehicle as explained in Fig. 1.

Each request is assigned to the vehicle based on its middle points and shipment points when the middle point refers to the x–y coordinate that lies in the middle of the pickup and delivery locations of each request while shipment points are denoted as a pair of pickup–delivery locations of each request.

Route construction starts by considering the priority of the request in the pickup–delivery priority list one by one. The candidate vehicle of the request is the vehicle that its coverage area covers middle point of the considered request or at least one of the shipment points of the request. The selection of vehicle for each request is based on the score which can be calculated by two different methods based on the situation of each vehicle. The first method is used when either pickup location is located at the vehicle's start location or delivery location is located at the same place as the vehicle's end location. The score is considered as a half of the distance between the other shipment point and the vehicle orientation point. Otherwise, the score is the Euclidean distance between the point of the request and the vehicle orientation point as explained in Fig. 2. All candidate vehicles are arranged by their scores in ascending order, and the request is then assigned to a candidate vehicle. After that, pickup and delivery operations of each request are inserted into the vehicle operation sequences according to a pickup–delivery priority list. Further, the problem constraints such as time window, vehicle capacity, and maximum route time are checked for the feasibility. If it is found that the insertion is infeasible, the next candidate vehicle will be considered to serve such request. Lastly, the assignment of the remaining requests is then performed to reduce the number of unassigned requests.

6 Experimental Results

The purpose of this work is to modify the DE algorithms to improve the effectiveness of the DE algorithm as solution method for GVRP-MDMPDR. Two different DE-based algorithms are developed, and the results are compared to those of

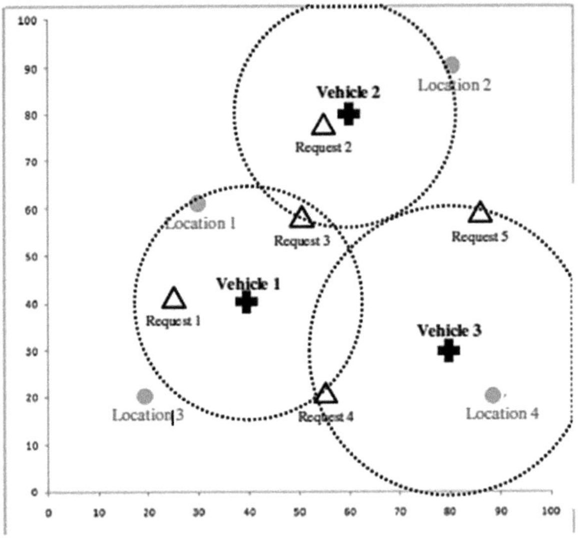

Vehicle	Start Station	End Station
1	1	1
2	2	2
3	4	4

Request no.	Pickup location	Delivery location
1	1	3
2	2	1
3	3	2
4	4	3
5	4	2

	Score of Vehicles		
Request No.	1	2	3
1	0.5 x (28.28)	-	-
2	0.5 x (64.03)	0.5 x (22.37)	-
3	18.03	0.5 x (72.11)	-
4	25	-	0.5 x (60.83)
5	-	0.5 x (67.08)	0.5 x (60.00)

	Candidate Vehicles	
Request No.	1st	2nd
1	Vehicle 1	-
2	Vehicle 2	Vehicle 1
3	Vehicle 1	Vehicle 2
4	Vehicle 1	Vehicle 3
5	Vehicle 3	Vehicle 1

Fig. 2 Construction of candidate vehicle list for each request

classical version of DE algorithm. All DE-based algorithms are implemented in C# programming language using Microsoft Visual Studio 2010. Parameter settings of all DE-based algorithms are obtained via experiments and are set to be the same. Population size and the maximum number of iteration are set as 200 and 500, respectively. Scaling factor is set as 2. The crossover rate is linearly increasing from 0.2 to 0.8. Thirty benchmark instances from Sombuntham and Kunnapapdeelert (2012) are used to evaluate the performance of the proposed algorithms. This set of instances composes five different groups with three different characteristics of location distribution of the requests as clustered, randomly distributed, and half-random-half-clustered scenarios. The percentage deviation from the solution obtained from classical DE algorithm is used for measuring the solution quality of the proposed algorithms which can be calculated by using the following formula:

$$\%\,\text{deviation from DE algorithm} = \left(\frac{\text{improved DE solution}_i - \text{DE solution}_i}{\text{DE solution}_i}\right) \times 100,$$

where i is the ith problem stance.

The comparison of the computational results of subgrouping DE algorithm and classical DE algorithm is presented in Table 1. This table composes 8 columns. The first column presents the name of benchmark instances used in the experiment. Columns 2, 3, and 4 contain the results obtained from classical DE algorithm, objective function (total cost), number of vehicle used, and minimum distance, respectively. The computational results from subgrouping DE algorithm are in columns 5, 6, and 7 for objective function, number of vehicles used, and total distance, respectively. The last column represents the percent deviation from the solution obtained from classical DE algorithm.

The results from Table 1 depicts that DE algorithm with subgrouping scheme is superior to the classical DE algorithm when applied to deal with GVRP-MDMPDR. The proposed algorithm improves the solution quality over those of classical DE algorithm 26 out of 30 problem instances with the average improvement of 4.10 % ranging between 0.11 and 12.16 %. The results also illustrate that the proposed algorithm works very well when the problem size is large.

The objective function used in this work composes fixed cost of the vehicles and variable cost of the traveling distance of each vehicle. Consequently, the number of vehicles used and the distance travelled by each vehicle are generally directly proportional to total transportation cost, which leads to high value of objective function. However, in problem Barc1 (60 requests and 16 vehicles), DE algorithm with subgrouping scheme found the better solution with shorter travelled distance but higher number of vehicles than those of DE algorithm. Although this seems to be common, upon further investigation, it is found that the vehicle capacity and fixed cost of each vehicle in problem instance Barc1 are not the same and different vehicle priority may lead to fewer vehicles if larger vehicle is used.

The performance comparison of classical DE algorithm and DE algorithm with switching scheme in crossover process is presented in Table 2 that composes 8 columns. Names of benchmark instances used in the experiment are in the first column. The results obtained from classical DE algorithm, objective function (total cost), number of vehicle used, and minimum distance are provided in columns 2, 3,

Table 1 Performance comparison of classical DE algorithm and DE algorithm with subgrouping crossover scheme for solving GVRP-MDMPDR

Instance	DE			Subgrouping crossover DE			% dev. from DE
	Objective	NV	Distance	Objective	NV	Distance	
Aac1	875.25	5	435.25	864.22	5	444.22	−1.26
Aac2	781.10	4	421.10	780.23	4	420.23	−0.11
Aar1	1976.56	6	1416.56	1972.90	6	1392.90	−0.18
Aar2	2227.32	6	1647.32	2287.01	6	1707.01	2.68
Aarc1	1455.28	4	1075.28	1465.98	5	985.98	0.74
Aarc2	1703.32	5	1243.32	1688.25	5	1208.25	−0.88
Bac1	1763.69	7	1123.69	1783.69	7	1143.69	1.13
Bac2	1927.72	7	1287.72	2022.64	8	1322.64	4.92
Bar1	4112.93	11	3072.93	4055.32	11	3015.32	−1.40
Bar2	8365.46	20	6565.46	7959.81	19	6239.81	−4.85
Barc1	2871.29	9	2031.29	2861.93	10	1941.93	−0.33
Barc2	3207.20	10	2287.20	2975.79	9	2175.79	−7.22
Cac1	3866.23	13	2686.23	3550.76	12	2490.76	−8.16
Cac2	4701.06	15	3341.06	4549.76	13	3389.76	−3.22
Car1	6750.80	17	5150.80	6428.79	17	4808.79	−4.77
Car2	8274.97	19	6554.97	8189.45	19	6429.45	−1.03
Carc1	5017.12	14	3757.12	4851.85	14	3571.85	−3.29
Carc2	5069.91	14	3849.91	4777.29	14	3557.29	−5.77
Dc1	2881.36	10	1981.36	2768.56	9	1968.56	−3.91
Dc2	3047.56	10	2127.56	2833.98	9	2033.98	−7.01
Dr1	3636.28	11	2656.28	3319.04	10	2439.04	−8.72
Dr2	3877.96	10	2997.96	3572.43	10	2692.43	−7.88
Drc1	4494.09	12	3414.09	4049.29	11	3009.29	−9.90
Drc2	5141.78	13	3961.78	4869.12	12	3729.12	−5.30
Ec1	3701.39	20	1861.39	3379.34	19	1659.34	−8.70
Ec2	3440.44	19	1720.44	3322.19	19	1542.19	−3.44
Er1	3444.67	14	2224.67	3186.14	12	2106.14	−7.51
Er2	3492.58	12	2392.58	3067.98	11	2047.98	−12.16
Erc1	4204.42	15	2764.42	3890.36	15	2530.36	−7.47
Erc2	7058.49	21	5178.49	6488.01	19	4768.01	−8.08

and 4, respectively. The computational results from DE algorithm with switching scheme are in columns 5, 6, and 7 for objective function, number of vehicle used, and total distance, respectively. The last column shows the percent deviation from the solution obtained from classical DE algorithm.

As the computational results from DE algorithm with switching scheme are compared to those of DE algorithm, the results indicate that when applying the proposed algorithm to deal with GVRP-MDMPDR, DE algorithm with switching scheme can find the better solution than that of DE algorithm especially in large-

Table 2 Performance comparison of classical DE algorithm and DE algorithm with switching crossover scheme for solving GVRP-MDMPDR

Instance	DE			Switching crossover DE			% dev. from DE
	Objective	NV	Distance	Objective	NV	Distance	
Aac1	875.25	5	435.25	893.31	5	433.31	2.06
Aac2	781.10	4	421.10	784.27	4	424.27	0.41
Aar1	1976.56	6	1416.56	2023.47	6	1463.47	2.37
Aar2	2227.32	6	1647.32	2364.35	7	1724.35	6.15
Aarc1	1455.28	4	1075.28	1479.91	5	1019.91	1.69
Aarc2	1703.32	5	1243.32	1728.80	6	1208.80	1.50
Bac1	1763.69	7	1123.69	1917.92	8	1177.92	8.74
Bac2	1927.72	7	1287.72	2125.53	8	1385.53	10.26
Bar1	4112.93	11	3072.93	4056.66	12	2936.66	−1.37
Bar2	8365.46	20	6565.46	9286.26	22	7346.26	11.01
Barc1	2871.29	9	2031.29	3547.63	11	2547.63	23.56
Barc2	3207.20	10	2287.20	2979.40	10	2059.40	−7.10
Cac1	3866.23	13	2686.23	3022.79	9	2202.79	−21.82
Cac2	4701.06	15	3341.06	3940.69	13	2760.69	−16.17
Car1	6750.80	17	5150.80	4615.76	14	3335.76	−31.63
Car2	8274.97	19	6554.97	6929.39	18	5249.39	−16.26
Carc1	5017.12	14	3757.12	4866.14	15	3546.14	−3.01
Carc2	5069.91	14	3849.91	5470.08	16	3990.08	7.89
Dc1	2881.36	10	1981.36	2636.00	9	1836.00	−8.52
Dc2	3047.56	10	2127.56	2945.98	9	2165.98	−3.33
Dr1	3636.28	11	2656.28	3414.53	10	2494.53	−6.10
Dr2	3877.96	10	2997.96	3733.89	11	2713.89	−3.72
Drc1	4494.09	12	3414.09	4294.45	12	3194.45	−4.44
Drc2	5141.78	13	3961.78	4877.83	13	3717.83	−5.13
Ec1	3701.39	20	1861.39	3881.68	22	1921.68	4.87
Ec2	3440.44	19	1720.44	3211.30	17	1631.30	−6.66
Er1	3444.67	14	2224.67	3024.17	12	2044.17	−12.21
Er2	3492.58	12	2392.58	3330.92	12	2270.92	−4.63
Erc1	4204.42	15	2764.42	4008.50	15	2668.50	−4.66
Erc2	7058.49	21	5178.49	7028.10	19	5128.10	−0.43

size problems. On average, the proposed algorithm provides the better solution than those obtained from DE algorithm by about 2.56 %. Considering problem Carc1 (100 requests and 30 vehicles), the solution obtained from DE algorithm with switching scheme provides the better solution with shorter travelled distance but higher number of vehicles than those of DE algorithm. Upon further investigation, the vehicle capacity and fixed cost of each vehicle in problem Carc1 are not the same and different vehicle priority may lead to fewer vehicles if larger vehicle is used.

According to the concept of the proposed algorithms, DE algorithm with subgrouping scheme allows the vector to search for the solution from more searching areas than DE algorithm with switching scheme. Comparing the performance of DE algorithm with subgrouping and switching schemes, they indicate that DE algorithm with subgrouping scheme provides about 1.54 % better solutions than that of DE algorithm with switching scheme. This confirms that DE algorithm with subgrouping scheme can enhance searching efficiency better than DE algorithm with switching scheme.

7 Summary and Conclusions

Two enhanced DE algorithms are compared and presented in this paper. The first DE-based algorithm is modified based on subgrouping concept in the crossover process to improve the efficiency of the search for solving GVRP-MDMPDR. Two crossover processes: exponential crossover and binomial crossover are used in the crossover process concurrently. The vector population is divided into two subgroups. Exponential crossover is used in the first subgroup, while binomial crossover is applied in the other subgroup. The second DE-based algorithm is developed based on the idea of switching in crossover process for enhancing the searching ability of DE algorithm. In this proposed algorithm, two crossover probabilities are used alternately. The crossover process will be switched to the other whenever the solution does not improve for 25 consecutive iterations. The proposed algorithms are implemented with solution representation SD3.

The results from 30 benchmark test problem instances show that both modified DE algorithms are effective for solving GVRP-MDMPDR. The results present that both proposed algorithms perform very well when applied to solve GVRP-MDMPDR. Comparing the solution quality of two proposed algorithms, the computational results indicate that DE algorithm with subgrouping scheme in crossover provide the better results than that of DE algorithm with switching scheme in crossover. The new route construction method, new solution representation, and decoding method should be further investigated for the future research.

References

Cao E, Lai M (2007) An improved differential evolution algorithm for the vehicle routing problem with simultaneous delivery and pick-up service. In: Proceedings of the 3rd international conference on natural computation (ICNC 2007)

Čičková Z, Števo S (2010) Flow shop scheduling using differential evolution. Manag Inf Syst 5 (2):8–13

Dantzig GB, Ramser JH (1959) The truck dispatching problem. Manag Sci 6(1):80–91

Das S, Suganthan PN (2011) Differential evolution: a survey of the state-of-the-art. IEEE Trans Evol Comput 15(1):4–31

Fard FA, Setak M (2011) Comparison between two algorithms for multi-depot vehicle routing problem with inventory transfer between depots in a three-echelon supply chain. Int J Comput Appl 28(6):39–45
Ho W, Ho GTS, Ji P, Lau HCW (2008) A hybrid genetic algorithm for the multi-depot vehicle routing problem. Eng Appl Artif Intell 21(4):548–557
Kunnapapdeelert S, Kachitvichyanukul V (2013) Solving multi-depot vehicle routing problem with pickup and delivery requests via differential evolution. In: Proceedings of the institute of industrial engineers Asian conference, pp 749–756
Rachman A, Dhini A, Mustafa N (2009) Vehicle routing problems with differential evolution algorithm to minimize cost. In: The 20th national conference of Australian society for operations research and the 5th international intelligent logistics system conference, 1 Sept 2009
Rahimi-Vahed A, Crainic TG, Gendreau M, Rei W (2015) Fleet-sizing for multi-depot and periodic vehicle routing problems using a modular heuristic algorithm. Comput Oper Res 53:9–23
Renaud J, Laporte G, Boctor FF (1996) A tabu search heuristics for the multi-depot vehicle routing problem. Comput Oper Res 23(3):229–235
Sombuntham P (2010) PSO algorithms for generalized multi-depot vehicle routing problems with pickup and delivery requests. Master's Thesis No. ISE-10-07, Asian Institute of Technology, Thailand
Sombuntham P, Kachitvichyanukul V (2010) Multi-depot vehicle routing problem with pickup and delivery requests. IAENG T Eng Technol 5:71–85
Sombuntham P, Kunnapapdeelert S (2012) Benchmark problem instances for generalized multi-depot vehicle routing problems with pickup and delivery requests. In: Proceedings of the Asia pacific industrial engineering and management systems conference, pp 290–297
Storn R, Price K (1997) Differential evolution—a simple and efficient heuristic for global optimization over continuous spaces. J Glob Optim 11:341–359
Surekha P, Sumathi S (2011) Solution to multi-depot vehicle routing problem using genetic algorithms. World Appl Prog 1(3):118–131
Wisittipanich W, Kachitvichyanukul V (2011) Differential evolution algorithm for job shop scheduling problem. Int J Ind Eng Manag Syst 10(3):203–208
Wisittipanich W, Kachitvichyanukul V (2012) Two enhanced differential evolution algorithms for job shop scheduling problems. Int J Prod Res 50(10):2757–2773

Part V
Metaheuristics and Artificial Intelligence Methods

Flexible Multistage Forward/Reverse Logistics Network Under Uncertain Demands with Hybrid Genetic Algorithm

Thitipong Jamrus, Chen-Fu Chien, Mitsuo Gen and Kanchana Sethanan

Abstract Logistics network is increasingly crucial because of shortened product life cycles, increasing competition, and uncertainty introduced by globalization. The logistics network distribution involves a multistage supply chain that consists of the flexible forward directions (i.e., factories, distribution centers, retailers, and various customers) and the flexible backward directions (i.e., re-manufacturing and reuse). Customer demands fluctuate and are unpredictable, thereby causing an imprecise customer quantity demand in each period in the production distribution model, and increasing inventory and related costs. Most studies have addressed the traditional multistage forward directions problem with certain demands or a single period. To fill the gap, this chapter proposes the hybrid genetic algorithm approaches for solving flexible, multiple periods, multiple stages, and forward/reverse logistics network. In particular, triangular fuzzy demands are considered to minimize the total cost, including transportation costs, inventory costs, shortage costs, and ordering costs, in the multistage and multi-time-period supply chain. The experimental results demonstrated practical viability for the proposed approaches.

T. Jamrus · C.-F. Chien (✉)
Department of Industrial Engineering and Engineering Management,
National Tsing Hua University, Hsinchu 30013, Taiwan
e-mail: cfchien@mx.nthu.edu.tw

T. Jamrus
e-mail: s101034881@m101.nthu.edu.tw

M. Gen
National Tsing Hua University, Hsinchu, Taiwan
e-mail: mitsuogen@gmail.com

M. Gen
Fuzzy Logic System Institute, Fukuoka 820-0067, Japan

K. Sethanan
Faculty of Engineering, Department of Industrial Engineering, Khon Kaen University,
Khon Kaen 40002, Thailand
e-mail: skanch@kku.ac.th

V. Kachitvichyanukul et al. (eds.), *Toward Sustainable Operations of Supply Chain and Logistics Systems*, EcoProduction,
DOI 10.1007/978-3-319-19006-8_26

Keywords Forward/reverse logistics network · Multistage and multi-time-period supply chain · Triangular fuzzy demand · Genetic algorithm

1 Introduction

Today's production distribution and demand fulfillment are increasingly challenging because of shortening product life cycles, rapid technological change, increasing competition, and uncertainty introduced by globalization (Chien et al. 2013). The success of a company will depend on their ability to achieve effective integration of organizational relationships within a supply chain. The integrated model for supply chain management includes not only the traditional forward flow from raw materials to products along the supply chain toward the customers, but also the reverse logistics network operations. This calls for the coordination and synchronization of groups of companies that work together with the added advantage that this in turn leads to strategic benefits, reduced costs, and improved competitiveness (Fuente et al. 2008). The reverse logistics system also is a new challenge to companies that retrieve all possible products and waste materials, addressing the economic value which still remains in the products returned.

In practice, customer demands are imprecise in each period of multi-time-period production distribution models, thus causing supplier production distribution complexity in meeting fluctuating demands in a multi-time period. The supply chain has to be prepared for demand unpredictability, interruption, or customer delivery delay, in which the multi-time period or infinite-horizon production distribution problem is unobtainable over the planning horizon until supply is insufficient to meet demand (Jamrus et al. 2015).

The data in supply chain networks are characteristically fuzzy, whereas customer demand is typically uncertainly quantified in supply chain procedures. Thus, traditional approaches with constant demand may not be appropriate for solving the uncertainty of supply chain networks that are necessary for effectively determining solutions. Moreover, one of the main difficulties associated with implementing forward and reverse logistics activities is the degree of uncertainty in terms of the timing and quantity of products. Previous research in the area of forward and/or reverse logistics network design often limited itself to only considering deterministic demand, single period, and one direction supply chain. Thus, we addressed the flexible multistage forward and backward logistics network with uncertain demands.

This study aims to develop a flexible multistage forward and reverse logistics network under uncertain demands. We propose hybrid genetic algorithm for solving this model. The approaches were solved under triangular fuzzy demands to satisfy demands with the minimal total cost by considering the transportation costs, inventory costs, and ordering costs of the forward supply chain network. In addition, considering a recovery costs of returned products and disposal costs in the reverse logistics network.

The remainder of this chapter is organized as follows: Sect. 2 reviews the literature on the forward and reverse supply chain problem and introduces the uncertain function of the triangular fuzzy numbers. Section 3 introduces the forward/reverse multistage and multi-time-period supply chain model. Section 4 describes the hybrid genetic algorithm for forward/reverse logistic model that consists of encoding chromosome and hybrid genetic algorithm. Section 5 presents the examples of experiments of the problems for comparing hybrid genetic algorithm with traditional genetic algorithm and particle swarm optimization. Section 6 concludes the chapter with a discussion on contributions and future research directions.

2 Literature Review

2.1 Forward and Reverse Supply Chain Problems

Various studies have considered different supply chain network classifications for each complexity of the problem that can be divided into two parts: a single-way supply chain network (forward or reverse supply chains) and an integrating forward and reverse supply chains. Several researchers have solved a single-way supply chain network problem. For example, Costa et al. (2010) proposed an innovative encoding and decoding procedure with the GA for determining the location and opening of facilities in a single-product three-stage forward supply chain network. They proposed a new, efficient chromosome representation procedure based on a parsimonious permutation decoding of the network string representation for reducing infeasible transportation trees. Zeballos et al. (2014) proposed mixed-integer linear programming for 10 layers (five stages) of closed-loop forward supply chains with uncertain supply and demand. Their objective function minimizes the expected costs of facilities, purchasing, storage, transportation, and emissions, minus the expected revenue of returned products. Sethanan et al. (2013) considered production planning, ordering, and inventory in analyzing costs for solving the forward supply chain problem. On the other hand, researchers consider the reverse and green supply chains such as Govindan et al. (2015) reviewed environmental, legal, social, and economic factors, reverse logistics and closed-loop supply chain issues that the most important extension in current objective functions is regarding green, sustainable, environmental, and resilience objectives. Xia et al. (2011) explained the definition of closed-loop supply chain and remanufacturing supply chain and reverse logistics management. In the reverse logistics that is an important means and methods to reduction, reuse, and recycle is also a supplementary. Also, Amin and Zhang (2013) proposed a mixed-integer linear programming model in closed-loop supply chain network that minimized the total cost under uncertain demand and return. Cardoso et al. (2013) used a mixed-integer linear programming (MILP) formulation is developed for the designing and planning of supply chains with reverse flows while considering simultaneously production, distribution, and reverse logistics activities under products' demand uncertainty.

However, actually the logistics network distribution involves a multistage supply chain that consists of the forward and reverse supply chain. Fuente et al. (2008) proposed an integrated model for supply chain management was aimed at redefining demand management procedures, order management procedures, manufacturing management procedures, procurement management procedures, distribution management procedures, and client management procedures. Kocabasoglu et al. (2007) focused the capture and exploitation of used products and materials liking forward and reverse supply chain. Easwaran and Uster (2010) considered a multi-product closed-loop logistics network design problem with hybrid manufacturing/remanufacturing facilities and finite-capacity hybrid distribution/collection centers to serve a set of retail locations. Moreover, few researchers used meta-heuristics to solve in forward and reverse supply chain problem. For example, Pishvaee et al. (2010) proposed solution memetic algorithm which used a new dynamic search strategy by employing three different local searches for integrated logistics network design to avoid the sub-optimality caused by a separate, sequential design of forward and reverse logistics networks. And Ko and Evan (2007) presented genetic algorithm-based heuristic for the dynamic integrate forward and reverse logistics network for third-party logistics that helped in the determination of various resource plans for capacities of material handling equipment and human resources.

Nevertheless, none of the above approaches can be used directly to solve the uncertainties that there are several uncertainties in the real manufacturing. The uncertainty may stem from a number of possible sources such as resources may become unavailable, interrupt activities, or due date may have to be changes. In this study, we propose the hybrid genetic algorithm approaches for solving flexible multistage forward/reverse logistics network in multiple periods and consider for minimizing the total cost, including transportation costs, inventory costs, shortage costs, and ordering costs.

2.2 Fuzzy Number

Fuzzy theory Zadeh (1965) presented mathematical techniques for deriving particular solutions. Chien et al. (2011) developed a comprehensive modular framework to derive various configurations of fuzzy numbers for fuzzy ranking. Also, Kumar (2012) summarized the definitions of the arithmetic operations of fuzzy numbers as follows.

If X is a collection of objects, then the fuzzy subset $\tilde{a}$ of X is defined as a set of ordered pairs:

$$\tilde{a} = \{(x, \mu_{\tilde{a}}(x)|x \in X\}$$

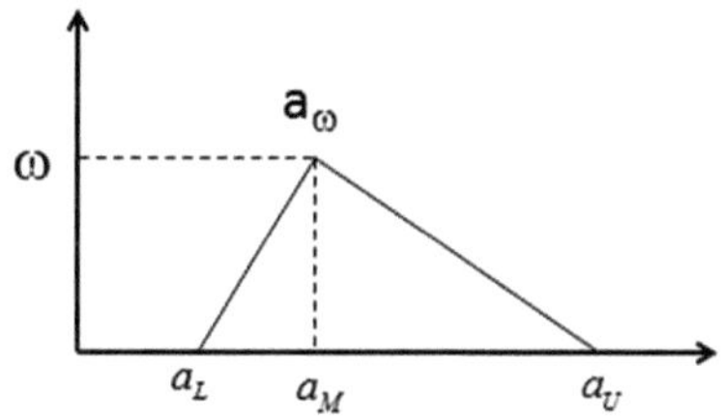

Fig. 1 A triangular fuzzy number

where $\mu_{\tilde{a}}(x)$ is the membership function for the fuzzy set $\tilde{a}$. The membership function maps each element of X to a membership grade or membership value between 0 and 1.

A real fuzzy number $\tilde{a} = [a_L\ a_M\ a_U]$, when defined as a triangular fuzzy number when $a_L \le a_M \le a_U$, is a fuzzy subset of the set of real numbers R, with membership function $\mu_{\tilde{a}}$ satisfying the following conditions:

$\mu_{\tilde{a}}$ is continuous from R to the closed interval [0, 1];
$\mu_{\tilde{a}}$ is strictly increasing and continuous $[a_L\ a_M]$.

The triangular fuzzy number is frequently used for practical purposes (Li et al. 2014). In particular, the membership function $\mu\tilde{a}(x)$ of the triangular fuzzy number $\tilde{a}$ is expressed as follows:

$$\mu_{\tilde{a}}(x) = \begin{cases} 0 & \text{for } x \le a_L \\ (x - a_L)/(a_M - a_L) & \text{for } a_L \le x \le a_M \\ 1 & \text{for } x = a_M \\ (a_U - x)/(a_U - a_M) & \text{for } a_M \le x \le a_U \\ 0 & \text{for } x \ge a_U \end{cases}$$

According to t definition of a triangular fuzzy number as shown in Fig. 1, let $\tilde{a} = (a_L(r), a_U(r))$, $(0 \le r \le 1)$ be a fuzzy number. Then, calculate the average value of the triangular fuzzy number $\bar{a}_0$ as follows:

$$\bar{a}_0 = \frac{1}{2}\int_0^1 \{a_L(r) + a_U(r)\}\mathrm{d}r = \frac{1}{4}[2a_M + a_L + a_U]$$

If $\bar{a}_\omega = (a_L(r), a_U(r)) = \{a_L + r(a_M - a_L)/\omega, a_U + r(a_M - a_U)/\omega\}$ is an arbitrary triangular fuzzy number at a decision level higher than "α" for $\alpha, \omega \in [0, 1]$, the value $\bar{a}_\omega$ can be calculated as follows:

If $\omega > \alpha$, then

$$\bar{a}_\omega = \frac{1}{2}\int_\alpha^\omega \{a_L(r) + a_U(r)\}\mathrm{d}r = \frac{1}{4\omega}\left[2a_M(\omega^2 - \alpha^2) + (a_L + a_U)(\omega - \alpha)^2\right]$$

3 Forward/Reverse Multistage and Multi-time-Period Supply Chain Model

The indices, parameters, and decision variables are listed as follows:

Indices

i index of plants ($i = 1, 2, \ldots, I$)
j index of distribution centers ($j = 1, 2, \ldots, J$)
k index of retailers ($k = 1, 2, \ldots, K$)
l index of customers ($l = 1, 2, \ldots, L$)
m index of collection centers ($m = 1, 2, \ldots, M$)
t index of time period ($t = 1, 2, \ldots, T$)

Parameters

I number of plants
J number of distribution centers
K number of retailers
L number of customers
M number of collection centers
T total time period
P_i plant i
D_m number of disposal center m
DC_j distribution center j
R_k retailer k
$\tilde{d}_l(t)$ customer uncertain demand l at time t
c^1_{ij} unit shipping cost of product from plant i to distribution center j
c^2_{ik} unit shipping cost of product from plant i to retailer k
c^3_{jk} unit shipping cost of product from distribution center j to retailer k
c^4_{jl} unit shipping cost of product from distribution center j to customer l
c^5_{kl} unit shipping cost of product from retailer k to customer l
c^6_{lm} unit shipping cost of product from customer l to collection center m
c^7_{mi} unit shipping cost of product from collection center m to plant i
f^1_j fix charge of products transportation in distribution center j
f^2_k fix charge of products transportation to retailer k
h^1_j unit holding cost of inventory per period at distribution center j
h^2_k unit holding cost of inventory per period at retailer k
$\mathrm{pr}_i(t)$ amount of products from plant i at time t
$S(t)$ cost saving of using product at time t
t^{L1}_{ij} lead time for transporting products from plant i to distribution center j
t^{L2}_{ik} lead time for transporting products from plant i to retailer k

t_{jk}^{L3} lead time for transporting products from distribution center j to retailer k

t_{jl}^{L4} lead time for transporting products from distribution center j to customer l

t_{kl}^{L5} lead time for transporting products from retailer k to customer l

α_x $= \begin{cases} 1 & x \neq 0 \\ 0 & \text{otherwise} \end{cases}$ α_x is the step function

Decision variables

$x_{ij}^{1}(t)$ amount of transporting product from plant i to distribution center j at time t

$x_{ik}^{2}(t)$ amount of transporting product from plant i to retailer k at time t

$x_{jk}^{3}(t)$ amount of transporting product from distribution center j to retailer k at time t

$x_{jl}^{4}(t)$ amount of transporting product from distribution center j to customer l at time t

$x_{kl}^{5}(t)$ amount of transporting product from retailer k to customer l at time t

$x_{lm}^{6}(t)$ amount of transporting product from customer l to collection center m at time t

$x_{mi}^{7}(t)$ amount of transporting product from collection center m to plant i at time t

$y_{j}^{1}(t)$ amount of inventory in distribution center j at time t

$y_{k}^{2}(t)$ amount of inventory in retailer k at time t

The proposed model was formulated for a forward/reverse multistage supply chain network. This problem involves two transfers—normal transfer from the first stage to the next closing stage and direct delivery—for transporting the product from distribution centers to customers instead of from retailers. Besides, the returned products are sent to collection centers that have the following responsibilities: collecting of used products from demand markets, determining the condition of returns by inspection and/or separation to find out whether they are revocable or not, sending recoverable returns to plants, and sending the unrecoverable returns to the disposal center (Fig. 2).

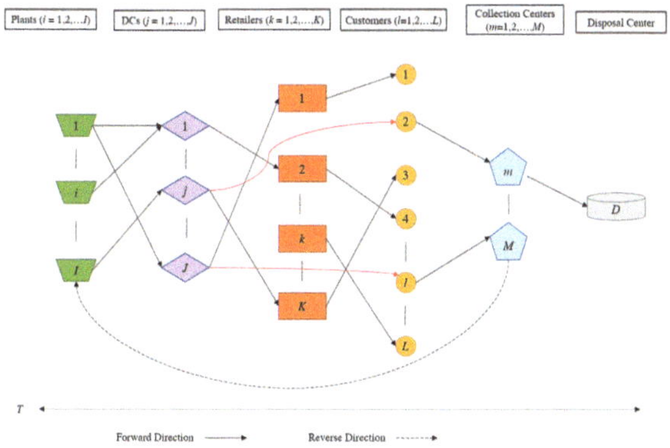

Fig. 2 Forward/reverse flexible multistage logistics network model

The total cost function of any of the following cost items is calculated as the sum of an item's costs over the planning horizon T. The objective function is to minimize the total costs are shown as follows:

$$
\begin{aligned}
\min z = \sum_{t=0}^{T} \Bigg[& \sum_{i=1}^{I}\sum_{j=1}^{J} c_{ij}^{1} x_{ij}^{1}(t) + \sum_{i=1}^{I}\sum_{k=1}^{K} c_{ik}^{2} x_{ik}^{2}(t) + \sum_{j=1}^{J}\sum_{k=1}^{K} c_{jk}^{3} x_{jk}^{3}(t) + \sum_{j=1}^{J}\sum_{l=1}^{L} c_{jl}^{4} x_{jl}^{4}(t) + \sum_{k=1}^{k}\sum_{l=1}^{L} c_{kl}^{5} x_{kl}^{5}(t) \\
& + \sum_{i=1}^{I}\sum_{j=1}^{J} f_{j}^{1} \alpha(x_{ij}^{1}(t)) + \sum_{i=1}^{I}\sum_{k=1}^{K} f_{k}^{2} \alpha(x_{ik}^{2}(t)) + \sum_{j=1}^{J}\sum_{k=1}^{K} f_{k}^{2} \alpha(x_{jk}^{3}(t)) + \sum_{j=1}^{J} h_{j}^{1} y_{j}^{1}(t) + \sum_{k=1}^{K} h_{k}^{2} y_{k}^{2}(t) \\
& - \sum_{k=1}^{K} S x_{mi}^{7}(t) + \sum_{k=1}^{k}\sum_{l=1}^{L} c_{lm}^{6} x_{lm}^{6}(t) \Bigg]
\end{aligned}
$$

The objective function is to minimize the total costs, which are shipping transportation costs (1st through 5th terms), fixed charge of product transportation costs (6th through 8th terms), holding costs (9th through 10th terms), recovery costs of returned products from collection centers to plants (11th term), and disposal costs (12th term).

4 Hybrid Genetic Algorithm Approach

4.1 Proposed Chromosome Encoding Procedure

The encoding representation designed in this chapter consists of the number of plants, distribution centers, retailers, customers, and collection centers. Figure 3 shows an example of encoding chromosomes with two plants, three distribution centers, three retailers, five customers, and two collection centers based on the random permutation, in which the priority of each gene is to allocate each stage so that the total number of genes is 13 in chromosome part 1. Chromosome part 2 is composed of assigning retailers to a plant or distribution center that has three genes. The value is 0 in the first gene (k_1) which means a direct shipment from plant i to the first retailer, and the value of 1 in the second and third gene means a direct delivery from distribution center j to k_2 and k_3. The final part of this chromosome that means customers reject products to collection centers, so that two integer numbers range from 1 to 2. In addition, there are random permutations that are the priority of collection centers for allocating in chromosome part 3.

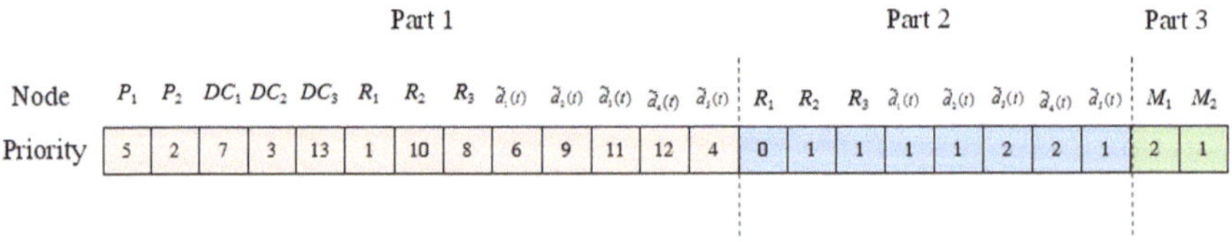

Fig. 3 Chromosome encoding

4.2 Hybrid Genetic Algorithm Operators

- **Particle swarm optimization (PSO)**

PSO is based on the simulation of social iterations proposed by Kennedy and Eberhart (1995) and is initialized with a population of random candidate solutions as particles. The position of each kth particle $x_k(t)$ is a potential result of the problem under study. Each particle remembers the best position that it has found thus far during the search process $h_{\text{best}}(t)$ and knows the global best position of swarm $g_{\text{best}}(t)$. We can calculate the particle's fitness by inputting its position into a designated objective function that is expressed according to the following equations:

$$v_k(t+1) = v_k(t) + b_1 r_1 (h_{\text{best}}(t) - x_k(t)) + b_2 r_2 (g_{\text{best}}(t) - x_k(t))$$
$$x_k(t+1) = x_k(t) + v_k(t+1)$$

where $v_k(t)$ is the kth particle's flying velocity at the tth iteration; b_1 and b_2 are the positive constants, called the acceleration constants; and r_1, $r_2 \in [0,1]$ are the uniform random numbers.

- **Crossover**

The crossover operators are exchanged between two chromosomes from one generation to the next generation. In this research, we used weight mapping crossover (WMX) in chromosome part 1 that extended one-cut point crossover, and two chromosomes would choose a random-cut point and generate the offspring by using a segment of its own parent to the left of the cut point, then remap the right segment based on the weight of other parent of right segment. Besides, chromosomes part 2 and 3 were used the uniform crossover that randomizes by uniform number then selects a locus and exchanges the subsequences before and after the locus between two parent chromosomes to two new offspring chromosomes if the random number less than 0.5 as shown in Fig. 4.

- **Mutation**

The mutation operator is applied to each child solution resulting from the crossover operator. In this study, we used two mutation operators, which are insert mutations, as shown in Fig. 5. The insert mutation operates in chromosome that selects a substring and insert within this chromosome to decrease the checking and repairing process of the offspring chromosome and to modify the gene.

- **Selection**

In the proposed algorithm, the well-known roulette wheel selection method is used for selecting parents from the old population, making the selection probability of a chromosome proportional to its fitness value.

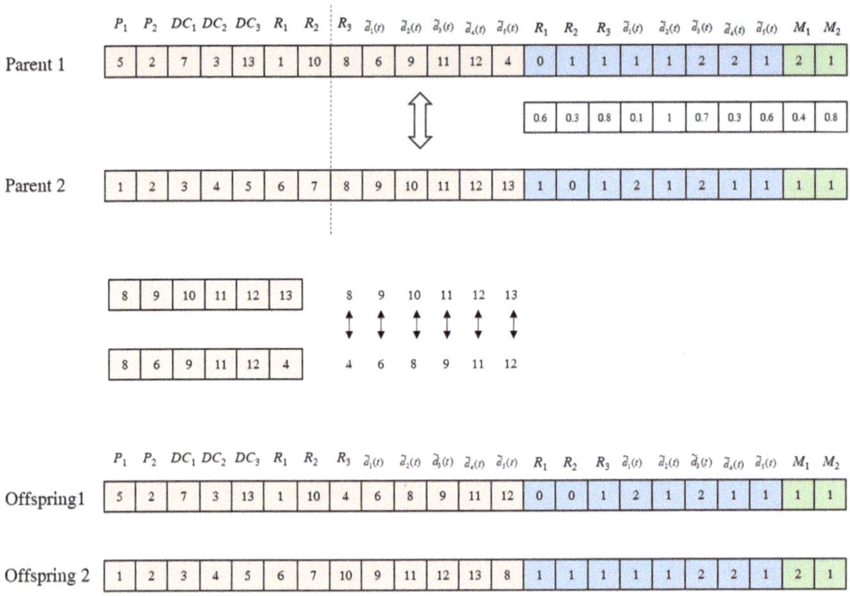

Fig. 4 Weight mapping crossover (WMX) and uniform crossover operators

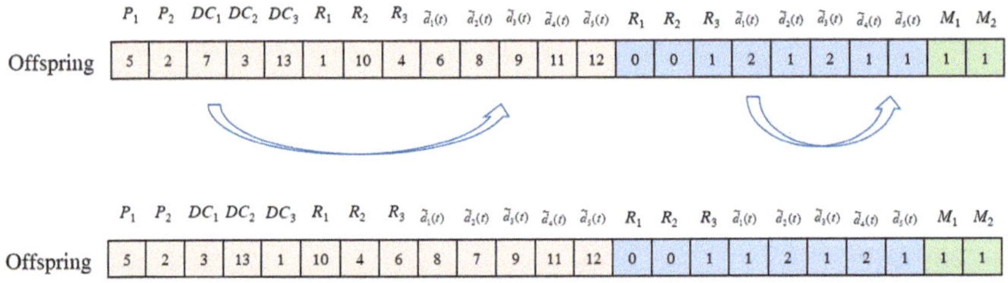

Fig. 5 Insert mutation operation

4.3 Proposed Chromosome Decoding Procedure

A description of the decoding procedure is detailed as follows: The first stage is the retailer or distribution center that delivers products for customer demand allocation in the period of triangular fuzzy customer demands; the values are obtained according to the proposed Sect. 2.2 ($\omega = 1$ and $\alpha = 0$). The allocation among retailers, distribution centers, customers, and collection centers is obtained by decoding chromosome part 1–3, which follows a priority-based procedure: Select the customer with the highest priority in chromosome part 1 and determine delivery from retailers or distribution centers according to chromosome part 2 (customers part). A gene number of 1 (Fig. 3) means a direct delivery from distribution center *j* to customer l, whereas a gene number of 2 means a direct delivery from retailer *k* to customer l; the highest priority of the customer corresponds to the minimum transportation cost; the customer allocation *t* continues until the maximum capacity has been reached and all customers are allocated; then, the throughput of each retailer is calculated.

The retailer allocation from the distribution center or plant, a stage at which each retailer lost stock of products from allocating customer demands; however, this stage is similar to the first stage, except that different retailers are positioned according to the chromosome and gene number. Then, select a suitable transportation cost to minimize costs; allocation continues until the maximum capacity has been reached and all retailers are allocated; and then, calculate the throughput of each distribution center.

The plants deliver direct shipment products to each distribution center because each distribution center lost stock of products from allocating customer demands and retailers. This step entails selecting the distribution *j* with the highest priority in chromosome part 1 and involves determining a direct shipment from plant *i* to DC *j*, in which plant *i* has sufficient capacity and the lowest transportation cost until all distribution centers have been assigned to the equal upper bound of distribution center capacity; then, the total cost from the encoding procedure is calculated. Finally, the returned products are sent to collection centers that have the following responsibilities: collecting of used products from demand markets, determining the condition of returns by inspection and/or separation to find out whether they are revocable or not, sending recoverable returns to plants, and sending the unrecoverable returns to the disposal center.

The overall procedure of this hybrid genetic algorithm approach is presented in the procedure in Fig. 6.

```
procedure : Hybrid genetic algorithm (minimize cost)
input : problem data and PSO (f(x), v_k(0), [hbest_k], gbest, b_1, b_2) and GA parameters (popSize, maxGen, p_M , p_C)
output : the best solution
begin
   t ← 0;
   initialize x_k(t) by operation and machine based encoding;//step 1 population P(t) =[x_k(t)]
   evaluate x_k(t) by operation and machine based decoding and keep the best solution;
   while (not terminating condition) do
       for each particle x_k   in swarm do
           update velocity v_k(t+1);
           update position x_k(t+1) using (11) and adjust x_k(t+1) by rounding routine;
        if f(x_k(t+1)) < f (hbest_k) then     // hbest_k : own best position of the
           particle k update hbest_k = x_k(t+1)
        end;
        if f(x_k(t+1)) < f(gbest) then //update gbest, gbest : the global best
            position of the swarm
           update gbest = x_k(t+1)
        end;
       create offspring C(t) from x_k(t+1) by WMX and uniform crossover routine;
       create offspring C(t) from x_k(t+1) by insertion mutation routine;
       check-and-repair C(t) for feasible solution;
       evaluate C(t) by decoding routine and update the best solution by comparing with gbest;
       select P(t+1) from P(t) and C(t) by selection routine;
       t ← t+1;
   end;
   output: the best solution
end;
```

Fig. 6 The procedure of the hybrid genetic algorithm

5 Experiments

To demonstrate the efficiency and effectiveness of the hybrid genetic algorithm in the fuzzy demand environment, the GA parameters are designed $pC(t) = 0.8$ and $pM(t) = 0.2$. The numerical experiments are 10 times the popSize = 50 and maxGen = 500. The proposed algorithms were run using MatLab on a 2.10 GHz PC, with 8 G-Byte of RAM, for testing and evaluation. For illustration, we generated three problems (Table 1) involving the various numbers of plants, distribution centers, retailers, customers, collection centers, and planning times. Uncertain demands generated in the a_L and a_U fuzzy values interval are [10, 20] for each problem. We assumed the fixed charge of product transportation and the unit holding cost of inventory per period at each node to be equal to 1. Table 2 provides information on the unit shipping cost for each node of a small-scaled MSMT problem (Problem 1).

We tested the hybrid genetic algorithm performance using three sizes of test problems. The computational test compared the best and average total cost of each solution for solving each problem, and traditional genetic algorithm and particle swarm optimization for forward/reverse supply chain and multi-time periods are shown in Table 1. Thus, the hybrid genetic algorithm and particle swarm optimization determined that total costs and have enhanced results compared with the traditional GA and PSO, whereas in the comparison results between the traditional GA and PSO for each problem, the PSO spends fewer number of iterations than the number of generations in the GA for finding the best solution for large problems. Time is the CPU time in seconds for every run, and the example of the hybrid genetic algorithm experiment facilitated determining the best solution at 43.9 s in

Table 1 Size of tested problems

Problem no.	No. of plants I	No. of DCs J	No. of retailers K	No. of customers L	No. of collection center M	Total planning time T
1	2	3	3	5	1	8
2	3	4	5	18	2	12
3	3	5	8	30	3	15

Table 2 Unit shipping cost of each node ($/unit)

C^1_{ij}	1	2	3	C^2_{ik}	1	2	3	C^6_{lm}	1	2	3	4	5		
1	9	2	6	1	5	8	10	1	2	3	3	5	4		
2	1	4	8	2	15	4	17								
C^3_{jk}	1	2	3	C^4_{jl}	1	2	3	4	5	C^5_{kl}	1	2	3	4	5
1	4	7	5	1	7	2	8	3	9	1	10	8	4	2	7
2	15	3	12	2	6	4	5	10	7	2	11	3	6	8	13
3	2	8	9	3	3	18	11	6	5	3	12	17	9	5	3

Table 3 Comparison of the best and average total cost of each solution from solving each problem [K · $]

Solutions problem no.	Traditional PSO		Traditional GA		Hybrid GA	
	Best solution	AV. solution	Best solution	AV. solution	Best solution	AV. solution
1	**1162**	1243.2	1162	1168.4	1162	1162.0
2	5535	5702.9	5420	5512.7	5113	5264.7
3	11,653	13,848.1	10,045	11,478.3	9416	9745.2

Problem 2. However, the CPU time of traditional approaches found a solution that is less than the hybrid genetic algorithm. The result of hybrid genetic algorithm outperformed the traditional approaches in all problems as shown in Table 3. However, they can determine which real-world demands must be optimized for long-term efficient operation of the entire supply chain.

6 Conclusion

Owing to globalization has introduced an extended competition and inconstant, factors, customer demands are typically vague in each period and uncertain customer quantity demands in production–distribution models affect inventory costs and related costs because of unsatisfactory planning. In this research, we formulated a forward/reverse supply chain network problem under uncertain demands as a hybrid genetic algorithm. The hybrid genetic algorithm approach combined between genetic algorithm and particle swarm optimization using a fuzzy number is frequently used for practical purposes and is relatively easy to employ with triangular fuzzy numbers for solving large-scale problems.

The model also examines the effects of uncertain demand and return on the network configuration. We determine delivery to customers in each period by using a triangular fuzzy number of customer demands. We selected the distribution center and retailers for each customer in the production–distribution network for satisfying uncertain customer demands. Moreover, returned products are considered to collection centers. The experimental results prove that the hybrid genetic algorithm can be applied for solving all problem types with the solution under uncertain customer demands. The proposed hybrid genetic algorithm outperforms the traditional genetic algorithm and particle swarm optimization. This research can extend to various supply chain environment problems by providing an enhanced method with large echelons and multi-objective function.

Acknowledgments This research is supported by the Ministry of Science and Technology, Taiwan (NSC102-2221-E-007-057-MY3; NSC103-2622-E-007-002; MOST103-2218-E-007-023), the Advanced Manufacturing and Service Management Research Center of National Tsing Hua University (102N2075E1), and Hsinchu Science Park (102A26).

References

Amin SH, Zhang G (2013) A multi-objective facility location model for closed-loop supply chain network under uncertain demand and return. Appl Math Model 37:4165–4176

Cardoso SR, Barbosa-Povoa APFD, Relvas S (2013) Design and planning of supply chains with integration of reverse logistics activities under demand uncertainty. Eur J Oper Res 226:436–451

Chien C-F, Chen J, Wei C (2011) Constructing a comprehensive modular fuzzy ranking framework and illustrations. J Qual 18:333–350

Chien C-F, Wu J-Z, Wu C-C (2013) A two-stage stochastic programming approach for new tape out allocation decisions for demand fulfillment planning in semiconductor manufacturing. Flex Serv Manuf J 25:286–309

Costa A, Celano G, Fichera S, Trovato E (2010) A new efficient encoding/decoding procedure for the design of a supply chain network with genetic algorithms. Comput Ind Eng 59:986–999

de la Fuente MV, Ros L, Cardos M (2008) Integrating forward and reverse supply chains: application to a metal-mechanic company. Int J Prod Econ 111:782–792

Easwaran G, Uster H (2010) A closed-loop supply chain network design problem with integrated forward and reverse channel decisions. IIE Trans 42:779–792

Govindan K, Soleimani H, Kannan D (2015) Reverse logistics and closed-loop supply chain: a comprehensive review to explore the future. Eur J Oper Res 240:603–626

Jamrus T, Chien C-F, Gen M, Sethanan K (2015) Multistage production distribution under uncertain demands with integrated discrete particle swarm optimization and extended priority-based hybrid genetic algorithm. Fuzzy Optim Decis Making 1–23

Kennedy J, Eberhart R (1995) Particle swarm optimization. In: Proceeding of the IEEE international conference on neural network, pp 1942–1948

Ko HJ, Evans GW (2007) A genetic algorithm-based heuristic for the dynamic integrated forward/reverse logistics network for 3PLs. Comput Oper Res 34:346–366

Kocabasoglu C, Prahinski C, Klassen RD (2007) Linking forward and reverse supply chain investments: the role of business uncertainty. J Oper Manage 25:1141–1160

Kumar BR (2012) On fuzzy transportation problem using triangular fuzzy numbers with modified revised simplex method. Int J Eng Sci Technol 4:285–294

Li X, Chien C-F, Yang L, Gao Z (2014) The train fueling cost minimization problem with fuzzy fuel prices. Flex Serv Manuf J 26:249–267

Pishvaee MS, Farahani RZ, Dullaert W (2010) A memetic algorithm for bi-objective integrated forward/reverse logistics network design. Comput Oper Res 37:1100–1112

Sethanan K, Jamrus T, Boonthavee P (2013) Solving production scheduling of egg production in hen farms with capacity of slaughterhouse by heuristic algorithms. IACSIT Int J Eng Technol 5:50–53

Xia W-h, Jia D-y, He Y-y (2011) The remanufacturing reverse logistics management based on Closed-loop supply chain management processes. In: 2nd international conference on challenges in environmental science and computer engineering, vol 11, pp 351–354

Zadeh LA (1965) Fuzzy sets. Inf Control 8:338–353

Zeballos LJ, Mendez CA, Barbosa-Povoa AP, Novais AQ (2014) Multi-period design and planning of closed-loop supply chains with uncertain supply and demand. Comput Chem Eng 66:151–164

Development of Heuristics in Sugarcane Harvest Scheduling for Mechanical Harvester in Sugarcane Supply Chain

Chuleeporn Kusoncum, Kanchana Sethanan and Chatnugrop Sangsawang

Abstract This chapter focuses on the inbound logistic section of the sugarcane industry, which is one of the important industries in Thailand. For inbound logistics, there are three major operation steps: cultivation, harvest, and transportation. From the three operations of inbound logistics, the highest inbound logistic cost is harvest. At present, there are three harvest patterns in Thailand: (1) labor cutting and loading, (2) labor cutting and trans-loader truck, and (3) mechanized harvester (cane harvester). The cane harvester usage tends to increase because of lack of labor and increasing cane production costs. In addition, the pattern of harvesting management depends on the experience and expertise of operators. So it is necessary to develop decision-making tools for high management efficiency and reduced risk of uncertainty. In this chapter, a genetic algorithm (GA) was developed to solve the harvest scheduling problem, which consists of 2 main parts: (1) sugarcane field clustering and (2) harvester routing. The objective is to reduce the harvesting cost by minimizing harvester travel distance. Experimental results show that the developed heuristic is quite effective and gives better result than the current practice from the real case of the sugarcane industry.

Keywords Planning · Sugarcane harvester · Heuristics · Increasing sugarcane quality

---

C. Kusoncum · K. Sethanan (✉) · C. Sangsawang
Research Unit on System Modeling for Industry, Department of Industrial Engineering, Faculty of Engineering, Khon Kaen University, Khon Kaen 40002, Thailand
e-mail: ksethanan@gmail.com

C. Kusoncum
e-mail: ckusoncum@gmail.com

C. Sangsawang
e-mail: s_chatnugrop@hotmail.com

V. Kachitvichyanukul et al. (eds.), *Toward Sustainable Operations of Supply Chain and Logistics Systems*, EcoProduction,
DOI 10.1007/978-3-319-19006-8_27

1 Introduction

Thailand is an agricultural country, with the majority of the population counted as farmers (41.1 %). The nature of agriculture in Thailand in the past was traditional agriculture, which focused on manufacturing to mainly meet the needs of household consumers. Later, there was an improvement in productivity, and it became more commercial agriculture. In this case, cane and sugarcane industries are vital to Thailand's economics at a very high level. There are 9.23 million rai of cane cultivation, with sugarcane products at around 103.66 million tons per year, and production of 11.29 million tons of sugar. Sugarcane consumption in Thailand in years 2013/2014 was 2.42 million tons. In sugarcane manufacturing, regardless of meeting the needs of the population inside the country, sugarcane produced from the mentioned industry was an important export product. In 2013/2014, 8.87 million tons of sugarcane were exported. This means that 32 % of the manufactured sugarcane was consumed within the country and the rest, around 68 %, exported to the international market. Thailand was the 2nd ranking sugarcane exporter, and Brazil was the first. Thailand holds 8.27 % of sugarcane in the world market, with the vital export markets of Indonesia, Japan, Cambodia, and India, in order (Fig. 1) (Office of the Cane and Sugar Board 2014). Moreover, the trend of the world market's needs is to increase consistently. There are 51 sugarcane factories located in different regions of Thailand, and this has truly contributed to employment and local income distribution. In such industries, there are many stakeholders such as farmers, labor, owners or entrepreneurs for machines, and equipment and trucks, as well as sugarcane factories. Hence, cane and sugarcane industries are also vital to significantly develop Thailand's economy and the quality of life of cane farmers.

However, the quantity of cane delivery in each period affects many stakeholders. During the first press, there is lack of labor for farmers to harvest canes as this period of time overlaps with the time to harvest other plants. Even though there were ready-to-deliver canes for the factory, the farmers could not cut canes and send to the factory because they did not have laborers to harvest canes. The result was that the quantity of input canes was less than the manufacturing capacity, and this led to an increasing production cost per unit. In the middle of the press,

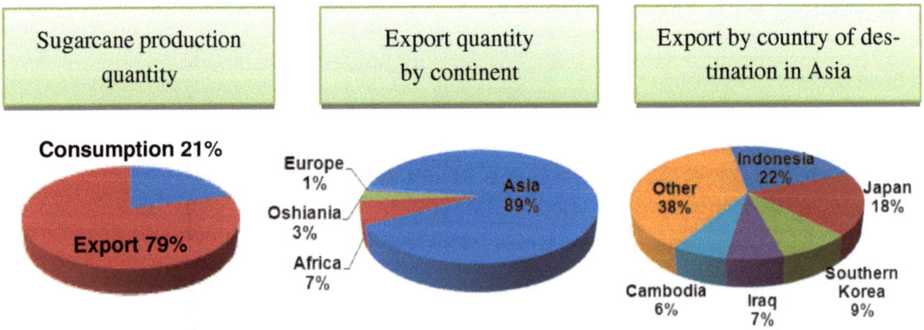

Fig. 1 Consumption quantity, export by continent and country of destination in 2014

sometimes the trucks would be in a long queue or there was no quota to take the canes, and this caused the problems that the labor had to slow down or stop harvesting canes in order to wait for the truck or wait for the quota to take the canes down. The farmers faced difficulties in managing cane-harvesting labor because the labor wanted a steady and inevitable income. If they could not cut the canes at that time, the labors would go to find other work to do, and it was so hard to find new laborers. At this time, it was the time to get large product, so canes were left on the front yard, and it lowered the quality of canes as well as decreasing CCS value for farmers and the purchasing price, including affecting continuity to the factory. In this case, it reduced the percent value of the sugarcane products as the amount per ton decreased. If the cost per unit was considered, it resulted in the increasing cost per unit for the factory, and at the final press, the farmers faced the problems of lack of cane-harvesting labor because at that time, some labors would go back to work in industry which provided consistent income and it was more convenient for them. Also, the weather was too hot, so the labors refused to harvest the fresh canes, which resulted in high productivity but low quantity of canes sent to the factory; therefore, the cost of production per unit was increased (Fig. 2). In the future, the problems of lack of cane-harvesting laborers will be more serious because the trend of the cane laborers now is to be mostly middle aged with an average age from 41 to 50 years old. The teenage laborers now focus more on industrial work as it is more convenient and it provides consistent income. Hence, using agricultural machines instead of human workforce is an interesting option in order to make a solution to the lack of labor, the complexity in labor management, and other problems in the cane pressing season. These reasons provide the answer why the costs of productivity for canes have increased. However, using agricultural machines such as cane-cutting mowers as another option has many limitations, such as the high price of the

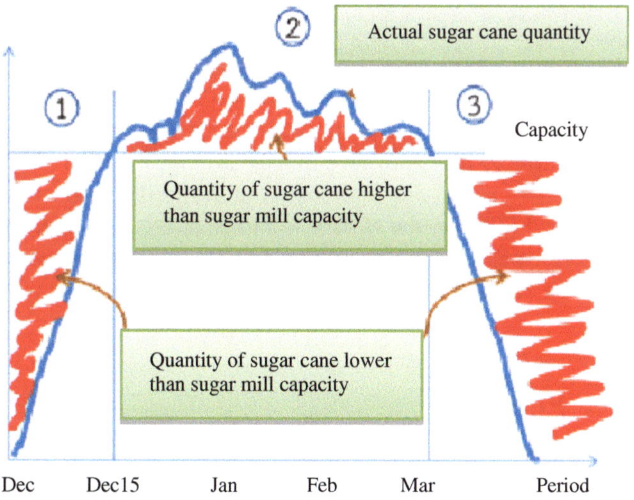

Fig. 2 Quantity of canes in each press

mowers, limitation on cutting per press, and other physical factors. Therefore, to use machines effectively and valuably, a well-managed method is required.

It is very difficult to manage the harvest by using good and effective cane mowers as there is complexity in the cane and sugarcane industries: (1) There are many stakeholders, leading to uncertainty, and (2) there are large plots of canes with different sizes and distributed in many places. The above factors result in ineffective management of the cane harvest because the scheduling of such actions requires skill, experience, and professionals as the main factors. In the meantime, if other conditions are considered, the mentioned scheduling should take time to produce a new plan, which might be less effective because of the limitations of human capacities in consideration, or to summarize, the harvest scheduling may be the best pattern or not. If a system of decision-making tools is developed, it should provide fast harvesting management and cane transportation, cost savings, and more effectiveness. If the system exists without any direct operators, the work can still be performed continuously, or it could take only a small amount of time to train workers who would take the responsibilities. It could be said that "the person who has knowledge, professionalism, and experience is the system maker, and the created system will develop operators to have more knowledge and professionalism."

Therefore, in order to make effective management of the cane-harvesting process and the transportation process to the factory, a decision-making tool for scheduling cane harvesting and transportation was proposed to be used in such an industry. The cane harvesting and transportation scheduling would divide the farmers into groups and include the quantity of cane and the cutting power of the cane-cutting mowers, as well as ordering the priority in cutting cane, by considering the distance in moving the cutting mowers to ensure the mowers are the most effective tools to cut cane, and since it is a waste of time to move to each field to cut cane. In addition, it was considered to harvest the sweetest canes to make the most profit.

The research led to the development of inbound logistics management in cane and sugarcane industries which focused on the management and allocation of limited resources to make the most benefit, by scheduling cane harvesting to make most benefit and make it conform to the production capacity of the factory, and the quality of the products would not decrease, so it led to increasing returns for the farmers. For the factory, it would have quality raw materials to the factory continuously, timely, adequately, and it should conform to the production activity. If the farmers had good lifestyle, the factory would be stable. The method for solving the problem is presented in Sect. 4. Section 5 outlines research results. Section 6 provides a research summary and conclusion.

2 Literature Review

In the supply chain and logistics of the cane and sugarcane industry, there are three important components: inbound logistics, Internal Logistics, and Outbound Logistics (see Table 1).

Table 1 Each component in the supply chain and logistics of cane and sugarcane industry

Inbound logistics	Internal logistics	Outbound logistics
• Cultivating area plan • Cultivation and care • Harvesting • Canes waiting spot management • Transportation • Platform management	• Production plan • Production and quality control • Material transfer • Packaging and storage • Information management	• Marketing communication • Product transportation and distribution • Warehouse management • Export management • Information management

Each activity in the supply chain will cause different costs. If any of the events in the system change, it affects other activities as well. Therefore, considering the resolutions should consider other parts as well. In this case, 60 % of the total cost of the activities mainly occurs from activities in the inbound logistics, and it was found that the cost of harvesting and transportation activities when sending canes to the factory was quite high, and it was counted as 40 % of the total cost of the system for all inbound logistics. In addition, the values were different in accordance with the harvest and transportation in each field. Such a harvest pattern not only affected the cost of transportation, but it also affected the long-term waiting at the front yard and CCS value for the farmers as well. The activities in the harvesting process consisted of cane cutting and loading.

From the review of the related literature on cane harvesting and transportation of cane to the factory, it was found that Higgins et al. (1998) had studied mathematical linear programming to find the best cane-harvesting day with the highest total sugarcane production quantity and the highest value for total returns depending on the conditions of location, differences of cane types, and cultivation characteristics. Later, Salassi et al. (2002) proposed mixed integer programming to find the best cane-harvesting day with the highest total sugarcane product quantity by estimating the growth of canes, and Jiao et al. (2005) proposed the second-order polynomial sweetness trend of canes and then developed it by using computer software for harvest scheduling, with the result of the proportion of the quantity of canes which would be harvested at each time in each cultivation field, so that the farmers could track the trend of sweetness value from cane harvesting. Grunow et al. (2007) created mathematical pattern and planning for the cane harvesting to reduce cost through the speed of the cane press, the harvested cane storage, the harvest of canes by using laborers and machines, as well as cane transportation to the factory. Moreover, the planning for cane harvest in order to lose the least sucrose was also proposed. Singh and Abeygoonawardana (1982) and Singh and Pathank (1994) have developed a computer program to simulate the process of sugarcane harvesting and transportation and also to calculate the cost of harvesting and transportation of sugarcane and the number of trucks required to perform the conversion with 1 cane-cutting mower. Next, Chamnanhlaw et al. (2004) developed a model to assist in the allocation of trucks to transport cane to the factory, in order to keep costs as low as possible for transport in the cane pressing season. Under the restriction of trucks used to transport, development of a mathematical model using linear programming and the application of a genetic algorithm (GA) model for allocating cane trucks did not consider the effectiveness of the performance of cane-cutting mowers which would vary according to the nature and size of canes. Higgins (2006) developed a mathematical model for truck scheduling to transport canes to the factory, to shorten the waiting time of trucks carrying canes in the queue in order to deliver canes to the factory and the unemployment period of cane pressing to be at the lowest cost by using 2 meta-heuristic approaches, including Tabu Search and Variable Neighborhood Search. Later on, Kaewtrakulpong et al. (2008) presented the development of cane-cutting mowers and cane truck allocation in each cane field to improve efficiency and reduce costs in the process of harvesting and transportation of canes, by applying the proper value for multi-objective optimization and then tested the result using computer simulation.

From the review of the related literature, it was found that most literature has no harvesting scheduling for using agricultural machines such as cane-cutting mowers with the principle that there should be sugarcane field clustering before routing, because fields in Thailand are quite small and social factors are different from other countries. Also, the trend of using cane-cutting mowers instead of laborers has increased more and more. Therefore, this research on Development of Heuristics in Sugarcane Harvest Scheduling for Mechanical Harvester in Sugarcane Supply Chain was important and should be focused on.

3 Problem Statement

The problem in this study was a case study in the cane and sugarcane industry which was in the process of harvesting management because of the complexities of the industry mentioned above, limitations, and conditions affecting the harvesting management in both quantity and CCS value of canes. Therefore, the study for development of scheduling for using cane-cutting mowers in harvesting was important under these conditions in order to produce continuously a flow of raw materials and use limited resources to produce the most benefit. As a result, it caused a lower cost for the total cost of the logistic system.

At present, the sampled factory harvested canes with fresh canes numbering 70–75 % and fired canes at 25–30 %. Normally, there were 3 harvesting methods which were (1) cut and load canes by labor, (2) cut and load canes by labor and cutting machine, and (3) cut and load canes by cutting machine. Mostly, farmers used method no. 2 (cut and load canes by labor and cutting machine) which was seen at 67 %, and no. 3 (cut and load canes by cutting machine) which was seen at 33 % of all canes cut, and there were 100 cutting machines in the system. About 68 cutting machines were from farmers, and the other 38 were from a factory. There were 2 types of cutting machine management. The first type was that the factory would manage the cutting machines, which was at 30 %. The raw material section of the factory would manage the cane cutting in both scheduling and ordering for each farmer. Major farmers mostly requested their need to the factory in order that the factory would provide cane cutting using machines, and the second type was that the factory would let farmers manage the cane cutting themselves, at 70 %. These were mostly small farmer groups that requested the need to the factory to provide cane-cutting machines. Then, the factory would allocate the cane-cutting machines for farmers in each group to conform to the production capacity of each machine. Next, the farmer groups should manage cane cutting by themselves. The factory identified that 1 cane-cutting machine should cut 15,000 ton/pressing season. The harvest managing of farmers may not be effective enough as such scheduling required skill, experience, or expertise in the work area. In the meantime, if it was considered necessary for other conditions, the new plan should be performed for such an order, and it may result in inadequacy because of the capacity limitation of human consideration. Moreover, it could not be concluded that such harvest scheduling was the best method or not. The most important points were that (1) many stakeholders were related to the cane and sugarcane industries, which resulted in high uncertainty, and (2) many cane cultivation plots were distributed in each area with differences in size, pattern, and physical factors. The above factors resulted in complexity and difficulty for the harvesting process, and this led to a bad effect on both product quality and cost of transportation in the inbound logistics system. Therefore, it was necessary to develop tools or systems to help in making decisions to support the uncertainty that might occur, in order to make decisions with human experience or expertise to give the most accurate and effective result. In this case, the canes should be in competition with competitors.

4 Methodology

In this paper, we propose and develop two GA algorithms (clustering and routing) for sugarcane harvester scheduling. From reviewing the literature, GA is a very efficient method for solving the clustering and scheduling problems, especially the NP class.

4.1 Genetic Algorithm for Field Clustering

Solving the sugarcane-harvesting problem starts with clustering types of sugarcane fields corresponding to the capacity of the harvester for minimizing the traveling distance. The steps for clustering using the GA are presented below.

4.1.1 Heuristic-Based Encoding and Decoding Routine

The essential factor to cluster the sugarcane fields is the maximum capacity of the mechanical harvester. In the clustering step, sugarcane fields are grouped corresponding to the capacity of the mechanical harvester with the objective to minimize the traveling distance. Hence, the process of generating the chromosome starts with the number of population using the heuristic-based encoding routine detailed in Figs. 3 and 4 shows the initial chromosome for the 9 sugarcane fields.

From Fig. 4, we can see that the sequence of fields in this cluster is 1-9-5-2-8-4-6-3-7. The objective of this chromosome can be determined using the heuristic-based decoding routine detailed in Fig. 5. For example, the parameters used to cluster the sugarcane fields are sugarcane quality in each field (see Table 2) and capacity of the three mechanical harvesters which are 15, 15, and 10 units, respectively. According to these parameters, the results of determining field clustering after applying the heuristic-based decoding routine are as follows: Group 1: Fields no. 1, 9, 5, and 2 with the total amount of sugarcane being 15 units, while Group 2 consists of fields no. 8, 4, 6, 3, and 7 containing the total amount of sugarcane at 10 units. Once the field clusters are obtained, the routes will be determined and the traveling distance can be calculated as detailed in Sect. 4.2.

4.1.2 Genetic Operator

Crossover is an operation to generate a new string (i.e., child) from two parent strings. It is the main operator of GA. In recent years, a number of crossover

Procedure: Heuristic based encoding routine for clustering
Input: Harvester scheduling data
Output: Chromosome v_k
Begin
    Select the largest sugarcane field
    While (no. of sugarcane fields≠0) **do**
        Sort the sequences of 4 neighborhood fields from the selected field
        Select the next field by random from the 4 fields
        Update the remaining sugarcane fields
    End
    Output Chromosome v_k
End

Fig. 3 Heuristic-based encoding routine for clustering

1	9	5	2	8	4	6	3	7

Fig. 4 Illustration of chromosome with heuristic-based encoding routine

```
Procedure: Heuristic based decoding routine for clustering
Input: Chromosome vk
Output: evaluated value (distance)
Begin
  For (field: first order to last order) do
    Harvester = 1
    While (Harvester capacity > 0) do
       Assign capacity for harvester
       Update capacity of harvester
    End
    Harvester = Harvester+1
   Calculate travel distance
   Output evaluate value
   End
End
```

Fig. 5 Heuristic-based decoding routine for clustering

Table 2 Amount of sugarcane in each field (units)

Field	1	2	3	4	5	6	7	8	9
Sugarcane Q'ty	5	4	1	2	4	2	3	2	2

operators have been presented, but the one-point crossover method has been selected for this study. For any pair of randomly selected points, the set of fields between them will always be inherited from the first parent to the child, while the other fields will appear in the order matching their appearance in the second parent. Furthermore, it is possible that when offspring are produced illegally in the form of a solution which is not feasible, conversion to a legal form might be possible through a repairing technique.

Mutation is another usually used operator of GA. Such an operation can be viewed as a transition from a current solution to its neighborhood solution in a local search algorithm. The main purpose is to ensure that premature solutions resulting in only a local optimum will not occur. This study applied 3 mutation strategies (swap, flip, and slide mutation) which are shown in Fig. 6a, b, c, respectively.

4.1.3 Selection Operation

One further factor which is especially important in GA is selection. It is a procedure to select offspring from parents to the next generation. The general definition holds that the probability of selecting a given chromosome should reflect the performance measure of that chromosome within the population. This ensures that a parent performing to a high level is more likely to be selected for the next generation. For the fitness function, in this clustering problem, the objective is to minimize distance

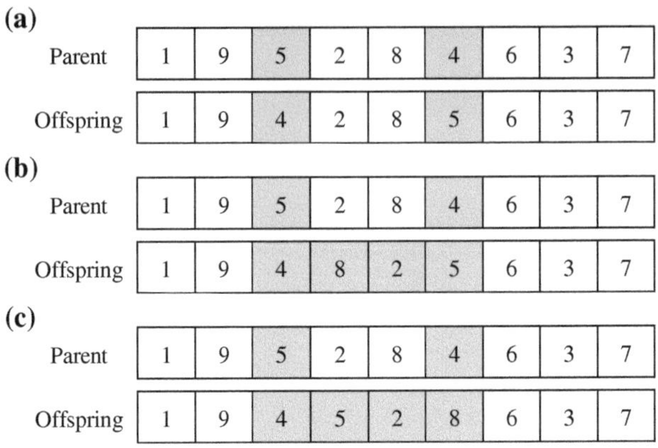

Fig. 6 Illustration of mutation operation. **a** Swap mutation, **b** flip mutation, **c** slide mutation

between fields. Therefore, the evaluation function used for the GA proposed is based on the distance (ΣD_{ij}) of the schedule using Eq. (1).

$$\text{Eval}(v_i) = 1/\Sigma D_{ij}; \quad i = 1, 2, 3, \ldots, \text{popsize} \tag{1}$$

This study implements the parent selection process through the common roulette wheel selection approach outlined by Goldberg.

4.2 Genetic Algorithm for Harvester Routing

Once the field cluster is obtained, the next step is to determine the mechanical harvester route using the GA detailed below.

4.2.1 Initial Population

The chromosome obtained from the clustering phase is used to determine the mechanical harvester route. For example, from Step 4.1.1, Group 1 consists of fields no. 1, 9, 5, and 2. The initial chromosome is presented in Fig. 7.

From this figure, we find that each chromosome is random for the sequence number corresponding to the number of fields. Additionally, the sequence number obtained from the randomization will get one initial chromosome, and the number of random numbers equals the number of population. Next, the chromosome generated for routing is evaluated for the objective function in order to maximize the total profit using Eq. (2),

Fig. 7 Illustration of chromosome encoding

Field	1	9	5	2
Random	1	4	2	3

Field	1	5	2	9
Order	1	2	3	4

$$\text{Total profit} = \text{Revenue} - \text{cost} \tag{2}$$

$$\text{Revenue} = Q \times S,\ Q = \text{Sugar cane quantity},\ \text{and } S = \text{Sugar cane price}$$
$$\text{Cost} = F \times TD \times CR,\ F = \text{Fuel price},\ TD = \text{Total travel distance},$$
$$CR = \text{Fuel consumption rate}$$

4.2.2 Genetic Operator

In order to improve the routing solution using GA, crossover, mutation, and selection processes are applied which is similar to the process for clustering. Hence, in this paper, one-point crossover, swap, flip, slide mutation, and roulette wheel selection are applied.

4.3 Genetic Parameter

The proposed algorithms were run with MATLAB on a 1.60 GHz PC, with 4 GB of RAM, for testing and evaluation. To solve the problem using the proposed GA, for clustering, the parameters used were set as follows: maximum generation maxGen = 5,000, population size popSize = 100, crossover probability $p_C = 0.8$, and mutation probability $p_M = 0.2$. For routing, the parameters used were set as follows: maximum generation maxGen = 10,000, population size popSize = 100, crossover probability $p_C = 0.8$, and mutation probability $p_M = 0.2$.

5 Results

From Development of Heuristics in Sugarcane Harvest Scheduling for Mechanical Harvester in Sugarcane Supply Chain, the researcher took the study results to apply in 2 sugarcane factories which were located in the central area of Thailand. The pattern of harvest management was divided into 2 cases which were (1) case of clustering and harvester routing pattern and (2) case of the pattern with only harvester routing as below.

Table 3 Comparison of effectiveness of cane harvest scheduling between current practice from a real case and the proposed pattern

Area	Distance (km)			Average capacity (ton/day)		
	Current practice	Program	Improvement (%)	Current practice	Program	Improvement (%)
1	4953.21	1513.87	69.44	88.13	102.07	15.81
2	3927.64	862.92	78.03	98.37	111.07	12.91
3	1353.45	812.41	39.97	87.79	106.03	20.78
4	613.78	560.04	8.76	48.64	84.97	74.69
5	840.83	308.53	63.31	129.77	134.07	3.32
6	610.45	308.51	49.46	92.92	122.16	31.46
Average	2049.89	727.71	51.49	90.94	110.06	26.49

5.1 Case of Sugarcane Field Clustering and Harvester Routing Pattern

From the study of data from a sample sugarcane factory in the central region, the area was divided into 6 regions, and clustering and plan for harvester routing in each area were carried out. The results from performance testing on cane harvest scheduling in current practice from the real-case sugarcane pattern and the developed pattern show clearly that the cane harvest scheduling developed pattern was more effective than that from current practice from the real case (present pattern used in the factory), and the average distance for the traditional pattern of the factory was 2049.89 km, while that from the newly developed method was 727.71 km. In this case, it reduced 1322.18 km or 51.49 %. Considering the average capacity of the cane harvester for the traditional pattern of the factory, it was 90.94 ton/day, whereas the developed pattern provided 110.06 ton/day, an average capacity increase of 20.12 ton/day, or 26.49 % as presented in Table 3.

5.2 Case of Pattern with Only Harvester Routing

In fact, most farmers have their own group due to physical, social, and cultural factors such as cane fields of farmers in each area being located closely to others who they are familiar with and provide confidence, respect, etc., so the order complied with the management of the current crop that farmer groups had at present. This study presents a scheduling of harvester routing only. By studying data of a sampled sugarcane factory in the central region, the area was divided into 9 regions, and the plan for harvester routing made for each vehicle. The results from performance testing on cane harvest scheduling in current practice from the real-case pattern and the developed pattern show clearly that the cane harvest scheduling developed pattern was more effective than that from current practice for the real

Table 4 Comparison of effectiveness of cane harvest scheduling with the harvester routing only pattern between current practice from the real-case sugarcane industry and the present program

Area	Distance (km)			Average capacity (ton/day)		
	Current practice	Program	Improvement (%)	Current practice	Program	Improvement (%)
1	492.40	224.61	54.38	143.91	220.37	53.13
2	446.39	323.60	27.51	264.50	225.32	−14.81
3	283.18	194.14	31.44	139.07	222.51	60.00
4	647.39	289.93	55.22	278.65	224.58	−19.40
5	371.05	259.24	30.13	225.27	222.46	−1.25
6	368.68	311.17	15.60	207.40	222.58	7.32
7	933.16	702.52	24.72	300.45	224.25	−25.36
8	357.07	276.89	22.45	277.64	224.78	−19.04
9	343.51	252.75	26.42	222.72	224.25	0.69
Average	471.43	314.98	31.99	228.85	223.46	4.59

case (present pattern used by factory). The average distance for the traditional pattern of the factory was 471.43 km and from the newly developed program was 314.98 km. In this case, it reduced 130.98 km or 31.99 %. After considering the average capacity of the cane harvester for the traditional pattern of the factory, it was 228.85 ton/day, whereas the developed pattern provided 223.46 ton/day, an increase of 5.39 ton/day, or 4.59 % (as shown in Table 4).

From the two patterns of harvest scheduling using cane-cutting mowers, it resulted in a shorter total distance to move mowers, and it could increase effectiveness of the cane-cutting mowers. Considering distance, the first pattern (clustering and harvester routing) was more effective than the second pattern (harvester routing only) as the second pattern has less flexibility in physical, social, and cultural aspects of farmers than the first pattern. Considering the average capacity of cane harvester for the first pattern, it was found that the cutting power of each cane-cutting mower on the field would have similar quantity, because the principle of clustering would try to equalize the cutting power of the mowers to have similar cutting quantity. For the second pattern, harvesting routing without clustering, the group was arranged according to the existing farmer groups. Therefore, pattern no. 1 would be more effective than pattern no. 2 because it was more flexible.

6 Summary and Conclusions

According to the study of Development of Heuristics in Sugarcane Harvest Scheduling for Mechanical Harvester in Sugarcane Supply Chain, after research on two-sampled sugarcane factories in the central area, it was seen that the harvest scheduling using GA gave a better pattern than the traditional pattern of the factory.

Considering routing, (1) the clustering and harvester routing pattern which was developed had shorter distance than the traditional pattern of the factory by 51.49 % and (2) the harvester routing alone pattern provided a shorter distance than the traditional pattern of the factory by 21.21 %. Pattern no. 1, clustering and harvester routing, was more effective than the case of pattern no. 2, harvester routing, as it was more flexible. By effective development of harvest management using the mentioned mowers, it gave a total distance shorter than before, so it resulted in reduction of power consumption, and it led to a reduction of inbound logistics capital. Moreover, it resulted in high efficiency of mowers use.

Acknowledgments This study was supported by collaboration with Research Unit on System Modeling for Industry (SMI), Department of Industrial Engineering, Faculty of Engineering, Khon Kaen University, Khon Kaen, Thailand, and the assistance and facilities from staff from sugar mill.

References

Chamnanhlaw C, Arnonkijpanich B, Pathumnakul S (2004) Solving truck allocation problem in sugar cane industry by genetic algorithms. In: 2004 IEEE international on engineering management conference. Proceedings, vol 3, pp 1324–1328. IEEE, Oct 2004

Grunow M, Günther HO, Westinner R (2007) Supply optimization for the production of raw sugar. Int J Prod Econ 110(1):224–239

Higgins A (2006) Scheduling of road vehicles in sugarcane transport: a case study at an Australian sugar mill. Eur J Oper Res 170(3):987–1000

Higgins AJ, Muchow RC, Rudd AV, Ford AW (1998) Optimizing harvest date in sugar production: a case study for the Mossman mill region in Australia. Field Crops Res 57:153–162

Jiao Z, Higgins AJ, Prestwidge DB (2005) An integrated statistical and optimisation approach to increasing sugar production within a mill region. Comput Electron Agric 48(2):170–181

Kaewtrakulpong K, Takigawa T, Koike M (2008) Truck allocation planning for cost reduction of mechanical sugarcane harvesting in Thailand: an application of multi-objective optimization. In: Operations research proceedings 2007, pp 331–336. Springer, Berlin

Office of the Cane and Sugar Board (2014) Report of plantation area in Thailand. Retrieved 8 Dec 2014. Website: http://www.ocsb.go.th/th/cms/ detail. php?ID = 923& SystemModuleKey = journal

Salassi ME, Breaux JB, Naquin CJ (2002) Modeling within-season sugarcane growth for optimal harvest system selection. Agric Syst 73(3):261–278

Singh G, Abeygoonawardana KAR (1982) Computer simulation of mechanical harvesting and transporting of sugarcane in Thailand. Agric Syst 8(2):105–114

Singh G, Pathak BK (1994) A decision support system for mechanical harvesting and transportation of sugarcane in Thailand. Comput Electron Agric 11(2):173–182

Territory Energy Management Performance Improvement

Gilles Dedeban and Paul-Eric Dossou

Abstract The lack of growth in Europe is one of the reasons behind the dire economic situation facing many European countries today. The question is how to address this situation at the same time meeting the expectations of both the general public and enterprises. European countries and enterprises have to resist the crisis and prepare for the future. They need to reorganize themselves in order to improve their performance. The research of ICAM (School of engineers) in industrial organization is destined to enterprises, public establishments and departments. It is based on GRAI Methodology, one of the three main methodologies for enterprise modeling. GRAIMOD is a software tool being developed in ICAM for supporting GRAI Methodology. The approach used is composed of the modeling of the existing system, the diagnosis, and the analysis of the obtained models for detecting inconsistencies and the design phase for improving the system. A general typology is proposed for enterprises and general public entities, then for facilitating the improvement during the design phase a reference model is proposed for each enterprise, public establishment, or department domain. The performance criteria used for improving countries and enterprises are quality (of products, system, process, supplying), cost, lead time, carbon management, social, societal, and environmental management (including energy management). According to the climate and energy package adopted in December 2008 by the European Parliament in the current context of energy transition, the department of Vendée recently passed a plan to improve energy self-sufficiency of its territories. Through modeling of consumption and potential local renewable energy production, this chapter offers several energy scenarios and assesses the efforts needed to achieve objectives. An illustration is given as an example in one of its urban communities, a territory of

G. Dedeban (✉)
ICAM VENDEE, La Roche-sur-Yon, France
e-mail: gilles.dedeban@icam.fr

P.-E. Dossou
ICAM Paris-Senart, Senart, France
e-mail: paul-eric.dossou@icam.fr

V. Kachitvichyanukul et al. (eds.), *Toward Sustainable Operations of Supply Chain and Logistics Systems*, EcoProduction,
DOI 10.1007/978-3-319-19006-8_28

30,000 inhabitants. This example is used for elaborating a reference model according to the concerned domain. The proximity of ICAM with enterprises and general public facilitates the elaboration of these different reference models and the validation of concepts elaborated.

Keywords Energy management · Sailboat and powerboat manufacturing · Energy reduction

1 Introduction

The study presented is a part of the Vendée Energies Nouvelles (VEN) Plan developed by the Vendee department. This aims to anticipate energy issues and seeing energy transition as an opportunity for promoting employment, boosting the local economy, and reducing energy dependence. To achieve the 50 % target electric range by 2025, two levers of action are required:

- control of energy demand: awareness on energy savings, use of alternative energy sources such as boilers—wood and taking into account the energy efficiency of buildings,
- local electricity production: facilities start or increase in the coming years in wind farms, solar fields, and biogas plants.

Actually, Vendee department produces only 10 % of his consumed electrical energy, by using renewable sources of energy. In France, between 4 and 5 % of the consumed electrical energy comes from a renewable source (Region of Pays de la Loire 2013).

In order to help local communities, an offshore wind project will be commissioned in 2021 and is expected to yield 25 % power autonomy. For the remaining 15 %, urban communities have an important role to play. They should be based on local and non-relocatable potential such as wind, solar, geothermal, and agricultural waste.

This study is the starting point for elaborating a reference model according to the GRAI Methodology.

2 Global Methodology

GRAI Methodology is one of the three main methodologies used for analyzing and designing enterprises and collectivities. The GRAI approach is composed of four phases: an initialization phase to start the study, a modeling phase where the existing system is described, an analysis phase to detect the inconsistencies of the studied system, and a design phase during which the inconsistencies detected are corrected and a new system is proposed. These concepts could be used to ensure the transformation of enterprises to meet real market needs (globalization, relocation, capacity to be proactive, cost optimization, lead time, quality, flexibility, etc.) and have to be adapted.

An enterprise or an organization is completely described according to GRAI Methodology by finding five models: functional (functions of the enterprise and their links), physical (the production system), informational (the net, tools, and informational flows), process (series of sequences or tasks), and decisional (structure of orders, hierarchic organization). Then these models could be improved for increasing enterprise performance.

GRAIMOD (Dossou and Mitchell 2013) is a new tool being developed by ICAM Engineer School for proposing concrete solutions to improve enterprises or organization according to new market evolutions. Nowadays, it contains five modules working around three submodules (Fig. 1).

This tool is composed of different modules such as GRAIXPERT, GRAIMANAGER, GRAIKERN, GRAITRANS, GRAIWORKER, GRAISUC, and GRAI_SSE. GRAI_SSE is the new module being developed specifically for integrating social, societal, and environmental dimensions in the improvement of enterprises and organization. It is composed of a submodule GRAICARB destined

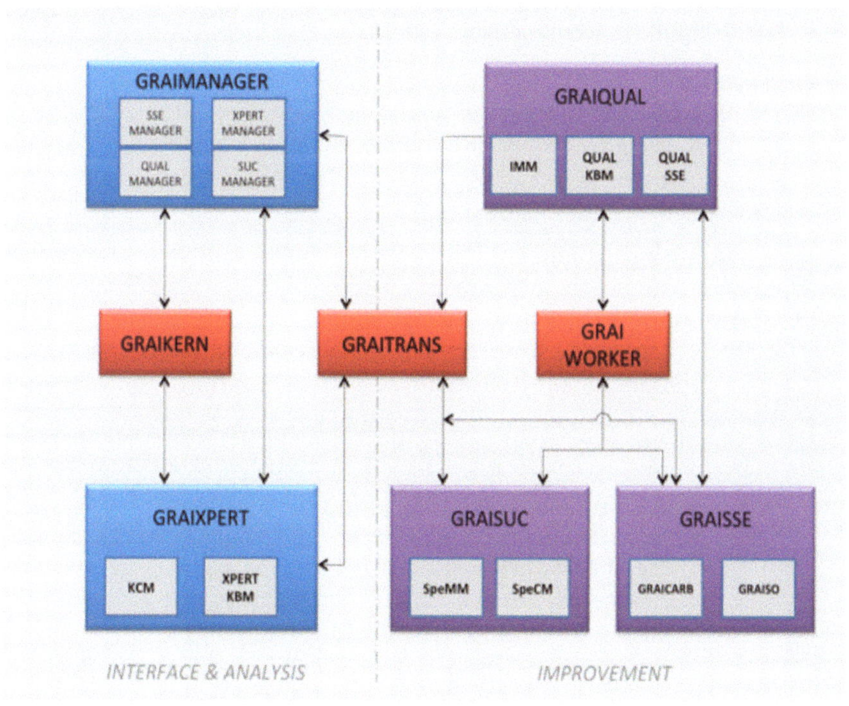

Fig. 1 Architecture of GRAIMOD

to manage carbon footprint and GRAI_SO being elaborated for improving the other aspects of environmental, social and societal dimensions, and energy management. These modules are detailed in Dossou and Mitchell (2013).

The objective was to integrate in the tool the best practices for improving enterprises or organizations. Reference models are being elaborated. This study is made in this goal.

3 Approach and Deliverables

The project is structured in several large stages according to the diagram given below (Fig. 2).

The final objective was to establish a dialogue and bring out projects at the local level.

Deliverables produced throughout the process are as follows:

1. Awareness media
   This presentation is the same for all communities. It is intended for:
   - introducing the topic of energy transition and increase awareness of the issues of energy,
   - explaining the framework and objectives of the mission, and
   - explaining the different possible actions.
2. Excel application
   This Excel application is a tool for formatting electricity production and power consumption data provided by ERDF (French Energy Company).

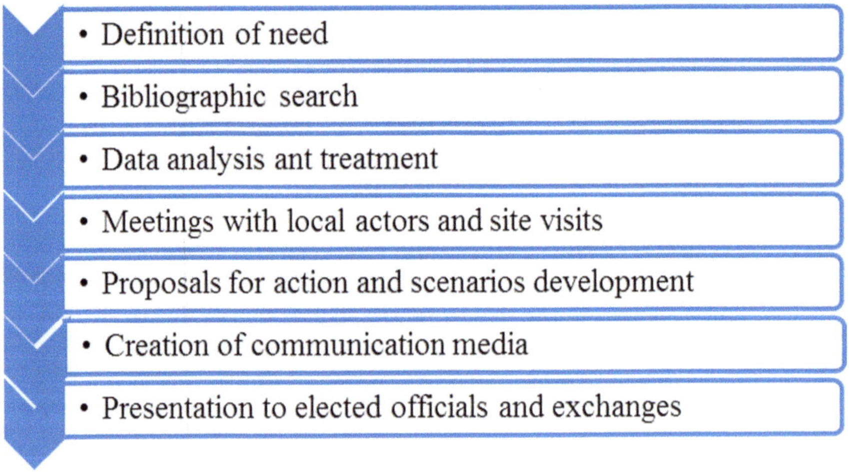

Fig. 2 Project stages

Socioeconomic data are similarly integrated there provided by Vendée Expansion.
From the database of local communities, the system produces raw communication documents.
This system allows now to ensure a follow-up of the evolution of data. Indeed, each year, data may be updated and the new consolidated reports are generated.

3. Territorial diagnostic energy report
The basis of this report was generated via the Excel application.
After data and bibliographic searches, several meetings with local actors, and site visits, we completed it with ideas of local actions in line with the identified potential and resources and we modeled three forward-looking scenarios according to the different actions.
4. Diagnostic presentation media
The first communication media for the "association of cities" is a slideshow, which incorporates the elements of the report.
Following are the objectives of the presentation with the officials:

- Make the data readable and exploitable.
- Encourage to invest in the Plan "Vendée Energies Nouvelles."

Finally, make a poster A3 to contain the set of important elements of territorial diagnosis (Fig. 3).

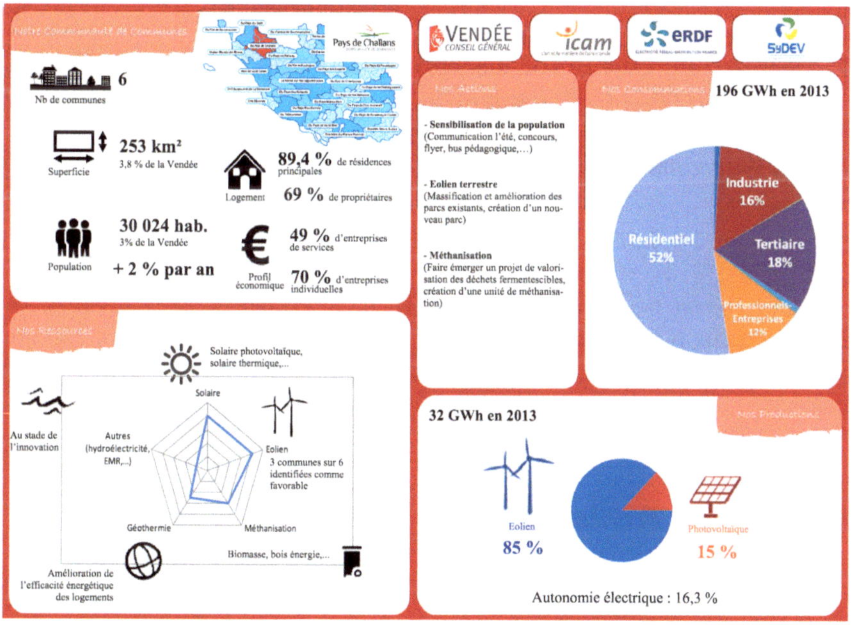

Fig. 3 Synthesis poster

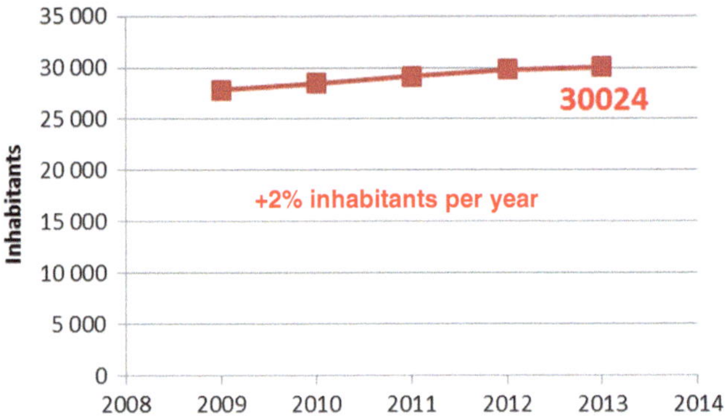

Fig. 4 Demographic evolution

This poster includes the following:

1. social and economical components of the territory (area, number of inhabitants, demographic evolution, rate of primary, secondary residences, and types of businesses),
2. evaluation of local energy resources,
3. current electrical consumption and local production, and
4. action plan to manage to achieve autonomy objectives.

4 Application: One of the Territories in French Department of Vendee (West of France)

4.1 Demographic Evolution

This represents an evolution of the population of 2 % per year, an increase of 543 inhabitants (Fig. 4).

In comparison, the French population evolves at the rate of 0.5 % per year. The territory density is 119 inhabitants per km^2. Below 150 inhabitants per km^2, the area can be classified as rural, although the population is unevenly distributed on the territory.

4.2 Economic Profile

The territory includes 2949 companies. The economic activity of the territory is characterized by many small-size companies for services (only 5 companies to more than 200 employees). However, there are a large number of shops. Industrial

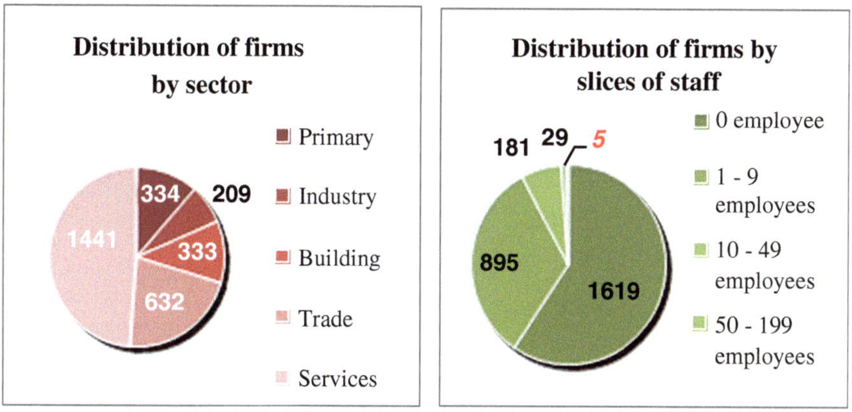

Fig. 5 Distribution of firms

companies are also well represented through 10 poles of activities present in the territory and proposing a dynamic economic environment (services, trade/handicraft, agriculture) (Fig. 5).

4.3 *Evolution of the Electric Consumption*

In 2013, electric consumption reached 196 GWh (Fig. 6).

It increased by 3.4 % per year over the last 5 years (French electric consumption increase was about 1.5 % per year in the same period). Variations on the displayed graph can be explained by the more or less severe winters (ERDF 2014).

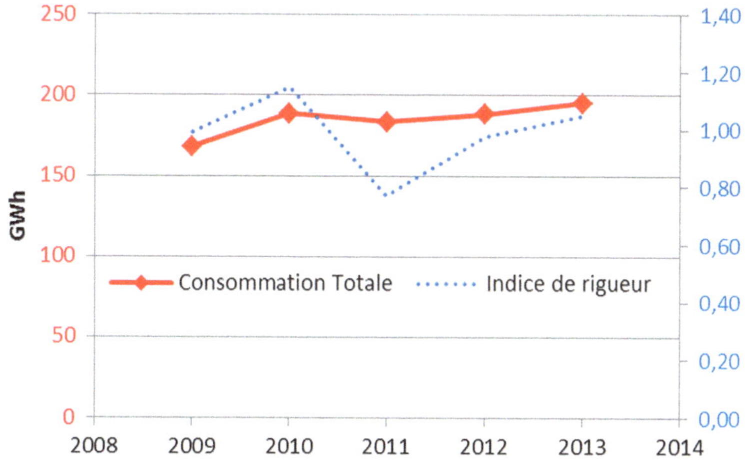

Fig. 6 Consumption evolution

We should also have in mind whether the study area is connected or not to the gas network in order to judge its power consumption profile.

In this case, the territory is well served in gas: 3 towns of 6 are connected to the gas network, which represents about 86 % of the population of the country.

4.4 Consumer Distribution

In the territory, the main sector of electricity consumption is residential to 52 % (Fig. 7).

The total power consumption per inhabitant is 6513 kWh in 2013.

Reduced to residential, this consumption amounted to 3418 kWh per inhabitant. Although it is always less than the French average that is 3569 kWh per inhabitant, there is an increase of 13 % over the last 5 years (Fig. 8).

Here are the five main activity sectors for electricity consumption in 2013 (Fig. 9).

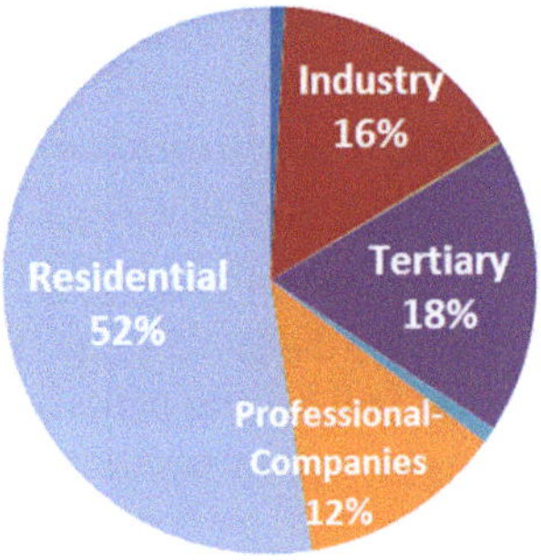

Fig. 7 Residential consumption 1

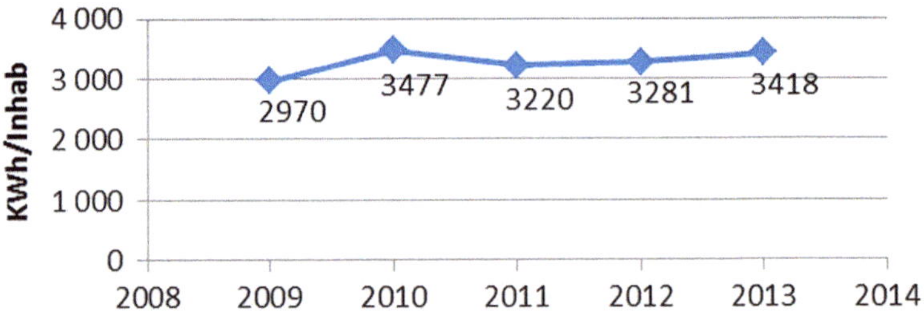

Fig. 8 Residential consumption 2

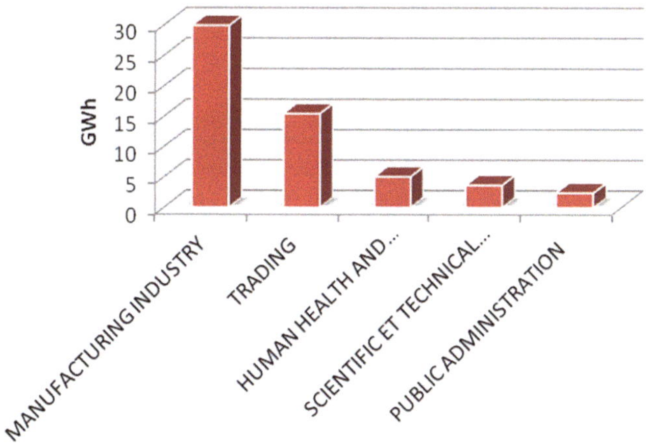

Fig. 9 The most consuming sectors

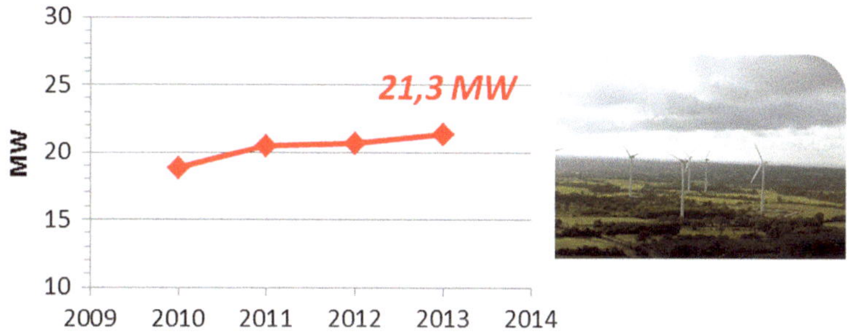

Fig. 10 Local electric production

4.5 Evolution of the Electric Production

Installed electrical power by the land was from 21,332 kW in 2013, distributed as shown in the chart below (Figs. 10 and 11).

4.6 Electric Autonomy

The same year, the country produces 32 GWh and consumes 196 GWh.

Energy independence thus amounts to 16 %. All of the electricity comes from renewable sources of energy (wind, photovoltaic). This situation places the territory as advance to the objective of the department.

Fig. 11 Installed electric power

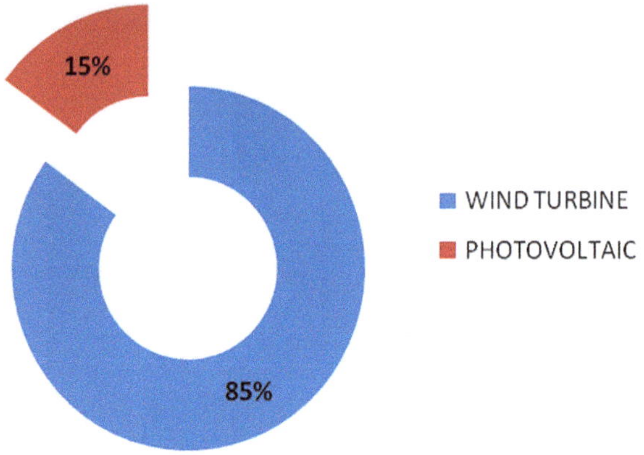

Indeed, the department wants to achieve a degree of power autonomy of 50 % by 2025. The objective will be achieved by an annual increase of production of 2.2 % (taking into account the growth of the current energy consumption).

4.7 Resources and Potential

Analysis of resources and potentials allows us to stock the following:

Sunshine:
The territory has a rather favorable photovoltaic potential (1260 kWh/m^2: comply with departmental and national average).

Wind:
3 towns of 6 are located in favorable areas according to the regional wind power scheme.

Biomass:
Deposits of liquid manure from farms of cattle and poultry and food waste (Table 1).

Table 1 Biomass estimated potential

Material	Potential methane (m^3 of CH_4/tonne of raw material)	Potential estimated of the territory
Cattle manure	20	3,997,296 m^3
Cattle manure	40	5,329,728 m^3
Green waste	125	500,000 m^3
Slaughterhouse grease	180	4 slaughterhouses

Geothermal energy:
The geographical area is potentially favorable for vertical geothermal.

The most favorable solution is to create an "energy mix" efficient and sustainable which takes into account complementarities of these energy sources in the territory and to ensure the development of renewable energy.

4.8 Local Actions Proposals

Here are three ideas for local actions proposed in city areas. These ideas are from the analysis of the diagnosis and the potential of the territory. Other actions are of course possible.

1. Public Awareness

The awareness aims to disseminate a common culture to the people to change behaviors and daily habits. Communication with locals helps make visible energy and energy transition that are not concepts mastered by all.

The Plan "Vendée Energies Nouvelles" requires all stakeholders to unite around the issue of the energy transition (Soleil 2007a, b). This is also an opportunity for the territory of communicating the current actions.

Here are several examples of implementation of awareness:

- Dissemination actions in municipal magazines,
- Vendée Electric Tower
  - Guide Distribution practice of good energy practices
  - Days "In town without my car"
  - Awareness of school through fun activities.

The territory is characterized by a dynamic trading sector. Therefore, one could imagine communication campaigns and games on the energy transition or other topics in supermarkets and places of high traffic (city center, commercial area).

Sensitization may also boost local employment by directing individuals to professionals and craftsmen of the area.

The various existing financing related to energy renovation can encourage individuals to citizen initiatives. The impact of awareness is measured on long-term energy independence.

2. Earth Wind Turbine

The regional wind scheme identified 3 towns favorable to the installation of wind farms in the territory.

There is already in the territory a 18 MW wind farm which consists of 9 wind turbines of 2 MW each which today accounts for almost 85 % of electricity production in the territory.

The extension of the park with the addition of seven new wind turbines is scheduled for 2016.

The impact is straightforward since the addition of a park directly drives the increase in electricity production, and therefore, autonomy.

For example:
The new fleet of 14 MW (7 wind turbines) will bring in 13.6 % more electric range under current conditions.

(a) Methane

- Methanation adds value to the fermentable waste planning. Waste can come from:
  - agriculture
  - food industry
  - collective canteen
  - waste disposal (green): 4000 t/year estimated here.

It is important to know the status of potential biogas. These studies can be carried out by the Chamber of Agriculture of the Vendée or other organizations.

For example:
3000 tons/year of waste are enough to power a biogas plant 75 kWel that can earn 0.25 % electric range under current conditions.

4.9 Prospective Scenarios

The current electric range of the territory is 16.3 %. However, to increase the electric range by 2025, the territory will have to make significant efforts in terms of control of water consumption and production of electricity.

Currently, there is a wind farm consisting of 8 turbines producing 28 GWh per year. It contributes significantly to the electric range. Further, a photovoltaic power is also in operation and production is 4 GWh annually.

The various scenarios below include the start of offshore wind farm of two islands from 2021.

Scenario 1: Consumption is under control, but production remains at current levels

⇒ **Insufficient electric autonomy** (Fig. 12).

Consumption: It increases of 3.4 %/year, reaching 196 GWh in 2013.

Consumption is controlled (green curve) through awareness on energy conservation, energy efficiency of buildings, and alternative energy such as wood energy and solar thermal panels.

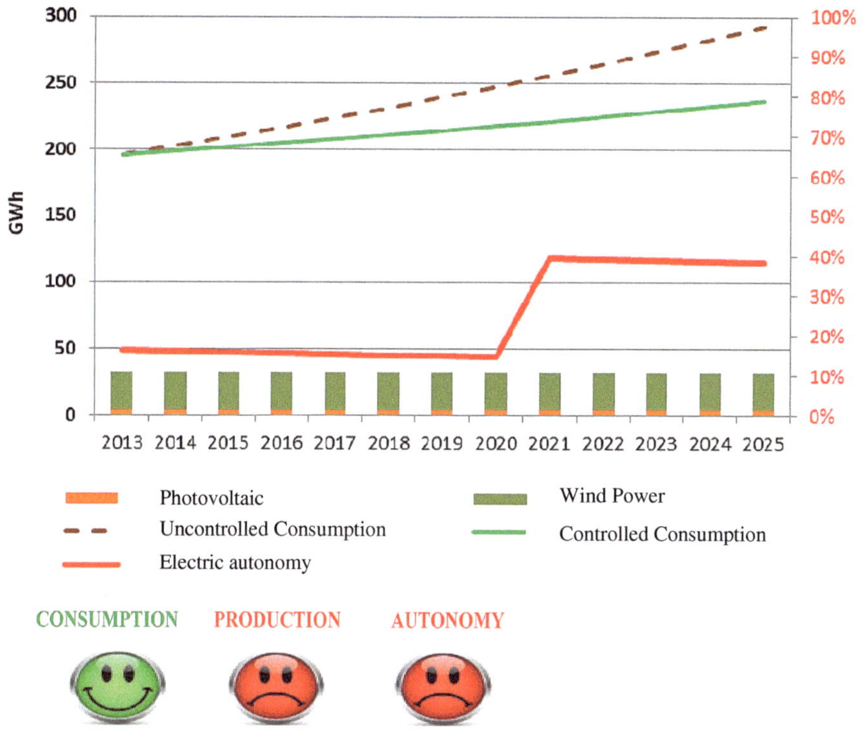

Fig. 12 Electrical scenario 1

This reduces almost 55 GWh of electricity consumption in 2025.

Production: It is 32 GWh in 2013.

In this scenario, there is no additional installed capacity and no new system is put in place.

Electrical life: It includes the contribution of offshore wind farm and its 25 % additional power autonomy.

With the commissioning of the offshore wind farm in 2021, autonomy should only reach 38.5 % in 2025.

Scenario 2: Consumption is not controlled, but renewable energy is developed

⇒ **Insufficient electric autonomy** (Fig. 13).

Consumption: In this scenario, consumption is not controlled (brown line), the current trend is continuing.

Production: To increase the production of green electricity, new facilities are in place.

- Methanation unit: three units starting in 2016, 2019, and 2022 for a unit producing 0.53 GWh (75 kW operating 7000 h per year)

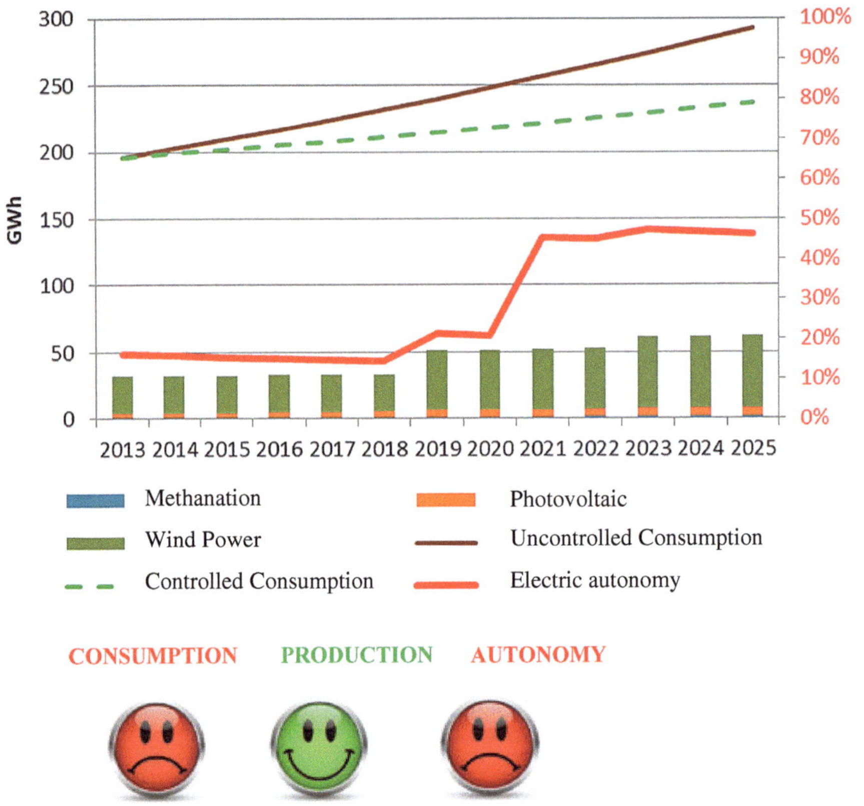

Fig. 13 Electrical scenario 2

Wind farm: start of a park in 2016 for a production of 16.8 GWh (4.2 MW wind turbines operated 2100 h per year),

- Wind farm: extension of a park in 2023 for a production of 8.4 GWh (2.2 MW wind turbines running 2100 h per year), and
- Photovoltaic: 1 house of 8 will be equipped with rooftop PV 2025 (1575 houses × 40 m^2). This corresponds to a production of 3.2 GWh. It is divided over 10 years, from 2016 to 2025 (+0.32 GWh per year).

Electrical autonomy: It includes the contribution of offshore wind farm and its 25 % additional power autonomy. It was 16.3 % in 2013. Despite large investments in electricity production, consumption increases too much and does not meet the energy needs.

The electric range should only reach 46.2 % in 2025 with the offshore fleet.

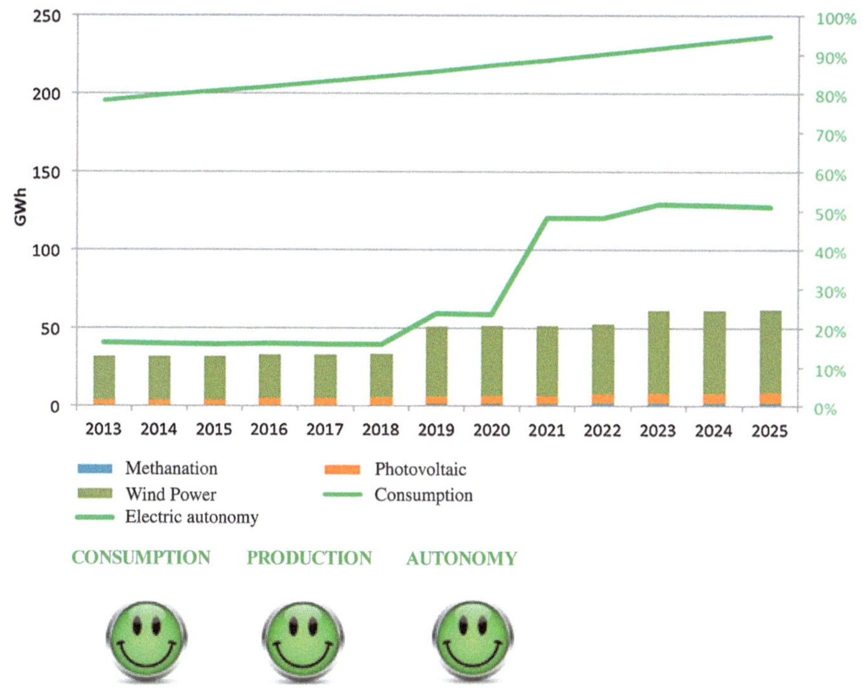

Fig. 14 Electrical scenario 3

Scenario 3: Consumption is controlled and developed renewable energy production

⇒ **Sufficient electrical autonomy** (Fig. 14).

Consommation: Same as scenario 1.
Production: Same as scenario 2.
Electrical autonomy: Through the efforts of the territory (in terms of energy efficiency, alternative energy, and awareness to energy saving) and the new offshore wind farm, the electric range is expected to reach 51.1 % in 2025.

It is important to act on two parameters to achieve the set goals: to control the consumption and production of electrical energy.

5 Conclusion

The presentation of this work with the elected officials has generated great interest and has given rise to the ideas and the following questions: *What about financial aid for individuals? What about concrete actions? What are objectives for reducing consumption in short term? What are the human resources of anaerobic digestion? How the photovoltaic recycling is?*

Different topics could be studied and developed for continuing this work:

- Make a study of the anaerobic digestion,
- Geothermal,
- Photovoltaic,
- Awareness on consumption, and
- Alternative energy.

Some ideas have to be improved: solar farm, methanation, development of wind power, development of photovoltaic, additional wind farm, and geothermal on the aquatic center.

Other improvements are also defined: prioritize the issues, ROI study on geothermal, and link with OPAH (Programmed operation Housing Improvement).

This first step of information and stimulation is performed, and the department of Vendée wishes to support the territory in second phase of feasibility study of the proposed actions.

For ICAM, this application gives a concrete case and data for elaborating a reference model for GRAIMOD by using GRAI methodology.

References

Dossou PE, Mitchell P (2013) Elaboration of reference models for improving enterprise performance. In: Advances in sustainable and competitive manufacturing systems. Lecture notes in mechanical engineering. Springer, Berlin, pp 899–910

ERDF (2014) Vendée. Data of electricity consumption and generation of electricity (years 2009–2013)

Soleil A (2007a) Alliance Soleil. Study of the Vendée potential and possible applications of solar thermal and photovoltaic

Soleil A (2007b) Alliance Soleil. Study of the Vendée potential and possible applications of geothermal and aerothermal

Region of Pays de la Loire (2013) Region of Pays de la Loire, Regional Scheme for Wind Power

http://agreste.agriculture.gouv.fr/recensement-agricole-2010/

www.vendee.fr

http://www.vendee-expansion.fr/

www.sydev.fr

http://www.developpement-durable.gouv.fr/-La-transition-energetique-pour-la-.html

http://www.ademe.fr/

http://www.lesenergiesnouvelles.com/

http://www.chambres-agriculture.fr/

http://www.agribiomethane.fr/

An Assessment of Customer Contentment for Ready-to-Drink Tea Flavor Notes Using Artificial Neural Networks

Athakorn Kengpol and Worrapon Wangkananon

Abstract Ready-to-drink (RTD) tea achieves estimated sales of 6.7 billion in 2012 (according to Chicago-based market) (Zegler in Tea and RTD Teas—US—Chicago-based. Mintel Consumer Market ResearchReport, 2013). The consumption of RTD tea is increasing rapidly in the USA (Cernivec in Food Chem 136, 1309–1315, 2015) and also in Thailand (Katenil, Bottled Beverage Market in Thailand, 2014), and the growth is projected to increase steadily for the next five years. The taste of beverage is the key success for achieving customer loyalty in which the flavor impact plays a crucial role to the taste. Traditional flavors such as lemon, peach, raspberry, citrus, and plain tea have survived the test of time; however, many beverage companies are seeking alternative flavors that are not typically associated with tea, such as pineapple, apple, mint, strawberry, chocolate, and herbal ingredients. In general, flavor consists of many compounds that make the odor notes. The chapter reports a study to assess which compound affects customer contentment for RTD tea flavor. The study selects the jasmine, lemon, peach, citrus, and plain tea flavors that are most wildly used in this market segment. In order to identify the hidden pattern of the customer's contentment, the artificial neural networks (ANNs) have been applied to classify the key volatile compound of the flavors. According to the input data of the 4 customer groups and 5 key volatile compounds (5 flavors) as the output, the results show that the best structure of ANNs is 4-7-5 with 1.54^{e-2} MSE and it can predict 75.5 % of accuracy. The compounds that carry the most effect on customer contentment are women–adult is jasmine; women–teen is lemon; men–teen is citrus; and men–adult is plain tea.

Keywords Ready-to-drink tea · Flavor notes · Artificial neural networks (ANNs)

A. Kengpol (✉) · W. Wangkananon
Advanced Industrial Engineering Management Systems Research Center, Department of Industrial Engineering, Faculty of Engineering, King Mongkut's University of Technology, North Bangkok 10800, Thailand
e-mail: athakorn@kmutnb.ac.th

V. Kachitvichyanukul et al. (eds.), *Toward Sustainable Operations of Supply Chain and Logistics Systems*, EcoProduction,
DOI 10.1007/978-3-319-19006-8_29

1 Introduction

Tea and ready-to-drink (RTD) tea achieved appraised sales of $6.7 billion in 2012, according to Chicago-based market research (Zegler 2013). The sales are expected to increase to nearly $8 billion by 2014. In order to support the increased interest of tea, beverage makers are producing a variety of tea drinks using different types of tea, such as black, green, oolong, and white. Consumer favorites for tea are becoming increasingly varied. The Tea Council in the UK (Khokhar and Magnusdottir 2002) estimates that there are now more than 1,500 varieties of the Camellia sinensis tea plant; nevertheless, black and green tea varieties stay the most popular. The market expects these varietals to continue to govern the segment for the next five years. White tea's flavor, however, seems to be dropping in the USS.

In the last few years, companies have been asking for a more authentic tasting tea. For iced tea drinks, stronger tea taste can become a higher preference for beverage producers in order to equip a point of difference between brands, to meet consumers' desires for tea products with a truer taste, and to obtain higher levels of the healthful ingredients found in tea, which is the main reason that many consumers drink tea.

Popular fruit flavors and a greater number of sweeteners also complement the smooth and mild taste of black tea. Traditional flavors such as lemon, peach, raspberry, citrus, and plain tea have remained alive through the test of time (Yu et al. 2014); however, many beverage companies are seeking alternative flavors that are not regularly associated with tea, such as pineapple, apple, mint, strawberry, chocolate, and herbal ingredients. The taste of the beverage is one of the key successes for achieving customer loyalty. However, the flavoring drink requires an expert to determine the type and amount of flavor to garnish the drink each species or individual scent. In addition, the flavor must be suitable for the consumer; for instance, young customers may enjoy the intense fruit flavors, while older people may like the smell of natural light. The odor language is a language that is difficult to describe. The key issue is that materials developed smells that are composed of many kinds of volatile flavor varieties. It is difficult to choose a scent to suit the product development. Thus, it is important to classify as volatile influencing satisfaction of each consumer group.

The objective of this study, therefore, is to assess which compound, which affects customer contentment for RTD tea flavor notes, would be successful in the market. In order to identify the hidden pattern of the customer's needs, the artificial neural networks (ANNs) have been applied to classify the key volatile compound of the flavors. In this study, the RTD tea segment is selected as a case study due to market trend. The 4 customer groups and 5 compounds of flavor are used as a data.

2 Literatures Review

In this part, we present 2 main ideas consisting of flavoring and ANNs that are related to the study.

2.1 Flavoring

Flavor (USA) or flavour (UK) is the sensory impression of a food or other substance and is defined mainly by the chemical senses of taste and smell (Ho et al. 2008). The "trigeminal senses," which detect chemical irritants in the mouth and throat as well as temperature and texture, are also very important to the overall Gestalt of flavor perception. The flavor of the beverage, as such, can be altered with natural or artificial flavorants, which influence these senses. Although the terms "*flavoring*" or "*flavorant*" in common language denote the combined chemical sensations of taste and smell, the same terms are usually used in the fragrance and flavors industry to refer to edible chemicals and extracts that alter the flavor of food and beverage through the sense of smell. Most artificial flavors are specific and often complex mixtures of singular, naturally occurring flavor compounds combined together to either mimic or improve a natural flavor. These mixtures are formulated by flavorists to give a food product a unique flavor and to maintain flavor consistency between different product batches or after formula changes. The list of known flavoring agents includes thousands of volatile compounds, and the flavor chemist (flavorist) can often mix these together to generate many of the common flavors. Many flavorants consist of esters, which are often described as being "sweet" or "fruity" (Hui et al. 2004). According to Table 1 (Maarse 1991) are illustrated some of the chemical compounds that produce odor.

The compounds used to generate artificial flavors are almost identical to those that occur naturally. It has been suggested that artificial flavors may be safer to consume than natural flavors due to the standards of purity and mixture consistency that are compelled either by the company or by law. Natural flavors in distinction may contain impurities from their sources, while artificial flavors are typically more pure and are required to tolerate more testing before being sold for consumption. Flavors from food products are usually the result of a blending of natural flavors, which set up the basic smell profile of a food product, while artificial flavors modify

Table 1 Some of the chemical compounds that produce odor

Chemical	Odor
Diacetyl	Buttery
Isoamyl acetate	Banana
Benzaldehyde	Bitter almond
Cinnamaldehyde	Cinnamon
Ethyl propionate	Fruity
Methyl anthranilate	Grape
Limonene	Orange
Ethyl decadienoate	Pear
Allyl hexanoate	Pineapple
Ethyl maltol	Sugar, cotton candy
Ethyl vanillin	Vanilla
Methyl salicylate	Wintergreen

the smell to accent it. Unlike smelling, which occurs upon inhalation, the sensing of flavors in the mouth occurs in the exhalation phase of breathing and is apperceived differently by an individual. In other words, the smell of food is different depending on when you smell it in front of you or whether it has already come in your mouth.

The flavorings related to previous study have illustrated the following. Auvray and Spence (2008) have reviewed the concept of perception of food (Flavor), which concluded that the smell of food affects the sense of touch in many aspects such as taste, smell sensory face, and feel. Sorenson et al. (2009) have proposed the development of functional beverage products. It is recommended that you create value by adding ingredients that help consumers feel healthier with fiber, such as collagen. In addition, the flavors of the drink are important factors that must be designed to meet the target audience. Harke et al. (2009) have studied the preferences of consumers in each group with the sweetness and flavor of the kiwi by a change in the method of cultivation. The results allow consumers to eat kiwi fruit which was satisfied with the different taste.

2.2 Artificial Neural Networks; ANNs

For the classification problems, there are many classification algorithms techniques that must be properly used. The examples of linearity problems are Fisher's linear discriminant, linear regression, naive Bayes classifier, support vector machines, etc. For nonlinearity problems, there are perceptron, kernel estimation, ANNs, etc. ANNs are algorithms that can be used to perform nonlinear statistical modeling and provide a new alternative to logistic regression, the most commonly used method for developing predictive models for dichotomous outcomes in medicine. Neural networks offer a number of advantages, including requiring less formal statistical training, ability to implicitly detect complex nonlinear relationships between dependent and independent variables, ability to detect all possible interactions between predictor variables, and the availability of multiple training algorithms. Disadvantages include its "black box" nature, greater computational burden, proneness to over fitting, and the empirical nature of model development (Vineis and Rainoldi 1997).

Artificial neural networks (ANNs) are a family of statistical learning algorithms infused by biological neural networks (the central nervous systems of animals, in particular the brain) and are used to estimate or approximate functions that can depend on a large number of inputs and are mainly unknown (Galushkin 2007). ANNs are generally presented as systems of interconnected "neurons" which can calculate values from inputs, and are capable of machine learning as well as pattern recognition thanks to their adaptive nature. For example, a neural network for handwriting recognition is identified by a set of input neurons which may be initiated by the pixels of an input image. After being weighted and transformed by a function (determined by the network's designer), the activations of these neurons are then passed on to other neurons. This process is repeated until finished: An output neuron is activated. This determines which character was read. Like other machine learning methods—systems that learn from data—neural networks have been used to resolve a wide variety of tasks that are hard to solve using ordinary rule-based programming, including computer vision and speech recognition.

A multilayer perceptron (MLP) is a feed-forward artificial neural network model that maps sets of input data onto a set of suitable outputs (Patterson 1998). A MLP consists of multiple layers of nodes in a directed graph, with each layer fully connected to the next one. Except for the input nodes, each node is a neuron (or processing element) with a nonlinear activation function. MLP employs a supervised learning technique called backpropagation for training the network. MLP is a modification of the standard linear perceptron and can differentiate data that are not linearly separable.

If a MLP has a linear activation function in all neurons, that is, a linear function that maps the weighted inputs to the output of each neuron, then it is easily justified with linear algebra that any number of layers can be eliminated to the standard two-layer input–output model (see perceptron). What makes a MLP different is that some neurons use a *nonlinear* activation function which was improved to model the frequency of action potentials, or firing, of biological neurons in the brain. This function is modeled in several ways.

The two main activation functions used in current applications are both sigmoids and are illuminated by

$$y(wi) = \tanh(wi)\,\text{and}\,y(wi) = (1 + e - wi) - 1 \quad (1)$$

in which the former function is a hyperbolic tangent which ranges from −1 to 1 and the latter, the logistic function, is similar in shape but ranges from 0 to 1. Here $\boldsymbol{y(w_i)}$ is the output of the $\boldsymbol{i}$th node (neuron) and $\boldsymbol{w_i}$ is the weighted sum of the input synapses. Alternative activation functions have been offered, including the rectifier and softplus functions. More specialized activation functions include radial basis functions which are used in another class of supervised neural network models.

A sample of MLP architecture (3 nodes in input layer, 8 nodes in hidden layer, 1 node in output layer) is demonstrated in Fig. 1. The MLP consists of three or more layers (an input and an output layer with one or more *hidden layers*) of nonlinearly activating nodes and is thus considered a deep neural network. Each node in one layer connects with a certain weight $\boldsymbol{w_{ij}}$ to every node in the following layer. Some people do not contain the input layer when counting the number of layers, and there is contradiction about whether w_{ij} should be interpreted as the weight from $\boldsymbol{i}$ to $\boldsymbol{j}$ or the other way around.

Learning appears in the perceptron by modifying connection weights after each piece of data is processed, based on the amount of error in the output compared to the expected result (mean square error; MSE). This is an example of supervised learning and is carried out through backpropagation, a generalization of the least mean squares algorithm in the linear perceptron. We illustrated the error in output node j in the $\boldsymbol{n}$th data point (training example) by

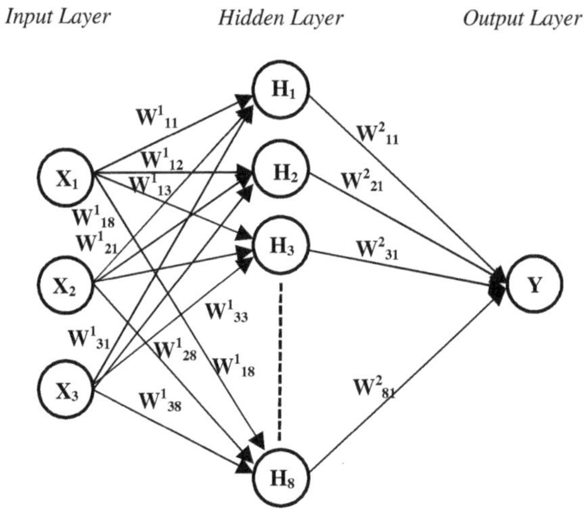

Fig. 1 Multilayer perceptron ANN architecture (3-8-1). Adapted from Galushkin (2007)

$$ej(n) = dj(n) - yj(n) \tag{2}$$

where $\boldsymbol{d}$ is the target value and $\boldsymbol{y}$ is the value generated by the perceptron. We then make corrections to the weights of the nodes based on those corrections which minimize the error in the entire output, given by

$$\sum(n) = 1/N \sum ej2(n) \tag{3}$$

Using a first-order optimization algorithm (gradient descent), we find our modification in each weight to be

$$Wji(n) = -l\eth \sum(n) yj(n)/\eth Wj(n) \tag{4}$$

where $\boldsymbol{y_j}$ is the output of the previous neuron and $\boldsymbol{l}$ is the *learning rate*, which is carefully selected to ensure that the weights converge to a reaction fast enough without producing oscillations. In programming applications, this parameter typically ranges from 0.2 to 0.8.

The derivative to be calculated depends on the influenced local field $\boldsymbol{w_j}$, in which it varies. In order to prove that for an output node, this derivative can be simplified to

$$\eth \sum(n)/\eth wj(n) = ej(n)\o(wj(n)) \tag{5}$$

where $\boldsymbol{\o}$ is the derivative of the activation function illustrated above, which itself does not vary. The analysis is heavier for the change in weights to a hidden node, but it can be shown that the relevant derivative is

$$-\eth\sum(n)/\eth wj(n) = \emptyset(wj(n))/\eth wj(n)\sum\left(-\eth\sum(n)wkj(n)\right) \quad (6)$$

This depends on the change in weights of the ***k***th nodes, which explain the output layer. In order to change the hidden layer weights, we must first change the output layer weights according to the derivative of the activation function, and so this algorithm represents a *backpropagation (BP) of the activation function*.

The ANNs related to previous study have illustrated the following. Boccorh and Paterson (2002) developed an artificial neural network for predicting flavor intensity using gas chromatography analysis (gas chromatography; GC) for black currant concentrates from three different seasonal cultivation. Cultivated area and extraction technology between estimated concentrations of aromas help manufacturers juice to determine the amount of flavor that is appropriate. Kengpol and Wangananon (2006) have developed an expert system to choose fragrance to suit groups of people. The application of ANN-type network includes multilayer perceptron (MLP) and training the networks with backpropagation (BP) to classify the fragrance notes. The smell was associated with satisfaction with the fragrance of consumer groups. The results of the expert system allow vendors to choose a perfume fragrance to suit a group of consumers. Also, Hanafizadeh et al. (2010) have developed an expert system to select a fragrance to suit the consumer as well. It has improved the techniques used in the teaching system. Using the technique of fuzzy Delphi with learning spread back (BP with fuzzy Delphi), the data analysis becomes faster and more precise. Ceballos-Magaña et al. (2013) classified the machine drink tech Avila (Tequila) into 4 categories using the results from gas chromatography analysis with Metz Spec Nitro Dmitry (gas chromatography—mass spectrometry; GC-MS). GC-MS was used to predict the type of beverage and used two types of comparison techniques including dyslexia discriminant out (discriminant analysis) technique with ANNs with multiple network layers. The results show that neural network techniques predict more accurately the type of drink.

3 Research Methodology

3.1 Conceptual Framework

The methodology to explore customer contentment on flavor notes by classifying the flavor notes to the key volatile compound is shown below.

Figure 2 shows the flavor notes in the middle of picture that each of the flavors (Formulas) included 5 volatile compounds in the right (with different proportions in Table 2). The customer groups in the left can select the contentment formula that has difference compounds. This model looks like networks that have complex

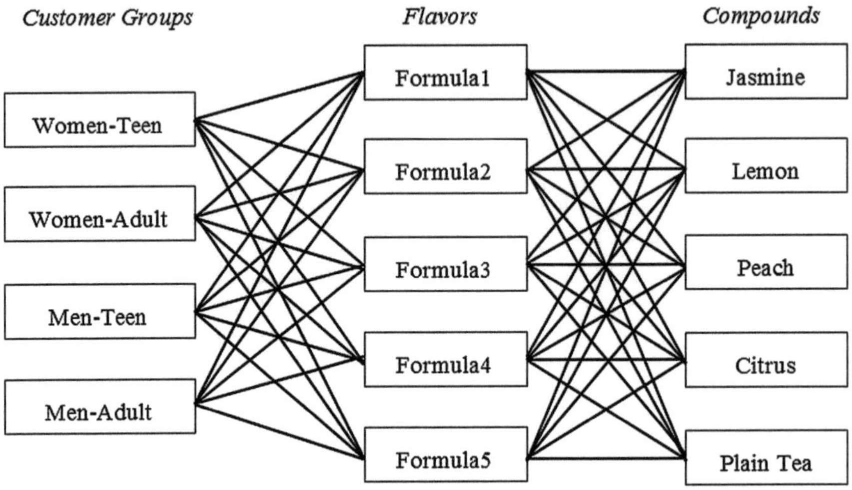

Fig. 2 Conceptual framework

Table 2 Chemical compounds of simulated data in percentage

Volatile compounds	Odors	Formula 1	Formula 2	Formula 3	Formula 4	Formula 5
2-Acetyl-1-pyrroline	Jasmine	0.1	0.6	0.1	0.1	0.1
Limonene	Lemon	0.6	0.1	0.1	0.1	0.1
Gamma-decalactone	Peach	0.1	0.1	0.6	0.1	0.1
Citral	Citrus	0.1	0.1	0.1	0.6	0.1
Cis-3-hexanol	Plain tea	0.1	0.1	0.1	0.1	0.6

Remarks
Formula 1 contains lemon 60 %, jasmine 10 %, peach 10 %, citrus 10 %, and plain tea 10 %
Formula 2 contains jasmine 60 %, lemon 10 %, peach 10 %, citrus 10 %, and plain tea 10 %
Formula 3 contains peach 60 %, lemon 10 %, jasmine 10 %, citrus 10 %, and plain tea 10 %
Formula 4 contains citrus 60 %, lemon 10 %, jasmine 10 %, peach 10 %, and plain tea 10 %
Formula 5 contains plain tea 60 %, lemon 10 %, jasmine 10 %, peach 10 %, and citrus 10 %

relationship so we apply the ANNs to classify the volatile compounds that are suitable for each of the customer groups.

3.2 Methodology

For this paper, the conceptual modeling has only been proposed. Therefore, the data of the flavor's compound and the customer's contentment are gathered from only

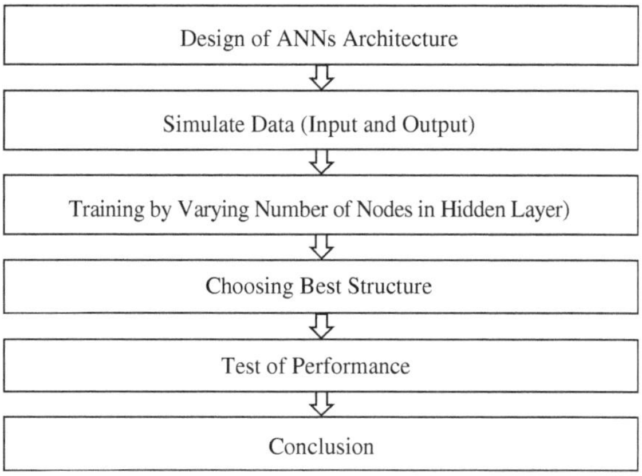

Fig. 3 Research methodology

one expert in the field. Firstly, the ANN architecture as multilayer feed-forward network has 4 nodes (customer group) at the input layer and 5 nodes (flavor compounds) at the output layer. In the case of training, BP training algorithms applied with MATLAB 2103R software. Then, the performance by the value of MSE is compared for choosing the best number of hidden layers (McCleskey et al. 2003). Finally, the model is tested by the prediction of the contentment of each customer group.

Figure 3 demonstrates the process chart of methodology of this study, beginning with the design of ANN structure, simulated data, training ANNs, choosing best structure, test of performance, and conclusion.

- Design of ANN structure; based on multilayer feed-forward network with 4-H-5 structure, by 4 nodes of customer group at the input layer, 5 nodes of flavor compounds at the output layer, and H nodes of hidden layer selected with minimum MSE by varying the number of nodes from 1 to 20.
- Simulated data; gathered from the expert by fixing the value of customer contentment as shown in Table 2.
- Training ANNs; by varying the number of nodes from 1 to 20 nodes.
- Choosing best structure; by selecting the number of nodes in hidden layer with minimum MSE.
- Test performance; by predicting the value of each customer's contentment and comparing with the data that expert fixed value at the beginning.
- Conclusion; which compound that affect each of the customer's contentment.

3.3 Input Data

Simple structure of flavor notes by 5 volatile compounds has designed and varies the percentage of compound concentration that is shown in Table 2.

Mixtures of odors are created to make the real like RTD tea flavors that have many compounds (odors) in one formulation; nevertheless, it should have only one odor that has strong impact in contrast with customers' contentment.

The 4 customer groups are created: women–teen, women–adult, men–teen, and men–adult. When the data have input, the binary value 0/1 is used in the input nodes and the concentrations of each formulation are used in the output nodes as shown in Fig. 4 (sample of data women–adult selected Formula 4). The value of customer contentment is shown in Table 3.

The data in Table 3 show the contentment of each customer group weight by the expert such as women–teen like Formula 2 for 80 % which means they like Formula 2 more than any other formulas (for 100 peoples of women–teen, 80 peoples like Formula 2). The trained data have fixed the amount of most customers' contentment between 60 and 80 % by the experience of the expert. So this table is used to test the performance of the model in the final conclusion.

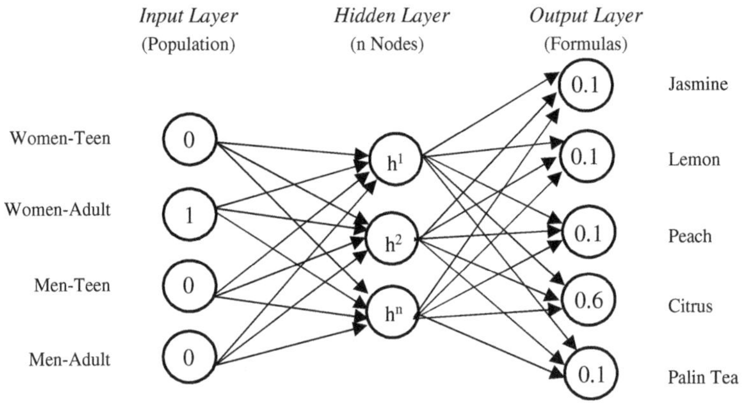

Fig. 4 Sample of data input to ANNs (women–adult—Formula 4)

Table 3 Customer contentment (gathered from only one expert in the field)

Contentment	Women–teen (%)	Women–adult (%)	Men–teen (%)	Men–adult (%)
Formula 1	5	75	5	20
Formula 2	80	10	10	5
Formula 3	5	5	10	5
Formula 4	5	5	70	10
Formula 5	5	5	5	60

4 Result

4.1 Training Results

For training data in ANNs, the number of nodes in a hidden layer starting from 1 to 15 versus the MSE value is shown in Fig. 5. The result shows that 7 nodes in the hidden layer have a minimum MSE value (1.54^{e-2}) so we select the 7 nodes as the best number of nodes in the hidden layer. Then, we used 4-7-5 ANN architecture to classify the flavor notes based on customer contentment.

According to Fig. 5, first node has highest value of MSE (3.52^{e-2}), then suddenly decreases to 1.77^{e-2} at 4th node, and rises again until 7th node that has minimum MSE value (1.54^{e-2}), and after that, it is slightly increased. The dash line at 7 nodes shows the minimum of MSE value.

After the 4-7-5 ANN architecture was applied, the model was running again. Figure 6 shows the MSE decreasing and convergence near to zero MSE. According to Fig. 6, at the beginning, the values of MSE are highest 0.55, then suddenly decrease to 0.054 at the second epochs, and slightly decreasing until the 5th epochs where the program was terminated because the MSE value was convergence. The MSE can be computed by Eq. (3).

4.2 Test of Performance

For the test of model performance, the data in the Table 3 are used to compare with the prediction of the model shown in Table 4.

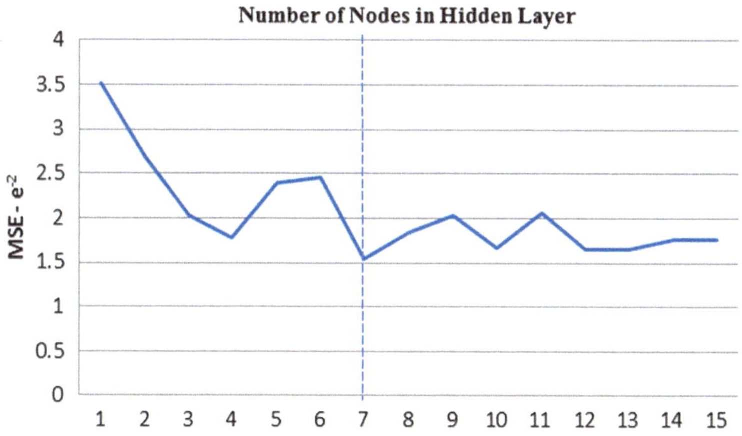

Fig. 5 MSE value versus number of nodes in hidden layer

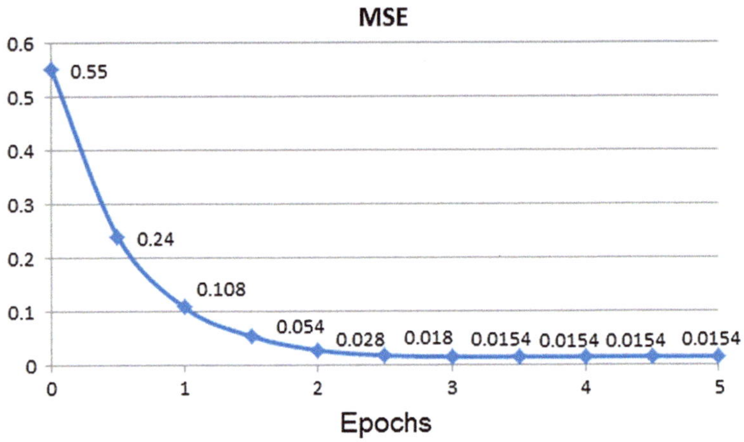

Fig. 6 MSE versus epochs

From Table 4, the total average of percentage difference between data and prediction of trained model was 0.245 (24.5 %). The prediction of compounds that has the most effect on customer contentment are:

Women–adult: Jasmine is the highest by 50.58 %, next are lemon, citrus, and plain tea; and peach is the lowest.

Women–teen: Lemon is the highest by 51.18 %; next are jasmine, plain tea, and peach; and citrus is the lowest.

Men-teen: Citrus is the highest by 48.73 %; next are jasmine, lemon, and peach; and plain tea is the lowest.

Men–adult: Plain tea is the highest by 47.5 %; next are jasmine and citrus; and lemon and peach are the lowest.

For Table 5, we forecast the unknown data or do not train directly where only women, only men, only teen, only adult, and overall. The forecasted result showed percent of each flavor notes that are suitable for each customer such as only

Table 4 Customer contentment (trained data)

Women–Teen			Women–Adult			Men–Teen			Men–Adult		
F.1	Predict	%D	F.2	Predict	%D	F.4	Predict	%D	F.5	Predict	%D
0.1	0.1294	0.294	0.6	**0.5058**	0.157	0.1	0.1352	0.352	0.1	0.1347	0.347
0.6	**0.5118**	0.147	0.1	0.1362	0.362	0.1	0.1282	0.282	0.1	0.1278	0.278
0.1	0.1221	0.221	0.1	0.1072	0.072	0.1	0.1282	0.282	0.1	0.1278	0.278
0.1	0.1074	0.074	0.1	0.1362	0.362	0.6	**0.4873**	0.1878	0.1	0.1347	0.347
0.1	0.1294	0.294	0.1	0.1145	0.145	0.1	0.1211	0.211	0.6	**0.475**	0.2083
Average		0.206			0.2196			0.2629			0.2916
Total average = 0.245 from [(0.206 + 0.2196 + 0.2629 + 0.2916)/4]											

Table 5 Customer contentment (forecast unknown data)

Odor	Only women	Only men	Only teen	Only adult	Overall
Jasmine	0.5442	−0.1144	0.4044	−0.0185	0.4258
Lemon	0.0564	0.3273	0.2224	0.4991	0.3179
Peach	0.2647	0.0739	0.3932	0.2503	0.3502
Citrus	0.1736	0.5514	**0.5311**	0.2756	0.4415
Plain tea	0.0503	0.2522	0.2186	0.1028	−0.2313

teen age. If we want to develop the RTD tea for only teen customers, we should put the citrus flavor up to 53.11 % (from Table 5, only teen in 4th row); on the other hand, we should not put the jasmine flavor in the formula.

5 Conclusions

The objective of this research is to assess which compound affects customer contentment on flavor notes of RTD tea. In order to identify the hidden pattern of the customer's needs, the ANNs have been applied to classify the key volatile compound of the flavors. The architecture of developed ANNs is multilayer feed-forward network with 4-7-5 structure, by 4 nodes of customer group at the input layer, 5 nodes of flavor compounds at the output layer, and 7 nodes of hidden layer selected with minimum MSE by varying the number of nodes from 1 to 20. In the training, we used BP training algorithm. The results of this structure show that the MSE was 1.54^{e-2}, and it could predict 75.5 % (24.5 % errors) of accuracy. The compounds that most affect customer contentment are as follows: women–adult is jasmine; women–teen is lemon; men–teen is citrus; and men–adult is plain tea.

The advantages of ANNs are high-accuracy prediction. The weaknesses of ANNs are time-consuming to achieve the best structures, and the solution may be local optimum; therefore, we should combine the optimization technique, for example, genetic algorithm (GA) or particle swarm optimization (PSO), to solve this problem.

References

Auvray M, Spence C (2008) The multisensory perception of flavor. Conscious Cognit 17(3):1016–1031

Boccorh RK, Paterson A (2002) An artificial neural network model for predicting flavour intensity in blackcurrant concentrates. Food Qual Prefer 13(2):117–128

Ceballos-Magaña SG, de Pablos F, Jurado JM, Martín MJ, Alcázar Á, Muñiz-Valencia R, Izquierdo-Hornillos R (2013) Characterisation of tequila according to their major volatile composition using multilayer perceptron neural networks. Food Chem 136(3):1309–1315

Cernivec S (2015) Trends in ready-to-drink tea. Dairy Food membrane technology forum, June 3-5 2015, Minneapolis, MN
Galushkin AI (2007) Neural networks theory. Springer Science+Business Media, Berlin
Hanafizadeh P, Ravasan AZ, Khaki HR (2010) An expert system for perfume selection using artificial neural network. Expert Syst Appl 37(12):8879–8887
Harker FR, Carr BT, Lenjo M, MacRae EA, Wismer WV, Marsh KB, Pereira RB (2009) Consumer liking for kiwifruit flavour: A meta-analysis of five studies on fruit quality. Food Qual Prefer 20(1):30–41
Ho CT, Lin JK, Shahidi F (eds) (2008) Tea and tea products: chemistry and health-promoting properties. CRC press, Boca Raton
Hui YH, Meunier-Goddik L, Josephsen J, Nip WK, Stanfield PS (eds) (2004) Handbook of food and beverage fermentation technology, vol 134. CRC Press, Boca Raton
Katenil N (2014) Bottled Beverage Market in Thailand. Thailand Food Market Report, May 2014, Food Intelligence Center, Food Institute
Kengpol A, Wangananon W (2006) The expert system for assessing customer satisfaction on fragrance notes: Using artificial neural networks. Comput Ind Eng 51(4):567–584
Khokhar S, Magnusdottir SGM (2002) Total phenol, catechin, and caffeine contents of teas commonly consumed in the United Kingdom. J Agric Food Chem 50(3):565–570
Maarse H (1991) Volatile compounds in foods and beverages, vol 44. CRC Press, Boca Raton
McCleskey SC, Floriano PN, Wiskur SL, Anslyn EV, McDevitt JT (2003) Citrate and calcium determination in flavored vodkas using artificial neural networks. Tetrahedron 59(50):10089–10092
Patterson DW (1998) Artificial neural networks: theory and applications. Prentice Hall PTR, New Jersey
Sorenson D, Bogue J, Paquin P (2009) Consumer-oriented development of functional beverages. In: Functional and speciality beverage technology, vol 200. Woodhead Publishing Limited/ CRC Press, Cambridge/UK, pp 421–450
Vineis P, Rainoldi A (1997) Neural networks and logistic regression: Analysis of a case-control study on myocardial infarction. J Clin Epidemiol 50(11):1309–1310
Yu P, Yeo ASL, Low MY, Zhou W (2014) Identifying key non-volatile compounds in ready-to-drink green tea and their impact on taste profile. Food Chem 155:9–16
Zegler J (2013) Tea and RTD Teas—US—Chicago-based. Mintel Consumer Market Research Report, July 2013

Particle Swarm Optimization with Multiple Learning Terms for Storage Location Assignment Problems Considering Three-Axis Traveling Distance

Warisa Wisittipanich and Pongsakorn Meesuk

Abstract This chapter presents an approach based on particle swarm optimization (PSO) for minimizing total traveling distance in warehouse storage location assignment problems (SLAP). The traveling distance in an order-picking process is considered with three-axis traveling distance: two horizontal axes and one vertical axis. A mathematical model is first presented, and LINGO optimization program is used to find optimal solutions for a set of generated problems. As the problem size increases, LINGO could not find solutions within reasonable time. Thus, particle swarm optimization (PSO) is applied to solve SLAP. The proposed algorithm employs multiple learning terms and utilizes the random key representation to generate a solution. The numerical experiments show that the proposed PSO is able to generate good solutions with relatively shorter computing time.

Keywords Particle swarm optimization · Warehouse storage · Storage location assignment

1 Introduction

Typical operations in warehouse management are comprised of receiving, storing, picking, and delivering. Warehouse management problem and, more specifically, storage location assignment problem (SLAP) is a critical subject in warehouse

W. Wisittipanich (✉) · P. Meesuk
Department of Industrial Engineering, Faculty of Engineering, Chiang Mai University, 239 Huay Kaew Road, Muang District, Chiang Mai 50200, Thailand
e-mail: warisa@eng.cmu.ac.th

P. Meesuk
e-mail: enjoyneering@hotmail.com

V. Kachitvichyanukul et al. (eds.), *Toward Sustainable Operations of Supply Chain and Logistics Systems*, EcoProduction,
DOI 10.1007/978-3-319-19006-8_30

operations because it is highly resource-consuming. Storage location assignment problem (SLAP) is a problem of assigning products to storage locations that aim to satisfy one or more objectives, i.e., space utilization, total transfer time, and total transportation distance.

The storage management system can be classified into three main policies: dedicated storage policy, random storage policy, and class-based storage policy. In dedicated storage policy, all items of a given type are stored in a particular area. In random storage policy, each stock-keeping unit (SKU) can be stored at any storage location. In class-based storage policy, a group of storage location is allocated to a class of SKUs and random storage is allowed within the group of storage locations.

Many researchers have proposed several solution techniques to solve SLAP. van den Berg and Zijm (1999) presented the hierarchy of decision problem for storage systems in the stage of designing, planning, and controlling. Roodbergen and Koster (2001) proposed heuristic methods for solving order-picking routing problem in warehouses where two or more aisles exist and random storage is used. Peterson and Aase (2004) applied simulation techniques to compare the principles for picking, storage, and routing in manual order-picking system. Hsu et al. (2005) presented a batching approach based on genetic algorithm (GA), which directly minimizes the total travel distance. Chen and He (2008) presented a mathematical model for warehouse assignment optimization with strategies for automatically storage systems and implemented particle swarm optimization with Pareto concept to overcome big-size warehouse assignment problems. Muppani and Adil (2008a) studied the integer programing of a storage system with class-based storage policy and developed the simulated annealing (SA) algorithm to solve storage assignment and form the classes. Then, Muppani and Adil (2008b) proposed a nonlinear integer programing for class-based storage system considering area decrease, handling costs, and storage area cost and used the brand and bound (B&B) algorithm for solving the developed nonlinear model. Sooksaksun et al. (2012) proposed a particle swarm optimization algorithm for class-based storage policy to determining the aisle layout and dimension while simultaneously assigning shelf spaces for storing the items based on item classes in order to minimize total distance in warehouse.

This study extends the work of Kasemset and Meesuk (2014) in which the mathematical model for storage location assignment for a particular warehouse layout considering three-axis travel distance has been proposed. To solve large-scale problem, this paper presents an implementation of particle swarm optimization (PSO) with multiple learning term to solve storage location assignment problems (SLAP). The remainder of this paper is organized as follows. The problem description and model specification of SLAP are given in Sect. 2. Section 3 describes the proposed PSO algorithm and its application to the problem. Experimental results are reported in Sect. 4. Finally, conclusion and further research are provided in Sect. 5.

2 Problem Description

The problem of storage location assignment is to assign each product to be stored at the appropriate storage location in order to obtain certain objectives subjected to the constraints. In this paper, the objective of storage location assignment is to minimize total traveling distance along three-axis traveling distance: two horizontal axes and one vertical axis.

As mentioned earlier, the warehouse layout used in this study is based on the work of Kasemset and Meesuk (2014). The assumptions of the layout are as follows: (1) Warehouse layout is assumed to be symmetric, (2) the input/output (I/O) point is located at one corner of the warehouse, (3) the number of storage blocks is limited, and (4) one storage block can be assigned for one product only. Figure 1 illustrates an example of a warehouse layout with four columns, two racks, three levels, and two rows, seen from the top view and the side view of storage rack.

The notation and variable used in this model are listed as follows:

D_x	Distance from I/O point to origin point along x-axis
D_y	Distance from I/O point to origin point along y-axis
W_a	Width of aisle
W_r	Width of storage row (equal to two times of storage block width)
D_{ijkr}	Total traveling distance of three axes
X	Total traveling distance along x-axis
Y	Total traveling distance along y-axis
Z	Total traveling distance along z-axis
L_s	Length of storage block
H	Height of storage block

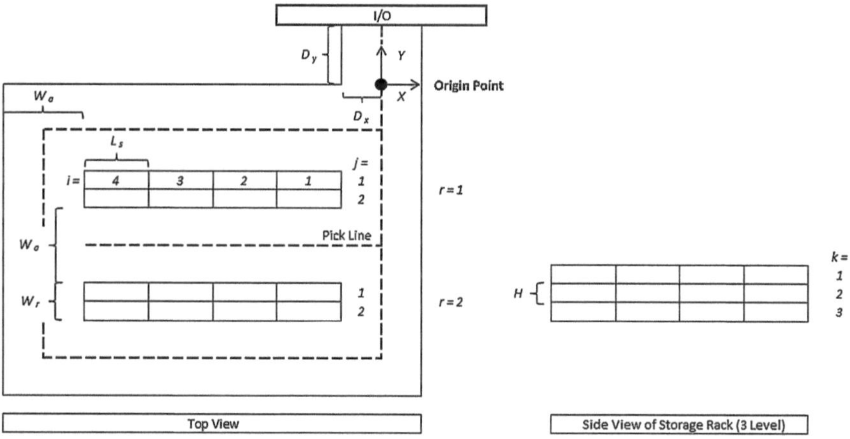

Fig. 1 A warehouse layout with four columns, two racks, three levels, and two rows

T_p Number of picking product p (p = 1, 2, 3,…, n)
S_p Number of storage block required for product p
I Order of storage block position i (i = 1, 2, 3, …, m)
J Order of storage rack j (j = 1, 2, 3, …, q)
K Order of level k (k = 1, 2, 3, …, k)
R Order of row r (r = 1, 2, 3, …, r)

$$X_{pijkr} = \begin{cases} 1, & \text{if product } p \text{ is assigned to storage position } i \text{ rack } j \text{ level } k \text{ row } r \\ 0, & \text{otherwise} \end{cases}$$

The mathematical model of the problem is formulated as follows:
Minimization of total traveling distance:

$$f_2 = \sum_{p=1}^{n} \sum_{i=1}^{m} \sum_{j=1}^{q} \sum_{k=1}^{k} \sum_{r=1}^{r} \left(\frac{Tp}{Sp}\right) X_{pijkr} D_{ijkr} \tag{1}$$

Subjected to constraints:

$$D_{ijkr} = X + Y + Z \tag{2}$$

$$X = D_x + (I - 0.5)L_s \tag{3}$$

$$Y = D_y + (0.5J\,W_a) + [(J - 1)(W_r + 0.5W_a)] + [(W_r + W_a)(R - 1)] \tag{4}$$

$$Z = H(K - 1) \tag{5}$$

$$\sum_{p=1}^{n} X_{pijkr} \leq 1 \quad \forall i, j, k, r \tag{6}$$

$$\sum_{i=1}^{m} \sum_{j=1}^{q} \sum_{k=1}^{k} \sum_{r=1}^{r} X_{pijkr} = S_p \quad \forall p \tag{7}$$

Equation (1) expresses the objective function of the model and minimization of the total traveling distance. It is important to note that traveling distances are measured along the aisle centerline and centerline of storage block, and horizontal traveling distance considered by the picker moved along the aisle floor and vertical distance considered by the height of rack shelf. Equations (2), (3), (4), and (5) explain the calculation of three-axis total traveling distance. Equation (6) guarantees that no more than one product is assigned to one storage unit. Equation (7) is to assure that each product stored in the storage units as equal to the number of storage needed.

3 Implementation of PSO to SLAP

3.1 Standard PSO

Particle swarm optimization (PSO), originally proposed by Kennedy and Eberhart in 1995, is a population-based random search approach that simulates the behavior of bird flocking or fish schooling. In the original model, each individual particle in a swarm learns and adapts its search behavior based on its own experience (cognitive term) and global experience (social term) during the search.

The standard PSO is formulated as Eqs. (8) and (9).

$$\omega_{id}(t+1) = \omega_{id}(t) + c_p u\left(\psi_{id}^p - q_{id}(t)\right) + c_g u\left(\psi_{id}^g - q_{id}(t)\right) \tag{8}$$

$$q_{id}(t+1) = q_{id}(t) + \omega_{id}(t) \tag{9}$$

q_{id} current position of dth dimension of ith particle
ω_{id} velocity of dth dimension of ith particle
ψ_{id}^p personal best position of dth dimension of ith particle
ψ_{id}^g global best position of dth dimension of ith particle
c_p weight of personal best position term
c_g weight of global best position term
u uniform random number in range [0, 1]

3.2 GLNPSO

One major drawback of the standard PSO is that the swarm tends to convergence prematurely. A swarm is frequently trapped in a local optimization and can no longer move. Several approaches have been proposed to deal with this problem. One of the approaches, introduced by Pongchairerks and Kachitvichyanukul (2005, 2009), utilizes multiple social learning terms and extends the concept of the standard PSO in GLNPSO. Instead of using only global best particle as reference, GLNPSO also incorporates the local best and near neighbor best as additional social learning reference terms. In GLNPSO, position and velocity in the dth dimension of particle i are updated by the following equations.

$$\begin{aligned}\omega_{id}(t+1) = w\omega_{id}(t) + c_p u\left(\psi_{id}^p - q_{id}(t)\right) + c_g u\left(\psi_{id}^g - q_{id}(t)\right) \\ + c_l u\left(\psi_{id}^l - q_{id}(t)\right) + c_n u\left(\psi_{id}^u - q_{id}(t)\right)\end{aligned} \tag{10}$$

$$q_{id}(t+1) = q_{id}(t) + \omega_{id}(t) \tag{11}$$

In this formula, w, introduced by Shi and Eberhart (1998), represents inertia weight to improve the search ability. The value of w is linearly decreasing so that the swarm can search the whole space aggressively at the early stage and gradually reduce the search space at the later stage. $\psi^q_d, \psi^l_{id}, \psi^n_{id}$ represent the dth dimension of the personal best, the global best, the local best, and the near neighbor best, respectively, and $q_{id}(t)$ is the current position of particle i at tth generation. The local best position of particle i, $\Psi^l_i = \{\psi^l_{i0}, \psi^l_{i1}, \ldots, \psi^l_{iD}\}$, can be determined as the best particle among K neighboring particles. It is equivalent to dividing the whole swarm into multiple sub-swarms with population of size K, and the local best particle is the best particle among the K neighbors, i.e., the particle gives the best fitness value in the sub-swarms.

The near neighbor best position of particle i, $\Psi^n_i = \{\psi^n_{i0}, \psi^n_{i1}, \ldots, \psi^n_{iD}\}$, represents the interaction between particles to achieve a better value of the objective function. The neighborhood is determined by the direct linear distance from the best particle. Each element of ψ^n_i is determined by Eq. (12).

$$\mathrm{FDR} = \frac{\mathrm{Fitness}(\Theta_\mathrm{i}) - \mathrm{Fitness}(\Psi_\mathrm{i})}{q_{id} - \psi_{id}} \tag{12}$$

The equations for updating position and inertia weight w in GLNPSO are the same as standard PSO formulas. Other parameters and working procedures are also similar to those in standard PSO.

3.3 Solution Representation

In this study, the number of particle dimension is set to be equal to the total number of storage blocks. Consider an example of a warehouse with two columns (i = 1, 2), two racks (j = 1, 2), two levels (k = 1, 2), and two rows (r = 1, 2). The warehouse stores three product types; A, B, and C. The value of Tp and Sp value of each product type are shown in Table 1.

The number of vector dimension is set to be equal to the number of storage units which is 16. Figure 2 illustrates a random key representation encoding scheme where each value in a vector dimension is initially generated with a uniform random number in a range [0, 1].

Table 1 Data information for product type A, B, and C

	Tp	*Sp*	*Tp*/*Sp*
Product type A	15	5	3.00
Product type B	18	3	6.00
Product type C	20	4	5.00

Dimension d	1	2	3	4	5	6	7	8	9	10	11	12	13	14	15	16
	0.29	0.04	0.56	0.73	0.71	0.94	0.60	0.38	0.63	0.73	0.68	0.42	0.44	0.24	0.75	0.90

Fig. 2 Encoding scheme for SLAP

Dimension d	2	14	1	8	12	13	3	7	9	11	5	4	10	15	16	6
	0.04	0.24	0.29	0.38	0.42	0.44	0.56	0.60	0.63	0.68	0.71	0.73	0.73	0.75	0.90	0.94
Product Type	B	B	B	C	C	C	C	A	A	A	A	A	-	-	-	-

Dimension d	1	2	3	4	5	6	7	8	9	10	11	12	13	14	15	16
	0.29	0.04	0.56	0.73	0.71	0.94	0.60	0.38	0.63	0.73	0.68	0.42	0.44	0.24	0.75	0.90
Product Type	B	B	C	A	A	-	A	C	A	-	A	C	C	B	-	-
Storage location	1	2	3	4	5	6	7	8	9	10	11	12	13	14	15	16

Fig. 3 Decoding scheme for SLAP

Next, the permutation of n-repetition of n jobs (Bierwirth 1995) with a sorting list rule is used in this study to assign each product unit to a storage location. In Table 1, product type B has the maximum value of movement ration (Tp/Sp), followed by product type C and A, respectively. Therefore, all product units of type B are first allocated to the dimension with sorted values, and then, product units of type C and A are allocated correspondingly. It is important to note that since the number of storage blocks is 16 and the number of total product units is 12, there are four storage blocks that are not assigned to store any product. This procedure results in completed storage location assignment as shown in Fig. 3.

4 Numerical Experiments

The experiments are implemented using the C# language of the Microsoft Visual Studio 9.0. The program runs on the platform of Intel®Core™ 2 Dou CPU 1.67 GHz with 3062 MRAM. The GLNPSO parameters used in this experiment are shown in Table 2.

Table 2 GLNPSO parameters

Parameter	Value
Inertia weight	Linearly decreases from 0.9 to 0.4
Constant acceleration	$c_p, c_g, c_l, c_n = 1.$
Swarm size	50 particles
Number of iterations	100

Table 3 Comparison of total traveling distances on a set of generated instances

Instance	Problem size	LINGO	Time (s)	GLNPSO	Time (s)
WH1	3 × 38 × 48	588.46	4	588.46	6.8
WH2	60 × 259 × 300	47,273.04	8	47,273.04	9.7
WH3	135 × 840 × 1000	112,705.19	240	112,705.19	16.0
WH4	270 × 1675 × 2000	204,283.57	1200	204,283.57	35.7
WH5	500 × 2825 × 3360	290,062.88	3600	290,062.88	70.4
WH6	500 × 2825 × 4992	–	–	269,784.52	300.9
WH7	500 × 2825 × 6300	–	–	260,944.12	590.6

The performance of GLNPSO is evaluated using a set of seven generated instances. Each instance is generated randomly based on a real-case warehouse and characterized by problem size: (number of product types) × (total number of products) × (total number of storage locations). Table 3 shows the experimental results of the traveling distance obtained by GLNPSO and those obtained by LINGO. It is noted that the best solution obtained by GLNPSO for each instance is obtained from five independent runs.

It can be easily seen from Table 3 that the GLNPSO shows its effectiveness in solving SLAP especially in the large-size problems. In small-size problem, GLNPSO is able to find optimal solutions equal to those obtained by LINGO. While solutions from LINGO are guaranteed to be optimal, when the problem size increases, computing time of LINGO increases rapidly, and eventually, the solution could not be found when the problem size becomes very large. On the other hand, for large-size problems, GLNPSO is able to provide good solution with relatively faster computing time.

5 Conclusions

This paper presents an implementation of particle swarm optimization with multiple learning terms, named as GLNPSO, for solving storage location assignment problem (SLAP). To encode and decode solutions, random key representation and permutation of n-job repetition are adopted to assign products to storage locations. The performance of GLNPSO is evaluated on a set of generated instances and compared with results obtained from LINGO optimization program. The experimental results show that GLNPSO can be used as an efficient alternative approach for solving SLAP as it yields optimal solutions in small-size problems and provides competitive solution quality, especially in the large-size problems, with fast computing time. The ongoing researches have been under investigation to improve algorithm performance and robustness and extend its applications for a wider range of combinatorial NP-hard problems.

References

Bierwirth C (1995) A generalized permutation approach to job shop scheduling with genetic algorithms. Oper Res Spectr 17(2–3):87–92

Chen Y, He F (2008) Research on particle swarm optimization in location assignment optimization. In: 7th World Congress on Intelligent Control and Automation, 2008. WCICA 2008, (pp 111–116). IEEE, New York

Hsu CM, Chen KY, Chen MC (2005) Batching orders in warehouses by minimizing travel distance with genetic algorithms. Comput Ind 56(2):169–178

Kasemset C, Meesuk P (2014) Storage location assignment considering three-axis traveling distance: a mathematical model. In: Logistics operations, supply chain management and sustainability. Springer International Publishing, Berlin, pp 499–506

Kennedy J, Eberhart RC (1995) Particle swarm optimization. In: Proceedings of IEEE International Conference on Neural Networks (Perth, Australia), IEEE Service Center, Piscataway, NJ, vol IV, pp 1942–1948

Muppani VR, Adil GK (2008a) Efficient formation of storage classes for warehouse storage location assignment: a simulated annealing approach. Omega 36(4):609–618

Muppani VR, Adil GK (2008b) A branch and bound algorithm for class based storage location assignment. Eur J Oper Res 189(2):492–507

Petersen CG, Aase G (2004) A comparison of picking, storage, and routing policies in manual order picking. Int J Prod Econ 92(1):11–19

Pongchairerks P, Kachitvichyanukul V (2005) Non-homogenous particle swarm optimization with multiple social structures

Pongchairerks P, Kachitvichyanukul V (2009) Particle swarm optimization algorithm with multiple social learning structures. Int J Oper Res 6(2):176–194

Roodbergen KJ, Koster R (2001) Routing methods for warehouses with multiple cross aisles. Int J Prod Res 39(9):1865–1883

Shi Y, Eberhart R (1998) A modified particle swarm optimizer. In: Evolutionary Computation Proceedings, 1998. IEEE World Congress on Computational Intelligence. The 1998 IEEE International Conference on, pp 69–73. IEEE, New York

Sooksaksun N, Kachitvichyanukul V, Gong DC (2012) A class–based storage warehouse design using a particle swarm optimisation algorithm. Int J Oper Res 13(2):219–237

van den Berg JP, Zijm WHM (1999) Models for warehouse management: Classification and examples. Int J Prod Econ 59(1):519–528

A Particle Swarm Optimization Approach for Solar Cell Industry Multi-stage Closed-Loop Supply Chain Model

Allen Wang, Li-Chih Wang and Yi-Wen Chen

Abstract This chapter presents the study of an integrated forward and reverse closed-loop supply chain (CLSC) network design problem with sustainable concerns in the solar energy industry. The focuses are in the logistics flows, capacity expansion, and technology investments of existing and potential facilities in the multi-stage CLSC. A deterministic multi-objective mixed integer programming model is formulated to capture the tradeoffs between the total cost and the carbon dioxide (CO_2) emission and to tackle the multi-stage CLSC design problem from both economic and environmental perspectives. Due to the multi-objective nature and the computational complexity, a multi-objective particle swarm optimization (MOPSO) with novel flow assignment algorithms is designed to search for non-dominated/Pareto CLSC design solutions. Finally, a case study of crystalline solar energy industry is illustrated to verify the proposed multi-objective CLSC design model and demonstrate the efficiency of the developed MOPSO algorithm in terms of computational time and solution quality.

Keywords Sustainable · Supply chain design · Multi-objective particle swarm optimization · Solar Energy Industry

A. Wang · L.-C. Wang (✉)
Department of Industrial Engineering and Enterprise information, Tunghai University, Taichung 40704, Taiwan, R.O.C
e-mail: wanglc@thu.edu.tw

Y.-W. Chen
China Medical University Hospital, 3D Printing Medical Research Center, Taichung, Taiwan, R.O.C

V. Kachitvichyanukul et al. (eds.), *Toward Sustainable Operations of Supply Chain and Logistics Systems*, EcoProduction,
DOI 10.1007/978-3-319-19006-8_31

1 Introduction

In past two decades, consumers are becoming more aware of the environmental and social implications of their day-to-day consumer decisions and are therefore beginning to make purchasing decisions related to their environmental and ethical concerns. To achieve a sustainable supply chain with low carbon, 3E (Effective, Efficient, Environmental; 3E), enterprises need to provide effective energy-saving and carbon-reduction means to meet the policies of Carbon Right and Carbon Trade, so that the waste reduction and recycling issues have been raised up. The new concept of end-of-life (EOL) product recycling, disassembling, repairing, and refurbishing is in vogue, which can be defined as a closed-loop supply chain (CLSC) for sustainable enterprises. In this scenario, forward and reverse logistics (FL/RL) have to be considered simultaneously in the network design of entire supply chain. Özceylan and Paksoy (2013) proposed a new mixed integer mathematical model for a CLSC network that includes FL/RL with multi-periods and multi-parts. Wang et al. (2011) studied a supply chain network design problem with environmental concerns. Fahimnia et al. (2009) found that variations in cost and environmental impacts occur over ranges of carbon pricing and, furthermore, built the foundation for optimization of carbon in CLSC environment.

This article particularly will build the CLSC model and solution suggestions based on particle swarm optimization (PSO) algorithm, a heuristic global optimization method. PSO was proposed by Eberhart and Kennedy (1995), and they claimed each particle movement is guided by their own best known position in the search space as well as the entire swarm's best known position. Many studies have used PSO to solve the NP-hard linear programming problems.

2 Multi-stage Closed-Loop Supply Chain

2.1 Problem Statement

This investigation assumes manufacturers will recycle, reuse, and refurbish the EOL products though the RC process. The EOL product collection includes product recycling from customers or used product markets. Any EOL products which cannot be used or recycled will leave the supply chain via disposal process. Thus, a multi-stage and multi-factory closed-loop supply chain structure will be formed as in Fig. 1.

Figure 1 includes the existing and potential components of forward logistics (FL) and reverse logistics (RL). Decision-makers have to determine the potential location and quantity of recycling units and, furthermore, design the capacity based on recycle quantity from customers. The production unit capacities may expand or shrink due to uncertain material supply, customer demand, and recycling. Reverse logistics and capacity increasing result in more carbon emission in closed-loop

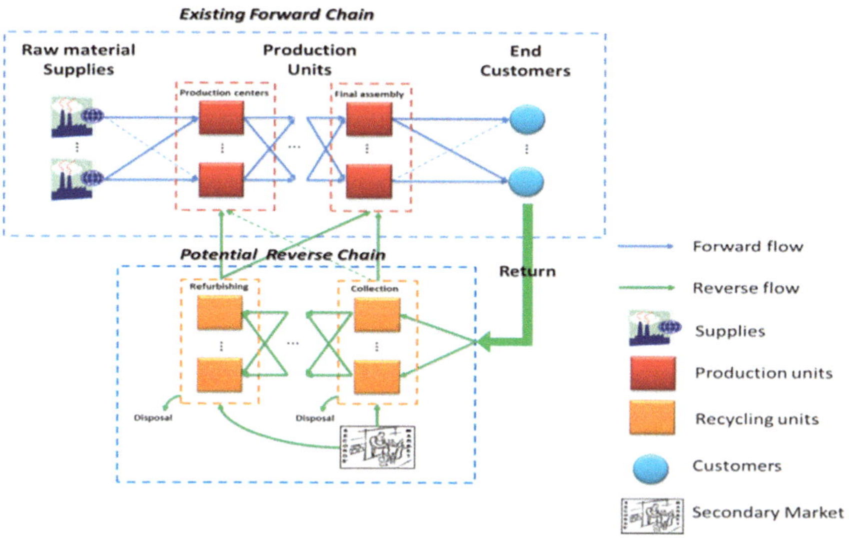

Fig. 1 The multi-stage and multi-factory structure of sustainable closed-loop supply chain

supply chain than open-loop supply chain. Effective means such as eco-technologies or lean management are necessary for manufacturers to reduce carbon emission at a lower cost.

2.2 *Multi-objective Closed-Loop Supply Chain Design (MCSCD) Model*

The economic objective, F_1, is measured by the total CLSC cost, including total material purchasing cost (PC), total installation cost (BC), total production cost (MC), total capacity expansion cost (CEC), total transportation cost (TC), and total disposal cost (DC). The environmental objective, F_2, is measured by the total carbon (CO_2) emission, including total production carbon emission (PCOE), total installation carbon emission (BCOE), and total transportation carbon emission (TCOE) in all the CLSC. For a MCSCD problem, we also need to consider material supply constraints, flow conservation constraints, capacity expansion and limitation constraints, and transportation constraints. Consequently, the MCSCD problem may be formulated as a multi-objective mixed integer programming model.

The assumptions used in this model are as follows: (1) The number of customers and suppliers and their demand are known, (2) second market is unique, (3) the demand of each customer must be satisfied, (4) the flow is only allowed to be transferred between two consecutive stages, (5) the number of facilities that can be opened and their capacities are both limited, and (6) the recovery and disposal percentages are given. The indices, input parameters, and decision variables of the multi-objective CLSC design model are defined as follows:

Indices

s	Material supplier index ($s = 1, \ldots, S$)
i, i'	Production stage index of FL ($i, i' = 1, \ldots, I$)
$\Phi(i)$	Set of existing production units at stage i of FL ($\Phi(i) = \{1, \ldots, \text{Num}(i)\}$)
$(i,k)(i',k')$	Existing production unit k, (k') at stage $i(i')$ of FL, where $k \in \Phi(i)\, k' \in \Phi(i')$
c	End customer index ($c = 1, \ldots, C$)
j, j'	Recycling stage index of RL ($j, j' = 1, \ldots, J$)
$\Psi(j)$	Set of potential recycling units in each stage j of RL ($\Psi(j) = \{1, \ldots, \text{Num}(j)\}$)
$(j,p)(j',p')$	Potential recycling unit $p(p')$ at stage $j(j')$ of RL, where $p \in \Phi(j) p' \in \Phi(j')$
L	The index of capacity expansion level ($l = 1, \ldots, L$)
T	The index of technology type ($t = 1, \ldots, T$)

Parameters

Cost-related parameters	
$\text{pc}_s^{\text{rm}}, \text{pc}_j^{\text{sm}}$	Purchasing cost from material supplier s in FL and from secondary market stage j in RL
$\text{dc}_{(j,p)}$	Disposal cost of potential recycling unit p at stage j in RL, which is denoted as RU(j, p)
$\text{sc}_{(i,k)(i',k')}$	Transportation cost from production unit k at stage i to production unit k' at stage i' of FL, which is denoted as PU(i, k) ($\text{sc}_{(0,s)(1,k)}$ represents the transportation cost from material supplier s to production unit k at 1st stage)
$\text{sc}_{(j,p)(j',p')}$	Transportation cost from RU(j, p) to RU(j', p') of RL ($\text{sc}_{(J+1,c)(J,p)}$ represents the transportation cost from customer c to recycling unit p at last stage J)
$\text{sc}_{(j,p)(i,k)}$	Transportation cost from RU(j, p) to PU(i, k)
$\text{mc}_{(i,k)\text{t}}, \text{mc}_{(i,p)\text{t}}$	Production cost using technology type t at potential PU (i, k) of FL and RU(j, p) of RL
$\text{bc}_{(i,k)}, \text{bc}_{(j,p)}$	Fixed installation cost of potential PU(i, k) and RU(j, p)
$\text{cc}_{(i,k)\text{tl}}, \text{cc}_{(j,p)\text{tl}}$	Capacity expansion cost of capacity level l using technology type t at existing PU(i, k) and potential RU(j, p)
Capacity-related parameters	
$\text{cl}_{(i,k)\text{tl}}, \text{cl}_{(j,p)\text{tl}}$	Expanded capacity amount of capacity level l using technology type t at existing PU(i, k) and potential RU (j, p)
Supply- and demand-related parameters	
$s_s^{\text{rm}}, s_j^{\text{sm}}$	The upper limit of supply quantity of material supplier s and secondary market stage j

(continued)

(continued)

d_c, r_c	Demand and recycling quantity of customer c
Environment-related parameters	
$\mathrm{pce}_{(i,k)t}, \mathrm{pce}_{(j,p)t}$	Fixed carbon emission quantity at PU(i, k) and RU(j, p)
$\mathrm{puce}_{(i,k)t}, \mathrm{puce}_{(i,p)t}$	Carbon emission quantity per unit of using technology type t at PU(i, k) and RU(j, p)
$\mathrm{tce}_{sk}, \mathrm{tce}_{(i,k)(i',k')}, \mathrm{tce}_{kc}, \mathrm{tce}_{cp}, \mathrm{tce}_{(j,p)(j',p')}, \mathrm{tce}_{(j,p)(i,k)}$	Carbon emission quantity per unit transporting from PU(i, k) to PU(i', k'), from RU(j, p) to RU(j', p'), and from RU(j, p) to PU(i, k)
Logistics-related parameters	
$t_{sk}, t_{(i,k)(i',k')}, t_{kc}; t_{cp}, t_{(j,p)(j',p')}, t_{(j,p)(i,k)}$	Transportation batch quantity from PU(i, k) to PU(i', k '), from RU(j, p) to RU(j', p') and from RU(j, p) to PU (i, k)
Ratio related parameters	
$\lambda_{(j,p)}^{\mathrm{dis}}$	Disposal percentage of potential RU(j, p)
$\gamma_{ji}^{\mathrm{ret}}$	recovery percentage from stage j of RL to stage i of FL

Decision variables

$\mathrm{TQ}_{sk}, \mathrm{TQ}_{(i,k)(i',k')}, \mathrm{TQ}_{kc}; \mathrm{TQ}_{cp}, \mathrm{TQ}_{(j,p)(j',p')}, \mathrm{TQ}_{(j,p)(i,k)}$	Transportation quantity from PU(i, k) to PU(i', k'), from RU (j, p) to RU(j', p'), and from RU(j, p) to PU(i, k)
$P_{(j,p)}$	Purchasing quantity of potential RU(j, p) from secondary market
$D_{(j,p)}$	Disposal quantity of potential RU(j, p)
	$X_{(j,p)}$ binary variable, $X_{(j,p)} = 1$, if potential is open; $X_{(j,p)} = 0$, otherwise
	$\mathrm{AC}_{(i,k)tl}$ binary variable, $\mathrm{AC}_{(i,k)tl} = 1$, if existing PU($i$, k) expands capacity level l using technology type t; $\mathrm{AC}_{(i,k)tl} = 0$, otherwise ($\mathrm{AC}_{(j,p)tl} = 0$, is defined similarly)
	$\mathrm{TA}_{(i,k)(i',k')}$ binary variable, $\mathrm{TA}_{(i,k)(i',k')} = 1$, if existing PU ($i$, k) transport goods to PU(i', k'); $\mathrm{TA}_{(i,k)(i',k')} = 0$, otherwise. ($\mathrm{TA}_{(j,p)(j',p')}$ is defined similarly)

Objective Function

$$\begin{aligned} &\text{Minimize}\, F_1 = \mathrm{PC} + \mathrm{BC} + \mathrm{MC} + \mathrm{CEC} + \mathrm{TC} + \mathrm{DC} \\ &\text{Minimize}\, F_2 = \mathrm{PCOE} + \mathrm{BCOE} + \mathrm{TCOE} \end{aligned}$$

Economic objective (F1):

$$\sum_{s \in S} \sum_{k \in \Phi(1)} \mathrm{pc}_s^{\mathrm{rm}} \times \mathrm{TQ}_{sk} + \sum_{j \in J} \sum_{p \in \Psi(j)} \mathrm{pc}_j^{\mathrm{sm}} \times P_{(j,p)} \tag{1}$$

$$\sum_{i\in I}\sum_{k\in\Phi(I)}\mathrm{bc}_{(i,k)}\times X_{(i,k)}+\sum_{j\in J}\sum_{p\in\Psi(J)}\mathrm{bc}_{(j,p)}\times X_{(j,p)} \tag{2}$$

$$\sum_{s\in S}\sum_{k\in\Phi(1)}\sum_{t\in T}\mathrm{mc}_{(1,k)t}\times \mathrm{TQ}_{sk}+\sum_{i=2}^{I}\sum_{k\in\Phi(i-1)}\sum_{k'\in\Phi(i)}\sum_{t\in T}\mathrm{mc}_{(i-1,k)(i,k')t}\times \mathrm{TQ}_{(i-1,k)(i,k')} \tag{3}$$

$$\sum_{c\in C}\sum_{p\in\Psi(1)}\sum_{t\in T}\mathrm{mc}_{(1,p)t}\times \mathrm{TQ}_{cp}+\sum_{j=2}^{J}\sum_{p\in\Psi(j-1)}\sum_{p'\in\Psi(j)}\sum_{tl\in T}\mathrm{mc}_{(j-1,p)(j,p')t}\times \mathrm{TQ}_{(j-1,p)(j,p')} \tag{4}$$

$$\sum_{j\in J}\sum_{p\in\Psi(j)}\sum_{i\in I}\sum_{k\in\Phi(i)}\sum_{t\in T}\mathrm{mc}_{(j,p)(i,k)t}\times \mathrm{TQ}_{(j,p)(i,k)} \tag{5}$$

$$\begin{aligned}&\sum_{i\in I}\sum_{k\in\Phi(j)}\sum_{t\in T}\sum_{l\in L}\mathrm{cc}_{(i,k)tl}\times \mathrm{cl}_{(i,k)tl}\times \mathrm{AC}_{(i,k)tl}\\&+\sum_{j\in J}\sum_{p\in\Psi(j)}\sum_{t\in T}\sum_{l\in L}\mathrm{cc}_{(j,p)tl}\times \mathrm{cl}_{(j,p)tl}\times \mathrm{AC}_{(j,p)tl}\end{aligned} \tag{6}$$

$$\begin{aligned}&\sum_{s\in S}\sum_{k\in\Phi(1)}\mathrm{sc}_{sk}\times \mathrm{TQ}_{sk}+\sum_{i=2}^{1}\sum_{k\in\Phi(i-1)}\sum_{k'\in\Phi(i)}\mathrm{sc}_{(i-1,k)(i,k')}\\&\times \mathrm{TQ}_{(i-1,k)(i,k')}+\sum_{k\in\Phi(i-1)}\mathrm{sc}_{sk}\times \mathrm{TQ}_{kc}\end{aligned} \tag{7}$$

$$\sum_{c\in C}\sum_{p\in\Psi(1)}\mathrm{sc}_{cp}\times \mathrm{TQ}_{cp}+\sum_{j=2}^{J}\sum_{p\in\Psi(j-1)}\sum_{p'\Psi(j)}\mathrm{sc}_{(j-1,p)(j,p')}\times \mathrm{TQ}_{(j-1,p)(j,p')} \tag{8}$$

$$\sum_{j\in J}\sum_{p\in\Psi(j)}\sum_{i\in I}\sum_{k\in\Phi(i)}\mathrm{sc}_{(j,p)(i,k)}\times \mathrm{TQ}_{(j,p)(i,k)} \tag{9}$$

$$\sum_{j\in J}\sum_{p\in\Psi(j)}\mathrm{dc}_{(j,k)}\times D_{(j,k)} \tag{10}$$

Equation (1) is the total material purchasing cost (PC) acquiring material from all suppliers in FL and recycling material from secondary market in RL. Equation (2) is the total fixed installation cost (BC) of all potential recycling units of all stages in RL. Equations (3)–(5) are the total production cost (MC) of FL and RL under different technology levels. Equation (6) states that total capacity expansion cost

(CEC) equals the sum of total capacity expansion costs in FL and RL. Equations (7) and (8) are total transportation cost (TC) between different production units of consecutive stages in FL and RL, respectively. Equation (9) is total transportation cost from RUs to PUs. Equation (10) is total disposal cost (DC) in RL due to the defect products generated from the recycling processes.

Environmental objective **(*F*2)**:

$$\sum_{s \in S} \sum_{k \in \Phi(1)} \sum_{t \in T} \mathrm{puce}_{(1,k)t} \times \mathrm{TQ}_{sk} + \sum_{i=2}^{I} \sum_{k \in \Phi(i-1)} \sum_{k' \in \Phi(i)} \sum_{t \in T} \mathrm{puce}_{(i-1,k)(i,k')t} \times \mathrm{TQ}_{(i-1,k)(i,k')} \tag{11}$$

$$\sum_{c \in C} \sum_{p \in \Psi(1)} \sum_{t \in T} \mathrm{puce}_{(1,p)t} \times \mathrm{TQ}_{cp} + \sum_{j=2}^{J} \sum_{p \in \Psi(j-1)} \sum_{p' \in \Psi(j)} \sum_{tl \in T} \mathrm{puce}_{(j-1,p)(j,p')t} \times \mathrm{TQ}_{(j-1,p)(j,p')} \tag{12}$$

$$\sum_{j \in J} \sum_{p \in \Psi(j)} \sum_{i \in I} \sum_{k \in \Phi(i)} \sum_{t \in T} \mathrm{puce}_{(j,p)(i,k)t} \times \mathrm{TQ}_{(j,p)(i,k)} \tag{13}$$

$$\sum_{i \in I} \sum_{k \in \Phi(I)} \mathrm{pce}_{(i,k)} \times X_{(i,k)} + \sum_{j \in J} \sum_{p \in \Psi(J)} \mathrm{pce}_{(j,p)} \times X_{(j,p)} \tag{14}$$

$$\sum_{s \in S} \sum_{k \in \Phi(1)} \mathrm{tce}_{sk} \times \mathrm{TQ}_{sk} + \sum_{i=2}^{I+1} \sum_{k \in \Phi(i-1)} \sum_{k' \in \Phi(i)} \mathrm{tce}_{(i-1,k)(i,k')} \times \mathrm{TQ}\,(i-1,k)(i,k's) \tag{15}$$

$$\sum_{c \in C} \sum_{p \in \Phi(1)} \mathrm{tce}_{cp} \times \mathrm{TQ}_{cp} + \sum_{j=2}^{J} \sum_{p \in \Phi(j-1)} \sum_{p' \in \Phi(j)} \mathrm{tce}_{(j-1,p)(j,p')} \times \mathrm{TQ}_{(j-1,p)(j,p')} \tag{16}$$

$$\sum_{j \in J} \sum_{p \in \Psi(j)} \sum_{i \in I} \sum_{k \in \Phi(i)} \mathrm{tce}_{(j,p)(i,k)} \times \mathrm{TQ}_{(j,p)(i,k)} \tag{17}$$

Equations (11)–(13) are total production carbon emission (PCOE) caused by the different capacity investment strategy (i.e., technology) in the closed-loop supply chain. Equation (14) represents total installation carbon emission (BCOE) of all PUs and RUs. Equations (15)–(17) are the total transportation carbon emission (TCOE) of FL and RL.

Constraints

- *Material supply constraints*

$$\sum_{k\in\Phi(1)} \mathrm{TQ}_{sk} \leq s_s^{rm}, \quad \forall\, s \in S \tag{18}$$

$$P_{(j,p)} \leq s_j^{sm}, \quad \forall\, j \in J, \quad \forall\, p \in \Phi(j) \tag{19}$$

Equation (18) shows the supply limitation purchasing from each material supplier s. Equation (19) shows the supply upper limit of each RU(j, p) purchasing from the secondary market in RL.

- *Flow conservation constraints*

$$\sum_{k'\in\Phi(i-1)} \mathrm{TQ}_{(i-1,k')(i,k)} + \sum_{j\in J}\sum_{p\in\Phi(j)} \mathrm{TQ}_{(j,p)(i,k)} = \sum_{k''\in\Phi(i+1)} \mathrm{TQ}_{(i,k)(i+1,k'')} \quad \forall\, i = \{1, 2, \ldots, I\}, \quad \forall\, k \in \Phi(i) \tag{20}$$

$$\sum_{k\in\Phi(I)} \mathrm{TQ}_{kc} = \sum_{c=1}^{C} d_c, \quad \forall\, c \in C \tag{21}$$

$$\sum_{p\in\Phi(J)} \mathrm{TQ}_{cp} = \sum_{c=1}^{C} d_c, \quad \forall\, c \in C \tag{22}$$

$$\sum_{p'\in\Phi(j-1)} \mathrm{TQ}_{(j-1,p')(j,p)} + P_{(j,p)} = \sum_{p'\in\Phi(j+1)} \mathrm{TQ}_{(j,p)(j+1,p')} + D_{(j,p)} + \sum_{i\in I}\sum_{k\in\Phi(i)} \mathrm{TQ}_{(j,p)(i,k)} \quad \forall\, j = \{1, 2, \ldots, J\}, \quad \forall\, p \in \Phi(j) \tag{23}$$

$$D_{(j,p)} = \lambda_{(j,p)}^{dis} \times \left(\sum_{p'\in\Phi(j-1)} \mathrm{TQ}_{(j-1,p')(j,p)} + P_{(j,p)} \right) \quad \forall\, j = \{2, \ldots, J\}, \quad \forall\, p \in \Phi(j) \tag{24}$$

$$\sum_{k\in\Phi(i)} \mathrm{TQ}_{(j,p)(i,k)} \leq \gamma_{ji}^{ret} \times \left(\sum_{p'\in\Phi(j-1)} \mathrm{TQ}_{(j-1,p')(j,p)} + P_{(j,p)} - D_{(j,p)} \right) \quad \forall\, i \in I, \quad \forall\, j = \{1, 2, \ldots, J\}, \quad \forall\, p \in \Phi(j) \tag{25}$$

Equation (20) satisfies the law of the flow conservation by inflow equal to outflow in each existing production unit of FL. It is to be noted that $\mathrm{TQ}_{sk} = \mathrm{TQ}_{(0,s)(1,k)}$ when $i = 1$, k' refers to supplier s, and $\mathrm{TQ}_{kc} = \mathrm{TQ}_{(I,k)(I+1,c)}$ when $i = \mathrm{I}$, k' refers to customer c. Equation (21) is to ensure that the customer demand may be satisfied. Equation (22) represents the recycling quantities from EOL products of customers in RL. Equation (23) satisfies the law of the flow conservation by inflow equal to outflow in each potential recycling unit of RL. It is to be noted that $\mathrm{TQ}_{cp} = \mathrm{TQ}_{(0,c)(1,p)}$ when $j = 1$, p' refers to customer c, and $\mathrm{TQ}_{(j,p)(j+1,p')} = 0$ when $j = J$.

Equation (24) calculates the disposal quantities of each potential recycling unit in RL, and Eq. (25) is the transportation limitation from each potential RU to PU.

- *Capacity expansion and limitation constraints*

$$\begin{aligned} &\sum_{k'\in\Phi(i-1)} \mathrm{TQ}_{(i=1,k')(i,k)} + \sum_{j\in J}\sum_{p\in\Phi(j)} \mathrm{TQ}_{(j,p)(i,k)} \le \mathrm{ca}_{(i,k)} \\ &\quad + \sum_{\mathrm{t}\in T}\sum_{\mathrm{l}=L} \left(\mathrm{cl}_{(i,k)\mathrm{tl}} \times \mathrm{AC}_{(i,k)\mathrm{tl}}\right) \quad \forall\, i = \{1,2,\ldots,I\}, \quad \forall\, k \in \Phi(i) \end{aligned} \tag{26}$$

$$\sum_{t\in T}\sum_{l\in L} \mathrm{AC}_{(i,k)\mathrm{tl}} \le 1 \quad \forall\, i \in I, \quad \forall\, k \in \Phi(i) \tag{27}$$

$$\begin{aligned} &\sum_{p'\in\Phi(j-1)} \mathrm{TQ}_{(j-1,p')(j,p)} + P_{(j,p)} \le \sum_{\mathrm{t}\in T}\sum_{\mathrm{l}\in L} \left(\mathrm{cl}_{(j,p)\mathrm{tl}} \times \mathrm{AC}_{(j,p)\mathrm{tl}}\right) \\ &\quad \forall\, j = \{1,2,\ldots,J\}, \quad \forall\, p \in \Phi(j) \end{aligned} \tag{28}$$

$$\sum_{t\in T}\sum_{l\in L} \mathrm{AC}_{(j,p)\mathrm{tl}} = X_{(j,p)}, \quad \forall\, j \in J, \quad \forall\, p \in \Psi(j) \tag{29}$$

Equation (26) represents the capacity limit of each PU, and Eq. (27) shows that only one capacity level of a certain technology is chosen when needing to expand capacity. Capacity expansion limit for RU can be obtained similarly in Eqs. (28) and (29).

- *Transportation constraints*

$$\begin{aligned} &t_{(i-1,k)(i,k')} \times \mathrm{TA}_{(i-1,k)(i,k')} \le \mathrm{TQ}_{(i-1,k)(i,k')} \le \mathrm{TA}_{(i-1,k)(i,k')} \times M \\ &\quad \forall\, i = \{1,2,\ldots,I\}, \quad \forall\, k \in \Phi(i-1), \quad \forall\, k' \in \Phi(i) \end{aligned} \tag{30}$$

$$t_{kc} \times \mathrm{TA}_{kc} \le \mathrm{TQ}_{kc} \le \mathrm{TA}_{kc} \times M, \quad \forall \quad k \in \Phi(I), \quad \forall \quad c \in C \tag{31}$$

$$t_{cp} \times \mathrm{TA}_{cp} \le \mathrm{TQ}_{cp} \le \mathrm{TA}_{cp} \times M, \quad \forall\, c \in C, \quad \forall\, p \in \Psi(1) \tag{32}$$

$$t_{(j-1,p)(j,p')} \times \mathrm{TA}_{(j-1,p)(j,p')} \leq \mathrm{TQ}_{(j-1,p)(j,p')} \leq \mathrm{TA}_{(j-1,p)(j,p')} \times M \\ \forall j = \{2,\ldots,J\}, \quad \forall p \in \Psi(j-1), \quad \forall p' \in \Psi(j) \tag{33}$$

$$t_{(j,p)(i,k)} \times \mathrm{TA}_{(j,p)(i,k)} \leq \mathrm{TQ}_{(j,p)(i,k)} \leq \mathrm{TA}_{(j,p)(i,k)} \times M \\ \forall j \in J, \quad \forall p \in \Psi(j), \quad \forall i \in I, \quad \forall k \in \Phi(i) \tag{34}$$

Equations (30) and (31) satisfy the minimum transportation quantities between the different PUs. Similarly, Eqs. (32)–(34) are for RUs' transportation capability constraint.

- *Domain constraints*

$$\mathrm{TQ}_{s(i,k)}, \mathrm{TQ}_{(i,k)(i',k')}, \mathrm{TQ}_{(i,k)c}, \mathrm{TQ}_{c(j,p)}, \mathrm{TQ}_{(j,p)(j',p')}, \mathrm{TQ}_{(j,p)(i,k)}, P_{(j,p)}, D_{(j,p)}, N_c \geq 0 \\ \forall s \in S, \quad \forall i \in I, \quad \forall k \in \Phi(i), \quad \forall c \in C, \quad \forall j \in J, \quad \forall P \in \Psi(j) \tag{35}$$

$$X_{(j,p)}, \mathrm{AC}_{(i,k)\mathrm{tl}}, \mathrm{AC}_{(j,p)\mathrm{tl}}, \mathrm{TA}_{s(i,k)}, \mathrm{TA}_{(i,k)(i',k')}, \mathrm{TA}_{(i,k)c}, \mathrm{TA}_{c(j,p)}, \mathrm{TA}_{(j,p)(j'p')}, \mathrm{TA}_{(j,p)(i,k)} \in \{0,1\} \\ \forall s \in S, \quad \forall i \in I, \quad \forall k \in \Phi(i), \quad \forall c \in C, \quad \forall j \in J, \quad \forall P \in \Psi(j) \tag{36}$$

3 Particle Swarm Optimization

In PSO algorithm evolution, Naka et al. (2003) introduced hybrid PSO (HPSO) which owns critical factors of W, R_1, and R_2, to converge the algorithm solutions for a practical distribution state estimation. This study will replace the random variables of R_1 and R_2 by using the value of C_r. Chuang et al. (2011) have declared that the C_r with the formula of $C_r(n+1) = k \times C_r(n) \times (1 - Cr(n))$ results in the better convergence for PSO algorithm. This study will adopt the nonlinear function to obtain the dynamic inertia weights which fit in with a high dimension of PSO problems. The formula of nonlinear function is shown in Eq. (37), which concluded that a reasonable selection of w should decrease gradually while the swarm search progresses (Fig. 2).

$$w(t) = (2/t)^{0.3} \tag{37}$$

$w(t)$ represents the inertia weight of t generations; t represents the t generation.

In this research, a novel multi-objective PSO with ideal-point non-dominated sorting is designed to find the optimal solution of the proposed MCSCD model. Figure 3 demonstrates the PSO algorithm and repair process for this MCSCD model, four major steps, described as follows, are employed to search for a feasible solution; then, a generic PSO process and a particle repair process are employed to

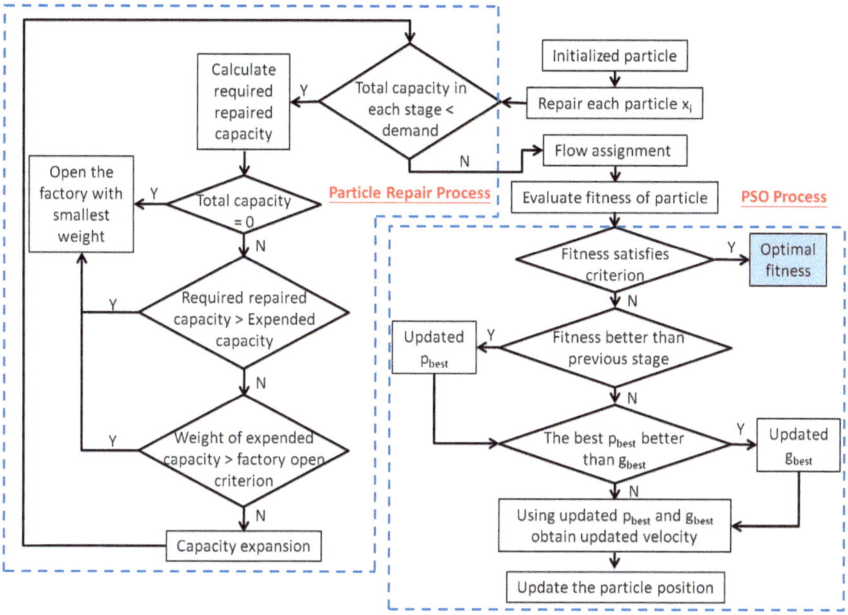

Fig. 2 PSO algorithm and repair process of MCSCD model

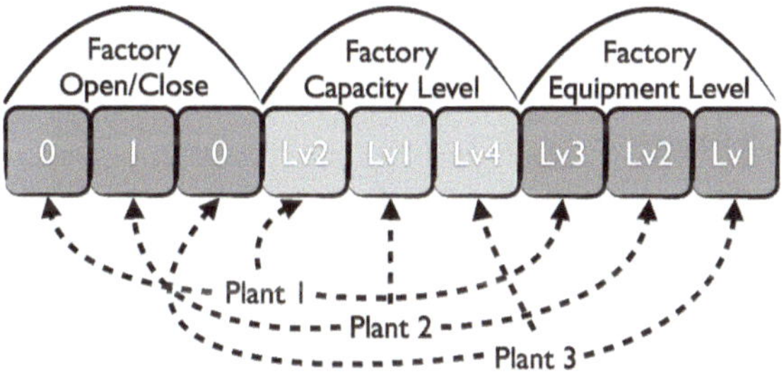

Fig. 3 The particle is encoded as the open/close decision, the capacity level, and the technology level of each factory

update the parameters of PSO algorithm and repair the value of each particle involved in step 2, respectively.

1. *Initialize the particle and the supply chain design decision*: The particle is encoded as decision of open/close of a factory, capacity level, and technology level for each level's each factory in the forward and reverse logistics, as shown in Fig. 4. Taking the second factory for example, factory 2 will be opened with level 1 capacity level and level 2 technology level.

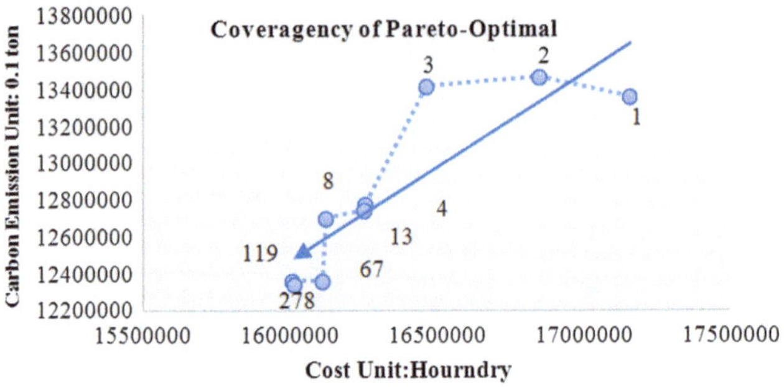

Fig. 4 Pareto track of 7 stages and 7 units

2. *Repair mechanism* is employed to first satisfy the aforementioned constraints and then to adjust the decision made in Step (1) by considering the factors such as the supply and demand balance.
3. *Flow assignment mechanism* is employed to determine the flow among all decision units by satisfying the total demand with lowest carbon emission and cost.
4. *Fitness computation, evaluation, and optimization*: (i) calculating the fitness of every particle by MCSCD model and obtain the *pBest*; (ii) updating the *gBest*, C_r, velocity, and position; (iii) optimization achievement.

To initialize a population is to randomly generate particles, the open/close decision particle is a 0 or 1 binary variable, and the capacity investment and technology level are continuous variables. The calculation of fitness value is divided into two parts: the economic objective function, F_1, is first calculated, and then, environmental objective function, F_2, is calculated. The *pBest* is the best position of each particle in its own searching process. During the iterations, the particle's fitness evaluation is compared with *pBest*. If the current value is better than *pBest*, then set *pBest* value equal to the current value. Then, crowding distance (CD) algorithm is employed to rank each particles' *pBest* in the Pareto set for each searching iteration in steps (2)–(4) in this multi-objective decision model, and the population's overall best, *gBest*, is updated based on the particle's crowd. Consequently, the velocity and position of the particles will be updated according to Eq. (38):

$$\begin{aligned} V_{id}(t+1) = w \times V_{id}(t) + c_1 \times r_1 \times (pbest_{id}(t) - X_{id}(t)) \\ + c_2 \times r_2 \times (gbest_d(t) - X_{id}(t)) \end{aligned} \tag{38}$$

4 Result and Analysis

4.1 Design of Experiment (DOE) and Analysis

This study used DOE to analyze the variable range and came out with the best composition of variables. Since this model is dealing with multi-objectives, we applied the ideal-point non-dominated sorting for non-dominated set to calculate F′ and the distance of $d(F)$ between $\left(F'_1, F'_2\right)$ and zero, where $(F_1 F_2)_{x_i}$ is an element in non-dominated set; $F_1^{\max}$ and $F_2^{\max}$ are the highest values of the first and second objectives among experiments; and $F_1^{\min}$ and $F_2^{\min}$ are lowest values of the first and second objectives among experiments. In the end, the distance will be sorted which leads to the proposed PSO.

$$F'{=}\left(F'_1, F'_2\right) = \left(\frac{F_1 - F_1^{\min} + \rho}{F_1^{\max} - F_1^{\min} + \rho}, \frac{F_2 - F_2^{\min} + \rho}{F_2^{\max} - F_2^{\min} + \rho}\right) \tag{39}$$

$$d(F) = d\left(F'_1, F'_2\right) = \sqrt{\left(F'_1, F_0\right)^2 + \left(F'_2, F_0\right)^2} \tag{40}$$

Based on ANOVA, we obtained the best parameter set of $C_1 = 6/C_2 = 6$; $w = 0.75$; C_r = varying parameters; MO method = ST; and particle number = 100. The suggested parameters were used to apply into PSO and compare the result to the optimal result of CPLEX based on the same parameter set. The algorithm logistic of CPLEX adopted the branch-and-bound (B&B) method. Three indicators are used in this study and validate the accuracy of PSO algorithm: (1) *C*-metric (CM) Zitzler and Thiele (1999); (2) maximum spread metric (*S*-metric; $S(A)$) Tan et al. (2006); and (3) maximum distance ($D_{\max}$). Results show that the PSO result is better when CPLEX converge value is close to 1. The converge ability of CPLEX and PSO for 2 stages and 2 units are 0.6 and 0 in CM indicator; furthermore, the converge percentage of CPLEX and PSO are 1 and 0.3737 in $S(A)$ indicator. The higher $S(A)$ value of PSO indicates the PSO solutions close to the optimal solutions.

In order to validate, this model can be converged in a complicated problem; this study provides the convergence trend of Pareto optimal of 7 stages and 7 units in Fig. 4. We observed that the Pareto optimal converged rapidly in the initial 4 generations. It also clearly shows the generations increased while the solution was close to the lower cost and CO_2 emission. The Pareto optimal and solution set optimal were achieved after 278 and 119 generations. These results showed that CLSC model can be optimal using PSO even though the problem is large and complicated.

Table 1 The optimal solution of case study in different scenarios

	Total cost	Carbon emission (ton)	Cost differential (%)	CO_2 emission differential (%)
Economic optimal	$788,341,600	635,378	–	–
Pareto optimal	$792,846,000	629,804	0.52	−0.63
Environmental optimal	$805,367,600	620,492	2.16	−2.34

4.2 Case Study

Since the solar cell can be recycled, refurbished, and reuse, it gave the potential to create the RL in supply chain design and, moreover, to achieve the economic and environmental benefits for solar enterprises. Solar cells will deliver to FL to refurbish or to RL for quality examination. Some solar cells after quality examination can extract the silicon materials and be reduced to silicon raw materials by silicon powder production factories, some solar cells that are of good quality will deliver to FL for reproduction, and few which did not pass the quality examination goes to disposal factories.

This experiment of applying MOPSO algorithm in solar cell industry case set up 500 iterations to obtain the optimal solution. Table 1 demonstrates the single optimization results and Pareto optimization results in which the decreasing carbon emission will cause the higher cost that can be observed. The Pareto results will need more 0.52 % cost, and in addition, the environmental model will need more 2.16 % cost to obtain the economic optimal. Figure 5 shows the Pareto optimal

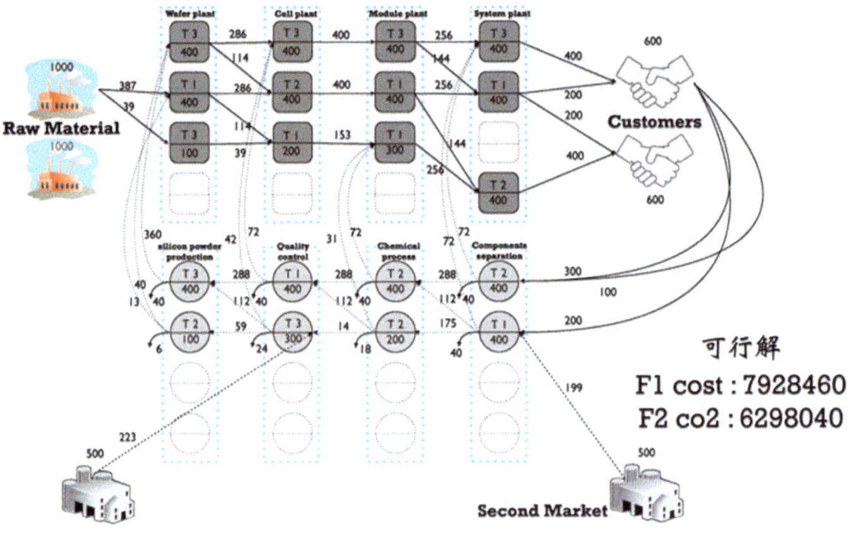

Fig. 5 Pareto optimal solution of 4 stages and 4 units

solution of 4 stages and 4 units with compromise of economic and environmental concerns. We observed that the solution contained all technologies which are 7 units for technology level 1, 6 units for technology level 2, and 7 units for technology level 3.

5 Conclusion

This paper studies an integrated CLSC network design problem with cost and environmental concerns in the solar energy industry. We adjusted the factory capacity based on selecting various manufacturing technology levels and, furthermore, diagnosed the economic and environmental impact based on various scenarios of FL/RL capacity and demand. While the scale of CLSC enlarging, more operated factories in FL/RL and carbon pollution will occur, in addition, the Pareto optimal solution which is provided by this investigation will give the suggestions of capacity expansion, technology selection, supply chain design, factory location options, and capacity allocation based on different local regulation for enterprises. This investigation also compared the PSO solution to CPLEX and studied the efficiency and performance. The application of MCSCD model in a solar cell industry shows that the solar enterprises are able to select the appropriated factories, technologies, and capacity to fulfill the economic and environmental concerns based on the enterprises' sustainable visions.

References

Chuang LY, Tsai SW, Yang CH (2011) Chaotic catfish particle swarm optimization for solving global numerical optimization problems. Appl Math Comput 217(16):6900–6916

Eberhart RC, Kennedy J (1995, Oct) A new optimizer using particle swarm theory. In: Proceedings of the sixth international symposium on micro machine and human science (vol 1, pp 39–43)

Fahimnia B, Marian R, Motevallian B (2009) Analysing the hindrances to the reduction of manufacturing lead-time and their associated environmental pollution. Int J Environ Technol Manage 10(1):16–25

Naka S, Genji T, Yura T, Fukuyama Y (2003) A hybrid particle swarm optimization for distribution state estimation. IEEE Trans Power Syst 18(1):60–68

Özceylan E, Paksoy T (2013) A mixed integer programming model for a closed-loop supply-chain network. Int J Prod Res 51(3):718–734

Tan KC, Goh CK, Yang YJ, Lee TH (2006) Evolving better population distribution and exploration in evolutionary multi-objective optimization. Eur J Oper Res 171(2):463–495

Wang F, Lai X, Shi N (2011) A multi-objective optimization for green supply chain network design. Dec Support Syst 51(2):262–269

Zitzler E, Thiele L (1999) Multiobjective evolutionary algorithms a comparative case study and the strength Pareto approach. IEEE Trans Evol Comput 3(4):257–271

A Combined Grey System Theory and Uncertainty Theory-Based Approach for Supplier Selection in Supply Chain Management

Muhammad Saad Memon, Young Hae Lee and Sonia Irshad Mari

Abstract The requirement of large sample size and the strong subject knowledge to build suitable membership function restrict the applicability of probability and fuzzy theories in supplier selection problem. Due to this fact, the combination of grey system theory and uncertainty theory-based approach is applied for supplier selection problem, which requires neither any exact probability distribution nor membership function. The proposed approach is able to consider stochastic and recognitive uncertainties associated with supplier selection decision in supply chain management. This chapter develops framework for reducing the risks associated with suppliers. The novelty of this research is to consider proper uncertainty approaches in different stages of proposed framework instead of solving the whole selection problem using the same uncertainty theory or grey system theory. The proposed framework is based on three stages: identification of supplier selection criteria, defining weights to goals and ratings to supplier attributes as grey linguistic variables, and supplier selection and order allocation using uncertain-goal programming considering uncertain demand and lead time. The proposed model will help the practitioners to effectively evaluate and select suitable set of suppliers and optimal order allocation.

Keywords Supplier selection · Grey system theory · Uncertainty theory · Supply chain management

1 Introduction

Many researchers have investigated supplier evaluation criteria and selection methods. Supplier selection is a multiple-attribute decision-making (MADM) problem (Karimi et al. 2014). Among various methods, the widely adopted methods

M.S. Memon · Y.H. Lee (✉) · S.I. Mari
Department of Industrial and Management Engineering, Hanyang University, Seoul, South Korea
e-mail: yhlee@hanyang.ac.kr

V. Kachitvichyanukul et al. (eds.), *Toward Sustainable Operations of Supply Chain and Logistics Systems*, EcoProduction,
DOI 10.1007/978-3-319-19006-8_32

are multi-criteria decision models (Wang et al. 2007) and linear weighting models (Sarkis et al. 2002), the analytic hierarchy process (Sipahi et al. 2010), and mathematical programming (MP) techniques (Talluri et al. 2003). These all above-described methods give little insights for supplier selection issue, because supplier selection is highly associated with uncertainties and depends on large amount of domain knowledge where expert's assessment plays an important role. Supplier selection problem is associated with *recognitive* and *stochastic* uncertainties. Therefore, it is necessary to develop a more effective supplier selection method, which can handle recognitive and stochastic uncertainties simultaneously (Deng et al. 2014). The involvement of decision maker's (DM) subjective judgment yields recognitive uncertainties, and therefore, traditional approaches may fail to handle this type of uncertainty. Another uncertainty embed in supplier selection decision is stochastic nature of various decision parameters. Various researchers proposed supplier selection model deal with stochastic parameters (Hu et al. 2012). These proposed stochastic models require large amount of data to handle the assumption of certain probability distribution, and this yields another problem for supplier selection decision as DMs have lack of knowledge or small availability of information for different set of suppliers. 'Probability and Statistics' theories can handle stochastic uncertainties, but it requires large amount of historical data to produce reliable results, and this is only possible when suppliers have long background and large sample data can be accessible to DMs, which is difficult in real situation. Recognitive uncertainty is largely due to lack of knowledge or incomplete information (Deng 1985). Fuzzy mathematics is widely used to handle the recognitive uncertainty. The problem with the fuzzy mathematics is that it requires strong knowledge about the subject to make membership functions. Recently, various researchers proposed fuzzy-based techniques to deal with supplier selection problem (Boran et al. 2009). The requirement of large sample size and the strong subject knowledge to build suitable membership function restrict the applicability of probability and fuzzy theories in supplier selection problem. Moreover, Liu (2012) showed that it is inappropriate to model belief degree by using probability theory because it may lead to counterintuitive results. Deng in 1982 developed the grey system theory to handle the problem of 'small sample size and poor information,' which requires neither any probability distribution nor membership function (Deng 1989). This theory attempts to solve above-discussed issues related to supplier selection problem.

The first model in supplier selection problem based on grey theory is proposed by Li et al. (2007). Later on Li et al. (2008) and Cao et al. (2012) extended the application of grey system theory in supplier selection problem. However, these proposed models also have lacking in finding out the optimal purchase quantity from each supplier in each period, which is the most important decision. This is due to lack of belief degree as exact value of grey parameter is unknown and decision makers have to choose the optimal purchase quantity from the given range, which is not possible in practical scenario. Decision maker may not order material quantity because he/she is usually not sure about belief degree of minimum and maximum limits of grey parameter. Kahneman and Tversky (1979) showed that decision makers usually overweight unlikely events. This problem limits the use of grey system theory in supplier selection problem. In this scenario, Liu (2007) proposed uncertainty theory to deal with belief degree. In order to overcome this problem, we proposed combined grey theory and uncertainty theory-based approach. Readers may consult Liu (2015) for recent development in uncertainty theory.

This paper developed a framework for supplier evaluation and selection in order to achieve both quantitative and qualitative objectives associated with suppliers using combined grey system theory and uncertainty theory. The proposed framework is based on three stages: (1) identification of supplier selection criteria, (2) defining weights to goals and ratings to supplier attributes as grey linguistic variables, and (3) supplier selection and order allocation using uncertain-goal programming considering uncertain demand and lead time. Goal programming technique is widely used by many researchers to handle the multi-attribute supplier selection problem (Chai et al. 2013).

2 Preliminaries

2.1 Grey System Theory

A 'grey number' $\otimes G$ is such number whose exact value is unknown, but a range within which the value lies is known. A grey number $\otimes G$ is defined as an interval with known lower limit and known upper limit as $\otimes G[\underline{G}, \bar{G}]$. Such a method supplements the expression of system uncertainties whenever the probability density and membership functions cannot be fully identified (Cai et al. 2011; Chang et al. 1996). A $\otimes G$ becomes a 'deterministic number' or 'white number' when $\underline{G} = \bar{G}$. Suppose $\otimes G[\underline{G}_1, \bar{G}_1]$ and $\otimes G[\underline{G}_2, \bar{G}_2]$ are two different grey numbers, the basic operations on these grey numbers are defined as follows (Li et al. 2007).

$$\otimes G_1 + \otimes G_2 = \left[\underline{G}_1 + \underline{G}_2, \overline{G}_1 + \overline{G}_2\right]$$

$$\otimes G_1 - \otimes G_2 = \left[\underline{G}_1 - \overline{G}_2, \overline{G}_1 + \underline{G}_2\right]$$

$$\otimes G_1 \times \otimes G_2 = [\min(\underline{G}_1\underline{G}_2, \underline{G}_1\bar{G}_2, \bar{G}_1\underline{G}_2, \bar{G}_1\bar{G}_2), \max(\underline{G}_1\underline{G}_2, \underline{G}_1\bar{G}_2, \bar{G}_1\underline{G}_2, \bar{G}_1\bar{G}_2)]$$

$$\otimes G_1 \div \otimes G_2 = \left[\underline{G}_1, \overline{G}_1\right] \times \left[\frac{1}{\underline{G}_2}, \frac{1}{\overline{G}_2}\right]$$

$$k.\otimes G = \left[k.\underline{G}, k.\overline{G}\right]$$

Most of the real-world processes in decision problems are in the grey stage due to lack of information and uncertainty (Mujumdar and Karmakar 2008). This is the reason why grey optimization provides a useful tool for decision-making problem under such uncertainties (Karmakar et al. 2006). Supplier selection problem also consists of uncertainties which may not be solved using probability or fuzzy theory. Hence, the grey system theory provides sufficient basis to handle recognitive uncertainty associated with supplier selection problem.

2.2 Uncertainty Theory

Let Γ be a nonempty set, and let A be a σ-algebra over Γ. Each element of A is called an event. According to Liu (2007), a set function is called an uncertain measure if and only if it satisfies the following four axioms:

Axiom 1 (Normality) $\mathcal{M}\{\Gamma\} = 1$;
Axiom 2 (Monotonicity) $\mathcal{M}\{A\} \leq \mathcal{M}\{B\}$ whenever $A \subseteq B$;
Axiom 3 (Self-Duality) $\mathcal{M}\{A\} + \mathcal{M}\ \{A^c\} = 1$ for any event A;
Axiom 4 (Countable Subadditivity) $\mathcal{M}\{\cup_i A_i\} \leq \sum_{i=1}^{\infty} \mathcal{M}\{A_i\}$ for any countable sequence of events $\{A_i\}$.

Definition 2.2.1 Liu (2007) Let Γ be a nonempty set, and let A be a σ-algebra over it. If $\mathcal{M}$ is an uncertain measure, then the triplet $(\Gamma, A, \mathcal{M})$ is called an uncertainty space.

Definition 2.2.2 Liu (2007) Uncertain variable ξ is defined as a measurable function from an uncertainty space $(\Gamma, A, \mathcal{M})$ to the set of real numbers $\Re$. That is, for any Borel set B, we have $\{\gamma \in \Gamma \mid \xi\ (\gamma) \in B\} \in A$.

Definition 2.2.3 Liu (2007) Let ξ be an uncertain variable. Then, the expected value of ξ is defined as below, provided that at least one of the two integrals is finite:

$$E[\xi] = \int_0^{+\infty} \mathbf{M}\{\xi \geq x\}\mathrm{d}x - \int_{-\infty}^{0} \mathbf{M}\{\xi \leq x\}\mathrm{d}x$$

Definition 2.2.4 Liu (2007) Let ξ be an uncertain variable with regular uncertainty distribution Φ and inverse uncertainty distribution Φ^{-1}. Then, the expected value of the uncertain variable ξ with regular uncertainty distribution Φ is

$$E[\xi] = \int_0^1 \Phi^{-1}(\alpha)\mathrm{d}\alpha.$$

3 Supplier Evaluation and Selection Criteria

Efficient evaluation criteria can help the enterprises to reduce the risks associated with suppliers. The key to success of manufacturing firms in today's fast-changing technology-based consumer market is to introduce innovative products, which is not possible without considering suitable suppliers (Kanagaraj et al. 2014). Evaluation criteria are also important to reduce the operational costs and increase the profit continuously (Wang et al. 2009). Supplier selection decision changes the global supply chain design problem in fundamental ways, because they are based on broad criteria and selected according to the buyer's perception (Yucenur et al. 2011). Based on extensive literature review related to supplier evaluation criteria from past researches (e.g. Amindoust et al. 2012; Awasthi et al. 2010; Chen 2011; Chou et al. 2008; Dickson 1966; Shaw et al. 2012; Weber et al. 1991), we develop important supplier evaluation criteria and their measuring principles as shown in Table 1.

4 The Proposed Supplier Selection Model

The combined grey system theory and uncertainty theory-based approach for supplier selection is proposed to solve the multi-attribute supplier selection problem. The grey system theory is employed because the selection of supplier attributes

Table 1 Criteria for supplier evaluation

Evaluation criteria	Description	Measuring principle
Quality	Poor-quality products are found during incoming inspection	Total number of rejected items are arrived from supplier in each batch
Delivery	The supplier capability to timely meet the demand	This can be analyzed by the percentage of demand meet in each period
Logistics service	Logistics service used by supplier and transportation time	This can be analyzed by the number of intermediate destinations from supplier to buyer. Also by analyzing the total number of rejected item due to poor logistics handling
Sustainability factor	The sustainability awareness of supplier	This can be analyzed by supplier awareness about sustainable performance such as their manufacturing methods, total embodied carbon footprints, waste management, social laws, and others
Risk factor	Vulnerability of suppliers due to risks such as natural/man-made disaster	This can be analyzed by the location of suppliers or any past disruption history of supplier

and preference for each type of procuring material belongs to DM's subjective judgment and it deals with recognitive uncertainty. As stated earlier, supplier selection problem also includes stochastic uncertainties which most of the time is unfeasible to use any exact probability distribution. The advantage of the proposed model over previous fuzzy and stochastic approaches is that it does not require any robust membership function nor require any probability distribution.

Based on the framework of the proposed model, initially both qualitative and quantitative data were collected based on experts' analysis along with available set of potential suppliers from procurement department. Then, the incomplete information collected from experts will be expressed as grey numbers, and grey system theory-based approach is used to obtain the weights and supplier's ratings for each attribute based on expert's opinions. Finally, uncertain-goal programming-based mathematical model is proposed to find the optimum set of suppliers and optimal purchase quantities.

4.1 Defining Weight to Objectives as Grey Linguistic Variables

In this research, the objective weights are considered as grey linguistic variables. Table 2 shows grey linguistics variables on seven scale numbers.

The procedure of defining weight to objectives is as follows. Assume that $w = \{w_1, w_2, \ldots, w_n\}$ is the vector of n objectives weights. Suppose that experts group consists of k persons, then attribute weight of O_j objective can be expressed according to geometric mean as:

$$\otimes w_j = \sqrt[k]{\otimes w_j^1 \times \otimes w_j^2 \times \cdots \times \otimes w_j^k} \tag{1}$$

where $\otimes w_j^k$ is the subjective judgment of the kth decision maker over jth objective.

4.2 Defining Attribute Ratings as Grey Linguistic Variables

The rating value to each supplier is also defined as linguistic variables. The scale of attribute ratings is shown in Table 3. Suppose that $S_i = \{S_1, S_2, S_3 \ldots S_n\}$ is set of potential suppliers. Then, the attribute rating of O_j objective can be expressed according to geometric mean of grey numbers as:

Table 2 Objective weights scale $\otimes w$

Scale	Very low (VL)	Low (L)	Medium low (ML)	Medium (M)	Medium high (MH)	High (H)	Very high (VH)
$\otimes w$	[0.0, 0.1]	[0.1, 0.3]	[0.3, 0.4]	[0.4, 0.5]	[0.5, 0.6]	[0.6, 0.9]	[0.9, 1.0]

Table 3 Attribute rating scale $\otimes G$

Scale	Very poor (VP)	Poor (P)	Medium poor (MP)	Fair (F)	Medium good (MG)	Good (G)	Very good (VG)
$\otimes G$	[0, 1]	[1, 3]	[3, 4]	[4, 5]	[5, 6]	[6, 9]	[9, 10]

$$\otimes G_{ij} = \sqrt[k]{\otimes G_{ij}^{1} \times \otimes G_{ij}^{2} \times \cdots \times \otimes G_{ij}^{k}} \tag{2}$$

where G_{ij}^{k} is the subjective judgment of the kth decision maker for ith supplier over jth objective.

The proposed mathematical model is based on combination of grey system theory and uncertainty theory. Therefore, we cannot use the grey parameter directly into mathematical model. Grey system theory provides useful solution to overcome this by applying *equal-weight mean whitenization*. Please refer to Liu et al. (2010) for further reading on whitenization of grey numbers.

After applying equal-weight mean whitenization to grey numbers, Eqs. (1) and (2) will be as follows, where $\tilde{\otimes}$ represents equal-weight mean whitenization value of grey parameter.

$$\tilde{\otimes} w_j = \frac{1}{2}\left(\underline{w_j} + \overline{w_j}\right) \tag{3}$$

$$\tilde{\otimes} G_{ij} = \frac{1}{2}\left(\underline{G_{ij}} + \overline{G_{ij}}\right). \tag{4}$$

4.3 Problem Formulation

Supplier selection problem concerns with how to choose the optimum supplier and how much quantity of material should be purchased. In the proposed model, both quantitative and qualitative attributes are included in the proposed model. The quantitative attributes are considered as uncertain purchase cost and lead time. The qualitative attributes are quality of material, delivery time, logistics service, sustainability factor, and risk. The information loss of these attributes due to lack of knowledge about suppliers and subjective judgment of DMs will be covered by using grey system theory. Notations used in the proposed model are as follows:

Sets:

i Set of suppliers $\{i = 1, 2\ldots n\}$

Parameters:

f_i	fixed cost associated with utilizing supplier i
p_i	unit purchase cost from supplier i
ξ	uncertain annual demand of material
η_i	uncertain lead time for supplier i
Φ_ξ	uncertain distribution of uncertain variable ξ
$\Phi_{\eta i}$	uncertain distribution of uncertain variable η_i
cs_i	capacity of supplier i
$\tilde{\otimes} ql_i$	quality level at supplier i as a grey parameter
$\tilde{\otimes} sl_i$	delivery service level of supplier i as a grey parameter
$\tilde{\otimes} ls_i$	quality of logistics service available at supplier i as a grey parameter
$\tilde{\otimes} sf_i$	level of sustainability awareness at supplier i as a grey parameter
$\tilde{\otimes} rl_i$	risk level at supplier i as a grey parameter
$\mathrm{TV}^{\mathrm{TC}}$	Threshold value of total cost
TV^{lt}	Threshold value of lead time
$\tilde{\otimes}\mathrm{TV}^{ql}$	Threshold value of quality level as a grey parameter
$\tilde{\otimes}\mathrm{TV}^{sl}$	Threshold value of delivery service level as a grey parameter
$\tilde{\otimes}\mathrm{TV}^{sl}$	Threshold value of logistics service level as a grey parameter
$\tilde{\otimes}\mathrm{TV}^{sf}$	Threshold value of sustainability factor level as a grey parameter
$\tilde{\otimes}\mathrm{TV}^{rl}$	Threshold value of risk level as a grey parameter

Decision variables:

$$x_i = \begin{cases} 1 & \text{If Supplier } i \text{ is selected, then 1, otherwise 0} \\ 0 & \end{cases}$$

y_i Proportion of demand fulfilled from supplier i

Total cost is the sum of annual purchase cost and fixed ordering cost. Where $y_i\xi$ represents quantity of material ordered to ith supplier.

$$\text{Purchase cost} = \sum_{i=1}^{n} p_i y_i \xi \tag{5}$$

$$\text{Fixed cost} = \sum_{i=1}^{n} f_i x_i \tag{6}$$

$$\text{Total cost} = c(y, \xi) = \sum_{i=1}^{n} f_i x_i + \sum_{i=1}^{n} p_i y_i \xi \tag{7}$$

Constraints of the model are estimated as follows.

$$\text{Total lead time} = Tlt(x, \eta) = \sum_{i=1}^{n} \eta_i x_i \tag{8}$$

$$\text{Total quality level} = Tql = \sum_{i=1}^{n} \tilde{\otimes} ql_i x_i \tag{9}$$

$$\text{Total quality level of logistics system} = Tls = \sum_{i=1}^{n} \tilde{\otimes} ls_i x_i \tag{10}$$

$$\text{Total delivery service level} = Tsl = \sum_{i=1}^{n} \tilde{\otimes} sl_i x_i \tag{11}$$

$$\text{Total sustainability level} = Tsf = \sum_{i=1}^{n} \tilde{\otimes} sf_i x_i \tag{12}$$

$$\text{Total risk} = Trl = \sum_{i=1}^{n} \tilde{\otimes} rl_i x_i \tag{13}$$

$$\text{Capacity constraint} = y_i \xi \leq cs_i \tag{14}$$

Constraints (15) and (16) show the minimization of goals for cost and lead time, respectively. It is worthwhile mentioning here that both constraints are uncertain variables since ξ and η are uncertain variable.

$$c(y, \xi) + d_1^+ = TV^{tc} \tag{15}$$

$$Tlt(x, \eta) + d_2^+ = TV^{lt} \tag{16}$$

Constraints (17)–(21) maximize the goals for all important evaluation criteria (see Table 1) by minimizing the total deviation of the goals from their threshold values.

$$Tql + d_3^+ = \tilde{\otimes} TV^{ql} \tag{17}$$

$$Tls + d_4^+ = \tilde{\otimes} TV^{ls} \tag{18}$$

$$Tsl + d_5^+ = \tilde{\otimes} TV^{sl} \tag{19}$$

$$Tsf + d_6^+ = \tilde{\otimes} TV^{sf} \tag{20}$$

$$Trl + d_7^+ = \tilde{\otimes} TV^{rl}. \tag{21}$$

4.4 Mathematical Model

In order to obtain the compromise solution of formulated problem, the expected value of goal programming model is developed as shown in Eq. (22). From mathematical viewpoint, there is no difference between deterministic and uncertain programming except uncertain programing contains uncertain variables.

$$\begin{cases} \min \left[\prod_{j=1}^{m} \tilde{\otimes} w_j d_j^+ \right] \\ \text{subject to} \begin{cases} E\left[\sum_{i=1}^{n} f_i x_i + \sum_{i=1}^{n} p_i y_i \xi \right] + d_1^+ = TV^{tc} \\ E\left[\sum_{i=1}^{n} \eta_i x_i \right] + d_2^+ = TV^{lt} \\ E[y_i \xi] \le cs_i \\ \sum_{i=1}^{n} E[y_i \xi] \ge E[\xi] \\ Tql + d_3^+ = \tilde{\otimes} TV^{ql} \\ Tls + d_4^+ = \tilde{\otimes} TV^{ls} \\ Tsl + d_5^+ = \tilde{\otimes} TV^{sl} \\ Tsf + d_6^+ = \tilde{\otimes} TV^{sf} \\ Trl + d_7^+ = \tilde{\otimes} TV^{rl} \\ y_i \ge 0,\ x_i \in \{0,1\},\ d_i^+ \ge 0, \quad i = 1, 2, \ldots, n \end{cases} \end{cases} \tag{22}$$

4.5 Equivalent Crisp Model

In order to find the optimum solution, we have to estimate the expected value or uncertain measure. Taking advantage of properties of uncertainty theory, the mathematical model (22) can be transformed into its crisp equivalent as shown in (23)

$$\begin{cases} \min\left[\prod_{j=1}^{m} \otimes \tilde{w}_j d_j^{+}\right] \\ \text{subject to} \begin{cases} \sum_{i=1}^{n} f_i x_i + \sum_{i=1}^{n} p_i y_i \int_{0}^{1} \Phi_{\xi}^{-1}(a)da + d_1^{+} = TV^{tc} \\ \sum_{i=1}^{n} x_i \int_0^1 \Phi_{\eta_i}^{-1}(a)da + d_2^{+} = TV^{lt} \\ y_i \int_0^1 \Phi_{\xi}^{-1}(a)da \leq cs_i \\ Tql + d_3^{+} = TV^{ql} \\ Tls + d_4^{+} = TV^{ls} \\ Tsl + d_5^{+} = TV^{sl} \\ Tsf + d_6^{+} = TV^{sf} \\ Trl + d_7^{+} = TV^{rl} \\ y_i \geq 0, x_i \in \{0,1\}, d_i^{+} \geq 0, \quad i = 1, 2, \ldots, n \end{cases} \end{cases} \tag{23}$$

5 Conclusion

This paper developed framework for reducing the risks associated with suppliers. The proposed framework is based on three stages: (1) identification of supplier selection criteria, (2) defining weights to goals and ratings to supplier attributes as a grey linguistic variable, and (3) supplier selection and order allocation using uncertain-goal programming. In real-world multi-criteria decision-making problems, the DM's judgments are uncertain and cannot be estimated exactly. We take advantage of grey system theory to tackle the stochastic and recognitive uncertainties associated with discussed problem. The advantage of using combined grey system theory and uncertainty theory over stochastic and fuzzy theory is that this approach requires neither large set of data points nor robust membership function. In this paper, various evaluation criteria are identified and described as grey linguistic variables which help the managers to evaluate their supplier performance more effectively. The uncertain-goal programming-based model is proposed to optimally select the suitable set of suppliers along with optimal order allocations. The proposed model will help the practitioners to understand the importance of handling recognitive and stochastic uncertainties simultaneously and give practitioners another tool to effectively evaluate and select suitable set of suppliers and optimal order allocation. The proposed methodology leads to future research directions such as the effectiveness of proposed model can be analyzed by applying the model to real case study. The performance of proposed methodology can further be compared with fuzzy-based techniques in order to provide significant effectiveness of this methodology.

Acknowledgments This research was supported by Basic Science Research Program through the National Research Foundation of Korea (NRF) funded by the Ministry of Education, Science and Technology (2012R1A1B3000631 and 2014R1A1A2056338).

References

Amindoust A, Ahmed S, Saghafinia A (2012) Supplier selection and performance evaluation of telecommunication company. Am J Eng Appl Sci 5(1):49–52

Awasthi A, Chauhan SS, Goyal SK (2010) A fuzzy multicriteria approach for evaluating environmental performance of suppliers. Int J Prod Econ 126(2):370–378

Boran FE, Genc S, Kurt M, Akay D (2009) A multi-criteria intuitionistic fuzzy group decision making for supplier selection with TOPSIS method. Expert Syst Appl 36(8):11363–11368

Cai Y, Huang G, Wang X, Li G, Tan Q (2011) An inexact programming approach for supporting ecologically sustainable water supply with the consideration of uncertain water demand by ecosystems. Stochastic Environ Res Risk Assess 25(5):721–735

Cao J, Cao G, Wang WW (2012) A hybrid model using analytic network process and gray relational analysis for bank's IT outsourcing vendor selection. Kybernetes 41(7–8):994–1013

Chai JY, Liu JNK, Ngai EWT (2013) Application of decision-making techniques in supplier selection: a systematic review of literature. Expert Syst Appl 40(10):3872–3885

Chang N-B, Wen C, Chen Y, Yong Y (1996) A grey fuzzy multiobjective programming approach for the optimal planning of a reservoir watershed. Part A: theoretical development. Water Res 30(10):2329–2334

Chen Y-J (2011) Structured methodology for supplier selection and evaluation in a supply chain. Inform Sci 181(9):1651–1670

Chou SY, Chang YH (2008) A decision support system for supplier selection based on a strategy-aligned fuzzy SMART approach. Expert Syst Appl 34(4):2241–2253

Deng J (1985) Special issue of grey system approach. Fuzzy Math 5(2):81–84

Deng J-L (1989) Introduction to grey system theory. J Grey Syst 1(1):1–24

Deng XY, Hu Y, Deng Y, Mahadevan S (2014) Supplier selection using AHP methodology extended by D numbers. Expert Syst Appl 41(1):156–167

Dickson GW (1966) An analysis of vendor selection systems and decisions. J Purchasing 2(1):5–17

Hu X, Munson CL, Fotopoulos SB (2012) Purchasing decisions under stochastic prices: approximate solutions for order time, order quantity and supplier selection. Ann Oper Res 201 (1):287–305

Kahneman D, Tversky A (1979) Prospect theory: an analysis of decision under risk. Econometrica: J Econometric Soc 263–291

Kanagaraj G, Ponnambalam S, Jawahar N (2014) Reliability-based total cost of ownership approach for supplier selection using cuckoo-inspired hybrid algorithm. Int J Adv Manuf Technol (online publication)

Karimi H, Rezaeinia A (2014) Supplier selection using revised multi-segment goal programming model. Int JAdv Manuf Technol 70(5–8):1227–1234

Karmakar S, Mujumdar P (2006) Grey fuzzy optimization model for water quality management of a river system. Adv Water Resour 29(7):1088–1105

Li GD, Yamaguchi D, Nagai M (2007) A grey-based decision-making approach to the supplier selection problem. Math Comput Model 46(3–4):573–581

Li GD, Yamaguchi D, Nagai M (2008) A grey-based rough decision-making approach to supplier selection. Int J Adv Manuf Technol 36(9–10):1032–1040

Liu B (2007) Uncertainty theory. Springer, Berlin

Liu B (2012) Why is there a need for uncertainty theory. J Uncertain Syst 6(1):3–10

Liu B (2015) Uncertainty theory, 4th edn. Springer, Berlin

Liu S, Lin Y, Forrest JYL (2010) Grey systems: theory and applications. Springer, Berlin (Vol 68)

Mujumdar P, Karmakar S (2008) Grey fuzzy multi-objective optimization. Fuzzy Multi-Criteria Decis Making 453–482 (Springer)

Sarkis J, Talluri S (2002) A model for strategic supplier selection. J Supply Chain Manage 38 (1):18–28

Shaw K, Shankar R, Yadav SS, Thakur LS (2012) Supplier selection using fuzzy AHP and fuzzy multi-objective linear programming for developing low carbon supply chain. Expert Syst Appl 39(9):8182–8192

Sipahi S, Timor M (2010) The analytic hierarchy process and analytic network process: an overview of applications. Manage Decis 48(5–6):775–808

Talluri S, Narasimhan R (2003) Vendor evaluation with performance variability: a max–min approach. Eur J Oper Res 146(3):543–552

Wang Y-J, Lee H-S (2007) Generalizing TOPSIS for fuzzy multiple-criteria group decision-making. Comput Math Appl 53(11):1762–1772

Wang H, Che Z, Wang M (2009) A three-phase integrated model for product configuration change problems. Expert Syst Appl 36(3):5491–5509

Weber CA, Current JR, Benton W (1991) Vendor selection criteria and methods. Eur J Oper Res 50(1):2–18

Yucenur GN, Vayvay O, Demirel NC (2011) Supplier selection problem in global supply chains by AHP and ANP approaches under fuzzy environment. Int J Adv Manuf Technol 56(5–8):823–833

A Fuzzy Multi-objective Model with MFCA Approach for Selecting Products Variety in a Textile Supply Chain

Amin Amid, Saeed Zolfaghari and Maryam Siyahpoush

Abstract Offering product variety is one of the strategies for today's competitive advantage. In spite of the fact that greater product variety allows manufacturers to meet a larger share of customer's satisfaction, it increases the complexity of the process and reduces efficiency of the operational performance in a supply chain. Increasing product variety also creates more wastes and more losses, and reduces efficiency of resources. In this chapter, a fuzzy multi-objective model is developed in an imprecise environment by using Material Flow Cost Accounting (MFCA) approach which concentrates on reducing resource consumption by reducing waste and losses as well as active response to environmental requirements. The cost objective function consists of waste management cost, system cost, material cost, and energy cost. The model considers different weights for each of the cost items. Due to uncertain seasonal market demand for textile productions, an imprecise production preparation time and vagueness of input data, a fuzzy weighted max–min multi-objective model is developed for this problem. In this model, the achieved level of objective functions will match the relative importance of objective functions. The model considers three objective functions: quality, product variety, and cost of negative products which focus on the appropriate variety of the products and help decision makers increase green productivity in a textile supply chain. A numerical example is developed to show the application of the model.

A. Amid (✉) · M. Siyahpoush
Industrial Engineering Department, Islamic Azad University North Tehran Branch, Tehran, Iran
e-mail: a_amid@iau-tnb.ac.ir

M. Siyahpoush
e-mail: m_siyahpoush@iau-tnb.ac.ir

A. Amid
Ryerson University, Toronto, Canada

S. Zolfaghari
Department of Mechanical and Industrial Engineering, Ryerson University, Toronto, Canada
e-mail: zolfaghari@ryerson.ca

V. Kachitvichyanukul et al. (eds.), *Toward Sustainable Operations of Supply Chain and Logistics Systems*, EcoProduction,
DOI 10.1007/978-3-319-19006-8_33

Keywords Material flow cost accounting (MFCA) · Product variety · Fuzzy multi-objective decision making · Weighted max–min

1 Introduction

Nowadays, product variety is a strategy that not only helps to increase market share and achieve customers' satisfaction but also is one of strategies for today's competitive advantage. Studies on the effects of increasing product variety from the aspects of marketing and operation indicate a trade-off relationship between the goals of increasing product variety and operational objectives of the organization or a supply chain (Ton and Raman 2010; Xia and Rajagopalan 2009; Thonemann and Bradley 2002).

Although an increase in product variety can increase customer satisfaction and improve the quality of social life, it makes the task of production managers more difficult and it also increases the rate of inventories, wastes, and losses in society. More product variety results in changing production strategies of mass production to mass customization which could reduce inventory problem. Besides many advantages of mass customization, implementation and planning is difficult and balance between variety and complexity should be considered (Daaboul et al. 2011; Berry and Cooper 1999).The other management practice is lean production. Lean production concentrates on continuous elimination of all wastes and reduces losses in the production process and can improve ability of a firm in high diversified range of products, at the lowest cost, with high levels of productivity and optimum quality (Arbós 2002).

On the other hand, the first impact of product variety would be lead to increase in diversity of raw material and resource consumption. Moreover, problem in after and during producing and disposal management should not be disregarded (Jayachandran et al. 2006). Since the competition was globalized, supply chains have been primarily focusing on design development as well as response expedition and reducing operational costs. However, decreasing the level of primary resources and increasing pollution have led to consideration of the environmental performance in the supply chain as the third criterion. Considering environmental performance in order to achieve a social, economic, and environmental balance as the basis of sustainable development is important in supply chain (Dyllick and Hockerts 2002). Therefore, another element which must be taken into account in product variety strategies is environmental considerations.

Furthermore, producing various products with short life cycles and low quality increases wastes and consumption of resources. Nonetheless, any effort in reducing resource consumption should not lead to deteriorating product quality (Niinimäki and Hassi 2011).

Taking these parameters into product and process design consideration needs information not only about how resources are spent and how their wastes is managed but also about how much cost losses. One of the environmental management system tools is material flow cost accounting (MFCA) which hereinafter is named (MFCA) and can avail this information. MFCA focuses on wastes and material losses both in monetary and physical units and can avail detail information which could not extract from other management systems. Under MFCA approach, whole losses especially input material and products are considered so that these items have not been considered in lean production (Tachikawa 2012). It means that wider range of process and product will be covered by MFCA.

Regarding above discussion, it seems that in product variety strategy, environmental impact and product quality should be taken into account. A key contribution of this study is to consider environmental factors with MFCA approach for choosing the optimal product variety by a fuzzy multi-objective model. Defined material losses and wastes in MFCA which have been named negative products (METI 2007) are regarded as environmental impact and are considered as an objective, but due to these data are both in monetary and physical units, we have considered different weight importance for each items costs in MFCA.

The model is developed in textile supply chain because textile supply chain is characterized by high product variety, high pollution and material and energy consumption, fashion trend, uncertainty in demand (Turker and Altuntas 2014; Allwood et al. 2008; Thomassey 2010), so it can be a good case for developing model. Accordingly, a weighted max–min fuzzy multi-objective model is represented in order to optimize the variety and quality of products with minimizing negative products cost (material losses and wastes) based on MFCA.

This paper is organized as follows. A review of the literature on product variety, product quality and MFCA, and textile supply chain are presented in Sect. 2. Objective functions are formulated in Sect. 3. The proposed fuzzy multi-objective model and weighted max–min method are developed in Sect. 4. A case study is presented in Sect. 5, followed by the concluding remarks in Sect. 6.

2 Literature Review

2.1 Product Variety

Higher product variety can meet the needs of heterogeneous customers. However, increasing variety of products increases the complexities particularly in the supply chain. Frizelle and Woodcock (1995) defined complexity results from variety and uncertainty in dynamical systems. Hu et al. (2008) presented a mix-model of product variety including manufacturing complexity for assembly systems and supply chain. Another area that is affected by product variety is in the resources with limited capacity such as material and energy, human resource, and equipment. Thus, product variety needs to be optimized. Susanto and Bhattacharya (2011) proposed a fuzzy multi-objective linear programming to optimize profit and product mix and considered the best feasible and optimal solution. Descending trend of performance can be balanced with reaching product variety to a high degree of similarity where it accurately forecasts the demand (Wan et al. 2012).

On the other hand, with increasing product variety, there will be a need for more changes in settings and processes, resulting in an escalated probability of creating wastes and reducing efficiency of resources. Therefore, the objective of minimizing consumption of resources and wastes in product variety in ecosystems must be considered as well (Tang and Yam 1996). Thus by focusing on these aspects, the importance of environmental considerations in selecting the optimized product variety will be more apparent.

2.2 *Material Flow Cost Accounting Approach*

Assessment and reducing environmental impacts and performance evaluation in a supply chain are issues which have received great deal of attention among researchers (Sarkis 2003; Canito et al. 2012; Clark and Clegg 2000).

Many studies also have been conducted on environmental costing in an organization or a supply chain (Roy et al. 2009). As Nikolopoulou and Ierapetritou (2012) have mentioned, there are many systematic methods which can be used as environmental performance indicators of product and process such as product life cycle assessment (LCA) and minimal environmental impact. Letmathe and Doost (2000) proposed cause and effect analysis to assign environmental cost. De Beer and Friend (2006) considered 5 types of costs including internal and external costs and also regarded environmental revenues.

Mangers often are interested in reduced adverse environmental impacts whether these activities lead to company profit. In order to meet these objectives, company needs to have useful and clear information about environmental costs. EMAs are calculation methods that connect environmental impacts to the economy (Higashida et al. 2013).

With respect to some hidden costs in environmental cost accounting that often are not considered, it seems that tracing flow of material could avail more information. Jasch (2003) evaluated environmental costs by using the concept of product life cycle and MFCA.

Nakano and Hirao (2011) used LCA and MFCA for data collection and proposed a supply chain collaboration model (SCCM) to promote improvement activity of product and environmental performance. Zhang and Huang (2013) have provided a fuzzy multi-objective model which optimizes economic. Hung et al. (2006) considered different qualitative and quantitative parameters and proposed a multi-objective model and used fuzzy AHP.

One of new methods that are recently introduced as one of environmental management tools is MFCA. MFCA method focuses on cost of material losses and wastes. Reducing wastes can lead to resource consumption reduction followed by adverse effects on its management. In fact, reducing adverse environmental impact throughout life cycle by environmental friendly and cost effective product could be result of applying MFCA (Jasch 2009). Integrating MFCA and enterprise resource planning (ERP) can provide useful information about quantity of wastes and other loss cost (Fakoya and van der Poll 2013).

In this approach, wastes and other material losses are defined as negative products. In MFCA analysis, at first, material flow in a quantity center is traced, the amount of material losses and wastes is identified, and then, the costs are calculated and allocated to losses (Kokubu and Kitada 2012). In MFCA, all cost items of negative product which are not calculated in conventional cost accounting separately are systemized and established (Nakajima 2010). Three types of costs, including energy and material costs, system costs, and wastes management costs, have been identified. Using an appropriate apportionment criteria based on mass balance of input and output, materials will be allocated to product and material losses and wastes (ISO 14051 2011; Kokubu and Kitada 2012). It provides possibility to detect other hidden and intangible costs in losses of material flow.

In order to reduce material losses and wastes besides improve resource productivity, efficiency corporation among related enterprises in a supply chain is needed (METI 2007; Nakajima 2010).

2.3 Product Quality

Quality is one of the distinction strategies that create prestige for manufacturers. High focus on environmental consideration in design phase must not lead to disregard customer expectation. Cristopher et al. (1996) developed Green QFD which utilizes the integration of the product life cycle and environmental considerations.

There are factors for measuring the quality of each product. Some examples of these features in the case of clothing and textiles are roughness and softness of handle, wrinkle-taking, fitness, color and light fastness, durability, suspension, and other related features. Kuo et al. (2009) integrated environmental considerations and the QFD by using the fuzzy systems and provided ECO-QFD function which is used in the product design. Yang et al. (2003) presented a QFD fuzzy system for buildability evaluation in design phase, and Chen and Ko (2008) employed a nonlinear fuzzy model of quality function by considering Kano concept. Taylan (2011) implemented fuzzy linguistic terms to state product quality and product grading for continuous quality assessment.

2.4 Textile Supply Chain

Textile supply chain includes suppliers, manufacturers, and customers. Manufacturing sector involves subsectors such as producers of fiber and yarn, producers of fabric inclusive knitting and weaving sectors, dyeing, finishing, printing, and clothing/sewing sectors (Seuring 2004; Allwood et al. 2008). Textile and clothing supply chains are usually a long chain which have many complexities due to the heterogeneous parts and the presence of small and medium enterprises (Salman 2006). In fact, efforts for searching lower production costs have led to chain moves toward far east (Bonacich et al. 1994; Thomassey 2010) which have lower labor and energy cost but instance consequent energy consumption for related transport has been observed.

Clothing and textile supply chains because of importance of quick responsibility and flexibility are called agile supply chains, and they represent a good example of a buyer-driven chain. High product variety will follow problem in production management, inventory, delivery time, responsively as well as increasing costs. High flexible manufacturing also is needed (Salman 2006; Turker and Altuntas 2014), so that the most items in each season and every new collections are not repeated (Thomassy 2010). Thus, unforeseen variation of season, fashion, and the other customer requirement can make significant impact on variety. In addition, in order to achieve high flexibility and responsibility, often environmental impact of this changes in textile industry is being disregarded (Turker and Altuntas 2014).

Toxic chemical material, water and energy consumption, emission and pollution, problem in disposal and solid wastes, distribution problem, etc., are some of negative environmental problems of textile industry (Allwook et al. 2000). Kalliala and Talvenmaa (2000) and Seuring (2004) addressed some details on environmental depiction of textile products in textile supply chain.

Producing low-quality textile products with short life cycles with the aim of continuous replacement of production leads to increased wastes and adverse environmental effects (Niinimäki and Hassi 2011).

Applying environmental management systems (EMSs) often put stress on consequent environmental impacts that in textile and fashion industry are in primary stages (Lo et al. 2012).

In this paper, a fuzzy multi-objective model which considers three objective functions, quality, product variety and cost of negative product by using MFCA approach, is developed. Next section will developed this issue.

3 Formulating the Model

In this paper, three different objective functions including product quality, cost, and variety are considered. Here, the cost refers to the total cost, and it is assigned to negative product by the definition in the MFCA approach. MFCA with monetary and physical tracing makes visible negative products and their aspects.

In this section, we primarily provide each of the objective functions and correspondingly the constraints. Then, we explain the solution methodology. The variables that are used in the model have been presented in Table 1.

Table 1 Description of variables

Variables	Description	Variables	Description
x_{ij}	Zero-one variable for model *i* to be allocated to design *j*	L	Minimum model constraints
a_{ij}	Number of available variety for model *i* design *j*	M, U	Minimum and maximum limitation for design
c_{ij}	Negative product costs ratio allocated to model *i* design *j*	Z_1	Objective function of variety
q_{ij}	Quality level	Z_2	Objective function of MFCA
w	Importance weight from AHP comparison	Z_3	Objective function of quality
UVI	Used variety indicator	b_{ijk}	Numerical value of comparison items by kth decision maker
α_i	Average weight of importance of variant *i* for customers	n	Number of respondents
V_i	Proposed variant	C_t	Total unit costs related to negative products
NV	Total number of all possible variants	β_i	Weight importance of each cost items
f_i	Each cost item in MFCA		

3.1 Objective Function of Product Variety

Generally, product variety can be resulted from changes in models, design, size, color, packaging, and distribution methods. Product variety is the number of different products which is provided to meet the explicit to inexplicit needs of the customers by the firm. Variety is an indicator which is highly effective on the value created for the customer, and the *used variety indicator* can be defined as follows (Daaboul et al. 2011):

$$UVI = \sum_{i=1}^{n} \alpha_i V_i / NV \tag{1}$$

where α_i stands for the average weight of importance of variant *i* for customers; V_i represents a proposed variant $1 \leq i \leq n$; $i \in Z$; and NV is the total number of all possible variants.

In this paper, the variety function is considered as follows:

$$Z_1 = \sum_{i=1}^{n} \sum_{j=1}^{m} a_{ij} x_{ij} \tag{2}$$

where $x_{ij} \in \{0,1\}$, it means that x_{ij} is a zero-one variable that is equal to 1 if model *i* is assigned to design *j*, and equal to 0 otherwise.

Weights of available variants are assumed to be equal as marketing expert distinguished.

3.2 MFCA Objective Function (Negative Product Cost Function)

In MFCA, after quantifying material losses and wastes (negative product), cost items are allocated to the products and wasted materials with a suitable criterion (METI 2007; ISO 14051 2011). Due to in MFCA, material losses and related costs are traced simultaneously, and overlapping costs can make incorrect decision. For example, high costs of materials compensate considerable wastes in energy, and the total cost of negative products does not provide accurate information for comparability of data especially where objectives are considered green. In green strategies, main idea is concentrated on reduction adverse environmental impact and losses as well as wastes rather than monetary impacts. Consequently, in practice, the importance and weight of each of these costs are not the same for different systems. In this paper, we consider different weights of importance for each of the cost items. Moreover, to specify the weight of items, the pair-wise comparison in analytic hierarchy process (AHP) is utilized.

Thus, primary function of the MFCA which is a cost function can be described as follow:

$$C_t = \sum_{i=1}^{4} \beta_i f_i \tag{3}$$

$$w = \sqrt{[n]\prod_{k=1}^{n} b_{ijk}} \quad w \geq 0 \tag{4}$$

$$\sum_{i=1}^{n} \beta_i = 1 \quad \beta_i \geq 0 \tag{5}$$

where f_1 is designated as the unit cost of material wasted, f_2 stands for the unit cost of energy wasted, f_3 represents unit system cost wasted, and f_4 is assigned as the unit wastes management cost (ISO 14051 2011).

C_t total unit costs of negative products,

β_i weight importance of each cost items which is calculated from Eq. 4,

b_{ijk} numerical value of comparison items (*ij*) by *k*th decision maker (DM), and

n number of respondents.

MFCA objective function which refers to cost of negative products is given as follows:

$$Z_2 = \sum_{i}^{n} \sum_{j}^{m} c_{ij} x_{ij} \tag{6}$$

3.3 *Quality Objective Function*

This section provides the quality objective function. As mentioned above, quality is an eligibility criterion of the product which is a key factor for product success in supply chain and is influenced by the quality of the input materials of downstream suppliers. Therefore, implementing strategies to reduce consumption of resources should not lead to reduction in the quality of the final product.

The quality objective function can be written as follows:

$$Z_3 = \sum_{i}^{n} \sum_{j}^{m} q_{ij} x_{ij} \tag{7}$$

4 Constraints and Solution Approach

While the demand for variety is theoretically unlimited, product diversification is commonly limited with supply side. Constraints in the supply side usually include resource limitations. Most of these restrictions have nowadays been solved with advances in technology (Tang and Yam 1996). Therefore, in the developed model, it is assumed that all constraints are related to production technology and equipment and they are crisp. In addition, it is possible to produce several products at the same time.

Given the above, the following model is proposed here:

$$\text{Max.}\, Z_1 = \sum_{i}^{n} \sum_{j}^{m} a_{ij} x_{ij} \tag{8}$$

$$\text{Min.}\, Z_2 = \sum_{i}^{n} \sum_{j}^{m} c_{ij} x_{ij} \tag{9}$$

$$\text{Max.}\, Z_3 = \sum_{i}^{n} \sum_{j}^{m} q_{ij} x_{ij} \tag{10}$$

Subject to:

$$\sum_{j=1}^{m} x_{ij} \geq L \quad i = 1, 2, \ldots, n \tag{11}$$

$$M \leq \sum_{i=1}^{n} x_{ij} \leq U \quad j = 1, 2, \ldots, m \tag{12}$$

Practical decision making is sometimes different than theoretical decision making. In practice, decision makers quite often choose the nearest solution to the desired objective, or they may make a decision in the absence of complete and accurate information on criteria and constraints. In such cases, fuzzy approaches are more effective. This may be more apparent in the case of quality where customers may use ambiguous terms which describe quality. For this reason, we use a fuzzy approach with weighted max–min to solve the model.

Fuzzy logic systems have been used extensively in decision-making problems especially supplier selection problems in a supply chain. Shaw et al. (2012) and Lee (2009) applied fuzzy models for supplier selection.

In this paper, different importance weight for each objective was considered. Weight of each objective can be achieved by using the method of AHP matrix of paired comparisons (using Eq. 4). To solve the problem with weighted max–min fuzzy method, we first need to set membership functions of fuzzy constraints and goals. Then, the minimum and maximum of each membership function are obtained under the constraints (Amid et al. 2006) (Table 2).

It is assumed that membership functions for minimization goal (z_k) and maximization goals (z_l) are expressed as follow:

$$\mu_{zk}(x) = \begin{cases} 1 & \text{for } z_k \leq z_k^- \\ (z_k^+ - z_k(x))/(z_k^+ - z_k^-) & \text{for } z_k^- \leq z_k(x) \leq z_k^+ \\ 0 & \text{for } z_k \geq z_k^+ \end{cases} \quad (k = 1, 2, \ldots, p) \tag{13}$$

Table 2 Variables for the fuzzy model

Variables	Description	Variables	Description
Z	Objective function	w_j	Importance of fuzzy objective function j
L	Index of max objectives	Z_j^+	Upper bound of objective function j
K	Index of min objectives	Z_j^-	Lower bound of objective function j
$\mu_{zj}(x)$	Membership function	λ_j	Achievement level for objective function j

$$\mu_{zl}(x) = \begin{cases} 1 & \text{for } z_l \leq z_l^+ \\ (z_l(x) - z_l^-)/(z_l^+ - z_l^-) & \text{for } z_l^- \leq z_l(x) \leq z_l^+ \\ 0 & \text{for } z_l \leq z_l^- \end{cases} \quad (l = p+1, p+2, \ldots, q) \tag{14}$$

In this model, the decision maker's preferences are considered in terms of weights in the model (Eq. 4). Therefore, the DM expects that achievement level of membership function should be close to the ratio of objective weights. This proportionality, however, may not always happen. If this proportionality is important to the DM, the weighted max–min method can be used. This model is formulated as follows (Amid et al. 2011; Dastkhan et al. 2011):

$$\text{Max.}\ \lambda \tag{15}$$

Subject to:

$$w_j \lambda \leq f_{\mu z_j}(x), \quad j = 1, \ldots, q \quad (\text{for all objective functions}) \tag{16}$$

$$g_r(x) \leq b_r, \quad r = 1, \ldots, m \tag{17}$$

$$0 \leq \lambda \leq 1 \tag{18}$$

$$\sum_{j=1}^{q} w_j = 1, \qquad w_j \geq 0 \qquad x_i \geq 0, \qquad i = 1, \ldots, n \tag{19}$$

New membership functions are defined as follows:

$$\mu'_{zk}(x) = \begin{cases} 1/w_k & \text{for } z_k \leq z_k^- \\ \frac{f_{\mu z_k}(x)}{w_k} & \text{for } z_k^- \leq z_k(x) \leq z_k^+ \\ 0 & \text{for } z_k \geq z_k^+ \end{cases} \quad (k = 1, 2, \ldots, p) \tag{20}$$

$$\mu'_{zl}(x) = \begin{cases} 1/w_l & \text{for } z_l \geq z_l^+ \\ \frac{f_{\mu z_l}(x)}{w_l} & \text{for } z_l^- \leq z_l(x) \leq z_l^+ (l = 1, 2, \ldots, q) \\ 0 & \text{for } z_l \leq z_l^- \end{cases} \tag{21}$$

5 Case Study

In this section, we describe the model by using a numerical example in textile industry. As mentioned above, textile industry mainly includes sub-units such as spinning, knitting/weaving, and wet processing that have high consumption of chemicals, energy, and especially water. The environmental impact of wastewater and wastes in this industry is significant. In this research, a textile firm is considered with spinning, weaving, dyeing, printing, finishing, and sewing units. Every unit in this firm has been considered as a quantity center and total cost of wastes, and material losses along whole quantity centers related to product variety have been calculated. Obtained data in quantity centers demonstrated that energy losses are more than material wastes but due to material cost was higher so, wastes of energy covered by material wastes. Negative product and associated cost have been increasing by raising the number of variety of products in colors and design due to changes in the setting as well as change in the final product type. In printing quantity center, all waste material such as dyestuff, auxiliary and chemical material, defects products, and other wastes have been calculated based on mass balance between inputs and outputs. Other costs including material, energy, system cost, and waste management cost (Dollars) are allocated by accountants and production experts. For quantifying, data in each quantity center have been considered as percent negative products cost ratio between 0 and 1. For example, 10 kg of negative product from 100 kg input product represents 0.1 (for more detail refers to METI 2007).

Quality level of a product can be taken into account by performance of a product on different features or criteria. Selected features of quality for the textile products in the case study are color fastness, light and washing fastness, strength, and staining which are measured by different standards and criteria such as gray and blue scales which are numbered from 1(low) to 5(high). Quality level of each product has been stated in level between 0 and 1. In practice, however, it may not be possible to select a precise value of each product in each criterion. In these cases, based on experts' opinion, the number between 1 and 5 is assigned to each product. This ratio is calculated from summation or total performance of a product on all criteria. For instance, quality ratio of a product with 20 out of 25 is 0.8.

The product family includes five different models. We assumed that in the set of products P_i where $i = 1, \ldots, 5$ by changing in design j where $j = 1, 2, 3$, different products can be produced. The variety is increased by increasing the number of design colors.

The data set for the values of quality level as well as the MFCA (Eq. 3) and available variety of each product are presented in Table 3.

Table 3 Data set for the problem

P	a_{ij}	c_{ij}	q_{ij}	P	a_{ij}	c_{ij}	q_{ij}	P	a_{ij}	c_{ij}	q_{ij}
P_{11}	2	0.22	0.8	P_{12}	4	0.35	0.6	P_{13}	1	0.29	0.8
P_{21}	1	0.16	0.7	P_{22}	5	0.38	0.4	P_{23}	1	0.20	0.6
P_{31}	1	0.17	0.5	P_{32}	6	0.24	0.4	P_{33}	8	0.35	0.7
P_{41}	5	0.22	0.4	P_{42}	1	0.28	0.5	P_{43}	3	0.20	0.7
P_{51}	6	0.26	0.6	P_{52}	1	0.20	0.65	P_{53}	7	0.40	0.5

Table 4 Max and min values for the objective functions under constraint in fuzzy environment

Objective function	w_j	Z_j^+	Z_j^-	Selected products
Z_1	0.45	46.00	8.00	$x_{11}, x_{12}, x_{22}, x_{32}, x_{33}, x_{41}, x_{43}, x_{51}, x_{53}$
Z_2	0.35	2.31	0.95	$x_{11}, x_{21}, x_{31}, x_{43}, x_{52}$
Z_3	0.20	6.00	3.35	$x_{11}, x_{12}, x_{13}, x_{21}, x_{33}, x_{42}, x_{43}, x_{51}, x_{52}$

We adopted linear programming to obtain the minimum and maximum of membership functions, and we used LINGO software to solve the model. The results are summarized in Table 4.

So the model will be provided as below:

$$\text{Max.}\ \lambda \tag{22}$$

Subject to:

$$0.45\lambda \leq (Z_1(x) - 8)/38 \tag{23}$$

$$0.35\lambda \leq (2.31 - Z_2(x))/1.36 \tag{24}$$

$$0.2\lambda \leq (Z_3(x) - 3.35)/2.65 \tag{25}$$

$$\sum_{j=1}^{3} x_{ij} \geq 1 \quad i = 1, \ldots, 5 \tag{26}$$

$$1 \leq \sum_{i=1}^{5} x_{ij} \leq 4 \quad j = 1, 2, 3 \tag{27}$$

$$x_{ij} \in \{0, 1\} \tag{28}$$

$$0 \leq \lambda \leq 1 \tag{29}$$

After solving the model in the form of a single-objective model, the following results are represented in Table 5.

Table 5 shows that Z_2 and Z_3 have been achieved simultaneously with product variety 26 which are obtained from 7 chosen products. Results from Tables 4 and 5 indicate that by implementing of the developed multi-objective model, average negative product decrease to 0.25 (Z_2 = 1.69 divided by 7) if the variety objective function is considered as a single objective under same constraints; 9 products will

Table 5 Solution of the numerical example

Method	Z_1	Z_2	Z_3	λ_1	λ_2	λ_3	Selected products
Weighted max–min	26	1.69	4.2	0.58	0.45	0.26	$(x_{11}, x_{23}, x_{32}, x_{33}, x_{41}, x_{43}, x_{51})$

be selected and average negative product costs will be increased to 0.3 ($Z_2^+ = 2.62$ divided by 9). In other words, average negative product costs have been reduced 16.67 % that subsequently will affect the amount of negative products.

The average of quality for 7 products with 26 varieties is 0.6 ($Z_3 = 4.2$ divided by 7). It represents that average quality level has been improved around 5.27 % rather than single-objective model with 9 products (0.57 = 5.1/9).

On the other hand, concentration on cost minimization as a single objective will lead to reduced variety to minimal mode (8 varieties) which is unexpected for managers and cannot meet the customer satisfaction and market demand. The results distinguish existent trade-off between objectives clearly and accentuate that product variety goal should be considered as multi-objective model. In fact, the model with optimization objectives under constraints has provided the best results inclusive fewer costs, higher product variety and quality in which aligned with decision maker's preferences.

It should be noted that the results hardly depend on decision makers' preferences. The obtained results represent that all managerial preferences have been met in developed model.

The main objective in Z_2 function is decreasing costs by MFCA approach which by reducing consumption energy and material and wastes reduces environmental impact.

6 Conclusion and Further Research

In any manufacturing firm, determining the firm's desirable product variety is among the most important management decisions. The product variety is affected by many parameters in terms of both market and technology and operational costs. With increasing attentions to sustainability in recent years, the amount of pollution levels, wastes, and resources are part of environmental considerations in supply chain. As such, selecting the optimal products that meet customers' expectations in terms of variety as well as managers' satisfaction in terms of the operational and environmental factors poses particular importance.

This paper presents a multi-objective model with choice of products that satisfies the desirable business goals and simultaneously with decrease resource consumption increases green productivity. What should be taken into consideration is deselecting a product does not mean rejecting it. As a matter of fact, to increase variety, decision makers can enter products into PDCA cycle in the MFCA and by reducing negative product and subsequently cost in supply chain can cause to select products and increase variety. Overviews of product design and process by choosing recyclable material as well as product strategy and supplier selection are effective in achieving these goals in manufacturing of these products. The model focuses on the appropriate variety of the products and helps decision makers to quantify how many varieties could be apply if green productivity and profit increased in a textile supply chain. Weight of the desired targets has a major role in selecting the products. In addition, since different products have different resource usages and wastes, considering the degree of differences is important in accounting negative product costs, which provides more clarity for decision making. In the weighted max–min method, the products selected can satisfy the objective functions with a higher level of satisfaction and provide more acceptable results based on decision makers' preference. However in this method, achievement levels for other objectives are closer to desired targets that show how this model can help to achieve the strategic objectives of a supply chain within reducing environmental negative impact of a textile supply chain.

Generally, available information on material flows provided by MFCA along upstream and downstream firms in a supply chain can ensure for reducing material losses and help to supply chain managers that have an effective decision on resource productivity as well as improvement in competitive strategies such as product variety and quality, while cost reduction is happened.

In real world, however, it may not be possible to select a precise value of each product in each criterion. For instance, the quality level of each product could be considered as a fuzzy variable. Further research is required to provide new fuzzy model and calculation.

MFCA objective function can be considered as an environmental impact index. While environmental negative impact reduces, green productivity increases. Designing of a measurement system that represents the relationships between green productivity and MFCA initiatives could be new area for researching.

Acknowledgments The authors gratefully acknowledge support of the Iran poplin textile company and Textile Industry Association of Iran in research process.

References

Allwood JM, Laursen SE, Russell SN, de Rodríguez CM, Bocken NMP (2008) An approach to scenario analysis of the sustainability of an industrial sector applied to clothing and textiles in the UK. J Clean Prod 16(12):1234–1246

Amid A, Ghodsypour SH, O'Brien C (2006) Fuzzy multiobjective linear model for supplier selection in a supply chain. Int J Prod Econ 104(2):394–407

Amid A, Ghodsypour SH, O'Brien C (2011) A weighted max–min model for fuzzy multi-objective supplier selection in a supply chain. Int J Prod Econ 131(1):139–145

Arbós LC (2002) Design of a rapid response and high efficiency service by lean production principles: methodology and evaluation of variability of performance. Int J Prod Econ 80 (2):169–183

Berry WL, Cooper MC (1999) Manufacturing flexibility: methods for measuring the impact of product variety on performance in process industries. J Oper Manag 17(2):163–178

Bonacich E, Cheng L, Chinchilla N, Hamilton N, Ong P (1994) Global pro-duction: the apparel industry in the Pacific Rim. Temple University Press, Pennsylvania

Caniato F, Caridi M, Crippa L, Moretto A (2012) Environmental sustainability in fashion supply chains: an exploratory case based research. Int J Prod Econ 135(2):659–670

Chen LH, Ko WC (2008) A fuzzy nonlinear model for quality function deployment considering Kano's concept. Math Comput Model 48(3):581–593

Clarke T, Clegg S (2000) Changing paradigms. Longman, Sydney
Cristopher M, Deshmukh A, Wang B (1996) Green quality function deployment. In: Proceeding of the 4th international conference on environmentally conscious design and manufacturing, pp 297–304
Daaboul J, Da Cunha C, Bernard A, Laroche F (2011) Design for mass customization: product variety vs. process variety. CIRP Ann Manufact Technol 60(1):169–174
Dastkhan H, Gharneh NS, Golmakani H (2011) A linguistic-based portfolio selection model using weighted max–min operator and hybrid genetic algorithm. Expert Syst Appl 38(9):11735–11743
De Beer P, Friend F (2006) Environmental accounting: a management tool for enhancing corporate environmental and economic performance. Ecol Econ 58(3):548–560
Dyllick T, Hockerts K (2002) Beyond the business case for corporate sustainability. Bus Strategy Environ 11(2):130–141
Fakoya MB, van der Poll HM (2013) Integrating ERP and MFCA systems for improved waste-reduction decisions in a brewery in South Africa. J Clean Prod 40:136–140
Frizelle G, Woodcock E (1995) Measuring complexity as an aid to developing operational strategy. Int J Oper Prod Manag 15(5):26–39
Higashida A, Kokubu K, Shinohara A (2013) Introducing material flow cost accounting and creating visibility–analyzing MFCA in practice based on a longitudinal case study. In: Proceedings of the seventh asia pacific interdisciplinary research in accounting conference, pp 26–28
Hu SJ, Zhu X, Wang H, Koren Y (2008) Product variety and manufacturing complexity in assembly systems and supply chains. CIRP Ann Manufact Technol 57(1):45–48
Hung ML, Yang WF, Ma HW, Yang YM (2006) A novel multiobjective programming approach dealing with qualitative and quantitative objectives for environmental management. Ecol Econ 56(4):584–593
International standard ISO 14051 (2011) Gerenal framework
Jasch C (2003) The use of environmental management accounting (EMA) for identifying environmental costs. J Clean Prod 11(6):667–676
Jasch C (2009) Environmental and material flow cost accounting. Springer, Berlin
Jayachandran R, Singh S, Goodyer J, Popplewell K (2006) The design of a sustainable manufacturing system: a case study of its importance to product variety manufacturing. In: 2nd virtual international conference on intelligent production machines and systems, Cardiff, UK
Kalliala E, Talvenmaa P (2000) Environmental profile of textile wet processing in Finland. J Clean Prod 8(2):143–154
Kokubu K, Kitada H (2012) Material flow cost accounting and conventional management thinking: introducing a new environmental management accounting tool into companies. In: Interdisciplinary perspectives on ac-counting conference, Cardiff
Kuo TC, Wu HH, Shieh JI (2009) Integration of environmental considerations in quality function deployment by using fuzzy logic. Expert Syst Appl 36(3):7148–7156
Lee AH (2009) A fuzzy supplier selection model with the consideration of benefits, opportunities, costs and risks. Expert Syst Appl 36(2):2879–2893
Letmathe P, Doost RK (2000) Environmental cost accounting and auditing. Manag Auditing J 15 (8):424–431
Lo CK, Yeung AC, Cheng TCE (2012) The impact of environmental management systems on financial performance in fashion and textiles industries. Int J Prod Econ 135(2):561–567
METI Ministry of Economy, Trade and Industry (2007) Guide for material flow cost accounting (Ver.1). In: Environmental industries office, environmental policy division industrial science and technology policy and environment bureau, Japan
Nakajima M, 中嶌道靖 (2010) Environmental management accounting for sustainable manufacturing: establishing management system of material flow cost accounting (MFCA)
Nakano K, Hirao M (2011) Collaborative activity with business partners for improvement of product environmental performance using LCA. J Clean Prod 19(11):1189–1197

Niinimäki K, Hassi L (2011) Emerging design strategies in sustainable production and consumption of textiles and clothing. J Clean Prod 19(16):1876–1883
Nikolopoulou A, Ierapetritou MG (2012) Optimal design of sustainable chemical processes and supply chains: a review. Comput Chem Eng 44:94–103
Roy P, Nei D, Orikasa T, Xu Q, Okadome H, Nakamura N, Shiina T (2009) A review of life cycle assessment (LCA) on some food products. J Food Eng 90(1):1–10
Salman M (2006) Global integration of supply chains in textile/apparel industry. The University of Nottingham, Nottingham
Sarkis J (2003) A strategic decision framework for green supply chain management. J Clean Prod 11(4):397–409
Seuring S (2004) Integrated chain management and supply chain management comparative analysis and illustrative cases. J Clean Prod 12(8):1059–1071
Shaw K, Shankar R, Yadav SS, Thakur LS (2012) Supplier selection using fuzzy AHP and fuzzy multi-objective linear programming for developing low carbon supply chain. Expert Syst Appl 39(9):8182–8192
Susanto S, Bhattacharya A (2011) Compromise fuzzy multi-objective linear programming (CFMOLP) heuristic for product-mix determination. Comput Ind Eng 61(3):582–590
Tachikawa H (2012) Conclusion/comments on MCSP—MFCA. In: Seminar on MFCA—its applications and impacts by Malaysian companies
Tang EP, Yam RC (1996) Product variety strategy-an environmental perspective. Integr Manufact Syst 7(6):24–29
Taylan O (2011) Estimating the quality of process yield by fuzzy sets and systems. Expert Syst Appl 38(10):12599–12607
Thomassey S (2010) Sales forecasts in clothing industry: the key success factor of the supply chain management. Int J Prod Econ 128(2):470–483
Thonemann UW, Bradley JR (2002) The effect of product variety on supply-chain performance. Eur J Oper Res 143(3):548–569
Ton Z, Raman A (2010) The effect of product variety and inventory levels on retail store sales: a longitudinal study. Prod Oper Manag 19(5):546–560
Turker D, Altuntas C (2014) Sustainable supply chain management in the fast fashion industry: an analysis of corporate reports. Eur Manag J 32(5):837–849
Wan X, Evers PT, Dresner ME (2012) Too much of a good thing: the impact of product variety on operations and sales performance. J Oper Manag 30(4):316–324
Xia N, Rajagopalan S (2009) Standard vs. custom products: variety, lead time, and price competition. Mark Sci 28(5):887–900
Yang YQ, Wang SQ, Dulaimi M, Low SP (2003) A fuzzy quality function deployment system for buildable design decision-makings. Autom Constr 12(4):381–393
Zhang X, Huang G (2013) Optimization of environmental management strategies through a dynamic stochastic possibilistic multiobjective program. J Hazard Mater 246:257–266

Part VI
Mathematical Programming, Operations Research and Statistical Techniques

Manpower Planning with Multiple Tasks for a Call Center in Healthcare Service

Sarusakorn Booranadiloak, Udom Janjarassuk, Kanokporn Rienkhemaniyom and Chumpol Yuangyai

Abstract Private hospitals offer an advanced appointment program that allows patients to receive medical care services at their convenient time. While the amount of callers has increased, many hospitals face difficulties to determine the number of operators to promptly respond the calls. Long waiting time may cause some callers to abandon their lines, which leads to the loss of opportunity. This chapter focuses on how to determine the optimal number of operators and their assignment in a service time horizon. An integrated framework is proposed using mixed-integer nonlinear programming to solve the staff planning and allocation problem. The result shows that the framework is viable.

Keywords Healthcare service · Mixed-integer nonlinear programming · Staff planning and allocation

1 Introduction

A call center has been playing a vital role in hospital as it is an initial contact point between customers and a hospital. The call center provides convenient services to customers, such as delivering hospital's service information and appointment rearrangement. While the demand of incoming calls has rapidly increased, many hospitals face a problem to determine the suitable number of operators to handle those calls. Several hospitals have no systematic approach to verify whether or not the current operators are capable of responding the calls (Nah and Kim 2013).

S. Booranadiloak · U. Janjarassuk · C. Yuangyai (✉)
Department of Industrial Engineering, Faculty of Engineering,
King Mongkut's Institute of Technology Ladkrabang, Bangkok 10520, Thailand
e-mail: chumpol.yuangyai@gmail.com

K. Rienkhemaniyom
Graduate School of Management and Innovation, King Mongkut's University of Technology,
Thonburi, Bangkok 10140, Thailand
e-mail: kychumpo@kmitl.ac.th

V. Kachitvichyanukul et al. (eds.), *Toward Sustainable Operations of Supply Chain and Logistics Systems*, EcoProduction,
DOI 10.1007/978-3-319-19006-8_34

Additionally, the hospital call center also suffered from the staff-arrangement issues due to performing multiple tasks, e.g., returning calls, and administrative work (Moore et al. 2001). The operators' competencies rely on experiences and intuition to prioritize when and which task has to be performed. The lack of these competencies results in customer's long waiting time and increased abandonment rate.

Much research focuses on considering only one type of inbound calls. In fact, there are several types of customers' needs leading to a variety of unpredictable types of inbound calls. For instance, some customers may want to get information and make an appointment, while other customers want to report and complain regarding services of hospitals. Other types of task include dialing outbound calls and other tedious jobs, such as collecting the complaints and comments from customers or other administrative jobs. These extra tasks make the operators not able to response customer calls effectively (Aksin et al. 2007; Atlason et al. 2008; Cezik and L'Ecuyer 2008; Ernst et al. 2004; Gans et al. 2003; Mehrotra 1997). To address the staff planning and allocation problem, several researchers focus on call center problem.

Specially, Nah and Kim (2013) have proposed a mixed-integer nonlinear model to determine the number of staffs and time slot allocation. Their objective function is to minimize the labor cost and penalty cost (both waiting time and abandon rate). However, in their study, a constant time is considered for each outbound calls, which could not represent a real situation where each outbound call has different characteristics, e.g, reminding customers of appointments regular of promotion announcement, keep close relation with special customers (for example, greeting on customer birthday), and respond to special requests or queries (Saltzman and Mehrotra 2007). Therefore, in our study, we will consider the additional constraints of different outbound characteristics as well as the administrative jobs.

The objective of this chapter is to develop a mathematical model to assist call center staff planner to determine a set of number of staffs and how to assign task to the staffs in each shift while observing labor cost, waiting time, and penalty.

The organization of this paper is as follows: We first present call center work process, then in Sect. 3, our mathematical model formation is explained. A numerical example is demonstrated in Sects. 4, and 5 provide conclusion.

2 Hospital Call Center Tasks

Figure 1 shows the hospital call center work process. When customers call, the inbound type is classified into three types, namely information inquiry, rearrangement requests, and complaints.

In the case of information inquiry, if the operator is capable, he/she will respond to the call immediately; otherwise, they will ask customer to wait in line until the customer receives information asked. However, if the customers wait for a long period of time, they may abandon calls and the operator will return call once he/she contact to other department for more information.

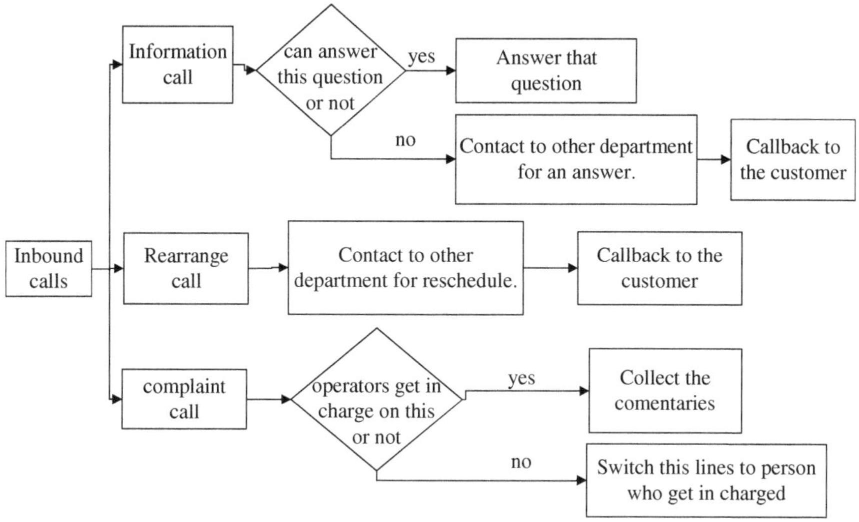

Fig. 1 Hospital call center work process

The second type of inbound call is rearranging an appointment. The operator does an online check in hospital database to make appointment changes for the customer. However, in some cases, the operator may require more time to contact other department before making changes. In this case, the operator will call back the customer once the issues have been resolved.

The last type of task is complaint/suggestion handling. If the operator is able to handle these tasks, he/she will collect all commentaries; otherwise, he/she will transfer the call to a responsible person.

3 Mathematical Programming Model Formulation

In this section, we present the mathematical model with all notations: indices, parameters, decision variables, objective functions, and constraints as follows:

3.1 Indices, Parameter, Decision Variable Description

See Table 1.

Let T be the set of 1 h time period where the call center department is open, and D denotes the set of working day. Then, t and d are the indices of the set T and D. Let m stand for the total number of operating days per week. We use k as a symbol for a shift in set K. Note that, there are several types of inbound call. Let I denote the set of call category; then, i is the indices of the set I.

Table 1 Notation for the model

T	Set of 1 h time period, $t \in T$
D	Set of working day, $d \in D$
K	Set of shift, $k = 1, 2, \ldots, K$
I	Set of category of inbound call, $i = 1, 2, \ldots, I$
P_i	Probability of call type i
F_i	Working time for administration of call type i
E_i	Operated time for outbound call type i
B_{td}	Volume of inbound call for (t, d)
α_{td}	Observed mean abandonment rate for (t, d)
u_{td}	Maximum allowable abandonment rate for (t, d)
g_d	Hour required for outbound calls for d
α	Mean daily wage per operator
β	Cost due to waiting time
μ	Cost due to abandonment line
z_{tdk}	Non-value-added period for (t, d, k)
l_{td}	Inbound load for (t, d)
W_{td}	Predicted waiting time for (t, d)
w_{td}	Observed mean waiting time (t, d)
A_{td}	Predicted abandonment rate for (t, d)
$f_1(\bullet)$	Relation between observed waiting time and inbound load
$f_2(\bullet)$	Relation between observed abandonment rate and inbound load
x_{tdk}	Total man-hour for inbound call for (t, d, k)
y_{tdk}	Total man-hour for outbound call for (t, d, k)
v_{tdk}	Total man-hour for administration task for (t, d, k)
N_k	Total number of operators on shift k

We assume that there are three kinds of calls, thus i is rearranging, information gathering, and complaints handling. Probability theory used in this research for classifying the sort of inbound calls. Let P_i denote the probability of call type i. The working time for administration is represented by F_i. Let E_i denote an operated time of outbound call in each call type i. The volume of inbound calls is denoted by B_{td}. We used w_{td} and α_{td} to denote observed mean waiting time per caller and observed abandonment rate, respectively. The w_{td} is used for determining the predicted waiting time (W_{td}) by formulating the relation between observed waiting time and inbound load. Similarly, predicted abandonment rate is denoted as A_{td} for the relation between observed abandonment rate and inbound load (Nah and Kim 2013). Let $f_1(\cdot)$ and $f_2(\cdot)$ stand for the mentioned relations.

In general, there are numerous expenses in the entire operations. Let α_k denote the mean daily wage per operator working on shift k. Nah and Kim (2013) present the method to evaluate the penalty cost due to waiting time and abandonment rate, and β and μ denote the penalty cost due to waiting cost per caller per time unit and the mean cost per lost call, respectively. The value of β can be estimated by the value of time based on the living expenses and the average wage. The amount of μ can be evaluated by comprehensively considering the effects of the lost calls, such as the current and potential mean profits generated by a patient being served, the percentage of inbound calls leading to appointments, and the percentage of lost calls resulting in lost customers.

Let N_k denote the number of operators in set k (k = 1, 2… K). The total man-hour of each tasks including, inbound call, outbound call, and administrations is indicated by x_{tdk}, $y_{tdk,}$ and v_{tdk}, respectively. The non-value-added time is z_{tdk}.

3.2 Mathematical Model

Here, we present a mixed-integer nonlinear mathematical model of the workforce planning and allocation for a hospital call center. Note that, the successful operations can be defined as a low total costs but maintain high quality of services. Therefore, the objective of the mathematical model can be stated as

Minimize

$$\Omega = \alpha m \sum_{k=1}^{K} N_i + \beta \sum_{d} \sum_{t} B_{td} W_{td} + \mu \sum_{d} \sum_{t} B_{td} A_{td} + \gamma \sum_{d} \sum_{t} B_{td} C_{td}$$

Subject to

$$l_{td} = \frac{B_{td}}{\sum_{k=1}^{K} X_{tdk}} \quad \forall t, d \tag{1}$$

$$W_{td} = f_1(l_{td}) \quad \forall t, d \tag{2}$$

$$A_{td} = f_2(l_{td}) \quad \forall t, d \tag{3}$$

$$A_{td} \leq u_{td} \quad \forall t, d \tag{4}$$

$$\frac{\sum_t B_{td} A_{td}}{\sum_t B_{td}} \leq u_{td} \quad \forall d \tag{5}$$

$$\frac{\sum_d \sum_t B_{td} A_{td}}{\sum_d \sum_t B_{td}} \leq u \tag{6}$$

$$g_d + P_i E_i \sum_{t} B_{td} \leq \sum_{t} \sum_{k} y_{tdk} \quad \forall d \tag{7}$$

$$P_i F_i \sum_{t} B_{td} \leq \sum_{k} \sum_{t} v_{tdk} \quad \forall d \tag{8}$$

$$N_k = x_{tdk} + y_{tdk} + z_{tdk} + v_{tdk} \quad \forall t, d \tag{9}$$

$$N_k \geq 0 \text{ and integer} \tag{10}$$

$$x_{tdk} \geq 0, y_{tdk} \geq 0, v_{tdk} \geq 0 \tag{11}$$

Our objective function is to minimize total cost which is composed of labor cost, waiting cost, abandonment cost, and loss of opportunity cost. The problem is subject to the following constraints.

The first constraint is an equation to determine workload which will be used to predict abandon rate in Eq. 2 and waiting time in Eq. 3. The expected abandonment rates should satisfy the hourly, daily, and weekly standard of the maximum allowable abandonment rate in Eqs. 4–6. Equation 7 indicates each type of demand of outbound calls. Equation 8 illustrates the calculation of man-hours for administration tasks.

The main purpose of this research is to find the optimal number of operators. We assume that the entire staffs can work at the same constant man-hour level which is 1 h per person per period. The number of operators (N_k) can be interpreted as total man-hours available during each one-hour period. Subsequently, as shown in Eq. 9, the number of operator in each shift can be calculated by the summation of total man-hour for inbound call, outbound call, administration tasks, and non-value added activities.

In Eq. 10, N_k is a nonnegativity integer. Also, we require positive man-hours for inbound call, outbound calls, and administrative hours as shown in Eq. 11.

4 A Numerical Example

To illustrate the mathematical model and provide number of results for the problem, we use a simple numerical example mainly based on the data represented in Nar and Kim (2013) as well as some data by consultation with our case study hospital call center.

The observed call center is operated from 8:00 to 18:00, and each interval represents a single hour. We consider only Monday through Friday. There are two shifts of operator which operate from 8:00 to 17:00 and 9:00 to 18:00, respectively. We also use the demand of inbound calls in each period as shown in Fig. 2. Tables 2 and 3 display model input parameters and maximum allowable abandonment rate in each day, respectively.

From the above information, the mathematical model is shown below; however, the loss of opportunity is not considered

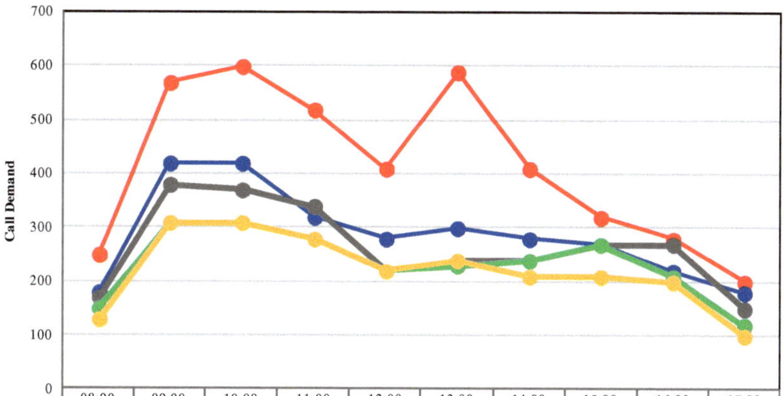

	08:00	09:00	10:00	11:00	12:00	13:00	14:00	15:00	16:00	17:00
Monday	250	570	600	520	410	590	410	320	280	200
Tuesday	180	420	420	320	280	300	280	270	220	180
Wednesday	170	380	370	340	220	240	240	270	270	150
Thursday	150	310	310	280	220	230	240	270	210	120
Friday	130	310	310	280	220	240	210	210	200	100

Fig. 2 Observing call demands each day

Table 2 Summary of input data

Parameter	Value
Probability of rearrangement call (P_{re})	0.4
Probability of information call (P_{in})	0.5
Probability of complaint call (P_{co})	0.1
Operated contact time of rearrangement (F_{re})	0.05 h
Operated contact time of information (F_{in})	0.05 h
Outbound operated time of rearrangement (E_{re})	0.05 h
Outbound operated time of information (E_{in})	0.05 h
Wage (α)	80 $
Penalty cost due to waiting (β)	7.2 $
Penalty cost due to lost call (μ)	8.7 $
Mean outbound call (g_d)	10 h

Minimize

$$\Omega = 80 \times 5 \sum_{1}^{2} N_i + 7.2 \times \sum_{t=1}^{10} \sum_{d=1}^{5} B_{td} W_{td} + 8.7 \times \sum_{t=1}^{10} \sum_{d=1}^{5} B_{td} A_{td}$$

Table 3 Maximum allowable abandonment rate

Time period	Monday	Tuesday	Wednesday	Thursday	Friday
8:00	22.5	16.5	15.0	13.5	12.0
9:00	22.5	16.5	15.0	13.5	12.0
10:00	22.5	16.5	15.0	13.5	12.0
11:00	33.8	24.8	22.5	20.3	18.0
12:00	45.0	33.0	30.0	27.0	24.0
13:00	33.8	24.8	22.5	20.3	18.0
14:00	16.9	12.4	11.3	10.1	9.0
15:00	16.9	12.4	11.3	10.1	9.0
16:00	16.9	12.4	11.3	10.1	9.0
17:00	16.9	12.4	11.3	10.1	9.0
u_d	22.5	16.5	15	13.5	12.0
u			15.0		

Subject to

$$l_{td} = \frac{B_{td}}{\sum_{k=1}^{2} X_{tdk}} \quad \forall t, d$$

$$W_{td} = 0.00249(l_{td})$$

$$A_{td} = \begin{cases} 0, & \text{if } l^{td} < 20 \\ 0.011457 l^{td} - 0.02261, & \text{otherwise} \end{cases}$$

$$A_{td} \le u_{td}$$

$$\frac{\sum_{t=1}^{10} B_{td} A_{td}}{\sum_{t=1}^{10} B_{td}} \le u_d$$

$$\frac{\sum_{d=1}^{5} \sum_{t=1}^{10} B_{td} A_{td}}{\sum_{d=1}^{5} \sum_{t=1}^{10} B_{td}} \le u$$

$$10 + 0.4 \times 0.05 \times \sum_{t=1}^{10} B_{td} + 0.5 \times 0.05 \times \sum_{t=1}^{10} B_{td} + 0.1 \times 0.05 \times \sum_{t=1}^{10} B_{td} \le \sum_{k=1}^{2} \sum_{t=1}^{10} y_{tdk}$$

$$0.4 \times 0.05 \times \sum_{t=1}^{10} B_{td} + 0.5 \times 0.05 \times \sum_{t=1}^{10} B_{td} + 0.1 \times 0.05 \times \sum_{t=1}^{10} B_{td} \le \sum_{k=1}^{2} \sum_{t=1}^{10} y_{tdk}$$

$$N_k = x_{tdk} + y_{tdk} + z_{tdk} + v_{tdk}$$

$$N_k \ge 0 \text{ and integer}$$

$$x_{tdk} \ge 0, y_{tdk} \ge 0, v_{tdk} \ge 0$$

The model is then solved using http://www.neos-server.org/. The results are shown in Tables 4, 5 and 6. It indicates that the optimal number of operators for shift 1 and shift 2 is 22 and 24, respectively.

Tables 4, 5 and 6 display the tasks that should be deployed in each time period. To illustrate the result, for example, during 10:00–11:00 on Wednesday of shift 1,

Table 4 Allocation of inbound tasks over weak

Time period	Shift 1					Shift 2				
	Mon	Tue	Wed	Thu	Fri	Mon	Tue	Wed	Thu	Fri
8:00	19.5	19.8	19.8	18.4	15.9	0.0	0.0	0.0	0.0	0.0
9:00	19.8	19.8	19.8	16.4	16.4	21.6	21.6	21.6	21.6	21.6
10:00	19.8	19.8	19.8	16.4	12.6	21.6	21.6	21.6	21.6	18.2
11:00	8.8	8.8	8.8	8.8	8.8	21.6	21.6	21.6	21.6	21.6
12:00	8.8	8.8	8.8	8.8	8.8	9.6	9.6	9.6	9.6	9.6
13:00	19.8	19.8	18.9	18.6	12.6	9.6	9.6	9.6	9.6	9.6
14:00	11.5	19.1	7.4	19.8	7.4	20.5	12.5	21.1	9.6	20.7
15:00	4.2	8.9	19.4	19.8	6.9	20.8	21.6	12.7	13.3	5.9
16:00	19.7	4.1	19.7	4.6	3.1	2.1	20.7	12.3	21.1	21.4
17:00	0.0	0.0	0.0	0.0	0.0	15.6	20.3	17.8	14.7	12.3

Table 5 Allocation of outbound task over weak

Time period	Shift 1					Shift 2				
	Mon	Tue	Wed	Thu	Fri	Mon	Tue	Wed	Thu	Fri
8:00	0.3	0.0	0.0	0.1	3.8	0.0	0.0	0.0	0.0	0.0
9:00	0.0	0.0	0.0	3.4	3.4	0.0	0.0	0.0	0.0	3.4
10:00	0.0	0.0	0.0	3.4	0.0	0.0	0.0	0.0	0.0	0.0
11:00	0.0	0.0	0.0	0.0	0.0	0.0	0.0	0.0	0.0	0.0
12:00	0.0	0.0	0.0	0.0	0.0	0.0	0.0	0.0	0.0	0.0
13:00	0.0	0.0	0.9	1.2	0.0	0.0	0.0	0.0	0.0	0.0
14:00	8.3	0.0	0.0	0.1	14.8	0.2	9.1	0.5	12.0	0.9
15:00	0.0	0.0	0.5	0.1	0.0	0.8	0.0	0.0	8.3	15.6
16:00	0.1	0.0	0.1	0.1	3.8	0.0	0.9	4.3	0.5	0.2
17:00	0.0	0.0	0.0	0.0	0.0	0.4	0.0	3.7	0.0	0.3

Table 6 Allocation of administration task over week

Time period	Shift 1					Shift 2				
	Mon	Tue	Wed	Thu	Fri	Mon	Tue	Wed	Thu	Fri
8:00	0.0	0.0	0.0	1.4	0.1	0.0	0.0	0.0	0.0	0.0
9:00	0.0	0.0	0.0	0.0	0.0	0.0	0.0	0.0	0.0	0.0
10:00	0.0	0.0	0.0	0.0	0.0	0.0	0.0	0.0	0.0	0.0
11:00	0.0	0.0	0.0	0.0	0.0	0.0	0.0	0.0	0.0	0.0
12:00	0.0	0.0	0.0	0.0	0.0	0.0	0.0	0.0	0.0	0.0
13:00	0.0	0.0	0.0	0.0	0.0	0.0	0.0	0.0	0.0	0.0
14:00	0.0	0.7	12.4	0.0	0.1	0.9	0.0	0.1	0.1	0.0
15:00	15.6	10.9	0.0	0.0	0.0	0.0	0.0	8.9	0.0	0.1
16:00	0.0	15.7	0.0	15.1	13.0	19.5	0.1	5.0	0.0	0.0
17:00	0.0	0.0	0.0	0.0	0.0	5.5	1.3	0.1	6.9	9.0

22 operators are required. According to Table 4, the man-hours for inbound call is 19.8 indicating that 19 operators should be assigned for handling inbound calls over this period and another one operator should be assigned for additional of 48 min. However, to assign for the whole shift during this period 2, operator will be idle as no more jobs waiting for handling.

The advantage of this guideline obtained from the results is that the hospital has systematic approach to determine the number of operators. In addition, those operators have an idea how to decide when to perform their tasks in each period as well they know which day they should take a leave for vocation or personal reason.

In contrast, this guideline may not be useful in some situations. For example, if there is high fluctuation on the nature of each calls resulting in the operators may not finish all tasks in time, e.g., on Thursday during 8:00–9:00, the result shows that there are 6 min to make outbound calls. In fact, the length of operating time for outbound call relies on the unpredicted behaviors of each customer rather than on the operators' skill. Thus, we may not guarantee that 6 min is sufficient to respond those unmanageable calls.

5 Conclusion and Future Work

The workforce planning and allocation is important in the hospital call center. In this paper, we present a mathematical model by using mixed-integer nonlinear programming to incorporate with some real situations. Experimental results are provided. Not only the number of operators required but also some insights for planner as a guideline for further planning are provided. Since it is time-consuming to solve the model, heuristic methods, genetic algorithm, particle swarm optimization, or differential evolution, will be used to help planning in practical way.

Furthermore, our model is not able to incorporate with the situation where the behavior of each customer is uncertain. The stochastic programming with the consideration of the planning results should be able to withstand the changes or differences of customers' behavior.

References

Aksin Z, Armony M, Mehrotra V (2007) The modern call-center: a multi-disciplinary per-spective on operations management research. Prod Oper Manage 16(6):665–688

Atlason J, Epelman MA, Henderson SG (2008) Optimizing call center staffing using simulation and analytic center cutting-plane methods. Manage Sci 54(2):295–309

Cezik MT, L'Ecuyer P (2008) Staffing multiskill call centers via linear programming and simulation. Manage Sci 54(2):310–323

Ernst AT, Jiang H, Krishnamoorthy M, Sier D (2004) Staff scheduling and rostering: a review of applications, method and models. Eur J Oper Res 153(1–4):3–27

Gans N, Koole G, Mandelbaum A (2003) Telephone call centers: Tutorial, review, and research prospects. Manuf Serv Oper Manage 5(2):79–141

Nah JE, Kim S (2013) Workforce planning and deployment for a hospital reservation call center with abandonment cost and multiple tasks. Comput Ind Eng 65:297–309

NEOS server (2014) NEOS Server: State-of-the-Art solvers for numerical optimization. http://www.neos-server.org/neos/. Accessed on 15 Jan 2015

Mehrotra V (1997) Ringing up big business. OR/MS Today 24(4):18–24

Moore CG, Wilson-Witherspoon P, Probst JC (2001) Time and money: effects of no-shows at a family practice residency clinic. Fam Med 33(7):522–527

Saltzman RM, Mehrotra V (2007) Managing trade-offs in call center agent scheduling: methodology and case study. In: Wainer GA, Vakilzadian H (eds) Proceedings of the 2007 summer computer simulation conference, San Diego, CA, pp 643–651

Developing Control Charts for Monitoring Time Interval Between Nonconforming Items in High-Quality Processes

Huynh Trung Luong and Zin Maung Maung Phyo Htet

Abstract High-quality processes with very low rate of nonconforming items can be found in many industries nowadays. For these processes, the traditional Shewhart control charts such as *p*-chart, *u*-chart, and *np*-chart are not applicable. Due to low fraction of nonconforming, it is better to establish the control chart to monitor the time between successive defectives of the process. In fact, time between successive defectives of high-quality processes usually follows exponential distribution. Due to the fact that the exponential distribution is highly skewed, some transformation techniques should be applied to help developing the control charts with unbiased in-control average run length (ARL). In this chapter, two different approaches have been applied using Weibull transformation and Cornish–Fisher expansion to develop unbiased ARL control charts in such a way that the probability of false alarm is at acceptable value.

Keywords Quality control · Weibull transformation · Cornish–Fisher expansion

1 Introduction

During the last years, many industries have been encountering problems in production management related to the distribution of goods through many networks to the customers in global emerging market. From several empirical studies, the firms can get greater benefits in terms of profitability and performance by focusing on quality management (Powell 1995; Easton and Jarrell 1998; Das et al. 2000;

H.T. Luong (✉) · Z.M.M.P. Htet
Asian Institute of Technology, 58 Moo 9 Paholyothin Highway, Klong Luang, Pathum Thani 12120, Thailand
e-mail: luong@ait.ac.th

Z.M.M.P. Htet
e-mail: zinmgphyohtet@gmail.com

V. Kachitvichyanukul et al. (eds.), *Toward Sustainable Operations of Supply Chain and Logistics Systems*, EcoProduction,
DOI 10.1007/978-3-319-19006-8_35

Douglas and Judge 2001; Kaynak 2003; Yeung et al. 2006; Mesut 2009; Kull and Narasimhan 2010).

Theoretically, quality assurance, a part of quality management to provide confidence that a product or service will satisfy given needs, relies on statistical analyses to deal with practices such as measuring, changing, and improving the process through continuous monitoring. Among statistical tools used for quality assurance, control charts are basically used for continuous monitoring and improving functions.

The use of control chart can help to determine whether the process is under control or out of the control. It also helps to determine the source of variation that results in degraded process performance. Many specific control charts have been developed for different practical implementations.

Nowadays, thanks to the technological advancement, many manufacturing companies like electronic companies can produce high-quality products with a very small amount of nonconforming (defective) fraction. The defective levels or count rates in the processes of these companies are extremely small (under 1000 occurrences per million). These processes are remarked as high-quality processes (Xie et al. 2002). In high-quality processes, most of the time, no defective items can be observed in the inspected sample because the average value of defective proportion is very low. This causes the range of the control limits to be narrowed down, and hence, the use of traditional Shewart control charts such as p-chart, np-chart, and u-chart is not appropriate to help detect the out of control situation quickly because most of the sample points to be plotted in the control chart may show nearly zero. Another possible problem is that the control chart in this situation can show significantly excess false alarm. The risk of false alarm can increase the cost unnecessarily and discredit the machines used. For high-quality processes, there is a need to develop improved control charts to replace the traditional ones. This is the aim of the research presented in this paper.

2 Literature Review

The drawbacks of applying conventional Shewhart control charts in high-quality processes motivated many research works toward establishing other alternatives without using the number of nonconforming items in the sample taken as the basis for the establishment of the control chart. Nelson (1994) introduced the chart based on the non-defective items. In this research, if the fraction of nonconforming item is very small, the number of defective items can be approximated by Poisson distribution and the number of non-defective items between these defective items will follow an exponential distribution.

A concept derived by Calvin (1983) and further studied by Goh (1987) is to use the distribution of the number of non-defective items between successive defective items. From this concept, a cumulative count of conforming control chart (CCC1) is developed to detect out of control signal based on cumulative count of conforming items. However, for small proportion of "p", it is better to use time-between-event (TBE) charts (Liu et al. 2004) which continuously monitor the number of conforming items between consecutive nonconforming items.

Several research studies related to TBE chart have been conducted for dealing with small "p" case. Yang et al. (2002) recognized that a geometric chart is quite slow to detect mild deteriorations of the system. Only if the "p" value rises, it shows out of control signals quickly. Also, rather than deciding the process should be stopped or not depending on the single failure, it is better to make decision until "r" failures have occurred.

Using negative binomial distribution, CCC-r chart, which uses the numbers of inspected items until "r" consecutive nonconforming items are observed, was first proposed by Bourke (1991). Besides, conforming run length (CRL) charts (Wu et al. 2001) and cumulative count of conforming (CCC) charts (Ohta et al. 2001) were also developed to deal with high-quality processes. Although there were different approaches, the main focused variable was the number of successes between failures. However, it should be noted that these charts are very sensitive, and if a large value of "r" is used, the charts become ineffective in detecting large upward shifts and need a relatively large number of items to be tracked down.

Tang and Cheong (2006) modified the existing cumulative count of conforming chart with the group inspection, while Albers (2010) considered the optimal design of negative binomial charts by choosing the value of "r" depending on the process. Chen et al. (2011) constructed the CCC chart and focused on the variable sampling intervals and control limits for high-quality process. All the charts mentioned above require 100 % inspection to the products.

Bersimis et al. (2014) proposed a new control chart to monitor the high-quality processes by using a compound rule based on the number of conforming units observed between the $(i-1)$th and the ith nonconforming unit and that observed between the $(i-2)$th and the ith nonconforming unit. Through comparison with the performance of other charts, this chart showed better results in most cases.

It should be noted that the cumulative sum (CUSUM) and exponential weighted moving average (EWMA) control charts have also been proposed for the use in high-quality processes (Yeh et al. 2008; Szarka and Woodall 2012). However, in some cases, they are very complex and complicated to understand for the practitioners.

Recently, Joekes and Barbosa (2013) introduced advanced p-chart using the Cornish–Fisher expansion. Cornish–Fisher expansion is a formula to approximate the quantiles of random variable based on its cumulants. The main purpose is to normalize the original value with one adjustment (adding the 3rd order cumulant to the limit) and two adjustments (adding the 3rd and the 4th order cumulants to the limit) to reduce the false alarm. As a result, the research presented a new rule for p-chart. If the value $np(1-p)$ is less than 5, it is suitable to use one or two adjustments. If the value $np(1-p)$ is between 0.08 and 0.25, only two adjustments should be used; however, if $np(1-p)$ is less than 0.08, another approach is needed.

3 Scope and Limitations of the Study

It is assumed that the fraction of defectives in high-quality process is very small and these rare events occur continuously and independently at a constant rate. Hence, the number of defective items in a fixed period of time will follow Poisson distribution, and therefore, the time between two consecutive defectives will follow exponential distribution. It is also assumed that the process is monitored continuously and 100 % inspection is applied so that the time at which a defective occurs can be recorded precisely.

Taken into consideration the fact that many processes are using continuous monitoring control system to check quality characteristics of the product, the control chart in this paper will be constructed based on the observed time between two consecutive defectives detected. Empirical analytical transformation technique as well as the use of cumulants of exponential distribution will be taken into consideration in the development of the control charts.

4 Control Charts for Exponential Distribution Using Transformation Method

For the exponential distribution with mean $\theta_0 = \frac{1}{\lambda_0}$, ($\lambda_0$ = poisson rate) which is used to model the time between two consecutive defectives, the main difficulty in constructing control chart is its high skewness. This skewness nature cannot be changed upon whatever the value of the mean as shown in the Fig. 1. However, the exponential distribution can be transformed into Weibull distribution which is less skew. The Weibull distribution is a continuous distribution which can be approximated by the normal distribution with appropriate shape parameter β. The effect of the shape parameter on skewness of the Weibull distribution is shown in Fig. 2.

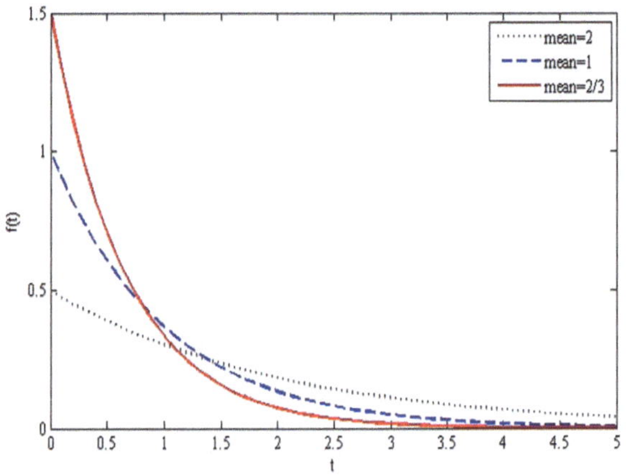

Fig. 1 Exponential distribution

Fig. 2 Weibull distribution

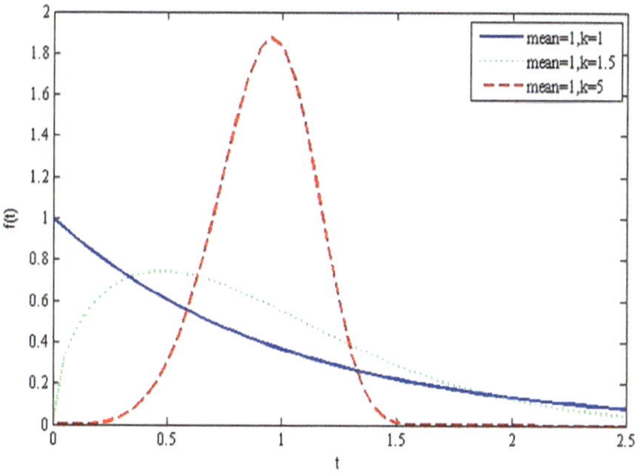

Assume that t is exponentially distributed with density function $f(t) = \frac{1}{\theta_0} * \mathrm{e}^{-t/\theta_0}, t \geq 0$, denote $T = t^{1/\beta}$ and using the transformation formula $f(T) = f(t) * \frac{\mathrm{d}t}{\mathrm{d}T}$, then the exponential distribution can be transformed into the Weibull distribution with density function

$$f(T) = \frac{\beta}{\theta_0^{1/\beta}} * \left(\frac{T}{\theta_0^{1/\beta}}\right)^{\beta-1} * \mathrm{e}^{-\left(\frac{T}{\theta_0^{1/\beta}}\right)^{\beta}}$$

The above expression is the probability density function of T which follows Weibull distribution with shape parameter β and scale $\theta^* = \theta_0^{1/\beta}$. From the research conducted by Nelson (1994), this Weibull distribution can be well approximated by a normal distribution using $\beta = 3.6$ as illustrated in Fig. 3.

Fig. 3 Normal approximation to Weibull distribution using shape parameter 3.6

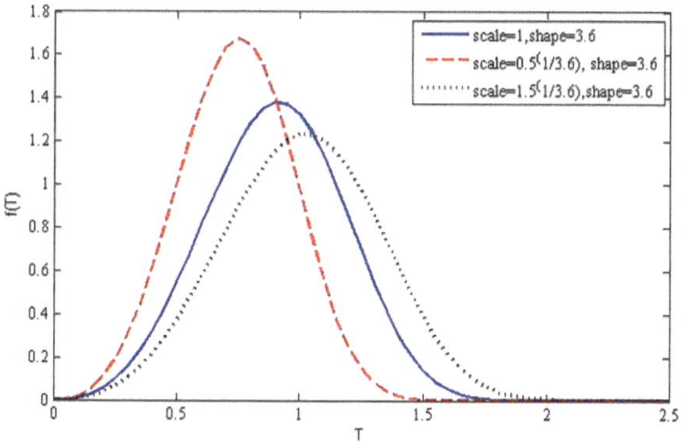

4.1 Control Limits Using $\pm 3\sigma$

Firstly, the control chart will be constructed using the standard limits derived for normal distribution because the exponential distribution has been approximated by the normal distribution through Weibull transformation. The standard control limits using $\pm 3\sigma$ from the mean of the Weibull distribution are

$$LCL = \mu_T - 3\sigma_T = \theta_0^{1/3.6} * \left[\Gamma\left(1 + \frac{1}{3.6}\right) - 3\sqrt{\Gamma\left(1 + \frac{2}{3.6}\right) - \left[\Gamma\left(1 + \frac{1}{3.6}\right)\right]^2} \right]$$

$$UCL = \mu_T + 3\sigma_T = \theta_0^{1/3.6} * \left[\Gamma\left(1 + \frac{1}{3.6}\right) + 3\sqrt{\Gamma\left(1 + \frac{2}{3.6}\right) - \left[\Gamma\left(1 + \frac{1}{3.6}\right)\right]^2} \right]$$

For normal distribution, $\pm 3\sigma$ is restricted from the mean as the standardized limits of allowing 0.27 % probability of false alarm. However, with the control limits showed above using Weibull distribution, the probability of false alarm $\alpha = P\{T < LCL\} + P\{T > UCL\} = 0.000754573$. This false alarm probability is much less than the desired value of 0.0027. Also, the in-control average run length value will be $ARL_0 = \frac{1}{\alpha} = 1325.25$ which is much greater than the normal value of 370. This control chart is, hence, needed to be adjusted.

4.2 Control Limits Using Correction Factor

In this section, the control limits will be adjusted as follows: $LCL = \mu_T + (-3 + c_1)\sigma_T, UCL = \mu_T + (3 + c_2)\sigma_T$, where c_1 and c_2 are correction factors to help get satisfactory results. Thus, the limits become

$$LCL = \theta_0^{1/3.6} * \left[\Gamma\left(1 + \frac{1}{3.6}\right) + (-3 + c_1) * \sqrt{\Gamma\left(1 + \frac{2}{3.6}\right) - \left[\Gamma\left(1 + \frac{1}{3.6}\right)\right]^2} \right]$$

$$UCL = \theta_0^{1/3.6} * \left[\Gamma\left(1 + \frac{1}{3.6}\right) + (3 + c_2) * \sqrt{\Gamma\left(1 + \frac{2}{3.6}\right) - \left[\Gamma\left(1 + \frac{1}{3.6}\right)\right]^2} \right]$$

The false alarm probability $\alpha = P\{T < LCL\} + P\{T > UCL\}$ can now be computed using $P(T < LCL) = 1 - e^{\frac{-(LCL)^{3.6}}{\theta_0}}, P(T > UCL) = e^{\frac{-(UCL)^{3.6}}{\theta_0}}$. Setting α to 0.0027, we have:

$$1 - e^{-(0.067044822+0.278020287c_1)^{3.6}} + e^{-(1.735166544+0.278020287c_2)^{3.6}} = 0.027 \quad (1)$$

It is also noted that for a change in the process mean from θ_0 to θ_1, the out of control average run length is $ARL_1 = \frac{1}{1-\beta}$, in which $\beta = P(LCL \leq T \leq UCL \backslash \theta = \theta_1)$. The most desirable fact is that ARL_1 should be maximized at $\theta_0 (ARL_0 > ARL_1)$, because any shift in the mean should be detected rapidly. Otherwise, the control chart will be unbiased and not be a better one. The above requirement is equivalent to the fact that $\frac{dARL_1}{d\theta_1}|_{\theta_1=\theta_0} = 0$, or

$$\begin{aligned}(1.735166544 + 0.278020287c_2)^{3.6} * e^{-(1.7355166544+0.278020287c_2)^{3.6}} \\ - (0.067044822 + 0.278020287c_1)^{3.6} e^{-(0.067044822+0.278020287c_1)^{3.6}} = 0\end{aligned} \quad (2)$$

By solving Eqs. (1) and (2), we can find $c_1 = 0.43271712$ and $c_2 = 0.19577971$.

4.3 Comparison of $\pm 3\sigma$ and Revised Control Charts

The two control charts can be compared by analyzing the operating characteristics curves (OC curves) and ARL curves. The OC curve plots the probability of not detecting the shift with respect to the shift of the mean. In Fig. 4, the OC curve of $\pm 3\sigma$ control chart shows better result, i.e., lower probability of not detecting the shift, than that of the revised control chart using correction factor if the shifted mean is more than 30 % of the original value. However, it should be noted that the cases when the shifted mean is less than the in-control mean are the main focus of the monitoring process. Taken this into consideration, it can be seen that the revised control chart performs better.

Also, Fig. 5 shows that the OC curve of the revised control chart has maximum value of ARL at the in-control mean value, while the OC curve of the $\pm 3\sigma$ control

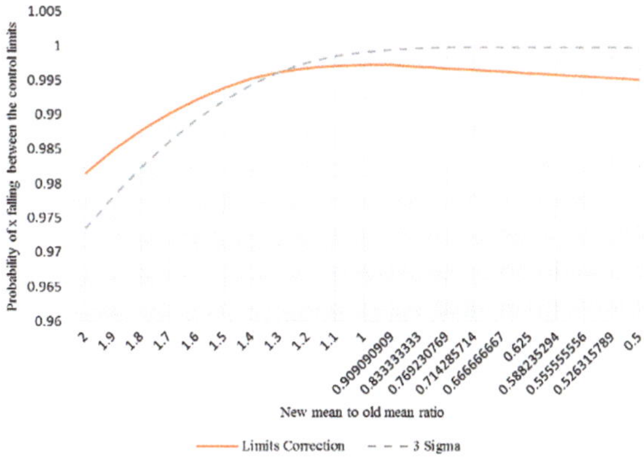

Fig. 4 Operating characteristics curve comparing two methods

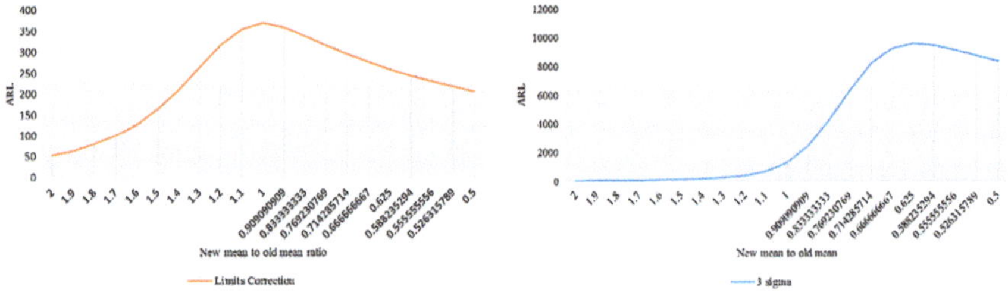

Fig. 5 ARL curves of the revised and $\pm 3\sigma$ control charts

chart shows that when the shifted mean is less than the in-control mean, the value of ARL will increase. The detection ability of the $\pm 3\sigma$ control chart is, therefore, not good.

5 Control Charts for Exponential Distribution Using Cumulants

In this section, an alternative way to build control chart using cumulants of exponential distribution is derived. At first, it is noted that if the exponential distribution is not transformed, the control chart constructed directly using $\mu = \theta = \frac{1}{\lambda}$ and $\sigma = \theta = \frac{1}{\lambda}$ will be

$$UCL = \frac{1}{\lambda} + 3\frac{1}{\lambda} = \frac{4}{\lambda} \quad LCL = \frac{1}{\lambda} - 3\frac{1}{\lambda} = -\frac{2}{\lambda} \to 0.$$

With the above control limit, the probability of false alarm is 0.0183, which is still much higher than 0.0027. Hence, the Cornish–Fisher expansion will be considered to help adjust the control limits.

5.1 Cornish–Fisher Expansion

The Cornish–Fisher expansion is a method to approximate the required quantiles of a distribution of random variable based on its cumulants. Cumulant is an alternative to provide the moment of the distribution. It determines the moment of the distribution. In order to apply the Cornish–Fisher expansion, the cumulants and moments of the exponential distribution are needed and can be found in the Appendix. The Cornish–Fisher expansion is

$$x'_\alpha = z_\alpha + \frac{1}{6}\left(z_\alpha^2 - 1\right) * K_3 + \frac{1}{24}\left(z_\alpha^3 - 3z_\alpha\right)$$
$$* K_4 - \frac{1}{36}\left(2z_\alpha^3 - 5z_\alpha\right) * K_3^2 - \frac{1}{24}\left(z_\alpha^4 - 5z_\alpha^2 + 2\right) * K_3 K_4 + \cdots$$

where z_α denotes the αth percentile value, K_3 is skewness, K_4 is kurtosis, and x'_α is the standardized αth percentile of the standard normal distribution which can be calculated as $x'_\alpha = \frac{x_\alpha - \mu}{\sigma}$.

5.2 Revised Control Limits Using Cornish–Fisher Expansion (Using Skewness Only)

Let T_α be an exponential random variable with parameter mean $\mu = \theta = \frac{1}{\lambda}$ and variance $\sigma = \theta = \frac{1}{\lambda}$. By considering the z-value associated with the αth percentile of the exponential distribution presented in Table 1 (using K_3 from Appendix), we can see that the percentile value of the exponential distribution becomes zero if the cumulative probability is not greater than 0.1587. This is because of the highly positive skewness of exponential distribution that leads to the fact that the 50th percentile of the normal distribution occurs at the z-value of zero while that percentile of the exponential distribution is at $0.6667/\lambda$.

So, by setting $z = -1$ for lower control limit and $z = 3$ for upper control limit, the control limits can be derived as follows: $UCL_1 = \frac{6.6667}{\lambda}$, $CL_1 = \frac{0.6667}{\lambda}$, and $LCL_1 = 0$. And, the in-control ARL of the control charts is $ARL_0 = \frac{1}{\alpha} = 785.79 (\gg 370.3)$.

It can be seen that the average run length is nearly double the standard value; therefore, the control limits need to be adjusted. To adjust the control limits, the Cornish–Fisher expansion with only skewness component is considered.

$$\text{control limit} = T_\alpha = \frac{1}{\lambda} + z_\alpha * \frac{1}{\lambda} + \frac{\left(z_\alpha^2 - 1\right)}{6} * K_3 * \frac{1}{\lambda}$$

Table 1 Required percentile value of exponential with each sigma limits

Cumulative probability	z-value	Percentile value of exponential distribution
0.9987	+3	6.6667/λ
0.9772	+2	4/λ
0.8413	+1	2/λ
0.5	0	0.6667/λ
0.1587	−1	0
0.0228	−2	0
0.0013	−3	–

In the above equation, z_α is the required percentile of the distribution and will be considered as a variable for adjustment. In details, to set the lower control limit and upper control limit, the *p*th percentile and the *q*th percentile are considered so as to give the preset alpha value of 0.0027.

Denote $z_p = -a$ and $z_q = b$, and substituting in Cornish–Fisher expansion, the control limits can be derived as follows:

$$\text{lower limit} = L = \frac{1}{\lambda} - a * \frac{1}{\lambda} + \frac{(a^2-1)}{6} * K_3 * \frac{1}{\lambda}$$

$$\text{Upper Limit} = U = \frac{1}{\lambda} + b * \frac{1}{\lambda} + \frac{(b^2-1)}{6} * K_3 * \frac{1}{\lambda}$$

Using the requirements that α value 0.0027 and ARL should be maximized when the process is in the control, the following two equations can be derived:

$$\alpha = 1 - e^{-\left(1-a+\frac{(a^2-1)}{6}K_3\right)} + e^{-\left(1+b+\frac{(b^2-1)}{6}K_3\right)} \tag{3}$$

$$\left(1+b+\frac{(b^2-1)}{6}K_3\right) * e^{-\left(1+b+\frac{(b^2-1)}{6}K_3\right)} = \left(1-a+\frac{(a^2-1)}{6}K_3\right) * e^{-\left(1-a+\frac{(a^2-1)}{6}K_3\right)} \tag{4}$$

By solving Eqs. (3) and (4), we can find

$$a = 0.99283$$
$$b = 3.46290$$

And the revised control limits will be

$$L = \frac{1}{\lambda} - a * \frac{1}{\lambda} + \frac{(a^2-1)}{6} * K_3 * \frac{1}{\lambda} = \frac{0.00241}{\lambda}$$

$$U = \frac{1}{\lambda} + b * \frac{1}{\lambda} + \frac{(b^2-1)}{6} * K_3 * \frac{1}{\lambda} = \frac{8.12676}{\lambda}$$

It is noted that the use of *a* and *b* above is equivalent to using the *z*-values of 3.463 and −0.993, respectively.

We also checked the application of Cornish–Fisher expansion with both skewness and kurtosis components, but the results are exactly the same and hence will not be presented here.

6 Comparison Between the Two Revised Control Charts

Noted that by the use of Weibull transformation with correction factor, the control limits of the revised control chart are

$$LCL_A = \theta_0^{1/3.6} * 0.187349$$
$$UCL_A = \theta_0^{1/3.6} * 1.78960$$

Also, by the use of Cornish–Fisher expansion, the control limits of the revised control chart are as follows:

$$LCL_C = \frac{0.00241}{\lambda} = 0.00241\theta_0$$
$$UCL_C = \frac{8.12676}{\lambda} = 8.12676\theta_0$$

It is interesting to note that the detection powers of the two revised control charts are exactly the same because the probability of not detecting the shift can be computed using the same formula

$$\beta = \exp\left\{-\frac{\theta_0}{\theta_1} * 0.00241\right\} - \exp\left\{-\frac{\theta_0}{\theta_1} * 8.12676\right\}$$

The main difference between the two revised control charts is that the time between two successive defectives t will be used for plotted point on the control chart developed using Cornish–Fisher expansion, while the transformed value $T = t^{1/3.6}$ will be used for the control chart developed using Weibull approximation.

7 Conclusions and Recommendations

In this research, control chart for monitoring the time elapsed between two consecutive defectives in high-quality processes, which is assumed to follow exponential distribution, is developed. To deal with skewness of the exponential distribution, two approaches have been considered, the first one uses Weibull transformation, and the second one uses cumulants by employing Cornish–Fisher expansion. Correction factors are determined in order to ensure that the revised control charts satisfy the basic requirements, i.e., the probability of false alarm is about 0.0027 and the average run length is maximized when the process is in control. It is interesting to note that the two revised control charts are equivalent. They have the same detection power, and the main difference is in the value plotted on the charts. The approaches used in this research can be easily expanded to deal with the cases when the in-control ARL value and/or false alarm rate are set at other desired values in some specific industries.

Appendix

The moment generating the function of the exponential distribution is $\phi(t) = \frac{\lambda}{\lambda - t} (\lambda > 0)$, where $\lambda = {}^{1}/_{\theta}$. So, the first four ordinary moments can be computed as $\mu'_1 = E(X) = \frac{1}{\lambda}, \mu'_2 = E(X^2) = \frac{2}{\lambda^2}, \mu'_3 = E(X^3) = \frac{6}{\lambda^3}, \mu'_4 = E(X^4) = \frac{24}{\lambda^4}$.

Also, the central moments of the exponential distribution can be calculated as $\mu_h = E(X - \mu)^h = \int_{D_x} (x - \mu)^h \cdot f(x) \mathrm{d}x$. And hence, $\mu_1 = 0\, \mu_2 = \mu'_2 - \mu^2 = \sigma_x^2 = \frac{1}{\lambda^2}, \mu_3 = \mu'_3 - 3\mu\mu'_2 + 2\mu^3 = \frac{2}{\lambda^3}, \mu_4 = \mu'_4 - 4\mu\mu'_3 + 6\mu^2\mu'_2 - 3\mu^4 = \frac{9}{\lambda^4}$.

The cumulants can then be obtained from the cumulants generating function (logarithm of the moments generating function) with the central moments to be standardized. The results are $K_1 = \frac{\mu}{\sigma} = 1, K_2 = \frac{\mu_2}{\sigma^2} = 1, K_3 = \frac{\mu_3}{\sigma^3} = 2$, $K_4 = \frac{\mu_4}{\sigma^4} - 3\left(\frac{\mu_2}{\sigma^2}\right)^2 = 6$.

References

Albers W (2010) The optimal choice of negative binomial charts for monitoring high-quality processes. J Statistical Plann Infer 140:214–225

Bersimis S, Koutras MV, Maravelakis PE (2014) A compound control chart for monitoring and controlling high quality processes. Eur J Oper Res 233:595–603

Bourke PD (1991) Detecting a shift in fraction nonconforming using run-length control charts with 100 % inspection. J Qual Technol 23:225–238

Calvin TW (1983) Quality control techniques for zero defects. IEEE Trans Compon Hybrids Manuf Technol CHMT-6:323–328

Chen YK, Chen CY, Chiou KC (2011) Cumulative conformance count chart with variable sampling intervals and control limits. Appl Stochast Models Bus Ind 27:410–420

Das A, Handfield RB, Calantone RJ, Ghosh S (2000) A contingent view of quality management-the impact of international competition on quality. Decis Sci 31(3):649–690

Douglas TJ, Judge WQ Jr (2001) Total quality management implementation and competitive advantage: the role of structural control and exploration. Acad Manag J 44(1):158–169

Easton GS, Jarrell SL (1998) The effects of total quality management on corporate performance: an empirical investigation. J Bus 71(2):253–307

Goh TN (1987) A charting technique for control of low-nonconformity production. Int J Qual Reliab Manage 5:53–62

Joekes S, Barbosa EP (2013) An improved attribute control chart for monitoring non-conforming proportion in high quality processes. Control Eng Pract 21:407–412

Kaynak H (2003) The relationship between total quality management practices and their effects of firm performance. J Oper Manage 21(4):405–435

Kull TJ, Narasimhan R (2010) Quality management and cooperative values: investigation of multilevel influences on work group performance. Decis Sci 41(1):81–113

Liu JY, Xie M, Goh TN, Ranjan P (2004) Time-between-events charts for on-line process monitoring. In: International engineering management conference, pp 1061–1065

Mesut A (2009) A multi-level examination of quality-focused human resource practices and firm performance: evidence from the US healthcare industry. Int J Human ResourManage 20(9):1945–1964

Nelson LS (1994) Control chart for parts per million conforming items. J Qual Technol 26:239–240

Ohta H, Kusukawa E, Rahim A (2001) A CCC-*r* chart for high-yield processes. Qual Reliab Eng Int 17:439–446

Powell T (1995) Total quality management as competitive advantage: a review and empirical study. Strateg Manage J 16(1):15–37

Szarka JL, Woodall WH (2012) On the equivalence of the bernoulli and geometric CUSUM charts. J Qual Technol 44:43–62

Tang LC, Cheong WT (2006) A control scheme for high-yield correlated production under group inspection. J Qual Technol 38:45–55

Wu Z, Zhang X, Yeo SH (2001) Design of the sum-of-conforming run length control charts. Eur J Oper Res 132:187–196

Xie M, Goh TN, Kuralmani V (2002) Statistical models and control charts for high quality processes. Kluwer Academic Publication, Massachusetts

Yang Z, Xie M, Kuralmani V, Tsui KL (2002) On the performance of geometric charts with estimated parameters. Qual Technique 34:448–458

Yeh AB, Mcgrath RN, Sembower MA, Shen Q (2008) EWMA control charts for monitoring high-yield processes based on non-transformed observations. Int J Prod Res 46:5679–5699

Yeung ACL, Edwin Cheng TC, Lai K (2006) An operational and institutional perspective on total quality management. Prod Oper Manage 15(1):156–170

Workforce Planning for Single Call Center with Service-Level Agreement

Thanyawan Chanpanit and Apinanthana Udomsakdigool

Abstract Workforce planning could be a critical process in call center operations. Due to arrival, calls are both time varying and uncertain, assigning the agents to the call center operations which responses to all calls might be not an easy task. This chapter demonstrates the workforce planning of a call center that services the number of calls required and satisfies the expected service level. The historical data are provided by a call center in a communication service company in Thailand. The data are forecasted by decomposition technique for the planning period. The queuing theory and the linear programming are used as the methodological tools. The call center operation is simplified to the *M/M/n* model and the Erlang C formulation is applied to solve for the number of agents required. In order to define shifts and assign agents to meet the needs, the model of Aykin is adapted to minimize the total of agents assigned. The results provide the number of agents to be assigned in the daily operations. The monthly workforce planning model is then formulated with the number of agents as decision variable and the objective is to minimize the total number of agent assignment. This problem can be solved by linear programming and the method could assist a manager to formulate monthly workforce plan that can satisfy the expected service level.

Keywords Call center operations · Workforce management · Decomposition technique · Erlang C formulation · Linear programming

T. Chanpanit (✉) · A. Udomsakdigool
King Mongkut's University of Techonology, Thonburi, Thailand
e-mail: thanyawan.c@mail.kmutt.ac.th

A. Udomsakdigool
e-mail: apinanthana.udo@kmutt.ac.th

V. Kachitvichyanukul et al. (eds.), *Toward Sustainable Operations of Supply Chain and Logistics Systems*, EcoProduction,
DOI 10.1007/978-3-319-19006-8_36

1 Introduction

The economic growth increased an importance of a call center in a service business. It also made call center operations more complicated with the diversities of client demands. Clients often had more expected to be served their calls as soon as possible. The service level had become a critical issue of call center performances. Actually, incoming calls might be not determined whenever they arrived. The assigning operators to response all calls impossibly be made. Thus, workforce planning had become an essential process of call center operations.

This paper aimed to examine workforce planning by setting the acceptable service level and determining the number of agents to satisfy the expected service level of a call center operation by the queuing theory. The linear programming was applied to perform the experiment. The paper was constructed to five sections including (i) introduction; (ii) literature review; (iii) methodology, model assumption, the descriptions of the model, and data analysis; (iv) experimental results; and (v) conclusions and recommendations. Eventually, the expected results could demonstrate the workforce planning for a call center operation with maintaining of the acceptable service level overtime.

2 Literature Review

A call center was a set of resources for communication between an organization and its customers over the phone (Buist et al. 2008). Aksin et al. (2007) stated staffing cost of call center was the main contribution by 60–80 % of the overall operation cost. It was very labor-intensive service industry. In the modern call center operations, multi-skill call center had been more concerned in order to response various customer's problems which increasing of complexity (Avramidis et al. 2010). Pot (2006) gave two reasons why a single call center can be inefficient. Firstly, it is very time-consuming to training agents with a considerable amount of useful information before starting work. Secondly, the gap between service time and productivity had great extent when agents handle call with a diversity of subjects. Although implementing multiple skills in a call center could reduce service handling time and increase productivity, the system required an advance technology to manage routing and staffing. This had brought an increasing of investment. Hence, to decide an implementation of a single or multi-skill, call centers need more strategic decision making with multi-criteria.

Typically, the core operation of a call center could be viewed by the concept of queuing (Dano 2004). It might be described the concept as a system including (i) an arrival process, (ii) a service process, (iii) a departure process, (iv) system capacity, and (v) queuing discipline (Dombacher 2010). The simplest and most widely used discipline of call center service was the *M/M/n* queue. The Kendall notation could define the characteristics of queue as a single-type single call center by *n* agents.

This also was known as Erlang C formula. The equation could be expressed by $C\ (\lambda,\ \mu,\ n)$ (Koole and Mandelbaum 2001). Noted the incoming calls randomly arrived at a constant rate λ, service process also follows an exponential distribution with fixed rate μ and operated by staffing level n (Aktekin 2014).

In the real world, however, the operation was more complex. Inbound calls did not arrive at fixed rate according to a stationary Poisson process; the call durations also were not exponential random variables, and their distribution may depend on the time of each day (Deslauriers et al. 2007). In case of the call center operation at Qwest Communications, Dietz (2011) presented Markovian queuing theory, quadratic programming, and variable-threshold rounding algorithms as a research tool to develop agent scheduling. This study was also performed by three parameters: (i) offered call volume, (ii) average handing time, and (iii) staff availability. It would appear that staffing cost was reduced. Li et al. (2011) studied the design of call center intelligent scheduling system in a large-scale call center. This research demonstrated the method for improving the precision of predicting historical data by neural network models. Then, the Erlang C formula was also applied to determine the number of agents assigned. The results gave productivity and standard of service had increased.

In operational management of a call center, workforce management was principally concerned with forecasting the workload and creating a set of schedules for the agents. There were three success factors in call center management: (i) improving customer satisfaction, services, and quality; (ii) increasing revenue; and (iii) reducing cost of providing excellent services (COPC CSP Standard 2014). Nevertheless, it was not easy to accomplish the success factors due to the allocating transactions in call center had come with uncertainty. There were considerable constraints that affected the operation such as workload, service level, and staff requirements (Liao et al. 2013). In addition, the effective scheduling had also to meet the service-level goals (Eveborn and Ronnqvis 2004). These factors were significant on overall cost. In order to improve the operation efficiently, call center optimization was applied. Goldberg et al. (2014), for example, demonstrated the prediction of the workload process, both directly and based on prediction of the arrival counts using the best linear unbiased predictor (BLUP). It was constructed for the continuation of a curve and applied the proposed estimator to real-world call center data. Aykin (1996) proposed the linear programming to determine the optimal number in every shift of agents needed with breaks assignment (Aykin 1996). Herowati (2005) also proposed the application of multi-shifts and break windows for helping shift manager to develop recruitments and selecting hiring process.

3 Methodology

The methodology of the paper divided into 2 parts. Firstly, the optimum number of agents could be computed by the concept of queuing model. To simplify the call center operations, the Erlang C formulation was applied. Second, the optimum number of agents was defined as the target number of agents at any working period. Then, the linear programming was used to determine the number of agents assigned to working shifts. Finally, the total number of agents daily required was set as constraints to schedule the monthly workforce. To simulate the models, the Erlang C formula and the programming were constructed in Excel spread sheet. It also has the availability of an optimization tool, Solver, to solve a mathematics program.

3.1 Model Assumption

To determine the number of agent requirements in any working period, the Erlang C formulation ($E_c(m,u)$) was applied. The queuing model of the call center was assumed as single-type single call center by *m* agents. The inputs of the model required an arrival rate (λ), service rate (μ), and the number of agents (m). An industry standard for telephone services seems to be the 80/20 rule, under which at least 80 % of the customers must wait no more than 20 s (Brezavscek and Baggia 2014; Koole 2007). The formulation could be expressed by $W(m, u, T_s, t)$ as shown in Eq. (1), and its components were detailed in Eqs. (2)–(5) (Hock 1996).

$$W(m,u,T_S,t) = P(\text{waiting time} \le t) = 1 - E_c(m,u) * \mathrm{e}^{-(m-u)\frac{t}{T_s}} \tag{1}$$

where

$$E_c(m,u) = P(\text{a call has to wait}) = \frac{\frac{u^m}{m!}}{\frac{u^m}{m!} + (1-\rho)\sum_{k=0}^{m-1}\frac{u^k}{k!}} \tag{2}$$

$$u = \frac{\lambda}{\mu} \tag{3}$$

$$\rho = \frac{u}{m} \tag{4}$$

$$\mu = \frac{1}{T_s} \tag{5}$$

Parameters

λ Call arrival rate,
T_s Average handling time (AHT),
m The number of agents, and
t Target waiting time.

3.2 Linear Programming

A linear programming problem could be defined as the problem of optimizing a linear function subject to linear constraints. To optimize efficiency of workforce planning, minimizing the total cost or the number of staff was often an objective function (Gomar et al. 2002; Liao 2011; De Brueker et al. 2015). The set of constraints could represent the number of agents required in any working period, such as consecutive working time or day-off, type of skills, and tasks (Dano 2004; Alfares 2003; Eun Nah and Kin 2013). At this state, there were two steps to determine an optimum workforce scheduling. Firstly, the linear programming was constructed to solve the shifting problem. The expect result could illustrate the number of shifts with an optimum agents that satisfy the number of agents needed daily. Secondly, another programming was built to figure out the feasible solution of daily agents needed under monthly workforce planning.

- Shifting and staff assignment problem

Formulating the linear programming to determine the number of shifts associated with the optimum agents, the model of Aykin was adapted (Aykin 1996). It could be described with the initial assumptions as follows: (i) the operation time was less than 24 h, (ii) every single period was of equal length, (iii) all of the agents had equally working periods and received a lunch break, and (iv) understaffing was not allowed.

To optimize efficiency of the operation, minimizing the total of agents assigned in any shift was an objective as shown in Eq. (6). The constraints (7) ensured that sufficient number of agents was presented in all periods. Every agent had to take exactly a lunch break by an hour and had of equal length of working hours. The model assumed that every single agent who assigned in shift k would take a break in period t as expressed in Eq. (8). The descriptions of the programming could be defined as follows:

- x_k was an integer decision variable. It represented the number of agents to be assigned in shift k.
- u_{kt} was the number of agents in shift k and started his/her lunch break in period t.
- m_t was a set of constraints. It was the number of agents needed in working period t.
- a_{kt} was a binary number where equal to 1 if period t was period for shift k and 0, otherwise.
- K was the set of all shifts.
- T was the set of working periods.
- BU_k was the set of working periods where agents in shift k could start his/her lunch break.
- TU_t was the set of shift where period t was a starting period for his/her lunch break.

Objective function

$$\text{Min z} = \sum_{k \in K} x_k \tag{6}$$

Subject to

$$\sum_{k \in K} a_{kt} x_k - \sum_{k \in TU_t} u_{kt} - \sum_{k \in TU_{t-1}} u_{kt-1} \geq m_t \tag{7}$$

$$x_k - \sum_{t \in BU_k} u_{kt} = 0 \tag{8}$$

Sign restricts

$$x_k, u_{kt}, m_t \geq 0, \text{Non-negative variables}$$
$$x_k, u_{kt}, m_t = \text{integer}$$

- Workforce scheduling problem

To develop the model, the initial assumptions included the following: (i) working week had 7 days that number 1–7 was marked from Sunday to Saturday; (ii) every single agent had 5 consecutive working days and 2 day-off equally; and (iii) understaffing is not allowed.

The objective function (9) was to minimize the total of agents assigned in every single day. Constraints (10)–(16) ensured that the number of agents assigned was at least equal to the number required for each day of the week. The variables could be described as follows:

- x_{ij} was an integer decision variable. It represented the number of agents assigned to day j in week i and
- b_{ij} was the number of agents required to day j in week i.

Objective function

$$\text{Min } z \sum_{i=1}^{n} \sum_{j=1}^{n} x_{ij} \tag{9}$$

Subject to

$$x_{i1} + x_{(i-1)4} + x_{(i-1)5} + x_{(i-1)6} + x_{(i-1)7} \geq b_{i1} \tag{10}$$

$$x_{i2} + x_{(i-1)5} + x_{(i-1)6} + x_{(i-1)7} + x_{i1} \geq b_{i2} \tag{11}$$

$$x_{i3} + x_{(i-1)6} + x_{(i-1)7} + x_{i1} + x_{i2} \geq b_{i3} \tag{12}$$

$$x_{i4} + x_{(i-1)7} + x_{i1} + x_{i2} + x_{i3} \geq b_{i4} \tag{13}$$

$$x_{i5} + x_{i1} + x_{i2} + x_{i3} + x_{i4} \geq b_{i5} \tag{14}$$

$$x_{i6} + x_{i2} + x_{i3} + x_{i4} + x_{i5} \geq b_{i6} \tag{15}$$

$$x_{i7} + x_{i3} + x_{i4} + x_{i5} + x_{i6} \geq b_{i7} \tag{16}$$

Sign restricts

$$\begin{gathered} x_{ij}, b_{ij} \geq 0, \text{ Non-negative variables} \\ x_{ij}, b_{ij} = \text{integer} \\ i = 1, 2, 3\ldots, n \quad \text{week } i \\ j = 1, 2, 3, \ldots, 7 \quad \text{day } j \end{gathered}$$

3.3 Data Collection

In order to perform the tests on monthly workforce planning of call center operations, the data were collected from a call center of a communication service company in Thailand during November 2014. Call arrivals and average handling time (AHT) were two main inputs used in the queuing model. The data had been provided during 9.00 a.m. to 10.00 p.m. It was total of 13 working hours.

To analyze the data, the working hours were divided into 13 periods of equal length by an hour. The first period started from 9.00 a.m. to 10.00 a.m., and the last period started from 9.00 p.m. to 10.00 p.m. Figure 1 was depicted the number of call arrivals in every single period every day in November 2014. The data could demonstrate that the number of call arrivals possibly had the same pattern over the same period on any day. Incoming calls dramatically increased and reached a peak point in the 2nd or 3rd period of the day. Then, the calls had a declined trend with slight fluctuation and went to the minimum point in the last period.

4 Experimental Results

4.1 The Number of Agents Required

In order to assign the workforce the next 4 weeks, decomposition of time series technique was applied to forecast the number of call offers. The components of a time series could be defined by trend, seasonality, cycles, and random variation (Heizer and Render 2011). Figure 1 illustrated that the historical data composed of

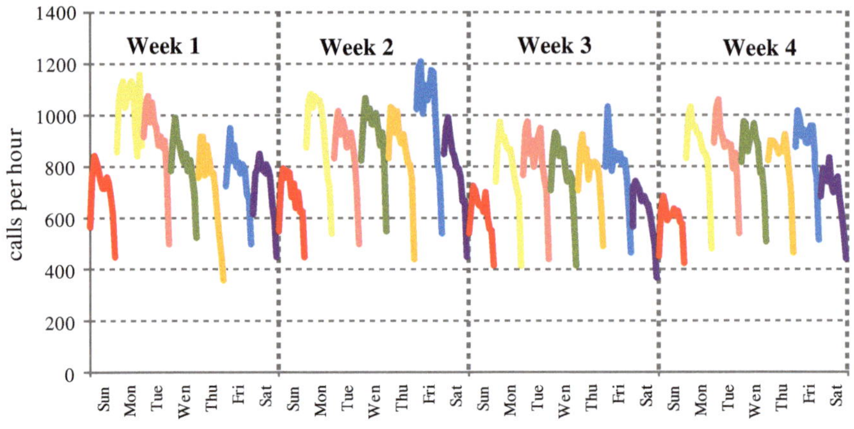

Fig. 1 The pattern of call arrivals in November 2014

trend and seasonality. Firstly, the number of call offers in the next 4 weeks would be determined by monthly trend projection. The regression line was determined by function of a least-squares method following equation $y = ax + b$. a was y-axis intercept. b was a slope of the regression line that expressed by $\sqrt{\sum xy - n\overline{xy}/\sum x^2 - n\bar{x}^2}$. y was the number of call offers to be projected, and x was time period. The results could be found as $y = 840.9000 - 0.2038x$. Then, a seasonal index would be computed to adjust the number of calls offers in daily period and hourly period. The seasonal index was determined by a fraction of the average number of call offers in each period and the total of call offers over the period. Finally, multiply the trend projection with the seasonal indexes. The 4-week forecasted number of call offered could be provided in Fig. 2. To predict the AHT

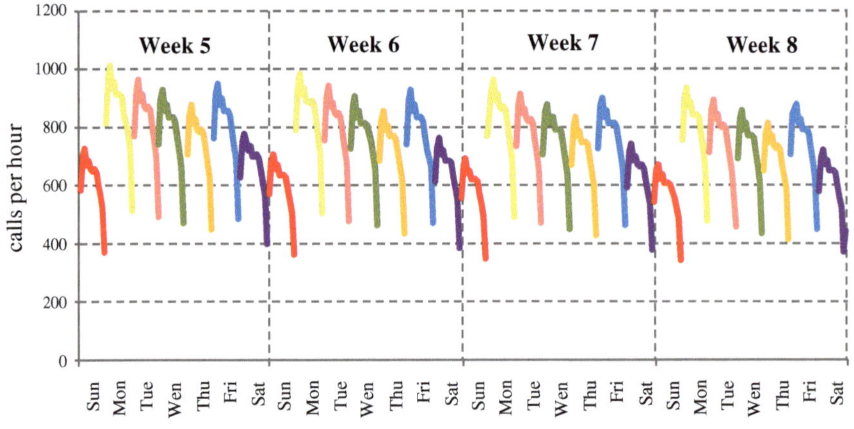

Fig. 2 The forecasted of call offers in the next 4 weeks

data, the forecasting technique was also applied. The linear regression could be found that $y = 761.8000 + 0.0079x$. Denote that y was AHT trend projection and x was hourly period. Additionally, a seasonal index would be computed to adjust the number of AHT in daily period and hourly period.

According to the assumptions of the call center operations, it was limited to the *M/M/n* model. The feasible number of required agents was determined by Erlang C formula subject to the expected service level at least or equal to 80 % with waiting time less than 20 s. The main inputs were arrival rate and service rate in every single hour over 13-working period entire the month. To determine arrival rate in period *ijk*, the hourly call arrivals were divided by 3600 s that equal to λ_{ijk} calls per second. In addition to the arrival rate, μ_{ijk} calls per second in period *ijk* was the fraction of AHT in the same period. Let $j = 1, 2, 3 \ldots 13$ represented to period 1–13, $i = 1, 2, 3 \ldots, 7$ marked from Sunday to Saturday, and $k = 1, 2, 3$ and 4 started from Week 5 to Week 8.

The results could be found in every single period and an example solution was shown in Table 1. The value illustrated the exact number of agents required that satisfied the expected service level by the rule of 80/20 in Week 5. The pattern of agent requirements performed in the same curve of the period $j = 1, 2, 3, \ldots, 13$. On Sunday, for example, it required 97 agents in the first period and sharply increased to a peak of 158 agents in the third period. Eventually, it decreased to the minimum value of 44 agents in the last period with slight fluctuation over time. As aforementioned, this pattern also performed in the same shape over the period every single day, but the requirement of agents was represented in the different numbers.

Table 1 The number of agents required in week 5

Period *j*	Day *i*						
	1	2	3	4	5	6	7
1	97	180	116	161	149	171	117
2	143	268	172	239	221	254	173
3	158	296	189	264	244	280	192
4	135	251	161	224	207	238	162
5	141	262	168	234	216	248	169
6	128	239	153	213	197	226	154
7	128	238	153	213	196	225	154
8	126	234	150	209	193	222	151
9	126	235	151	210	194	223	152
10	108	201	129	180	166	191	130
11	93	172	111	154	142	163	112
12	78	145	93	130	120	137	95
13	44	80	52	72	66	76	52

4.2 Shifting and Staff Assignment

The total working period was considered by 13 h between 9.00 a.m. and 10.00 p.m. It could be divided the daily operations into 13 periods equally and every shift had 9 working hours. Let shift $k = 1$ denoted the work shift that begins at 9.00 a.m. and finishes at 6.00 p.m. The shift $k = 2, 3, 4, \ldots n$ started next to the previous shift with equal length of 9 h until the last hour of the last shift finished at 10.00 p.m. All of the agents would be allowed to have a break at the 5th hour. Thus, there were 5 nominal shifts over the 13-h working period including: (i) 9.00 a.m. to 6.00 p.m., (ii) 10.00 a.m. to 7.00 p.m., (iii) 11.00 a.m. to 8.00 p.m., (iv) 12.00 p.m. to 9.00 p.m., and (v) 01.00 p.m. to 10.00 p.m.

In order to solve shifting and staff assignment problem, the linear programming as seen in Eqs. (6)–(8) was applied. The results in Sect. 4.1 gave the number of agents required in all periods (m_t) put in the program. The shifts could be made up entire the month. On Sunday in Week 5, for example, there were 5 shifts as shown in Table 2. The results suggested that the number of agents assigned to the shift could satisfy all of constraints.

4.3 Workforce Planning

In this section, the optimal solution of workforce allocations could be determined by using the linear programming as shown in Eqs. (9)–(16). The total daily agents assigned in the previous section would be put in the model as the target agents required (b_{ij}). The model assumptions could be described in Sect. 3.2 workforce scheduling problem. The result could be shown as the number of agents assigned each day over the month of planning periods in Table 3. This could also imply to the monthly plan of workforce in the call center. Eventually, the feasible solution might give the number of assigned agents that satisfied all of the requirements in every single period entire the month of planning period.

5 Conclusions and Recommendations

5.1 Conclusions

To conclude, the client satisfaction could be a critical issue in any service provider organizations. In case of a call center, the service level is also the considerable issues. Actually, the modern call center operations possibly comprise of multi-tasks in order to response to diversity of client needs. In addition to incoming calls possibly arrive with uncertainty, it might not predict accurately. Managing

Table 2 The number of agents assigned to each shift on sunday in week 5

Period t		1	2	3	4	5	6	7	8	9	10	11	12	13
Shift k		a.m.			p.m.									
		09–10	10–11	11–12	12–1	1–2	2–3	3–4	4–5	5–6	6–7	8-Jul	8–9	9–10
1	09.00 a.m. to 06.00 p.m.	97	97	97	97	B	97	97	97	97				
2	10.00 a.m. to 07.00 p.m.		46	46	46	46	B	46	46	46	46			
3	11.00 a.m. to 08.00 p.m.			61	61	61	61	B	61	61	61	61		
4	12.00 p.m. to 09.00 p.m.				34	34	34	34	B	34	34	34	34	
5	01.00 p.m. to 10.00 p.m.					44	44	44	44	44	B	44	44	44
Total of agents assigned		**97**	**143**	**204**	**238**	**185**	**236**	**221**	**248**	**238**	**185**	**139**	**78**	**44**
Number of agents required		**97**	**143**	**158**	**135**	**141**	**128**	**128**	**126**	**126**	**108**	**93**	**78**	**44**
Number of surplus agents		**0**	**0**	**46**	**103**	**44**	**108**	**93**	**122**	**112**	**77**	**46**	**0**	**0**

B Breaking period

Table 3 The number of agents to be assigned each day in the planning period

	Sun	Mon	Tue	Wed	Thu	Fri	Sat
Week I	0	430	9	0	0	56	273
Week II	0	181	0	2	238	63	27
Week III	0	172	59	190	0	52	21
Week IV	0	414	0	0	0	47	269

workforce in a call center could not be easy tasks. It hardly balances workforce and workload to satisfy the expected service level every working period.

The objective of this paper is to demonstrate the optimal solution of monthly workforce planning to assign call center agents to meet the number of calls offers which also satisfy the acceptable service level. The queuing model and the linear programming were used as experimental tools. The data used in the experiment were collected from a call center of a communication service company in Thailand. The methodology was demonstrated in 4 steps. Firstly, the data were forecasted in the next 4 weeks by decomposition of time series technique. Then, the call center operations were simplified to the model of *M/M/n* and determined the number of agents required by the Erlang C formulation. Next, the requirements of agents were set up to the target value in the linear programming used to determine the optimal shifts. The number of agents assigned in the shifts would meet the requirements in every single period. The model was simulated entire the month of planning period. Workforce scheduling problem was solved by the linear programming. The result, finally, gave us the feasible number of the agents assigned in the monthly workforce plan.

5.2 Recommendations

Even though the solution for workforce planning and deployment was obtained by using the Erlang C formulation, the expect service level also was satisfied, but the other system performances, for example, agent occupancy level, seem not to be taken into account. Agent occupancy level, for example, was probably less than 50 % in some periods because this experiment might be quite restrictive with steady-state environments. Therefore, the actual implementation probably adjusted in the number of agents assigned based on the simulation results.

In the real world, a call center is more complicated with multi-tasks and the diversity of client needs. The process also has become increasingly difficult since internal and external variables are to be added. Thus, in order to improve call center performances, the prospect research should more concern several aspects in call center optimizations, forecasting method, for example, more complex of the queuing model, routing problem, staff scheduling problems, multi-skill scheduling, and real-time adherence.

References

Alfares H (2003) Four-day workweek scheduling with two or three consecutive days off. J Math Model Algorithms 2:67–80

Aksin Z, Armony M, Mehortra V (2007) The modern call center: a multi-disciplinary perspective on operations management research. Prod Oper Manage 6(6):665–688

Aktekin T (2014) Call center service process analysis: Bayesian parametric and semi-parametric mixture modeling. Eur J Oper Res 234:709–719

Aykin K (1996) Optimal shift scheduling with multiple break windows. Manage Sci 42(4)

Avramidis A, Chan W, Gendreau M, L'Ecuyer P, Pisacane O (2010) Optimizing daily agent scheduling in a multiskill call center. Eur J Oper Res 200:822–832

Brezavscek A, Baggia A (2014) Optimization of a call center performance using the sto-chastic queuing models. Bus Syst Res 5(3)

Buist E, Chan W, L'Ecuyer P (2008) Speeding up call center simulation and optimization by Markov chain Uniformation. In: Proceeding of the 2008 winter simulation conference

COPC CSP Standard (2014) The performance management system for customer contact operations

Dano G (2004) Workforce scheduling with logical constraints: theory and applications in call centers, Master thesis, Department of Mathematics, Vrije Univesity, Amsterdam

De Brueker P, Van den Bergh J, Belien J, Demeulemeester E (2015) Workforce planning incorporating skills: state of the art. Eur J Oper Res 243:1–16

Deslauriers A, L'Ecuyer P, Pichittlamken J, Ingolfsson A, Avramidis A (2007) Markov Chain models of a telephone call center with call blending. Comput Oper Res 24:1616–1645

Dietz D (2011) Practical scheduling for call center operations. Omega 39(2011):550–557

Dombacher C (2010) Queuing models for call centers. Download from http://www.telecomm.at/documents/Queueing_Models_CC.pdf. Accessed 20th Dec 2014

Eun Nah J, Kin S (2013) Workforce planning and deployment for a hospital reservation call center with abandonment cost and multiple tasks. Comput Ind Eng 66(2):239–297

Eveborn V, Ronnqvis M (2004) Scheduler—a system for staff planning. Ann Oper Res 128:21–45

Goldberg Y, Ritov Y, Mandelbaum A (2014) Predicting the continuation of a function with applications to call center data. J Statistical Plann Infer 147:53–65

Gomar J, Haas T, Morton D (2002) Assignment and allocation optimization of partially multiskilled workforce. J Constr Eng Manage 103–109

Heizer J, Render B (2011) Operations management, global edition, 10th edn. Pearson Education, New Jersey

Herowati E (2005) The application of multi shifts and break windows in employees scheduling. Int J Comput Internet Manage 13(2)

Hock N (1996) Oueueing modeling fundamentals. Wiley, England, pp 114–115

Koole G (2007) Call center mathematics: a scientific method for understanding and improving contact centers. Version of July 2007. Download from http://inf.unisul.br/~davalos/materialpoii/book%20calcenter.pdf. Accessed 10th September 2014

Koole G, Mandelbaum A (2001) Queing models of call centers: an introduction. Download from http://www.columbia.edu/~ww2040/cc_review.pdf. Accessed 30th Nov 2014

Li M, Huang L, Gong L (2011) Research on the challenges and solutions of design large-scale call center. Procedia Eng 5:2359–2363

Liao S, van Delft C, Vial J (2013) Distributionally robust workforce scheduling in call centres with uncertain arrival. Optim Methods Softw 28(3):501–522

Liao S (2011) Staffing and shift-scheduling of call centers under call arrival rate uncertainty. Master Thesis, Engineering Sciences, Ecole Centrale, Paris

Pot A (2006) Planning and routing algorithms for multi-skill contact centers. Doctoral dissertation, Department of Mathematics, Vrije Univesity, Amsterdam

Image Analysis for High-Dimensional Control Chart in Sausage Color Consistency Detection

Piraya Kaewsuwan, Chen-Yang Cheng and Chumpol Yuangyai

Abstract A quality control activity is one of the key elements in a food logistic system. It plays an important role to reduce cost and customer-claim due to unsatisfactory food products. Typically, the food inspection is performed by human, and this leads to inconsistent inspection results due to fatigue and tediousness. Therefore, an automated quality inspection program is required to ensure that the system is effective. This chapter presents an approach to improve consistency of sausage color inspection after packaging using an integrated framework of image processing techniques and high-dimensional control charts. The results indicate that sausage color consistency can be improved resulting in a more stable process.

Keywords Color inspection · Image processing · Multivariate control chart · Sausage production

1 Introduction

A quality control activity is one of the key elements in a food logistic system (Soysal et al. 2012). It plays an important role to reduce food product rejection rate and to improve process for cost reduction as well as customer satisfaction (Essien 2003).

P. Kaewsuwan (✉) · C. Yuangyai
Department of Industrial Engineering, Faculty of Engineering,
King Mongkut's Institute of Technology Ladkrabang, Bangkok 10520, Thailand
e-mail: piraya.ks@gmail.com

C. Yuangyai
e-mail: chumpol.yuangyai@gmail.com

C.-Y. Cheng
Department of Industrial Engineering and Management, Taipei University of Technology,
Taipei 10608, Taiwan, ROC
e-mail: cycheng@ntut.edu.tw

V. Kachitvichyanukul et al. (eds.), *Toward Sustainable Operations of Supply Chain and Logistics Systems*, EcoProduction,
DOI 10.1007/978-3-319-19006-8_37

Therefore, a capable automated quality control system is required to ensure that the logistics system is effective.

One of the most important characteristics of quality perception and consumer preferences is food appearance, for example, sausage product (Francis 1995). Typically, the food appearance inspection is performed by human judgments. In the sausage production, color is important quality parameters of sausage. Its color characteristic depends on the natural color of components used and must be uniform and homogeneous. It means that no unusual colors such as black or dark green are allowed and overall perception in color must be similar (Essien 2003).

However, the sausage abnormality inspections by human are limited as the natures of these jobs are tedious, laborious, costly, and time-consuming (Gunasekarn 1996). This leads to the measurement system is incapable of distinguishing the food colors. Therefore, the inspection program requires an automate system using image processing technique to improve the inspection effectiveness (Du and Sun 2004; Brosnan and Sun 2004; Wu and Sun 2013).

Once the measurement system is complete, the use of available data is also crucial for further process improvement. Much research has shown the effective use of control charts for food inspection (Lim et al. 2014). For example, Yuangyai et al. (2012) proposed using Hotelling T^2 control chart associated with X-MR chart for color sausage production control in $L^*a^*b^*$ system. The proposed control charts are able to detect abnormalities that are noticeable between packages (consider as between subgroup variation). However, they do not consider the color homogeneity within each sausage package (considered as within subgroup variation).

The objective of the study is to improve a consistency of sausage color inspection after packaging. Typically, $L^*a^*b^*$ system is used to measure color; however, it may not be sufficient to inspect all possible abnormalities. The integrated framework of image processing techniques and high-dimensional/auto-correlated control charts is proposed. We will additionally use hue (h) and chroma (C^*) (Girolami et al. 2014) and use standard deviation of each value to detect color homogeneity (within subgroup variation). Then, high-dimensional control chart, a Hotelling T^2 control chart *on* L^*, a^*,b^*, h, C^*, S_{L*}, S_{a*}, S_{b*}, S_h, and S_{C*}, is constructed.

This paper is organized as follows: We will present image processing technique, color measurement, statistical process control, and our proposed framework in Sect. 2. Then, demonstration of our framework, results, and discussion is presented in Sect. 3, followed by Sect. 4, conclusion.

2 Materials and Methods

In this section, we introduce image processing technique, color measurement, statistical process control, and a proposed framework as the following subsections.

Fig. 1 Image inspection equipment

2.1 Image Processing Technique

Image processing involves a construction of meaningful description of physical objects, and then, the description is transformed to serve predefined proposes. There are three steps of interest within image processing: (1) image acquisition, (2) image preprocessing, and (3) object measurement. Image analysis has been commonly used in food quality inspection (Du and Sun 2004; Brosnan and Sun 2004).

Our inspection equipment is constructed based on Pedreschi et al. (2006) and Yuangyai et al. (2013) as shown in Fig. 1. The chamber consists of three main elements: (1) Four florescence lamps are attached at square 30 cm. above the sample and set at an angle of 45° to nugget location. (2) A Canon EOS 550D camera is located vertically above the sample. The angle between the camera lens axis and the lighting sources is at 45°; the image format is JPEG with resolution 3456 × 2304 and connected to a computer by the USB port. (3) Illuminators and the CDC are placed inside a mat acrylic box to prevent light and reflection from outside sources, and the box size is 50 × 50 × 50 cm. The frame is made of aluminum and is located above the ground at 100 cm.

2.2 Color Measurement

For the sausage color inspection, the image processing algorithm is coded using MATLAB. The color of sausage is measured in the $L^*a^*b^*$ system, as well as hue and chroma. $L^*a^*b^*$ system is an international standard for color measurements, adopted by the Commission Internationale d'Eclairage (CIE) in 1976. This color model is a device-independent color space (Johnson 1996). L^* is the luminance or lightness component, which ranges from 0 to 100, and a^* (from green to red) and b^* (from blue to yellow) are the two chromatic components, which range from −120 to 120 (Papadakis et al. 2000).

The hue is a color attribute associated with the dominant wavelength in a mixture of light waves that represent the dominant color as perceived by an observer; when an object is considered to be red, orange, or yellow, the hue is being specified. The chroma defines the color's strength. It is measured in numbered steps starting at one with weak colors having low chroma values (Plataniotis and Venetsanopoulos 2000).

The sausages images are acquired using a digital camera. In the pixel pre-processing, the raw image is converted into $L^*a^*b^*$ system using MATLAB to code all the pixels. hue $= \tan^{-1}\left(\frac{b^*}{a^*}\right)$ and chroma $= (a^{*2} + b^{*2})^{\frac{1}{2}}$ were calculated for each pixel. Then, the calculated average and standard deviation of each L^*, a^*, b^*, h, and C^* values were from all image pixel data by MATLAB.

2.3 Statistical Process Control

Importantly, there are two main assumptions underlying Shewhart's chart family: (1) Normality: Normal probability plot is constructed by MINITAB 17. (2) Independency: The correlations among L^*, a^*, b^*, h, C^*, S_{L*}, S_{a*}, S_{b*}, S_h, and S_{C*} are analyzed using matrix plots, a two-dimensional matrix of individual plots and Pearson coefficient generated by MINITAB 17.

If dependency exists, Hotelling T^2 control chart is commonly used procedure for a multivariate process monitoring and controlling. For individual observations, suppose that there are m samples, each of size $n = 1$, and p is the number of quality characteristics observed in each sample. Let $\bar{x}$ be the sample mean vector and S be covariance matrix of these observations. The Hotelling T^2 statistic is calculated as $T^2 = (x - \bar{x})'S^{-1}(x - \bar{x})$. The phase I limits should be based on a beta distribution. This would lead to phase I limit UCL given by UCL $= \frac{(m-1)^2}{m}\beta_{\alpha,p/2,(m-p-1)/2}$ and LCL = 0, where $\beta_{\alpha,p/2,(m-p-1)/2}$ is the upper α percentage point of a beta distribution with parameters $p/2$ and $(m - p - 1)/2$ (Montgomery 2009).

2.4 The Proposed Framework

The framework of image processing techniques and high-dimensional control charts is shown in Fig. 2. The food process control is begun by image processing and analysis, and then, we study the characteristic of color whether or not the data are

Fig. 2 Proposed framework

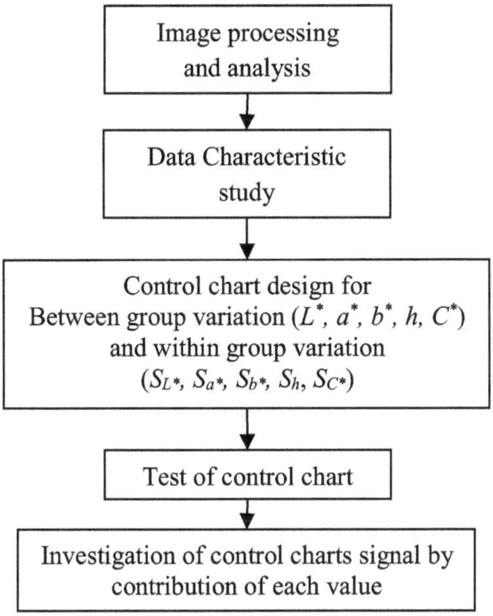

normally distributed and independent using normal probability plot and correlation plot. Next, we design and investigate control charts for homogeneity of sausage color. Finally, we analyze control chart signal by difference contribution of each characteristic.

3 Results and Discussion

3.1 The Study of Data Characteristics

To investigate the characteristics of the data, the sausage 30 packs were selected randomly from a local market, and we record the image individually overtime in values L^*, a^*, b^*, h, C^*, S_{L*}, S_{a*}, S_{b*}, S_h, and S_{C*}. The image data collected are investigated for their characteristics: normality and independency.

3.1.1 Normality

Normal probability plots of L^*, a^*, b^*, h, C^*, S_{L*}, S_{a*}, S_{b*}, S_h, and S_{C*} are shown in Fig. 3. The normal probability plot of L^* suggested that it is not normally distributed at 5 % significance level. We found that there was an abnormality and decided to remove this value.

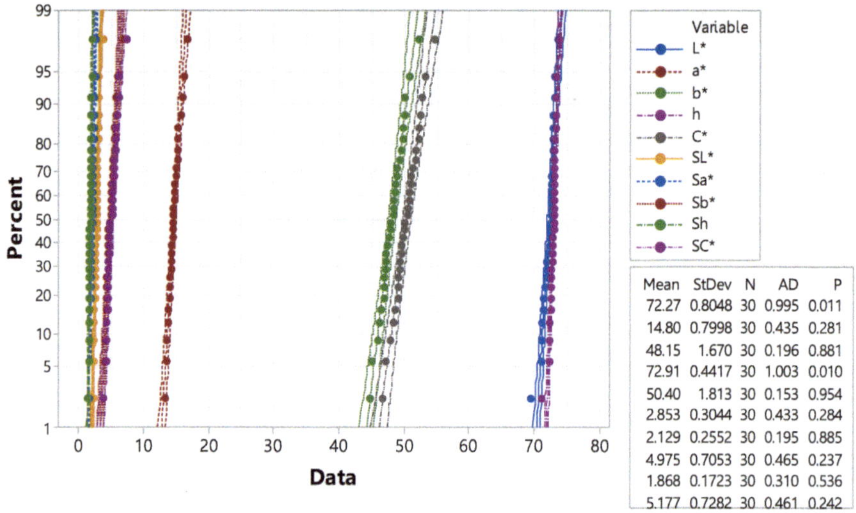

Fig. 3 Normal probability plot of L^{*}, a^{*}, b^{*}, h, C^{*}, S_{L*}, S_{a*}, S_{b*}, S_{h}, and S_{C*}

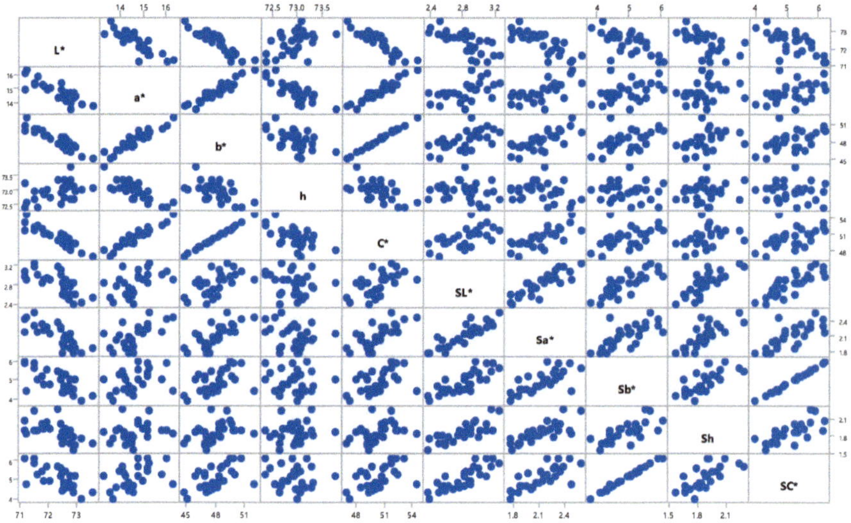

Fig. 4 Matrix plot of , $L^{*}a^{*}$, b^{*}, h, C^{*}, S_{L*}, S_{a*}, S_{b*}, S_{h}, and S_{C*}

3.1.2 Independency

The matrix plots and Pearson correlation are used to investigate independence assumption. The matrix plots and Pearson correlation of L^{*}, a^{*}, b^{*}, h, C^{*}, S_{L*}, S_{a*}, S_{b*}, S_{h}, and S_{C*} are shown in Fig. 4 and Table 1, respectively. They indicate that the S_h and a^{*}, S_{L*} and h, S_{a*} and h, S_{b*} and h, S_h and h, and S_{C*} and h have low

Table 1 Pearson correlation

Variable	L^*	a^*	b^*	h	C^*	S_{L*}	S_{a*}	S_{b*}	S_h
a^*	−0.829								
b^*	−0.914	0.932							
h	0.409	−0.775	−0.493						
C^*	−0.911	0.946	0.999	−0.529					
S_{L*}	−0.72	0.504	0.598	−0.177	0.593				
S_{a*}	−0.785	0.633	0.745	−0.212	0.738	0.866			
S_{b*}	−0.672	0.382	0.502	−0.079	0.493	0.799	0.704		
S_h	−0.572	0.223	0.393	0.129	0.378	0.798	0.789	0.686	
S_{C*}	−0.696	0.419	0.538	−0.101	0.529	0.819	0.745	0.998	0.692

correlation, but the rest of other pairs are highly correlated. Therefore, univariate shewhart's charts are not appropriate to monitor abnormalities.

3.2 Test of Sensitivity of Proposed Control Chart

We then construct a Hotelling T^2 control chart that compared Hotelling T^2 control chart of L^*, a^*, b^* (based on Yuangyai et al. 2012) with Hotelling T^2 control chart of L^*, a^*, b^*, h, C^*, S_{L*}, S_{a*}, S_{b*}, S_h, and S_{C*}; to analyze the color homogeneity of sausage color, 27 sausages are randomly selected from a local market which is represented as products from in-control process.

Next, we assumed that there are three packages considered as bad product (heterogeneity of color) as it is seen obviously by human inspection. Figure 5a–c shows that the color sausage within pack is not homogenous as marked in circle. The three packages are indicated as sample number 1, 3, and 17, respectively.

Figure 6 shows the Hotelling T^2 control charts of $L^*a^*b^*$. The upper control limit is 8.1029 with $\alpha = 0.027$. Notice that no points exceed this limit. The chart shows no abnormality as expected. Figure 7 shows the Hotelling T^2 control charts of L^*, a^*, b^*, h, C^*, S_{L*}, S_{a*}, S_{b*}, S_h, and S_{C*}. The upper control limit is 16.1937 with $\alpha = 0.027$.

Fig. 5 Sample sausage within pack is not homogenous obvious: **a** sample number 1, **b** sample number 3, **c** sample number 17

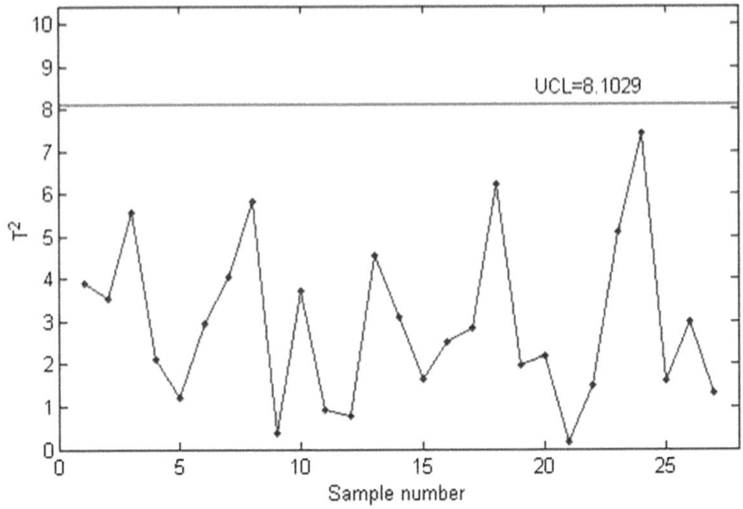

Fig. 6 The Hotelling T^2 control chart of $L^*a^*b^*$

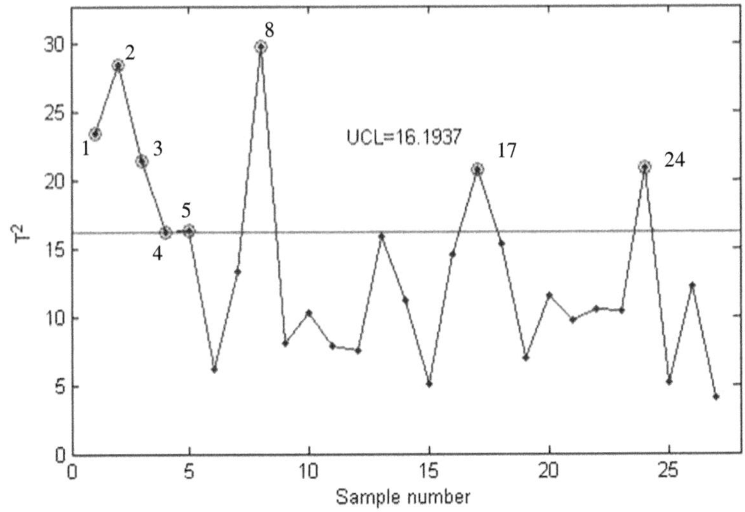

Fig. 7 The Hotelling T^2 control chart of L^*, a^*, b^*, h, C^*, S_{L*}, S_{a*}, S_{b*}, S_h, and S_{C*}

This control chart contributed to the identification of eight abnormalities, i.e., sample 1, 2, 3, 4, 5, 8, 17, 24. Sample 1, 3, and 17 (Fig. 5) suffered from color inhomogeneity and thereby detected by the proposed control chart. Sample 2, 4, 5, 8, and 24 (Fig. 8) undetected by the visual inspection but the control chart was capable of detecting the color inhomogeneity. These show that Hotelling T^2 control charts of L^*, a^*, b^*, h, C^*, S_{L*}, S_{a*}, S_{b*}, S_h, and S_{C*} are more sensitive.

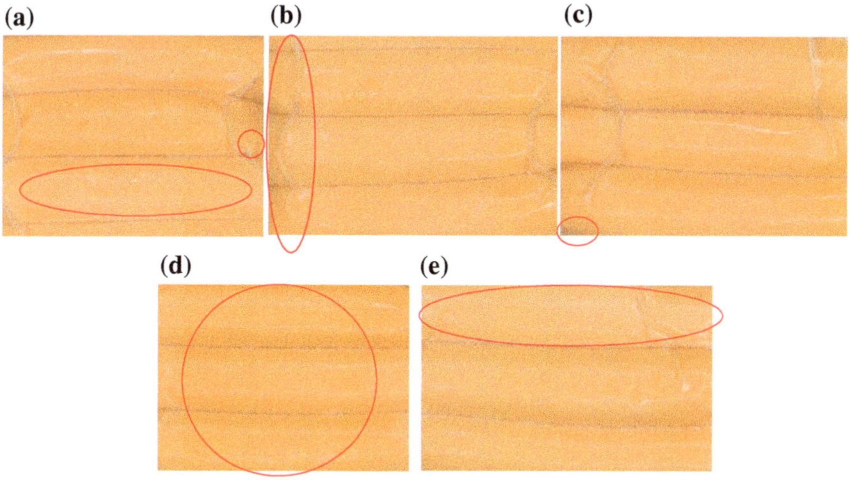

Fig. 8 Sample sausage within pack is not homogenous unobvious: **a** sample number 2, **b** sample number 4, **c** sample number 5, **d** sample number 8, **e** sample number 24

The most important step of process control is to identify the special cause of the abnormality. We then analyze control chart signal by using difference contribution of each value in mean vector to explain causes of unusual and interpreted using color attributes. The color difference between a sample (S) color with color values L^*, a^*, b^*, h, C^*, S_{L*}, S_{a*}, S_{b*}, S_h, and S_{C*} and an average (A) color with average color values L^*, a^*, b^*, h, C^*, S_{L*}, S_{a*}, S_{b*}, S_h, and S_{C*} (27 sausage from in process control) is given by the corresponding difference contributions in Eq. (1).

$$\begin{aligned} &\Delta L^* = L_S^* - L_A^*, \Delta a^* = a_S^* - a_A^*, \Delta b^* = b_S^* - b_A^*, \Delta h = h_S - h_A, \Delta C^* = C_S^* - C_A^*, \\ &\Delta S_{L^*} = S_{L^*,S} - S_{L^*,A}, \Delta S_{a^*} = S_{a^*,S} - S_{a^*,A}, \Delta S_{b^*} = S_{b^*,S} - S_{b^*,A}, \\ &\Delta S_h = S_{h,S} - S_{h,A} \quad \text{and} \quad \Delta S_{C^*} = S_{C^*,S} - S_{C^*,A} \end{aligned} \tag{1}$$

Table 2 shows the estimated difference contribution of each sample that contributed to the out-of-control signal. We consider the maximum contribution of each sample as shown in Table 3. Maximum contribution was to identify the possible causes of the out-of-control samples. From Table 2, the highest contribution of samples 1, 2, 3, 4, 5, 8, 17, and 24 is 2.062, 0.726, −3.392, −0.981, 0.563, 4.499, 0.666, and −2.271, respectively.

Maximum contribution of sample 1, 2, 3, 4, 5, 8, and 24 comes from chroma C^*. The chroma of sample 1, 2, 5, and 8 is more than zero that means its color attribute is more brilliant or clearer (Klein 2010). On the other hand, the chroma of sample 3, 4, and 24 is less than zero. It means that color attribute is duller or less chromatic. Maximum contribution of sample 17 is −0.488 from S_{b*} (standard deviation of b^*). The difference is more than zero, it means high variability in yellow color.

Table 2 Contributions of L^*, a^*, b^*, h, C^*, S_{L^*}, S_{a^*}, S_{b^*}, S_h, and S_{C^*}

Sample	ΔL^*	Δa^*	Δb^*	Δh	ΔC^*	ΔS_{L^*}	ΔS_{a^*}	ΔS_{b^*}	ΔS_h	ΔS_{C^*}
1	−0.869	0.579	1.980	0.009	2.062	0.369	0.208	1.046	0.034	1.073
2	0.061	0.597	0.577	−0.455	0.726	0.187	0.147	0.506	−0.074	0.553
3	1.239	−1.115	−3.200	0.161	−3.392	−0.302	−0.263	−0.717	−0.196	−0.730
4	0.563	−0.449	−0.885	0.190	−0.981	0.021	−0.160	−0.354	−0.084	−0.379
5	0.064	0.183	0.529	−0.022	0.563	0.289	0.178	0.123	0.171	0.144
8	−1.162	1.683	4.186	−0.351	4.499	0.099	0.357	−0.472	−0.028	−0.333
17	−0.048	−0.506	−0.368	0.406	−0.488	0.359	0.278	0.666	0.439	0.653
24	0.405	−1.378	−1.958	0.903	−2.271	0.014	0.049	0.132	0.023	0.159

Table 3 Maximum contribution for sample sausage out of control

Sample	Maximum contribution	Maximum contribution (Color values)	Possible causes in sausage production
1	2.062	C^* (Chroma)	Smoking process (placing sausage overlap, color setting)
2	0.726	C^* (Chroma)	Smoking process (color setting)
3	−3.392	C^* (Chroma)	Smoking process (placing sausage overlap, color setting)
4	−0.981	C^* (Chroma)	Smoking process (color setting)
5	0.563	C^* (Chroma)	Smoking process (color setting)
8	4.499	C^* (Chroma)	Smoking process (color setting)
17	0.666	S_{b*} (Standard deviation of b^*)	Mixing process (insufficient mixing time)
24	−2.271	C^* (Chroma)	Smoking process (color setting)

When considering Fig. 5a–c, they show some of color skin sausages are not homogeneous that can be detected by the visual inspection. Figure 8a–e shows some of color skin sausages are not homogeneous and they are not obviously detected by the visual inspection.

3.3 The Implication of Control Chart Signal for Sausage Production

When control chart signals an out of control alarm (e.g., color inhomogeneity, unusual colors such as black or dark green), it is difficult for process owners to determine its root causes because the Hotelling T^2 does not provide in-depth information. Therefore, we recommend using the contribution of color values L^*, a^*, b^*, h, C^*, S_{L*}, S_{a*}, S_{b*}, S_h, and S_{C*} and estimate maximum contribution of each sample in mean vector.

When ΔC^* value is more than zero, the color of the sausage is more brilliant or clearer. While ΔC^* value is less than zero, the color of the sausage is duller or less chromatic (Klein 2010). This indicates that surface of sausage is not homogeneous. It may be caused by smoking process. The most critical characteristic in smoke color is determined by moisture (lack of moisture on the surface and color setting immediately) after smoking to prevent the development of a mottled, streaky surface appearance (Essien 2003; Hansen 2000). Also, placing sausage overlap may cause surface of sausage is not homogeneous.

When ΔS_{b*}: standard deviation of b^* difference is more than zero, indicating the inhomogeneity of the yellow color on the surface of sausage that may be caused by insufficient mixing time (Essien 2003).

4 Conclusions

In this study, we proposed a framework for using image analysis and high–dimensional control chart for sausage production control. Image analysis includes color of sausage for ten characteristics, namely L^*, a^*, b^*, h, C^*, S_{L*}, S_{a*}, S_{b*}, S_h and S_{C*}. These values represent not only between subgroup variations (color different among packages) but also within subgroup variation (color homogeneity within package).

Then, the data characteristics are found that they are not independent that lead to the violation of the Shewhart's chart family assumption. The Hotelling T^2 control chart was constructed to identify the out-of-control signals. Due to control chart signals that is difficult for process owners to determine its root causes because the Hotelling T^2 does not provide in-depth information. Therefore, the Hotelling T^2 is associated with the analysis of maximum contribution of color values to identify out-of-control root causes for process improvement.

In this study, we only focus on within-group variation (homogeneity within package); however, in process control, not only within-group but also between-group variation has to be considered to maintain the process stability. In addition, the performance measure, e.g., average run length (ARL) of our control charts will be studied.

References

Brosnan T, Sun DW (2004) Improving quality inspection of food products by computer vision—a review. J Food Eng 61(1):3–16

Du C, Sun DW (2004) Recent developments in the applications of image processing techniques for food quality evaluation. Trends Food Sci Technol 15(5):230–249

Essien E (2003) Sausage manufacture. CRC Press, New York

Francis FJ (1995) Quality as influenced by color. Food Qual Prefer 6(3):149–155

Girolami A, Napolitano F, Faraone D, Bello GD, Braghieri A (2014) Image analysis with the computer vision system and the consumer test in evaluating the appearance of Lucanian dry sausage. Meat Sci 96:610–616

Gunasekarn S (1996) Computer vision technology for food quality assurance. Trends Food Sci Technol 7:245–256

Hansen R (2000) Seeing through the smoke screen. Meat Int 10(8):18–20

Johnson T (1996) Methods for characterizing colour scanners and digital cameras. Displays 16 (4):183–191

Klein GA (2010) Industrial color physics. Springer, New York

Lim SAH, Antony J, Albliwi S (2014) Statistical process control (SPC) in the food industry—a systematic review and future research agenda. Trends Food Sci Technol 37:37–151

Montgomery DC (2009) Statistical quality control: a modern introduction, 6th edn. Wiley, New York

Papadakis SE, Abdul-Malek S, Kamdem RE, Yam KL (2000) A versatile and inexpensive technique for measuring color of foods. Food Technol 5(12):48–51

Pedreschi F, LeÓn J, Mery D, Moyano P (2006) Development of a computer vision system to measure the color of potato chips. Food Res Int 39(10):1092–1098

Plataniotis KN, Venetsanopoulos AN (2000) Color image processing and applications. Springer, New York\

Soysal M, Bloemhof-Ruwaard JM, Meuwissen MPM, Van Der Vorst JGAJ (2012) A review on quantitative models for sustainable food logistics management. Int J Food Syst Dyn 3(2):136–155

Wu D, Sun DW (2013) Colour measurements by computer vision for food quality control—a review. Trends Food Sci Technol 29:5–20

Yuangyai C, Kaewsuwan P, Cheng CY (2012) Color-based image analysis for statistical sausage production control. Paper presented at the 13th Asia pacific industrial engineering and management systems conference, Phuket, Thailand, 2–5 Dec 2012

Yuangyai C, Matvises P, Janjarassuk U (2013) Image-based analysis for characterization of chicken nugget quality. Jurnal Teknik Industri 15(2):25–130

Author Index

V. Kachitvichyanukul et al. (eds.), *Toward Sustainable Operations of Supply Chain and Logistics Systems*, EcoProduction,
DOI 10.1007/978-3-319-19006-8

MIX
Papier aus verantwortungsvollen Quellen
Paper from responsible sources
FSC® C105338
www.fsc.org

If you have any concerns about our products,
you can contact us on
ProductSafety@springernature.com

In case Publisher is established outside the EU,
the EU authorized representative is:
Springer Nature Customer Service Center GmbH
Europaplatz 3, 69115 Heidelberg, Germany

Printed by Libri Plureos GmbH
in Hamburg, Germany